Communications in Computer and Information Science

2297

Series Editors

Gang Li, *School of Information Technology, Deakin University, Burwood, VIC, Australia*

Joaquim Filipe, *Polytechnic Institute of Setúbal, Setúbal, Portugal*

Zhiwei Xu, *Chinese Academy of Sciences, Beijing, China*

Rationale

The CCIS series is devoted to the publication of proceedings of computer science conferences. Its aim is to efficiently disseminate original research results in informatics in printed and electronic form. While the focus is on publication of peer-reviewed full papers presenting mature work, inclusion of reviewed short papers reporting on work in progress is welcome, too. Besides globally relevant meetings with internationally representative program committees guaranteeing a strict peer-reviewing and paper selection process, conferences run by societies or of high regional or national relevance are also considered for publication.

Topics

The topical scope of CCIS spans the entire spectrum of informatics ranging from foundational topics in the theory of computing to information and communications science and technology and a broad variety of interdisciplinary application fields.

Information for Volume Editors and Authors

Publication in CCIS is free of charge. No royalties are paid, however, we offer registered conference participants temporary free access to the online version of the conference proceedings on SpringerLink (http://link.springer.com) by means of an http referrer from the conference website and/or a number of complimentary printed copies, as specified in the official acceptance email of the event.

CCIS proceedings can be published in time for distribution at conferences or as post-proceedings, and delivered in the form of printed books and/or electronically as USBs and/or e-content licenses for accessing proceedings at SpringerLink. Furthermore, CCIS proceedings are included in the CCIS electronic book series hosted in the SpringerLink digital library at http://link.springer.com/bookseries/7899. Conferences publishing in CCIS are allowed to use Online Conference Service (OCS) for managing the whole proceedings lifecycle (from submission and reviewing to preparing for publication) free of charge.

Publication process

The language of publication is exclusively English. Authors publishing in CCIS have to sign the Springer CCIS copyright transfer form, however, they are free to use their material published in CCIS for substantially changed, more elaborate subsequent publications elsewhere. For the preparation of the camera-ready papers/files, authors have to strictly adhere to the Springer CCIS Authors' Instructions and are strongly encouraged to use the CCIS LaTeX style files or templates.

Abstracting/Indexing

CCIS is abstracted/indexed in DBLP, Google Scholar, EI-Compendex, Mathematical Reviews, SCImago, Scopus. CCIS volumes are also submitted for the inclusion in ISI Proceedings.

How to start

To start the evaluation of your proposal for inclusion in the CCIS series, please send an e-mail to ccis@springer.com

Mufti Mahmud · Maryam Doborjeh ·
Kevin Wong · Andrew Chi Sing Leung ·
Zohreh Doborjeh · M. Tanveer
Editors

Neural Information Processing

31st International Conference, ICONIP 2024
Auckland, New Zealand, December 2–6, 2024
Proceedings, Part XVI

 Springer

Editors
Mufti Mahmud
King Fahd University of Petroleum
and Minerals
Dhahran, Saudi Arabia

Kevin Wong
Murdoch University
Murdoch, WA, Australia

Zohreh Doborjeh
Auckland University of Technology
Auckland, New Zealand

Maryam Doborjeh
Auckland University of Technology
Auckland, New Zealand

Andrew Chi Sing Leung
City University of Hong Kong
Kowloon, Hong Kong

M. Tanveer
Indian Institute of Technology Indore
Indore, Madhya Pradesh, India

ISSN 1865-0929   ISSN 1865-0937 (electronic)
Communications in Computer and Information Science
ISBN 978-981-96-7035-2   ISBN 978-981-96-7036-9 (eBook)
https://doi.org/10.1007/978-981-96-7036-9

Preface

Welcome to the 31st International Conference on Neural Information Processing (ICONIP 2024) of the Asia-Pacific Neural Network Society (APNNS), held in Auckland, New Zealand, December 2–6, 2024.

The mission of the Asia-Pacific Neural Network Society is to promote active interactions among researchers, scientists, and industry professionals who are working in neural networks and related fields in the Asia-Pacific region. APNNS has Governing Board Members from 13 countries/regions - Australia, China, Hong Kong, India, Japan, Malaysia, New Zealand, Singapore, South Korea, Qatar, Taiwan, Thailand, and Turkey. The society's flagship annual conference is the International Conference on Neural Information Processing (ICONIP). The ICONIP conference aims to provide a leading international forum for researchers, scientists, and industry professionals who are working in neuroscience, neural networks, deep learning, and related fields to share their new ideas, progress, and achievements.

ICONIP 2024 received 1301 papers, of which 318 papers were accepted for publication in Lecture Notes in Computer Science (LNCS), representing an acceptance rate of 24.44% and reflecting the increasingly high quality of research in neural networks and related areas. From the remaining 983 papers, 472 papers (48.01%) were accepted for publication in Communications in Computer and Information Science (CCIS) series volumes. The conference focused on four main areas, i.e., "Theory and Algorithms", "Cognitive Neuroscience", "Human-Centered Computing", and "Applications". All the submissions were rigorously reviewed by the conference Programme Committee (PC), comprising 767 PC members, and they ensured that every paper had at least three high-quality single-blind reviews. Each paper was further reviewed by at least one chair.

We would like to take this opportunity to thank all the authors for submitting their papers to our conference, and our great appreciation goes to the PC members and the reviewers who devoted their time and effort to our rigorous peer-review process; their insightful reviews and timely feedback ensured the high quality of the papers accepted for publication. We hope you enjoyed the research program at the conference.

March 2025

Mufti Mahmud
Maryam Doborjeh
Kevin Wong
Andrew Chi Sing Leung
Zohreh Doborjeh
M. Tanveer

Organisation

Honorary Chairs

Nikola Kasabov	Auckland University of Technology, New Zealand
Derong Liu	Southern University of Science and Technology, China
Seiichi Ozawa	Kobe University, Japan
Jonathan Chan	King Mongkut's University of Technology Thonburi, Thailand

Conference Chairs

Maryam Doborjeh	Auckland University of Technology, New Zealand
Mufti Mahmud	King Fahd University of Petroleum and Minerals, Saudi Arabia
Michael Witbrock	University of Auckland, New Zealand

Program Chairs

Kevin Wong	Murdoch University, Australia
Andrew Chi Sing Leung	City University of Hong Kong, China
Zohreh Doborjeh	University of Auckland, New Zealand
M. Tanveer	IIT Indore, India

Finance Chairs

Saori Tanaka	University of Electro-Communications, Japan
Terry Brydon	Auckland University of Technology, New Zealand

Competition Chairs

Paul S. Pang	Federation University Australia, Australia
Goh Wen Bin Wilson	Nanyang Technological University, Singapore

Special Session Chairs

Tingwen Huang	Texas A&M University at Qatar, Qatar
Wei-Neng Chen	South China University of Technology, China
Edmund Lai	Auckland University of Technology, New Zealand

Tutorial and Workshop Chairs

Alex Sumich	Nottingham Trent University, UK
Kenneth Johnson	Auckland University of Technology, New Zealand
Marzia Hoque Tania	University of New South Wales, Australia
Akbar Ghobakhlou	Auckland University of Technology, New Zealand

Award Chairs

Kenji Doya	Okinawa Institute of Science and Technology, Japan
Tanja Mitrovic	University of Canterbury, New Zealand

Keynote Chairs

Zeng-Guang Hou	Chinese Academy of Sciences, China
Amir H. Gandomi	University of Technology Sydney, Australia
Stephen Thorpe	Auckland University of Technology, New Zealand
Leanne Bint	Auckland University of Technology, New Zealand
Jing Ma	Auckland University of Technology, New Zealand
Mahsa Mohaghegh	Auckland University of Technology, New Zealand
Wei Qi Yan	Auckland University of Technology, New Zealand
Amir Hussain	Edinburgh Napier University, UK
Roopak Sinha	Deakin University, Australia

Publicity Chairs

El-Sayed M. El-Alfy	King Fahd University of Petroleum and Minerals, Saudi Arabia
David Brown	Nottingham Trent University, UK
Siddhartha Bhattacharyya	VSB Technical University of Ostrava, Czech Republic

Hyeyoung Park	Kyungpook National University, South Korea
Qinmin Yang	Zhejiang University, China
Chu Kiong Loo	University of Malaya, Malaysia
Cosimo Ieracitano	Mediterranean University of Reggio Calabria, Italy
Bernhard Pfahringer	University of Waikato, New Zealand

Local Organising Chairs

Tek Tjing Lie	Auckland University of Technology, New Zealand
Felix Tan	Auckland University of Technology, New Zealand
Minh Nguyen	Auckland University of Technology, New Zealand
Yvonne Chan Cashmore	Auckland University of Technology, New Zealand
Peter Chong	Auckland University of Technology, New Zealand
Reza Enayatollahi	Toi Ohomai Institute of Technology, New Zealand
Selina Nihalani-Sharma	Auckland University of Technology, New Zealand
Janie van Woerden	Auckland University of Technology, New Zealand

International Advisory Committee

Giancarlo Fortino	University of Calabria, Italy
Hamido Fujita	Iwate Prefectural University, Japan
Shariful Islam	Deakin University, Australia
Tianzi Jiang	Institute of Automation, CAS, China
M. Shamim Kaiser	Jahangirnagar University, Bangladesh
Joarder Kamruzzaman	Federation University Australia, Australia
Francesco Carlo Morabito	University Mediterranea of Reggio Calabria, Italy
Stefano Panzeri	University Medical Center Hamburg-Eppendorf, Germany
Hanchuan Peng	SEU-Allen Institute for Brain & Intelligence, China
Nelishia Pillay	University of Pretoria, South Africa

Technical Program Committee

Hamid Abbasi	University of Auckland, New Zealand
Abdullah Abdullah	Universiti Teknologi PETRONAS, Malaysia
Hazel Abraham	Auckland University of Technology, New Zealand
Sonali Agarwal	IIIT Allahabad, India

Nehal Ahmad   Indian Institute of Technology Indore, India
Saad Bin Ahmed   RPTU, Germany
Toshiaki Aida   Okayama University, Japan
Aisha Ajmal   Open Polytechnic, New Zealand
Mushir Akhtar   IIT Indore, India
Zaid Al-Tameemi   Toi Ohomai Institute of Technology, New Zealand
Shafiq Alam   Massey University, New Zealand
Wajahat Ali   Aligarh Muslim University, India
Jumabek Alikhanov   HumbleBeeAI, South Korea
Fares Alkhawaja   Concordia University, Canada
Nathan Allen   Auckland University of Technology, New Zealand
Fahad Alvi   University of York, UK
Goran Andonovski   University of Ljubljana, Slovenia
Yuki Aoki   Osaka University, Japan
Sabri Arik   Istanbul University, Turkey
Tetsuya Asai   Hokkaido University, Japan
Mohammad Asif   IIIT Allahabad, India
G. Avinash   National Institute of Occupational Health, India
Helena Bahrami   Auckland University of Technology, New Zealand
Muhammad Zeeshan Baig   Macquarie University, Australia
Tao Ban   National Institute of Information and
                                   Communications Technology, Japan

Sourasekhar Banerjee   Umeå University, Sweden
Qiming Bao   University of Auckland, New Zealand
Soubhagya Barpanda   VIT-AP University, India
Arindam Basu   University of Canterbury, New Zealand
Boris Bačić   Auckland University of Technology, New Zealand
Aymen Ben Said   University of Regina, Canada
Krishan Berwal   Military College of Telecommunication
                                   Engineering, India

M. Agni Catur Bhakti   Sampoerna University, Indonesia
Raghavendra Bhalerao   IITRAM, India
Siddhartha Bhattacharyya   RCC Institute of Information Technology, India
Dasari Bhulakshmi   VIT, India
Ying Bi   Victoria University of Wellington, New Zealand
Jacques Blanc-Talon   DGA TA, France
Christian Bohn   University of Wuppertal, Germany
Phil Bones   University of Canterbury, New Zealand
Yee Ling Boo   RMIT University, Australia
Larbi Boubchir   University of Paris 8, France
Sally Britnell   Auckland University of Technology, New Zealand
Chenyang Bu   Hefei University of Technology, China

Sugam Budhraja	Auckland University of Technology, New Zealand
Michael Bunn	Auckland University of Technology, New Zealand
Jérémie Cabessa	Panthéon-Assas University Paris II, France
Wanqiang Cai	Nanjing University of Finance & Economics, China
Weidong Cai	University of Sydney, Australia
Cristian S. Calude	University of Auckland, New Zealand
Yuan Cao	Beijing University of Posts and Telecommunications, China
Elisa Capecci	Auckland University of Technology, New Zealand
Debasrita Chakraborty	Indian Statistical Institute, Kolkata, India
Srinivasa Chakravarthy	IIT Madras, India
Stephan Chalup	University of Newcastle, Australia
Victor Chan	Tsinghua University, China
Ritesh Chandra	IIIT Allahabad, India
Siripinyo Chantamunee	Walailak University, Thailand
Gu Chaochen	Shanghai Jiao Tong University, China
Nisha Chauhan	Indian Institute of Technology Roorkee, India
Zaineb Chelly Dagdia	Université Paris-Saclay, France
Bang Chen	Ningbo University, China
Dongjie Chen	University of Chinese Academy of Sciences, China
Jianyong Chen	Shenzhen University, China
Jinpeng Chen	Beijing University of Posts & Telecommunications, China
Kecheng Chen	University of Science and Technology of China, China
Ling Chen	Southwest University, China
Lyu Chen	Shandong Normal University, China
Wei-Neng Chen	South China University of Technology, China
Xiaoqing Chen	University of Chinese Academy of Sciences, China
Xinrui Chen	Tsinghua University, China
Zhewei Chen	Australian National University, Australia
Zhi Chen	University of Electronic Science and Technology of China, China
Long Cheng	Institute of Automation, CAS, China
Xingfu Cheng	Qufu Normal University, China
Ziqi Cheng	Chongqing University, China
Sung-Bae Cho	Yonsei University, South Korea
Chak Fong Chong	Macao Polytechnic University, China
Peter Han Joo Chong	Auckland University of Technology, New Zealand
Eashita Chowdhury	Indian Institute of Technology Kharagpur, India

Mamadou B. H. Cissoko	University of Strasbourg, France
Raphael Couturier	Marie and Louis Pasteur University, France
Stephen Cox	Residio, New Zealand
Brian Cusack	Auckland University of Technology, New Zealand
Jianhua Dai	Hunan Normal University, China
Shaoxiang Dang	Nagoya University, Japan
Anik Das	St. Francis Xavier University, Canada
Santosh Das	OmDayal Group of Institutions, India
Subham Das	Indian Institute of Technology Madras, India
Sudhansu Bala Das	NIT Rourkela, India
Marcílio De Souto	University of Orléans, France
Neda Deljavan	Institute for Advanced Studies in Basic Sciences, Iran
Jeremiah D. Deng	University of Otago, New Zealand
Nagaraj V. Dharwadkar	National Institute of Technology, Warangal, India
Kumar Dheenadayalan	Qualcomm, USA
Johannes Dimyadi	University of Auckland, New Zealand
Yuxin Ding	Harbin Institute of Technology, China
Maryam Doborjeh	Auckland University of Technology, New Zealand
Zohreh Doborjeh	University of Auckland, New Zealand
Bin Dong	Ricoh Software Research Center, China
Ninghua Dong	Beijing Jiaotong University, China
Ruiqi Dong	South China Normal University, China
Jordan Douglas	University of Auckland, New Zealand
Kenji Doya	OIST, Japan
Haiwen Du	Harbin Institute of Technology, China
Fuqing Duan	Beijing Normal University, China
Piotr Duda	Częstochowa University of Technology, Poland
Thao Duong	Murdoch University, Australia
Varun Dutt	Indian Institute of Technology Mandi, India
Pratik Dutta	Indian Institute of Technology Patna, India
Mansoor Ebrahim	Iqra University, Pakistan
El-Sayed M. El-Alfy	King Fahd University of Petroleum and Minerals, Saudi Arabia
Reza Enayatollahi	Toi Ohomai Institute of Technology, New Zealand
Farnoush Falahatraftar	Polytechnique Montréal, Canada
Chenyou Fan	South China Normal University, China
Jiangtao Fan	Durham University, UK
Junyuan Fang	City University of Hong Kong, China
Sen Fang	University of Victoria, Australia
Yifei Fang	University of Chinese Academy of Sciences, China

Zhiyuan Fang	University of Waikato, New Zealand
Muhammad Farhan	Glocal University, India
Jiaxuan Feng	Southwest University of Science and Technology, China
Yong Feng	Chongqing University, China
Jaco Fourie	Lincoln Agritech Ltd, UK
Frieyadie Frieyadie	STMIK Nusa Mandiri, Indonesia
Junjie Fu	Southeast University, China
Taoran Fu	Hunan University, China
Xiping Fu	University of Otago, New Zealand
Yulin Fu	University of Auckland, New Zealand
Fumiyo Fukumoto	University of Yamanashi, Japan
Haitao Gan	Hubei University of Technology, China
Mudasir Ganaie	Indian Institute of Technology Indore, India
Varun Ganjigunte Prakash	CogniAble, India
Qian Gao	Qilu University of Technology, China
Terry Gao	Moreton Bay City Council, Australia
Xinyi Gao	Auckland University of Technology, New Zealand
Xizhan Gao	University of Jinan, China
Chandan Gautam	I2R, A*STAR, Singapore
Liang Ge	Chongqing University, China
Mengmeng Ge	University of Canterbury, New Zealand
Trevor Gee	University of Auckland, New Zealand
Mingyang Geng	National University of Defense Technology, China
Thanawit Gerdprasert	Yamaguchi University, Japan
Akbar Ghobakhlou	Auckland University of Technology, New Zealand
Hamid Gholamhosseini	Auckland University of Technology, New Zealand
Ashish Ghosh	Indian Statistical Institute, Kolkata, India
Tripti Goel	National Institute of Technology Silchar, India
Zbigniew Gomolka	University of Rzeszow, Poland
Jiaying Gong	Virginia Tech, USA
Rui Gong	Chinese Academy of Sciences, China
Yingjie Gong	Zhejiang University, China
Maanvik Gounder	University of the South Pacific, Fiji
Richard Green	University of Canterbury, New Zealand
Jessica Gu	University of Auckland, New Zealand
Tamil Selvan Gunasekaran	University of Auckland, New Zealand
Long Guo	Sichuan University, China
Ping Guo	Beijing Normal University, China
Qin Guo	Peking University, China
Deepak Gupta	MNNIT Allahabad, India

Harshit Gupta	IIIT Allahabad, India
Tomasz Hachaj	Pedagogical University of Krakow, Poland
Katsuyuki Hagiwara	Mie University, Japan
Masafumi Hagiwara	Keio University, Japan
Luke Hallum	University of Auckland, New Zealand
Chansu Han	National Institute of Information and Communications Technology, Japan
Gao Han	Xidian University, China
Zhang Haoyu	Chinese Academy of Sciences, China
Md. Tarek Hasan	United International University, Bangladesh
Tatsuhito Hasegawa	University of Fukui, Japan
Hao He	Chinese Academy of Sciences, China
Xing He	Southwest University, China
Xinyang He	UCAS, China
Yangliu He	Beijing University of Posts and Telecommunications, China
Yuting He	University of Nottingham Ningbo, China
Mahshid Helali Moghadam	RISE Research Institutes of Sweden, Sweden
Hugo Hernault	Playtika, Japan
Akira Hirose	University of Tokyo, Japan
Yvonne Hong	Victoria University of Wellington, New Zealand
Yin Hongwei	Huzhou University, China
Adrian Horzyk	University of Krakow, Poland
Amanda Horzyk	University of Edinburgh, UK
Md Zakir Hossain	Australian National University, Australia
Zengguang Hou	Chinese Academy of Sciences, China
Menghao Hu	Pengcheng Laboratory, China
Yan Hu	University of New South Wales, Australia
Zhongyun Hua	Harbin Institute of Technology, China
He Huang	Soochow University, China
Kaizhu Huang	Duke Kunshan University, China
Sheng Huang	Chongqing University, China
Siyu Huang	Nanjing University of Aeronautics and Astronautics, China
Zhen Huang	Shenyang Institute of Computing Technology, China
Carolynne Hultquist	Pennsylvania State University, USA
Aarij Hussaan	Iqra University, Pakistan
David Iclanzan	Sapientia University, Romania
Cosimo Ieracitano	University "Mediterranea" of Reggio Calabria, Italy
Kazushi Ikeda	Nara Institute of Science and Technology, Japan

Radu Tudor Ionescu	University of Bucharest, Romania
Noriyuki Iwane	Hiroshima City University, Japan
Sana Jabbar	Lahore University of Management Sciences, Pakistan
Sapna Jaidka	University of Waikato, New Zealand
Debesh Jha	Northwestern University, USA
Ningning Jia	Beijing University of Posts and Telecommunications, China
Yan Jia	National University of Defense Technology, China
Peng Jiang	Wuhan University, China
Xianwei Jiang	Southwest University of Science and Technology, China
Xuesong Jiang	Qilu University of Technology, China
Chen Jiaqi	Dongguan University of Technology, China
Jin	Xi'an Jiaotong Liverpool University, China
Chunzhen Jin	Northeastern University, USA
Xiaozheng Jin	Qilu University of Technology, China
Vijay John	RIKEN, Japan
Kenneth Johnson	Auckland University of Technology, New Zealand
Seul Jung	Chungnam National University, China
Akbar K.	BITS Pilani, Goa Campus, India
Sweta Kaman	IIT Jodhpur, India
Keiji Kamei	Nishinippon Institute of Technology, Japan
Keisuke Kameyama	University of Tsukuba, Japan
Yoshimi Kamiyama	Aichi Prefectural University, Japan
Joarder Kamruzzaman	Federation University Australia, Australia
Bhavik Kanekar	IIT Mandi, India
Tomoyuki Kaneko	University of Tokyo, Japan
Jun-Su Kang	Kyungpook National University, South Korea
Shin'Ichiro Kanoh	Shibaura Institute of Technology, Japan
Nikola Kasabov	Auckland University of Technology, New Zealand
Arshpreet Kaur	NIT Jalandhar, India
Yoshinobu Kawahara	Osaka University/RIKEN, Japan
Hideaki Kawano	Kyushu Institute of Technology, Japan
Kostiantyn Khabarlak	Dnipro University of Technology, Ukraine
Mehshan Khan	Deakin University, Australia
Atikant Khanna	Auckland University of Technology, New Zealand
Daegyeom Kim	Korea University, South Korea
Jonghong Kim	Kyungpook National University, South Korea
Mutsumi Kimura	Ryukoku University, Japan
Irwin King	Chinese University of Hong Kong, China

Gisela Klette	Auckland University of Technology, New Zealand
Mallika Kliangkhlao	Walailak University, Thailand
Alistair Knott	Victoria University of Wellington, New Zealand
Kunikazu Kobayashi	Aichi Prefectural University, Japan
Rangachary Kommanduri	Indian Institute of Information Technology, Sri City, India
Aneesh Krishna	Curtin University, Australia
Rita Krishnamurthi	Auckland University of Technology, New Zealand
Adam Krzyzak	Concordia University, Canada
Sumant Kulkarni	Zenlabs, Zensar Technologies, India
Estine Kumar	University of the South Pacific, Fiji
Neetesh Kumar	IITR, India
Praveen Kumar	Banaras Hindu University, India
Rakesh Kumar	IIT BHU, India
Swaroop Kumar	L&T Technology Services, India
Chithrangi Kumarasinghe	University of Moratuwa, Sri Lanka
Anuradha Kumari	Indian Institute of Technology Indore, India
Naga Jyothi Kunchala	Massey University, New Zealand
Souraja Kundu	Indian Institute of Technology Guwahati, India
Hiroki Kurashige	Tokai University, Japan
Tomoki Kurikawa	Future University Hakodate, Japan
Kurnianingsih	Politeknik Negeri Semarang, Indonesia
Fatih Kurugollu	University of Sharjah, UAE
Thomas Lacombe	University of Auckland, New Zealand
Hamid Laga	Murdoch University, Australia
Edmund Lai	Auckland University of Technology, New Zealand
Leon Lange	University of California, San Diego, USA
Walter Langelaar	Victoria University of Wellington, New Zealand
Sang Hun Lee	Kookmin University, South Korea
Tet Chuan Lee	Auckland University of Technology, New Zealand
Yuan Lei	University of Chinese Academy of Sciences, China
Chi Sing Leung	City University of Hong Kong, China
Bingxian Li	Heilongjiang University, China
Bo Li	Baidu Inc, China
Fengyu Li	Beijing University of Posts and Telecommunication, China
Hongfei Li	Xinjiang University, China
Jiale Li	University of Auckland, New Zealand
Jun Li	Nanjing Normal University, China
Linjing Li	CASIA, China
Maodong Li	Soochow University, China

Mengmeng Li	Zhengzhou University, China
Mengshu Li	University of Toronto, Canada
Mengting Li	Beijing University of Posts and Telecommunications, China
Ming Li	Wuhan University of Technology, China
Peifeng Li	Soochow University, China
Ruifan Li	Beijing University of Posts and Telecommunications, China
Sirui Li	Murdoch University, China
Tieshan Li	Dalian Maritime University, China
Weiwei Li	UESTC, China
Xiaohong Li	Northwest Normal University, China
Yantao Li	Chongqing University, China
Yi Li	Lancaster University, UK
Yinbao Li	Baidu Tech., China
Yun Li	Nanjing University of Posts and Telecommunications, China
Ziwei Li	UCAS, China
Jiawen Liang	City University of Hong Kong, China
Xiaokun Liang	Shenzhen Institute of Advanced Technology, China
Xiaoyi Liang	Tongji University, China
Xu Liang	Harbin Institute of Technology, China
Zhuonan Liang	University of Sydney, Australia
Junfeng Liao	Shanghai University of International Business and Economics, China
Xiao-Cheng Liao	South China University of Technology, China
Jianpeng Lin	Guangdong University of Technology, China
Qiuhua Lin	Dalian University of Technology, China
Yang Lin	University of Sydney, Australia
Baodi Liu	China University of Petroleum, China
Binghao Liu	Beihang University, China
Gangli Liu	Tsinghua University, China
Hanyuan Liu	City University of Hong Kong, China
Jian-Wei Liu	China University of Petroleum, China
Juan Liu	Wuhan University, China
Lei Liu	Northeastern University, USA
Renyang Liu	National University of Singapore, Singapore
Shang Liu	UCAS, China
Weiteng Liu	China University of Petroleum, China
Wen Liu	Chinese University of Hong Kong, China
Xiaoyang Liu	Huazhong University Science & Technology, China

Yang Liu	Fudan University, China
Yanming Liu	Georgia Institute of Technology, USA
Yaozhong Liu	Australian National University, Australia
Zhaoyi Liu	KU Leuven, Belgium
Zhe Liu	Jiangsu University, China
Han Long	National University of Defense Technology, China
Jieting Long	University of Sydney, Australia
Chu Kiong Loo	University of Malaya, Malaysia
Andrew Lowe	Auckland University of Technology, New Zealand
Gewei Lu	Shanghai Jiao Tong University, China
Heng-Yang Lu	Nanjing University, China
Hongtao Lu	Shanghai Jiao Tong University, China
Weihai Lu	Peking University, China
Wenhao Lu	Nanyang Technological University, Singapore
Yu Lu	Shenzhen Technology University, China
Shijie Luan	Chinese Academy of Sciences, China
Fangzhou Luo	McMaster University, Canada
Róisín Luo	University of Galway, Ireland
Siwen Luo	University of Western Australia, Australia
Yuchuan Luo	National University of Defense Technology, China
Raymond Lutui	Auckland University of Technology, New Zealand
Jiancheng Lv	Sichuan University, China
Yuezu Lv	Beijing Institute of Technology, China
Liangfu Lyu	Federation University Australia, Australia
Qingguo Lü	Southwest University, China
Jing Ma	Auckland University of Technology, New Zealand
Jinwen Ma	Peking University, China
Ming Ma	Yeshiva University, USA
Yan Ma	Fudan University, China
Yuqi Ma	Chinese University of Hong Kong, China
Alexei Machado	Pontifical Catholic University of Minas Gerais, Brazil
Jyoti Maggu	Thapar Institute of Engineering and Technology, India
Tariq Mahmood	COMSATS Institute of Information Technology, Pakistan
Mufti Mahmud	King Fahd University of Petroleum and Minerals, Saudi Arabia
Snehashis Majhi	Inria Sophia Antipolis, France
Mishaim Malik	University of Auckland, New Zealand

Mamta	IIT Patna, India
Raquel Marasigan	University of Asia and the Pacific, Philippines
Gerard M. Freixas	Fudan University, China
Stefan Marks	Auckland University of Technology, New Zealand
Archana Mathur	Nitte Meenakshi Institute of Technology, India
Yoshitatsu Matsuda	Seikei University, Japan
Tomas Maul	University of Nottingham Malaysia Campus, Malaysia
Jacek Mańdziuk	Warsaw University of Technology, Poland
Yogendra Meena	Delhi University, India
Yi Mei	Victoria University of Wellington, New Zealand
Erik Meijering	University of New South Wales, Australia
Qing-Xin Meng	China University of Petroleum, China
Alexander Merkin	AUT, New Zealand
Cheng Miao	Guangxi Normal University, China
Ashish Mishra	University of Nevada Las Vegas, USA
Sajib Mistry	Curtin University, Australia
Tanja Mitrovic	University of Canterbury, New Zealand
Seiji Miyoshi	Kansai University, Japan
Mohammadreza Montazerijouybari	Polytechnique Montréal, Canada
Francesco C. Morabito	University of Reggio Calabria, Italy
Satoru Morita	Yamaguchi University, Japan
Mpatisi Moyo	AiTonomy, UK
Taslim Murad	Georgia State University, USA
Shingo Murakami	Chuo University, Japan
Dharmalingam Muthusamy	Bharathiar University, India
Chitrakala Muthuveerappan	Victoria University of Wellington, New Zealand
Muhan Na	Inner Mongolia University, China
Isao Nambu	Nagaoka University of Technology, Japan
Parma Nand	Auckland University of Technology, New Zealand
Ajit Narayanan	Auckland University of Technology, New Zealand
Gokulmuthu Narayanaswamy	IIT Kharagpur, India
Kiyohisa Natsume	Kyushu Institute of Technology, Japan
Azadeh Nazemi	Norwood Systems, Australia
Usman Nazir	Lahore University of Management Sciences, Pakistan
Elena Nechita	Vasile Alecsandri University of Bacau, Romania
Chandra Mohan Singh Negi	Siemens, India
Mehdi Neshat	University of Adelaide, Australia
Kourosh Neshatian	University of Canterbury, New Zealand
Frank Neumann	University of Adelaide, Australia

Bach Nguyen	Victoria University of Wellington, New Zealand
Bao Sinh Nguyen	HUST, Vietnam
Binh P. Nguyen	Victoria University of Wellington, New Zealand
Minh Nguyen	Auckland University of Technology, New Zealand
Quang Vinh Nguyen	Western Sydney University, Australia
Mukku Nisanth Kartheek	National Institute of Technology Warangal, India
Sou Nobukawa	Chiba Institute of Technology, Japan
Anupiya Nugaliyadde	Murdoch University, Australia
Jeongbin Ok	Victoria University of Wellington, New Zealand
Diego Oliva	Universidad de Guadalajara, Spain
Toshiaki Omori	Kobe University, Japan
Sihem Omri	Higher School of Communication of Tunis, Tunisia
Hideaki Orii	Fukuoka University, Japan
Seiichi Ozawa	Kobe University, Japan
Achmad Pahlevi	Auckland University of Technology, New Zealand
Susmita Palmal	Ramgarh Engineering College, India
Guangyuan Pan	Linyi University, China
Wenkai Pan	Linyi University, China
Pankaj Pandey	Indian Institute of Technology, Delhi, India
Jason Pang	Lincoln University, UK
Paresh Kumar Panigrahi	VIT-AP University, India
Dipendra Pant	Norwegian University of Science and Technology, Norway
Hyeyoung Park	Kyungpook National University, South Korea
Dipanjyoti Paul	Indian Institute of Technology Patna, India
Mangor Pedersen	Auckland University of Technology, New Zealand
Siamak Pedrammehr	Deakin University, Australia
Anjie Peng	Southwest University of Science and Technology, China
Nasca Peng	Statistics New Zealand, New Zealand
Yong Peng	Hangzhou Dianzi University, China
Joao Pereira	Imperial College London, UK
Krassie Petrova	Auckland University of Technology, New Zealand
Somnuk Phon-Amnuaisuk	Universiti Teknologi Brunei, Brunei Darussalam
Pasu Poonpakdee	Walailak University, Thailand
Vijay Prakash	Thapar University, India
Narinder Singh Punn	Mayo Clinic, Arizona, USA
Junjie Qi	Guangxi Normal University, China
Weihua Qiang	Tianjin University, China
Yu Qiao	Shanghai Jiao Tong University, China
Sitian Qin	Harbin Institute of Technology at Weihai, China

Xiaoyang Qu	Huazhong University of Science and Technology, China
Abdul Quadir	IIT Indore, India
Uday Kiran Rage	University of Aizu, Japan
Krishna Raghuwaiya	University of the South Pacific, Fiji
Ibrahim Rahman	Open Polytechnic of New Zealand, New Zealand
Jessica Rahman	University of Dhaka, Bangladesh
Shri Rai	Murdoch University, Australia
Surbhi Raj	Indian Institute of Technology Patna, India
K. Ramakrishnan	Auckland University of Technology, New Zealand
R. Kanesaraj Ramasamy	Multimedia University, Malaysia
Deepak Ranjan Nayak	Malaviya National Institute of Technology, Australia
Rabia Naseer Rao	University of Auckland, New Zealand
Munish Rathee	Auckland University of Technology, New Zealand
Ramesh Rayudu	Victoria University of Wellington, New Zealand
Khalid Raza	Jamia Millia Islamia, India
Jianfeng Ren	University of Nottingham Ningbo, China
Minsi Ren	Chinese Academy of Sciences, China
Shao Renrong	East China Normal University, China
Tobias Rettenmeier	University of Applied Sciences Heilbronn, Germany
Young Ju Rho	Tech University of Korea, South Korea
Daniel Riccio	University of Naples Federico II, Italy
Rishabh	University of Delhi, India
Horacio Gonzalez	Universidad de Guanajuato, Mexico
Gargi Roy	Brunel University London, UK
Kaushik Roy	West Bengal State University, India
Muhammad Fakhrur Rozi	National Institute of Information and Communications Technology, Japan
Ji Ruan	Auckland University of Technology, New Zealand
Khairun Saddami	Universitas Syiah Kuala, Indonesia
Michał Sadowski	Jagiellonian University, Poland
Amit Kumar Sah	South Asian University, India
Pranab Sahoo	Indian Institute of Technology Patna, India
Naveen Saini	Indian Institute of Information Technology Allahabad, India
Ken Saito	Nihon University, Japan
Toshimichi Saito	Hosei University, Japan
Md Sajid	IIT Indore, India
Ko Sakai	University of Tsukuba, Japan
Nazmus Sakib	Ahsanullah University of Science & Technology, Bangladesh

Rohit Salgotra	AGH University of Krakow, Poland
Michel Salomon	IUT Belfort-Montbéliard, France
Toshikazu Samura	Yamaguchi University, Japan
Xue Sang	Northwestern Polytechnical University, China
Takashi Sano	Tokyo University, Japan
Yassine Saoudi	Faculté des Sciences de Tunis, Tunisia
Naoyuki Sato	Future University Hakodate, Japan
Eri Sato-Shimokawara	Tokyo Metropolitan University, Japan
Seiya Satoh	Tokyo Institute of Technology, Japan
Wojciech Sałabun	West Pomeranian University of Technology, Poland
Erich Schikuta	University of Vienna, Austria
Eric Scott	MITRE Corporation, USA
Mahdi Setayesh	Microsoft Inc., USA
Noushath Shaffi	Sultan Qaboos University, Oman
Nida Shahab	Auckland University of Technology, New Zealand
Reza Shahamiri	University of Auckland, New Zealand
Jiaxing Shang	Chongqing University, China
Peng Shao	Jiangxi Agricultural University, China
Zhenzhou Shao	Capital Normal University, China
Sourabh Sharma	Avantika University, India
Megha Sharma	Indian Institute of Technology Mandi, India
Swakkhar Shatabda	BRAC University, Bangladesh
Hualei Shen	Henan Normal University, China
Zhixiang Shen	UESTC, China
Yin Sheng	Huazhong University of Science and Technology, China
Yongpan Sheng	Southwest University, China
Pumeng Shi	University of British Columbia, Canada
Qiushi Shi	NTU, Singapore
Xinxin Shi	Changchun University of Science and Technology, China
Hiroki Shibata	Tokyo Metropolitan University, Japan
Hayaru Shouno	University of Electro-Communications, Japan
Jiang Shuyu	Sichuan University, India
Shijing Si	Duke University, USA
Jiten Sidhpura	Sardar Patel Institute of Technology, India
Rangika Silva	Queensland University of Technology, Australia
Simeon Simoff	Western Sydney University, Australia
David Simpson	Massey University, New Zealand
Balkaran Singh	Auckland University of Technology, New Zealand
Om Singh	National Institute of Technology Patna, India

Shashank Singh	Kanpur Institute of Technology, India
Roopak Sinha	Deakin University, Australia
Soumen Sinha	Mahindra University, India
Stephen Skalicky	Victoria University of Wellington, New Zealand
Ferdous Sohel	Murdoch University, Australia
Upeka Somaratne	Murdoch University, Australia
Aiguo Song	Southeast University, China
Chao Song	Zhejiang Gongshang University, China
Haohao Song	Xiamen University, China
Liang Song	Fudan University, China
Xianfeng Song	South China University of Technology, China
Laxmi Soni	AKS University, India
Paul Sowman	Auckland University of Technology, New Zealand
Aleksei Staroverov	MIPT, Russia
Amy Stewart	University of Waikato, New Zealand
Martin Stommel	Auckland University of Technology, New Zealand
Jianbo Su	Shanghai Jiao Tong University, China
Xinyan Su	University of Chinese Academy of Sciences, China
Yila Su	Inner Mongolia University of Technology, China
Zhixun Su	Dalian University of Technology, China
Badri Narayan Subudhi	Indian Institute of Technology Jammu, India
Toshiharu Sugawara	Waseda University, Japan
John Sum	National Chung Hsing University, China
Alexander Sumich	Nottingham Trent University, UK
Hao Sun	Nankai University, China
Shiwen Sun	Inner Mongolia University, China
Shuo Sun	Inner Mongolia University, China
Tao Sun	Beihang University, China
Zhanquan Sun	University of Shanghai for Science and Technology, China
Suryavardan Suresh	New York University, USA
Kanata Suzuki	Fujitsu Limited, Japan
Satoshi Suzuki	NTT Yokosuka, Japan
Yoshimi Suzuki	University of Yamanashi, Japan
Mikołaj Słupiński	University of Wrocław, Poland
Izak Tait	Auckland University of Technology, New Zealand
Murtaza Taj	Lahore University of Management Sciences, Pakistan
Norikazu Takahashi	Okayama University, Japan
Hiroshige Takeichi	RIKEN, Japan
Takashi Takekawa	Kogakuin University, Japan

Hiroshi Tamura	Osaka University, Japan
Christine Nya-Ling Tan	Massey University, New Zealand
Chunyu Tan	Anhui University, China
Renzo Roel Tan	Nara Institute of Science and Technology, Japan
Ying Tan	Peking University, China
Zhengguang Tan	Guangdong University of Technology, China
Takuma Tanaka	Shiga University, China
Fengxiao Tang	Tohoku University, Japan
Haoran Tang	Beijing University of Posts and Telecommunications, China
Maolin Tang	Queensland University of Technology, Australia
Yang Tang	East China University of Science and Technology, China
Yihang Tang	Chongqing University of Posts and Telecommunications, China
Zerui Tang	Xiamen University, China
M. Tanveer	Indian Institute of Technology, Indore, India
Zerui Tao	Tokyo University of Agriculture and Technology, China
Jules-Raymond Tapamo	University of KwaZulu-Natal, South Africa
Prithvi Tarale	University of Massachusetts Amherst, USA
Shuhei Tarashima	NTT Communications Corporation, Japan
Yassin Terraf	Mohammed VI Polytechnic University, Morocco
Putthiporn Thanathamathee	Walailak University, Thailand
Veerakumar Thangaraj	National Institute of Technology Goa, India
Chuan Tian	University of Canterbury, New Zealand
Hao Tian	Zhejiang University of Technology, China
Aruna Tiwari	IIT Indore, India
Sadhana Tiwari	Galgotias College of Engineering & Technology, India
Ewaryst Tkacz	Silesian University of Technology, Poland
Kar-Ann Toh	Yonsei University, South Korea
Faranak Tohidi	Charles Sturt University, Australia
Cong Tran	QUOC, Vietnam
Chidentree Treesatayapun	Walailak University, Thailand
Richa Tripathi	Washington University in St. Louis, USA
Enmei Tu	Shanghai Jiao Tong University, China
Nitin Tyagi	IIT Roorkee, India
Hamid Usefi	Memorial University of Newfoundland, Canada
Alireza Valizadeh	Universitat de les Illes Balears, Spain
Manju Vallayil	Auckland University of Technology, New Zealand
Maryam Var Naseri	Victoria University of Wellington, New Zealand

Matthieu Vignes	Massey University, New Zealand
Nobuhiko Wagatsuma	Toho University, Japan
Shanchuan Wan	University of Tokyo, Japan
Tao Wan	Beihang University, China
Wang	City University of Hong Kong, China
Bin Wang	Nanjing University of Finance & Economics, China
Binqiang Wang	Chinese Academy of Sciences, China
Chao Wang	China Academy of Railway Sciences Corporation Limited, China
Chen Wang	Chinese Academy of Sciences, China
Chengliang Wang	Chongqing University, China
Chunshi Wang	Guilin University of Electronic Technology, China
Guanjin Wang	Murdoch University, Australia
Haizhou Wang	Sichuan University, China
Han Wang	Xidian University, China
Hao Wang	Shenzhen University, China
Haowen Wang	Alipay, Ant Group, China
Jiale Wang	Zhejiang Sci-Tech University, China
Jianzong Wang	Ping An Technology Co., Ltd., China
Jun-Wei Wang	University of Science and Technology Beijing, China
Kaier Wang	Volpara Health Technologies Ltd., New Zealand
Liang Wang	Beijing University of Technology, China
Liang Wang	Huazhong University of Science and Technology, China
Liantao Wang	Hohai University, China
Ming Hui Wang	China University of Petroleum, China
Nana Wang	Jiangsu Normal University, China
Peijun Wang	Anhui Normal University, China
Rui Wang	Ningbo University, China
Ruiying Wang	Southwest University of Science and Technology, China
Xianzhi Wang	University of Technology Sydney, Australia
Xiao Wang	Chongqing Normal University, China
Xiulin Wang	Dalian University of Technology, China
Yadi Wang	Henan University, China
Ye Wang	National University of Defense Technology, China
Yong Wang	Southeast University, China
Yongyu Wang	JD Logistics, China
Zhenni Wang	City University of Hong Kong, China

Zhongsheng Wang	University of Auckland, New Zealand
Zi-Peng Wang	University of Jinan, New Zealand
Ziwei Wang	Huazhong University of Science and Technology, China
Yoshikazu Washizawa	University of Electro-Communications, China
Fengchen Wei	University of Sussex, UK
Hongxi Wei	Inner Mongolia University, China
Alastair Wells	Auckland University of Technology, New Zealand
Guanghui Wen	RMIT University, Australia
Junjie Wen	Chinese University of Hong Kong, China
Jinta Weng	Chinese Academy of Sciences, China
Lei Wenjie	City University of Hong Kong, China
David White	Auckland University of Technology, New Zealand
Tom White	Victoria University of Wellington, New Zealand
Savindi Wijenayaka	University of Auckland, New Zealand
Michael Witbrock	University of Auckland, New Zealand
Hiutung Wong	City University of Hong Kong, China
Ka-Chun Wong	City University of Hong Kong, China
Kevin Wong	Murdoch University, Australia
Boqi Wu	University of Wuppertal, Germany
Chao Wu	Zhejiang University, China
Chengkun Wu	National University of Defense Technology, China
Haha Wu	Xinjiang University, China
Song Wu	Hainan Tropical Ocean University, China
Xianze Wu	Shanghai Jiao Tong University, China
Zipeng Wu	University of Birmingham, UK
Zhiqiu Xia	Rutgers University, USA
Ziwei Xiang	Chinese Academy of Sciences, China
Jinying Xiao	Changsha University of Science & Technology, China
Qiang Xiao	Huazhong University of Science and Technology, China
Xi Xiao	University of Alabama at Birmingham, USA
Shiwen Xie	Central South University, China
Zaipeng Xie	Hohai University, China
Yucheng Xing	Stony Brook University, USA
Wang Xinshen	Guangdong University of Foreign Studies, China
Wenxin Xiong	City University of Hong Kong, China
Chao Xu	Changsha University of Science and Technology, China
Fanchao Xu	University of Science and Technology of China, China

Jianhua Xu	Nanjing Normal University, China
Jinhua Xu	East China Normal University, China
Lele Xu	Southeast University, China
Qing Xu	Tianjin University, China
Xinyue Xu	Hong Kong University of Science and Technology, China
Junyu Xuan	University of Technology Sydney, Australia
Felix Yan	Victoria University of Wellington, New Zealand
Shankai Yan	Hainan University, China
Teng Yan	Shenzhen University, China
Weiqi Yan	AUT, New Zealand
Da Yang	Beijing University of Aeronautics and Astronautics, China
Haitian Yang	Chinese Academy of Sciences, China
Jie Yang	Shanghai Jiao Tong University, China
Mengyu Yang	Beijing University of Posts and Telecommunications, China
Minghao Yang	Chinese Academy of Sciences, China
Peipei Yang	Chinese Academy of Science, China
Peng Yang	Chongqing Three Gorges University, China
Qinmin Yang	Zhejiang University, China
Wei Yang	University of Chinese Academy of Sciences, China
Yanfeng Yang	South China University of Technology, China
Zhan Yang	Central South University, China
Zhikai Yang	KTH Royal Institute of Technology, Sweden
Wangshu Yao	Soochow University, China
Yifeng Yao	University of South China, China
Yun Ye	Intel, USA
Wang Yingying	Shenyang Aerospace University, China
Yongmin Yoo	Macquarie University, Australia
Shuyuan You	Tianjin University, China
Dianzhi Yu	Chinese University of Hong Kong, China
Na Yu	Zhejiang University, China
Ping Yu	Nanjing University of Science and Technology, China
Wenxin Yu	Southwest University of Science and Technology, China
Zhibin Yu	Ocean University of China, China
Zhiwen Yu	South China University of Technology, China
Fangfang Yuan	Chinese Academy of Sciences, China
Xin Yuan	Southeast University, China

Dmitry Yudin	Moscow Institute of Physics and Technology, Russia
Chao Yue	University of Chinese Academy of Sciences, China
Xiaodong Yue	Shanghai University, China
Farzana Zahid	University of Waikato, New Zealand
Ruizhe Zeng	Chinese Academy of Sciences, China
Weixin Zeng	National University of Defense Technology, China
Xinhua Zeng	Fudan University, China
Zekeng Zeng	Chinese Academy of Sciences, China
Ewa Zeslawska	University of Rzeszow, Poland
Zhiyuan Zha	Renmin University of China, China
Chao Zhang	Shanxi University, China
Chao Zhang	South China Normal University, China
Chenyi Zhang	University of Canterbury, New Zealand
Chong Zhang	Xi'an Jiaotong-Liverpool University, China
Chufan Zhang	Shanghai Jiao Tong University, China
Congwei Zhang	Southeast University, China
Fan Zhang	Shandong Technology and Business University, China
Francis X. Zhang	Durham University, UK
Gaoyan Zhang	Tianjin University, China
Han Zhang	Northwest University, China
Haoyang Zhang	Chongqing University of Posts and Telecommunications, China
Hongtao Zhang	Kochi University of Technology, Japan
Jiahui Zhang	Beijing University of Technology, China
Jinchuan Zhang	University of Electronic Science and Technology of China, China
Kai Zhang	East China Normal University, China
Kang Zhang	Kyushu University, Japan
Li Zhang	Soochow University, China
Qian Zhang	Jiangsu Open University, China
Ruixiao Zhang	University of Southampton, UK
Ting Zhang	Central China Normal University, China
Tong Zhang	China Mobile Research Institute, China
Weili Zhang	Xi'an Jiaotong University, China
Weilun Zhang	Tianjin University, China
Wendy Zhang	University of Canterbury, New Zealand
Xiaowei Zhang	Qingdao University, China
Xu Zhang	Jiangsu Normal University, China

Xulong Zhang	Ping An Technology (Shenzhen) Co., Ltd., China
Xunhui Zhang	National University of Defense Technology, China
Yang Zhang	City University of Hong Kong, China
Yangsong Zhang	Southwest University of Science and Technology, China
Zijing Zhang	University of Waikato, New Zealand
Bo Zhao	Beijing Normal University, China
Hui Zhao	University of Jinan, China
Jianhui Zhao	Wuhan University, China
Jigui Zhao	Xinjiang University, China
Jing Zhao	Qilu University of Technology, China
Keer Zhao	Zhejiang University of Technology, China
Ming Zhao	Central South University, China
Rui Zhao	University of Technology Sydney, Australia
Runjie Zhao	Zhejiang University, China
Xujian Zhao	Southwest University of Science and Technology, China
Yan Zheng	Kyushu University, Japan
Yingtao Zheng	University of Auckland, New Zealand
Yuchen Zheng	Shihezi University, China
Guoqiang Zhong	Ocean University of China, China
Jinghui Zhong	South China University of Technology, China
Ping Zhong	Central South University, China
Guangchong Zhou	Chinese Academy of Science, China
Jinjia Zhou	Hosei University, China
Shihua Zhou	Dalian University, China
Xiao-Hu Zhou	Chinese Academy of Sciences, China
Xinyu Zhou	Jiangxi Normal University, China
Yu Zhou	National University of Defense Technology, China
Zhenxiong Zhou	National University of Defense Technology, China
Leyi Zhu	University of Macau, China
Qiaoming Zhu	Soochow University, China
Qiyuan Zhu	Swinburne University of Technology, Australia
Xuanying Zhu	Australian National University, Australia
Yuesheng Zhu	Peking University, China
Wang Ziling	Sichuan University, China
Yuan Zong	Southeast University, China
Xu Zou	Zhejiang University, China

Contents – Part XVI

Optimisation of Fibre Selection for Tubes Production in Manufacturing of Optic Cables

Zbigniew Gomolka[1]([✉])(iD), Ewa Zeslawska[1](iD), and Lukasz Olbrot[2]

[1] College of Natural Sciences, University of Rzeszow, Rejtana Street 16C, 35–959 Rzeszow, Poland
{zgomolka,ezeslawska}@ur.edu.pl
[2] FIBRAIN Sp. z.o.o., Zaczernie 190F, 36–062 Zaczernie, Poland
l.olbrot@fibrain.pl

Abstract. The large–scale industrial production of fiber optic cables is a multi–stage, complex process that requires the use of raw materials and semi–finished products at various stages of their manufacturing. One of the key problems in cable production is tube extrusion. Fiber is used for its production. Depending on customer orders, the production process is carried out in appropriate stages. Minimizing production costs necessitates time–consuming planning work to select the proper capacity of fiber spools so that the amount of waste generated is low or zero. In this work, an intelligent algorithm for selecting fiber optic strands based on warehouse stock analysis and declared order segments is proposed. Due to the strongly NP–hard nature of the problem, algebraic–logical meta-modeling technology was applied. A model was constructed that enables the search for feasible solutions to the spool matching problem within a finite time, meeting technological requirements and minimizing the amount of fiber scraps as well as retooling resulting from the generated production scenario. The simulation part of the work proposes the use of a parallel allocation mechanism of dedicated computational molecules, enabling the realization of this task in a parallel environment, thereby reducing the solution search time.

Keywords: selection of fibers for tube production · algebraic–logic meta modelling · NP–hard problem · parallel computation execution

1 Introduction

The production of fiber optic cables is a technologically advanced process that brings significant benefits to the field of data transmission technology. Despite the challenges and high costs, fiber optic cables are a key component of modern telecommunication networks, offering high bandwidth and reliable data transmission. Fiber optic cables consist of one or more optical fibers, which are capable of transmitting data using light. These fibers are made from very pure glass or plastic and have a diameter of about $8 \div 10\,\mu m$. The production process of fiber optic cables typically is divided into two main stages:

M. Mahmud et al. (Eds.): ICONIP 2024, CCIS 2297, pp. 1–15, 2026.
https://doi.org/10.1007/978-981-96-7036-9_1

- The stage of fiber optic production, where the process of creating a preform, a glass cylinder that serves as the basis for drawing the actual fiber, is carried out. The result is a semi–finished product in the form of spools with fiber of a specified length (usually around 56 km) and a specified color of the coating covering the fiber.
- The actual production of fiber optic cables, which begins with the arrangement of the fibers into loose tubes. Next, the cable core is assembled, with the loose tubes arranged in a specific pattern around a central reinforcing element. The number of fibers in each tube and the number of tubes selected are determined by the technological specifications and transmission parameters of the cable being produced. This stage of the production process typically occurs in segments, which are dictated by the specifications of individual customer orders.

The lengths of the produced fibers, which constitute a semi–finished product, are typically limited by the capacity of the spools on which they are wound. The selection of the length of individual fibers for the production of individual tube segments must meet the requirement of continuity in each section (fibers are not welded due to the high time consumption and cost of this process). Therefore, the optimization problem considered in this work for selecting fibers for the production of tubes for fiber optic cables involves choosing individual fibers in such a way that eliminates the need for splicing during the tube production stage and simultaneously minimizes the amount of waste generated at this stage. Additionally, the selection of so–called spool proposals for tube production should reduce the number of necessary retoolings of the feeding module associated with each change of the required sets of spools needed to produce sets of tubes for a given cable segment or a set of such segments. In the currently used tube production technologies, the fiber selection and production planning stage is carried out by a dedicated expert due to the lack of available algorithmic solutions that would allow for the automation of this process. This requires both additional financial expenditures and extra time. Typically, such a manually conducted selection process is invariably associated with large amounts of waste, further increasing production costs [1,4,7–11,13,14].

In previous scientific research conducted by the authors, it has been demonstrated that applying mathematical modeling of real production processes using algebraic–logic technology allows for the construction of intelligent algorithms that ensure the attainment of optimal or acceptable solutions within a given system [12]. The conducted studies confirmed the feasibility of using algebraic–logic metamodelling technology to optimize processes with high complexity and variability of system resources. In works [2,3,5,6,12], it has been shown that processes described using algebraic-logic technology can be effectively optimized regardless of disturbances influencing that system.

In this work, an intelligent system for fiber selection during the production of tubes for fiber optic cables has been proposed. This system will involve optimizing the selection of spool sets to minimize fiber waste and the number of necessary retoolings of the production line feeding modules. Additionally, to reduce the time spent searching for acceptable solutions, the use of a parallel allocation mechanism for dedicated computational resources has been proposed. This allows the task to be executed in a parallel environment, thereby shortening the solution search time.

2 Materials and Methods

The considered problem of optimizing the management of the warehouse of raw materials and semi-finished products in the production of fiber optic cables is a key element that can bring significant benefits in the form of cost reduction, minimization of losses and improvement of operational efficiency in the production process of fiber optic cables. For the research purposes of this study, real data was obtained from FIBRAIN, the largest Polish manufacturer specializing in the production and delivery of fiber optic technology solutions. Considering the specifics of the production planning and monitoring process for fiber optic cables, a solution has been proposed consisting of two synchronized intelligent subsystems 2 and 3 depicted in Fig. 1, which perform two key tasks of production process planning and control respectively. In the further part of this work, a solution regarding the automation and optimization of managing the selection of raw materials and semi-finished products for a sample task related to the fiber optic cable production process implemented by module 2 will be presented.

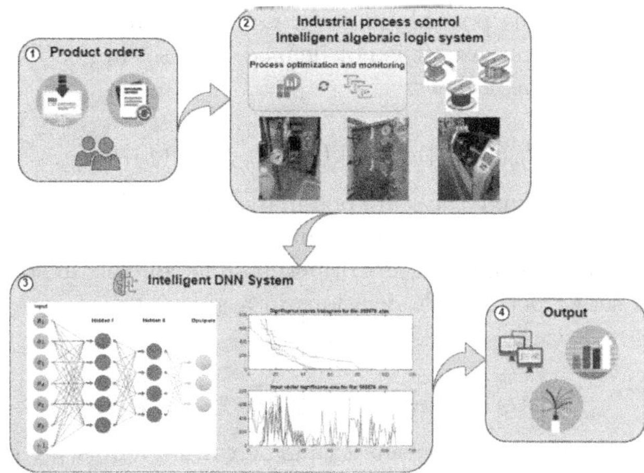

Fig. 1. The structure of the production planning and monitoring system for fiber optic cables.

2.1 ALMM Technology for Single Processor Environment

ALMM (Algebraic Logical Meta Model) is the name of a mathematical modeling technology used to formally describe a multi–stage decision-making process [2, 5].

An example optimization problem, before applying the algebraic-logical technology, requires the creation of a mathematical model that includes definitions in algebraic-logical terminology of permissible states, state transition functions, sets of possible and permissible decisions, forbidden states, and target states. The classic ALMM model for a single-processor computational environment assumes that the process P of selecting optical fiber for the production of tubes for fiber optic cables is uniquely determined by the sextuple:

$$P = (U, S, s_0, f, S_N, S_G) \tag{1}$$

where: U – the set of all decisions defining possible scenarios for fiber selection, $S = X \times T$ – the set of generalized states, X – the set of proper states, $s_0 = (x, t)$– the generalized initial state, $s_0 \in S$, $T \subset \mathbb{R}^+ \cup 0$ – a subset of non-negative real numbers representing time instances, $f : U \times S \to S$ – the transition function, defined using functions $f = (f_x, f_t)$, $f_x : U \times X \times T \to X$ specifies the next proper state, $f_t : U \times X \times T \to T$ specifies the next time moment and meets the condition $\Delta t = f_t(u, x, t) - t > 0$ and assumes a finite value. The sets U, X, T are non–empty sets, $S_N \subset S$ – the set of unacceptable generalized states, $S_G \subset S$ – a non–empty set of target generalized states, i.e., the states in which the process should end up as a result of making proper decisions. Usually, the transition function is defined as a partial function (defined only for certain pairs $(u, s) \in U \times S$) , which allows for considering all constraints regarding fiber selection decisions using so-called sets of possible decisions in state s, denoted as $U_p(s)$. If decision u is possible in state s, the transition function is defined for that pair (u, s). Otherwise, it is not defined. The set of possible decisions in state $U_p(s)$, is defined as follows:

$$U_p(s) = \{u \in U : (u, s) \in Dom(f)\} \tag{2}$$

The definition specifying the set of possible states is an essential component of the transition function definition. Constraints on generalized states defining S_N, can also be considered using the sets of permissible controls in state s, denoted as $U_d(s)$. The set of permissible decisions in a given state $U_d(s)$ is defined as follows:

$$U_d(s) = \{u \in U_d : f(u, s) \notin S_N\} \tag{3}$$

In the most general case, the sets U and X can be represented as Cartesian products $U = U^1 \times U^2 \times \cdots \times U^m$, $X = X^1 \times X^2 \times \cdots \times X^m$ (upper indices denote successive sets, not set powers). Generally, both decision u and state x are vectors $u = (u^1, u^2, \ldots u^m)$, $x = (x^1, x^2, \ldots x^n)$, where the individual coordinates of vector u represent separate decisions. It is important to note that, unlike control using physical signals, in the adopted simplified model, the decision-making moment does not have to coincide with the moment of its implementation. For the considered problem of selecting raw materials, the cardinality of the set $U_d(s)$ directly depends on the state of the raw material warehouse.

2.2 Description of the Problem at Hand

Figure 2 illustrates the modular diagram of the optimization system for selecting spool proposals, taking into account the minimization of fiber waste during the production of fiber optic cables.

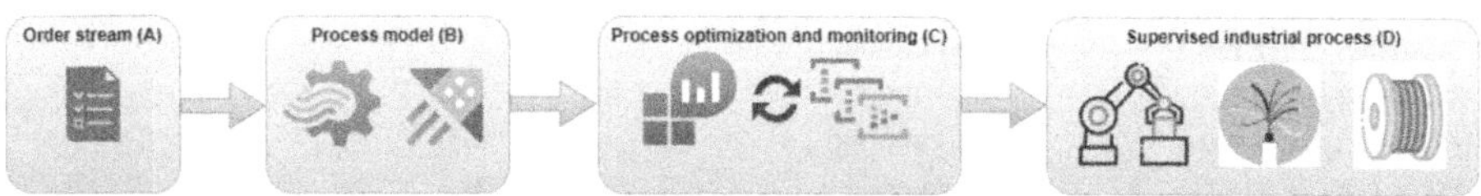

Fig. 2. The modular diagram of the fiber selection process implemented in block 2 of Fig. 1

Coming production tasks, i.e. specific orders of fiber optic sets, can appear in the system according to the properties of the process state vector (block A). Block B is responsible for extracting the properties of the input data necessary for the operation of the algebraic-logical model, and is responsible for inputting the properties of the input data provided to the optimization and process control system (C). The process monitoring system allows dynamically (division of the solved problem into subproblems with respect to selected variables, which is not disjoint, but must be characterized by the property of optimal substructure) to determine the quantities: U, S, s_0, f, S_N, S_G. These quantities are necessary for on–line determination of subsequent decisions controlling the process. Block D is a supervised industrial process, which, in the present case, implements the selection and optimization of fiber storage resources to a specific decision sequence. Taking into account the peculiarities of the computational problem in question, an analysis of the numerical complexity of the process of finding optimal or feasible solutions for hypothetical sample orders was carried out. It can be shown that the number of elements in the permutation matrix forming the decision set can be estimated as: the number of fiber colors available in the spool resources of the warehouse (N_{CS}) raised to the power of the average amount of unique spools (N_{aaus}) in a given color: $(N_{CS})^{N_{aaus}}$. For example, 12 fiber colors with an average of 50 spools of varying lengths in each color: $(N_{CS})^{N_{aaus}} = 50^{12} = 2.4414e + 20$ defines the number of permutations determining the power of the decision set (U_p). The cardinality of the decision set must be reduced by seeking algebraic-logical constraints concerning certain specific features, such as common multiples, the technologically permissible maximum length of a uniform order segment, acceptable scrap size, etc. Assuming allocation constraints for data structures appropriate for x64 architecture, amounting to 2.8147e+14, it was assumed that the modified ALMM model should have the ability to adapt to multiprocessor architectures enabling parallel search for solution trajectories. This way, it will be possible to search for an optimal or feasible solution in an acceptable time. Examples of actual orders flowing into the designed system are shown in Table 1.

Table 1. Overview of selected orders.

Order number	Item number	Commodity code Quantity [lm]	Number of cable sections	Section length
ZW-7/3/21/EKS	1	4000.0000	2	2000
ZW-22/3/21/EKS	1	4000.0000	1	4000
ZW-51/3/21/EKS	1	38000.0000	9	4000
ZW-51/3/21/EKS	1	38000.0000	1	2000
ZW-20/3/21/EKS	1	20000.0000	5	4000
ZW-20/3/21/EKS	2	60000.0000	60	1000
ZW-56/3/21/EKS	1	10000.0000	5	2000
ZW-61/3/21/EKS	1	4000.0000	1	4000
ZW-61/3/21/EKS	2	4000.0000	1	4000
ZW-61/3/21/EKS	3	4000.0000	1	4000

The order entered into the system is sequentially processed in an reasonable time by the module searching for optimal spool proposals, which takes into account the adopted components of the optimization criterion in the calculations.

2.3   Parallelisation of ALMM Technology for Fibre Selection

Considering the computational limitations regarding the search for solution trajectories in the state space, a parallel mechanism for dividing the problem P into subproblems P^n was introduced, where the solution for each subproblem is sought in a separate computational node:

$$P^{\parallel} = \left[P^1, P^2, ..., P^n \right] \tag{4}$$

where $n \in \{1, 2, ..., N\}$, N denotes the number of available computational nodes, whereas $P^{\parallel}$ is the global optimization process implemented on a parallel platform. In the notation of parallel computations, this is a *Fork-Join* model. Assuming that the set of spools with fibers in specific colors has an order $c = 1, ..., N_{CS}$, we can then divide the decision set U into subsets:

$$U = \bigcup_{n=1}^{N_{CS}/N} U^{(n)} \wedge \forall_{i,j \in (1,...,N_{CS}/N)} \left(U^{(i)} \cap U^{(j)} \right) = \varnothing, (i \neq j) \tag{5}$$

Thus, for each computational node, we have:

$$P^n = \left(U^{(n)}, S^{(n)}, s_0^{(n)}, f, S_N^{(n)}, S_G^{(n)} \right) \tag{6}$$

It should be emphasized here that the transition function f remains the same for all computational nodes. Moreover, due to the technological aspects of the process and the construction of the fiber feeding device, as well as the even number of available core variants, the expression N_{CS}/N takes only integer values. Accordingly, we have:

$$U_p^{(n)}(s) = \left\{ u^{(n)} \in U^{(n)} : \left(u^{(n)}, s^{(n)} \right) \in Dom(f) \right\} \tag{7}$$

Utilizing the properties of the multiprocessor environment, we can simultaneously define the subsets of permissible decisions as follows:

$$U_d^{(n)}\left(s^{(n)}\right) = \left\{u^{(n)} \in U_d^{(n)} : f\left(u^{(n)}, s^{(n)}\right) \notin S_N^{(n)}\right\} \tag{8}$$

It should be noted that the sets $S_N^{(n)}$, depending on the properties of the process, can be defined independently of the computational node or may have separate definitions related to the properties of $U_d^{(n)}$. The ranges of successive groups of fibers in specific colors, which will determine the cardinality of the individual decision sets in the respective computational nodes, can be expressed in the form of a sliding window:

$$u^{(n)} = \left[u_{i+k\cdot s'}^{(n)}, u_{i+k\cdot s'+1}^{(n)}, \ldots, u_{i+k\cdot s'+w'-1}^{(n)}\right] \tag{9}$$

where: s' – the window shift size, w' – the window width, k – the shift index. Therefore, in the general case, we have:

$$m' = \left\lfloor \frac{(N_{CS}/N) - i - w'}{s'} \right\rfloor + 1 \tag{10}$$

In the example order (see Fig. 3), which involves producing a 12-fiber tube, a solution is sought on a cluster of 4 computational nodes, forming 4 sets of window groups, each containing 3 colors $c \in \{[1\ 2\ 3], [4\ 5\ 6], [7\ 8\ 9], [10\ 11\ 12]\}$.

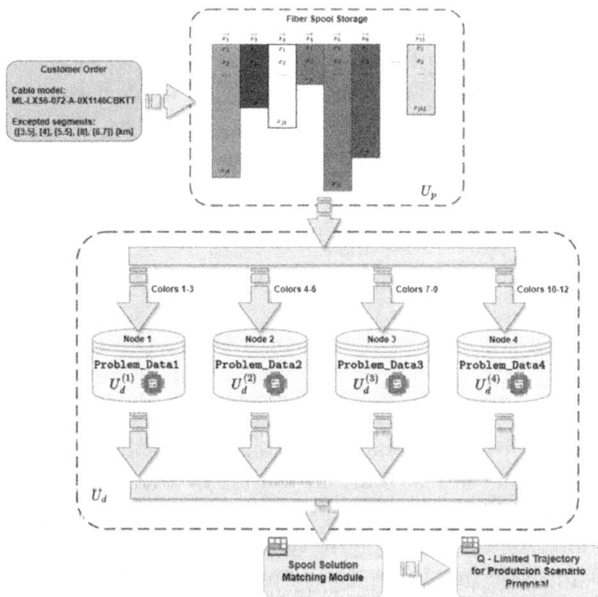

Fig. 3. Imposing fiber spool storage data into parallel structure of computing nodes

Assuming the selected group of spools with fibers in chosen colors $c(n)$, the space of possible spool configurations will take the form:

$$\forall_{n=1,\ldots,N}, \exists_{c(n)} : X^{(n)} = x^{c(n)(1)} \times x^{c(n)(2)} \times x^{c(n)(3)} \tag{11}$$

where:

$$x^{c(n)(i)} = \left\{ x_{(1,i)}, x_{(2,i)}, \ldots, x_{(j^{c(n)},i)} \right\} \tag{12}$$

2.4   Optimization Criterion of the Problem

In considered case parallelised ALMM model is equivalent to a suitable multi-stage decision problem of the process $P^{\|}$ denoted as a $\left(P^{\|}, Q^{\|}\right)$ pair where $Q^{\|}$, is a parallelised optimization criterion. The objective of optimizing the considered problem is to select successive spool proposals to minimize the criterion $Q^{\|}$, where its value determines the total scrap amount and the number of necessary changeovers. Therefore, we are looking for a finite sequence of permissible decisions $\tilde{u}^{(n)*} \in \tilde{U}_d^{(n)}$ in each of the n nodes such that:

$$Q^{\|} = \sum_{n=1}^{N} Q^{(n)} \left(\tilde{u}^{(n)*} \right) \tag{13}$$

and:

$$Q^{(n)} \left(\tilde{u}^{(n)*} \right) = \min_{\tilde{u}^{(n)*} \in \tilde{U}_d^{(n)}} Q^{(n)} \left(\tilde{u}^{(n)} \right) \tag{14}$$

where: $\tilde{u}^{(n)*}$ – denotes the optimal finite sequence of decisions, $\tilde{U}_d^{(n)}$ – denotes the set of all permissible decision sequences within a given computational node, which define permissible trajectories belonging to the set $S^{(n)}$. The total value of the objective function within a given node n is the sum of increments of the objective function in successive states:

$$Q^{(n)} \left(\tilde{u}^{(n)} \right) = \Delta Q^{(n)} \left(u_0^{(n)}, x_0^{(n)}, t_0 \right) + \cdots + \Delta Q \left(u_i^{(n)}, x_i^{(n)}, t_i \right) + \ldots$$
$$+ \Delta Q^{(n)} \left(u_{\omega-1}^{(n)}, x_{\omega-1}^{(n)}, t_{\omega-1} \right) \tag{15}$$

where: $\Delta Q^{(n)}(u_i^{(n)}, x_i^{(n)}, t_i)$ represents the cost increment in the tube production model resulting from implementing decision $u_i^{(n)}$ in state $s_i^{(n)} = (x_i^{(n)}, t_i)$. ω denotes the index of the last state in the trajectory. In the considered problem, this is the final segment of tube production in the adopted sequence of decisions. The form of the quality criterion affects the model's state structure, decisions, and transition functions. The sequence of decisions sought to minimize the criterion $Q^{\|}$ is determined based on a appropriately designed optimization algorithm.

3   Results and Discussion

In the context of the considered problem, variants have been studied where the selection of raw materials required by the BOM (Bill of Materials) for this semi–finished product is based on the stock status for a test set with different raw material lengths and selection for 12 different fiber colors. This variant presents both the NP–hard problem of selecting color fibers and includes a complex production cycle (a large number of possible segments for continuous or combined production with changeovers). In addition, threshold conditions have been set for the algorithm to be in compliance with:

- minimum length – the residue of the fiber, which is treated as a scrap due to its uneconomical reuse;
- cost variables resulting from the dependence of changeovers and rearming of lines within a given selection;
- set lengths – the residual fiber lengths that the selection process strives for, so that reuse coincides with the most frequently ordered lengths by customers.

To achieve this, a data structure named `Problem_Data{p,i}` was prepared to store the data of the considered problem, which can be replicated to individual computational nodes operating in the MIMD (Multiple Instruction, Multiple Data) model. The structure of the computational molecule is as follows:

```
Problem_Data{p,n}
    Segments
    Colors_Parts
    Span_type
    Indexes
    Spool_proposal
    Scrap
    Retoolings
    Estimated_cost
```

Listing 1.1. Structure that stores data

where p denotes the problem under consideration (specific order), while n is the index of the computational node where the local solution trajectory is sought. Examples of the values taken by each computational molecule:

```
Problem_Data{p,n}
          Segments: [20.2000  7.5000]
      Colors_Parts: [1  2  3]
         Span_type: {[1  2  3]    [4  5]}
           Indexes: [43,  1]
    Spool_proposal: [37800  37800  37800;  20400  12200  20400]
             Scrap: [17200  17200  17200;  12500  4300  12500]
        Retoolings: 1
    Estimated_cost: Scrap length:  0   Retoolings:  1 and
        estimated total cost: 12 PLN
```

Listing 1.2. Exemplary local results within node $n = 1$

For the above example, the number of hypothetical changeovers `Retoolings=1` (the production line start inherent in each order start is not included in the total number of changeovers). On the other hand, the total Scrap satisfying the condition mentioned above equals 0. Taking a test order, consisting of five sections with lengths $\{[3.5000]\}$ $\{[4]\}$ $\{[5.5000]\}$ $\{[8]\}$ $\{[6.7000]\}$, we get proposals to manufacture the tube on production lines in the scenarios shown in Fig. 4.

```
Exemplary order content in km's:
     {[3.5000]}     {[4]}     {[5.5000]}      {[8]}     {[6.7000]}
===============================================================================
Set of possible segments order:
===============================================================================
  1.  {[6.7000]}   {[    8]}   {[5.5000]}   {[    4]}   {[3.5000]}
  2.  {[6.7000]}   {[    8]}   {[5.5000]}   {[3.5000]}   {[    4]}
  3.  {[6.7000]}   {[    8]}   {[    4]}   {[5.5000]}   {[3.5000]}
  4.  {[6.7000]}   {[    8]}   {[    4]}   {[3.5000]}   {[5.5000]}
        . . .         . . .         . . .         . . .         . . .
 25.  {[    8]}   {[6.7000]}   {[5.5000]}   {[    4]}   {[3.5000]}
 26.  {[    8]}   {[6.7000]}   {[5.5000]}   {[3.5000]}   {[    4]}
 27.  {[    8]}   {[6.7000]}   {[    4]}   {[5.5000]}   {[3.5000]}
 28.  {[    8]}   {[6.7000]}   {[    4]}   {[3.5000]}   {[5.5000]}
        . . .         . . .         . . .         . . .         . . .
 49.  {[5.5000]}   {[6.7000]}   {[    8]}   {[    4]}   {[3.5000]}
 50.  {[5.5000]}   {[6.7000]}   {[    8]}   {[3.5000]}   {[    4]}
 51.  {[5.5000]}   {[6.7000]}   {[    4]}   {[    8]}   {[3.5000]}
 52.  {[5.5000]}   {[6.7000]}   {[    4]}   {[3.5000]}   {[    8]}
        . . .         . . .         . . .         . . .         . . .
 73.  {[    4]}   {[6.7000]}   {[    8]}   {[5.5000]}   {[3.5000]}
 74.  {[    4]}   {[6.7000]}   {[    8]}   {[3.5000]}   {[5.5000]}
 75.  {[    4]}   {[6.7000]}   {[5.5000]}   {[    8]}   {[3.5000]}
 76.  {[    4]}   {[6.7000]}   {[5.5000]}   {[3.5000]}   {[    8]}
        . . .         . . .         . . .         . . .         . . .
 97.  {[3.5000]}   {[6.7000]}   {[    8]}   {[5.5000]}   {[    4]}
 98.  {[3.5000]}   {[6.7000]}   {[    8]}   {[    4]}   {[5.5000]}
 99.  {[3.5000]}   {[6.7000]}   {[5.5000]}   {[    8]}   {[    4]}
100.  {[3.5000]}   {[6.7000]}   {[5.5000]}   {[    4]}   {[    8]}
        . . .         . . .         . . .         . . .         . . .
120.  {[3.5000]}   {[    4]}   {[5.5000]}   {[    8]}   {[6.7000]}
```

Fig. 4. An excerpt of the obtained scenarios for the considered problem

For such a number of combinations of production sequence for individual segments, permissible mergers are possible, which must adhere to additional algebraic-logical conditions limiting the decision set's power. These conditions stem from technological constraints and potential additional requirements associated with constraints related to the power of the available set, which is determined by the inventory state. Table 2 presents a list of all possible segment spans.

For example, the notation $\{[1\ 2\ 3\ 4]\}$ $\{[5]\}$ means, respectively, the sequence of production of the combined segments 1:4 and the change of the line and continuation of production of segment 5. The product for which the raw material selection trajectory is constructed is a 6–tube fiber optic cable containing 12–fiber tubes (technology designation MK–LXS6–072–A–0X1146CBKTT). On the demonstrated example, four processes of solution trajectory search are performed, which is the scenario of producing a sequence of tube sections with the optimization criterion $Q^{\|}$: final scrape size S_{CR}. Below is an example of a fiber

Table 2. List of all possible segments spans:.

No.	List	No.	List
1	{[1 2 3 4 5]}	9	{[1]} {[2 3 4 5]}
2	{[1 2 3 4]} {[5]}	10	{[1]} {[2 3 4]} {[5]} 4
3	{[1 2 3]} {[4 5]}	11	{[1]} {[2 3]} {[4 5]}
4	{[1 2 3]} {[4]} {[5]}	12	{[1]} {[2 3]} {[4]} {[5]}
5	{[1 2]} {[3 4 5]}	13	{[1]} {[2]} {[3 4 5]}
6	{[1 2]} {[3 4]} {[5]}	14	{[1]} {[2]} {[3 4]} {[5]}
7	{[1 2]} {[3]} {[4 5]}	15	{[1]} {[2]} {[3]} {[4 5]}
8	{[1 2]}{[3]} {[4]} {[5]}	16	{[1]} {[2]} {[3]} {[4]} {[5]}

selection that takes into account the specific content of the stock, when the number of spools of the selected length is significantly larger than the rest of the stock items and should naturally be considered as the "first choice" in the production sequence planning process. This variant proposes the scenario, for which the following amount of scrap is produced:

$$Q^{\|}(S_{CR}) = \sum_{j=1}^{Seg} \sum_{i=1}^{N_{CS}} s_{cr_j}(i) \tag{16}$$

where the number of fibers of the produced tube $N_{CS} = 12$; and the number of anticipated segments $Scg = 2$. In the final stsage of the simulation of the selection of raw materials for the implementation of the order variant in question, the scenario of the production strategy proposal is shown, which takes into account different values of the optimization factor, which is the obtained scrap $Q^{\|}(S_{CR})$. Table 3 shows the 3 exemplary spools set proposal, within I, II and III row.

Table 3. The proposed spools set for given production stage

Elapsed time is 31.252617 seconds.				
Finally proposed tube production sequence.				
Indexes of spool proposals in **Problem_Data{1,1:4}** and Scrapes:				
6	4	4	4	0
1	1	1	1	3 900
4	3	3	3	6 300
Total scrap caused by assumed production scenario:				10 200

Assuming above variant in which operator decided for technological and process reasons to execute scenario of tubes to be produced, the ratio S_{CR} reaches a final value of $Q^{\|}(S_{CR}) = 10.2\,[km]$. The individual spools statements that would enable such a production scenario for stages I–III are shown in Tables 4, 5, 6 and 7.

Table 4. Proposed spools set for molecule – `Problem_Data{1,1}`

No.	Segments [km]	Color_Parts	Span_type	Indexes	Spool_proposal [m]	Scrap [m]	Retoolings	Estimated_cost
1	[20.2000,7.5000]	[1,2,3]	[1,2,3];[4,5]	[43;1]	[35300,35300,35300;9200,9200,9200]	[14700,14700,14700;1300,1300,1300]	1	'Scrap length: 3900 Retoolings:1 and estimated total cost: 99.75 PLN'
2	[20.2000,4,3.5000]	[1,2,3]	[1,2,3];[4];[5]	[43;1;3]	[35300,35300,35300;9200,9200,9200;9200,9200,35300]	[14700,14700,14700;4800,4800,4800;5300,5300,31400]	2	'Scrap length: 0 Retoolings: 2 and estimated total cost: 24 PLN'
3	[14.7000,5.5000,7.5000]	[1,2,3]	[1,2];[3];[4,5]	[43;1;3]	[35300,35300,35300;9200,9200,9200;9200,9200,35300]	[20200,20200,20200;3300,3300,3300;1300,1300,27400]	2	'Scrap length: 12500 Retoolings: 2 and estimated total cost: 305.25 PLN'
4	[6.7000,21]	[1,2,3]	[1];[2,3,4,5]	[1;43]	[9200,9200,9200;35300,35300,35300]	[2100,2100,2100;13900,13900,13900]	1	'Scrap length: 6300 Retoolings: 1 and estimated total cost: 153.75 PLN'
5	[6.7000,8,13]	[1,2,3]	[1];[2];[3,4,5]	[1;3;43]	[9200,9200,9200;9200,9200,35300;35300,35300,35300]	[2100,2100,2100;800,800,26900;21900,21900,21900]	2	'Scrap length: 7900 Retoolings: 2 and estimated total cost: 201.75 PLN'
6	[23.7000,4]	[1,2,3]	[1,2,3,4];[5]	[43;1]	[35300,35300,35300;9200,9200,9200]	[11200,11200,11200;4800,4800,4800]	1	'Scrap length: 0 Retoolings: 1 and estimated total cost: 12 PLN'
⋮	⋮	⋮	⋮	⋮	⋮	⋮	⋮	
504	[3.5000,4,20.2000]	[1,2,3]	[1];[2];[3,4,5]	[1;3;43]	[9200,9200,9200;9200,9200,35300;35300,35300,35300]	[5300,5300,5300;4800,4800,30900;14700,14700,14700]	2	'Scrap length: 0 Retoolings: 2 and estimated total cost: 24 PLN'

Table 5. Proposed spools set for molecule – `Problem_Data{1,2}`

No.	Segments [km]	Color_Parts	Span_type	Indexes	Spool_proposal [m]	Scrap [m]	Retoolings	Estimated_cost
1	[20.2000,7.5000]	[4,5,6]	[1,2,3];[4,5]	[1;3]	[29800,29800,29800;29800,29800,35300]	[9200,9200,9200;21900,21900,27400]	1	'Scrap length: 0 Retoolings: 1 and estimated total cost: 12 PLN'
2	[14.7000,13]	[4,5,6]	[1,2];[3,4,5]	[1;3]	[29800,29800,29800;29800,29800,35300]	[14700,14700,14700;16400,16400,21900]	1	'Scrap length: 0 Retoolings: 1 and estimated total cost: 12 PLN'
3	[6.7000,21]	[4,5,6]	[1];[2,3,4,5]	[1;3]	[29800,29800,29800;29800,29800,35300]	[22700,22700,22700;8400,8400,13900]	1	'Scrap length: 0 Retoolings: 1 and estimated total cost: 12 PLN'
4	[23.7000,4]	[4,5,6]	[1,2,3,4];[5]	[1;3]	[29800,29800,29800;29800,29800,35300]	[5700,5700,5700;25400,25400,30900]	1	'Scrap length: 0 Retoolings: 1 and estimated total cost: 12 PLN'
5	[20.2000,7.5000]	[4,5,6]	[1,2,3];[4,5]	[1;3]	[29800,29800,29800;29800,29800,35300]	[9200,9200,9200;21900,21900,27400]	1	'Scrap length: 0 Retoolings: 1 and estimated total cost: 12 PLN'
6	[14.7000,13]	[4,5,6]	[1,2];[3,4,5]	[1;3]	[29800,29800,29800;29800,29800,35300]	[14700,14700,14700;16400,16400,21900]	1	'Scrap length: 0 Retoolings: 1 and estimated total cost: 12 PLN'
⋮	⋮	⋮	⋮	⋮	⋮	⋮	⋮	
432	[7.5000,20.2000]	[4,5,6]	[1,2];[3,4,5]	[1;3]	[29800,29800,29800;29800,29800,35300]	[21900,21900,21900;9200,9200,14700]	1	'Scrap length: 0 Retoolings: 1 and estimated total cost: 12 PLN'

The column `Span_type` in consecutive tables: Tables 4, 5, 6 and 7 provides a details of the structure of the combined segments of the fiber production, which takes into account how each segment is joined. Correspondingly, the `Spool_proposal` column illustrates spool set proposals that reflect the current method of connecting `Span_type` segments.

Due to the strongly NP-hard nature of the considered optimization problem, there are many factors that can significantly affect its performance and

Table 6. Proposed spools set for molecule – `Problem_Data{1,3}`

No.	Segments [km]	Color_Parts	Span_type	Indexes	Spool_proposal [m]	Scrap [m]	Retoolings	Estimated_cost
1	[20.2000,7.5000]	[7,8,9]	[1,2,3];[4,5]	[1;3]	[29800,29800,29800; 29800,29800,35300]	[9200,9200,9200; 21900,21900,27400]	1	'Scrap length: 0 Retoolings: 1 and estimated total cost: 12 PLN'
2	[14.7000,13]	[7,8,9]	[1,2];[3,4,5]	[1;3]	[29800,29800,29800; 29800,29800,35300]	[14700,14700,14700; 16400,16400,21900]	1	'Scrap length: 0 Retoolings: 1 and estimated total cost: 12 PLN'
3	[6.7000,21]	[7,8,9]	[1];[2,3,4,5]	[1;3]	[29800,29800,29800; 29800,29800,35300]	[22700,22700,22700; 8400,8400,13900]	1	'Scrap length: 0 Retoolings: 1 and estimated total cost: 12 PLN'
4	[23.7000,4]	[7,8,9]	[1,2,3,4];[5]	[1;3]	[29800,29800,29800; 29800,29800,35300]	[5700,5700,5700; 25400,25400,30900]	1	'Scrap length: 0 Retoolings: 1 and estimated total cost: 12 PLN'
5	[20.2000,7.5000]	[7,8,9]	[1,2,3];[4,5]	[1;3]	[29800,29800,29800; 29800,29800,35300]	[9200,9200,9200; 21900,21900,27400]	1	'Scrap length: 0 Retoolings: 1 and estimated total cost: 12 PLN'
6	[14.7000,13]	[7,8,9]	[1,2];[3,4,5]	[1;3]	[29800,29800,29800; 29800,29800,35300]	[14700,14700,14700; 16400,16400,21900]	1	'Scrap length: 0 Retoolings: 1 and estimated total cost: 12 PLN'
⋮	⋮	⋮	⋮	⋮	⋮	⋮	⋮	⋮
432	[7.5000,20.2000]	[7,8,9]	[1,2];[3,4,5]	[1;3]	[29800,29800,29800; 29800,29800,35300]	[21900,21900,21900; 9200,9200,14700]	1	'Scrap length: 0 Retoolings: 1 and estimated total cost: 12 PLN'

Table 7. Proposed spools set for molecule – `Problem_Data{1,4}`

No.	Segments [km]	Color_Parts	Span_type	Indexes	Spool_proposal [m]	Scrap [m]	Retoolings	Estimated_cost
1	[20.2000,7.50000]	[10,11,12]	[1,2,3];[4,5]	[1;3]	[29800,29800,29800; 29800,29800,35300]	[9200,9200,9200; 21900,21900,27400]	1	'Scrap length: 0 Retoolings: 1 and estimated total cost: 12 PLN'
2	[14.7000,13]	[10,11,12]	[1,2];[3,4,5]	[1;3]	[29800,29800,29800; 29800,29800,35300]	[14700,14700,14700; 16400,16400,21900]	1	'Scrap length: 0 Retoolings: 1 and estimated total cost: 12 PLN'
3	[6.7000,21]	[10,11,12]	[1];[2,3,4,5]	[1;3]	[29800,29800,29800; 29800,29800,35300]	[22700,22700,22700; 8400,8400,13900]	1	'Scrap length: 0 Retoolings: 1 and estimated total cost: 12 PLN'
4	[23.7000,4]	[10,11,12]	[1,2,3,4];[5]	[1;3]	[29800,29800,29800; 29800,29800,35300]	[5700,5700,5700; 25400,25400,30900]	1	'Scrap length: 0 Retoolings: 1 and estimated total cost: 12 PLN'
5	[20.2000,7.5000]	[10,11,12]	[1,2,3];[4,5]	[1;3]	[29800,29800,29800; 29800,29800,35300]	[9200,9200,9200; 21900,21900,27400]	1	'Scrap length: 0 Retoolings: 1 and estimated total cost: 12 PLN'
6	[14.7000,13]	[10,11,12]	[1,2];[3,4,5]	[1;3]	[29800,29800,29800; 29800,29800,35300]	[14700,14700,14700; 16400,16400,21900]	1	'Scrap length: 0 Retoolings: 1 and estimated total cost: 12 PLN'
⋮	⋮	⋮	⋮	⋮	⋮	⋮	⋮	⋮
432	[7.5000,20.2000]	[10,11,12]	[1,2];[3,4,5]	[1;3]	[29800,29800,29800; 29800,29800,35300]	[21900,21900,21900; 9200,9200,14700]	1	'Scrap length: 0 Retoolings: 1 and estimated total cost: 12 PLN'

the quality of the solutions obtained. First and foremost, it should be noted that the proposed algorithm is subject to speed limitations resulting from the architecture of the available parallel computing platform. The computational performance of individual nodes and the data transmission speed between these nodes necessitate the search for an equilibrium that accounts for both constraints simultaneously. Additionally, the authors suggest that there is a possibility of introducing an additional algebraic-logical condition into the proposed model, which would reduce the size of the decision set by utilizing histogram analysis of the storage state. In this approach, it would be possible to assign spool selection priorities in the form of a weighted function that considers the local extremes of the storage state function.

4   Conslusion

The paper presents the hybrid application: the use of ALMM technology and its parallel execution in the search for spool sets of optical fiber to manufacture tubes for optical cables. Due to its NP–hard nature, this task cannot be accomplished in a single x64 computing environment. The proposed mechanism of decomposing problem P into n sub–problems overcomes this limitation and allows for a solution within an acceptable timeframe using the technology employed in this process. Computational molecules `Problem_Data{p,n}` used in the experimental part enable obtaining sub–solutions, which are subsequently integrated using the *Fork–Join* model. In a generalized scenario, both the number of computing nodes and the variability of order segments and the number of fibers in a tube can vary. According to ours current best knowledge, this is the first solution of its kind applied in optical cable production. At the same time, the authors suggest that future research should expand the ALMM model to include an intelligent mechanism for selecting the number of computing nodes, depending on warehouse status and the specific nature of the order being processed. Additionally, the proposed distribution model of sub–problems should take into account both the computing speeds of individual clusters and the level of required transmission necessary for this purpose.

Acknowledgments. We gratefully acknowledge the support of the NCBR, as part of the competition: 6/1.1.1/2020 6/1.1.1/2020 SS Duze/MSP/JN 4, project number POIR.01.01.01-00-1425/20 and Fibrain Sp. z.o.o, which made this research possible.

References

1. Dorota, D., Smutnicki, C.: On-line scheduling multiprocessor tasks in the non-predictive environment. In: Zamojski, W., Mazurkiewicz, J., Sugier, J., Walkowiak, T., Kacprzyk, J. (eds.) System Dependability - Theory and Applications: Proceedings of the Nineteenth International Conference on Dependability of Computer Systems DepCoS-RELCOMEX. July 1–5, 2024, Brunów, Poland, pp. 59–68. Springer Nature Switzerland, Cham (2024). https://doi.org/10.1007/978-3-031-61857-4_6

2. Dudek-Dyduch, E., Gomolka, Z., Twarog, B., Zeslawska, E.: The concept of the ALMM solver knowledge base retrieval using protégé environment. In: Zamojski, W., Mazurkiewicz, J., Sugier, J., Walkowiak, T., Kacprzyk, J. (eds.) Engineering in Dependability of Computer Systems and Networks: Proceedings of the Fourteenth International Conference on Dependability of Computer Systems DepCoS-RELCOMEX, July 1–5, 2019, Brunów, Poland, pp. 177–185. Springer International Publishing, Cham (2020). https://doi.org/10.1007/978-3-030-19501-4_17

3. Dudek-Dyduch, E., Gomolka, Z., Twarog, B., Zeslawska, E.: Intelligent ALMM system - implementation assumptions for its knowledge base. ITM Web Conf. **21**, 00002 (2018)

4. Gaifutdinov, A., Andrianova, K., Amirova, L., Amirov, R.: Optimizing the manufacturing technology of high-strength fiber reinforced composites based on aluminophosphates. Compos. Appl. Sci. Manuf. **185**, 108310 (2024)

5. Gomolka, Z., Dudek-Dyduch, E., Zeslawska, E.: Generalization of ALMM based learning method for planning and scheduling. Appl. Sci. **12**(24) (2022)

6. Gomolka, Z., Twarog, B., Zeslawska, E., Dudek-Dyduch, E.: Knowledge base component of intelligent ALMM system based on the ontology approach. Expert Syst. Appl. **199**, 116975 (2022)

7. He, S., Ma, P.C., Duan, M.: Continuous fiber path optimization in additive manufacturing: a gradient-based b-spline finite element approach. Addit. Manuf. **86**, 104155 (2024)

8. Penkov, D., Koprinkova-Hristova, P., Kasabov, N., Nedelcheva, S., Ivanovska, S., Yordanov, S.: Grid search optimization of novel SNN-ESN classifier on a supercomputer platform. In. Lirkov, I., Margenov, S. (eds.) Large-Scale Scientific Computations: 14th International Conference, LSSC 2023, Sozopol, Bulgaria, June 5–9, 2023, Revised Selected Papers, pp. 435–443. Springer Nature Switzerland, Cham (2024). https://doi.org/10.1007/978-3-031-56208-2_45

9. Shah, N.K., Ierapetritou, M.G.: Integrated production planning and scheduling optimization of multi-site, multi-product process industry. In: Pistikopoulos, E., Georgiadis, M., Kokossis, A. (eds.) 21st European Symposium on Computer Aided Process Engineering, Computer Aided Chemical Engineering, vol. 29, pp. 1015–1019. Elsevier (2011)

10. Szederkenyi, B., Kovacs, N.K., Czigany, T.: A comprehensive review of fiber–reinforced topology optimization for advanced polymer composites produced by automated manufacturing. Adv. Ind. Eng. Polymer Res. (2024)

11. Xie, F., et al.: Topology optimization for fiber-reinforced plastic (FRP) composite for frequency responses. Comput. Methods Appl. Mech. Eng. **428**, 117114 (2024)

12. Zeslawska, E., Gomolka, Z., Dydek-Dyduch, E.: Application of ALMM technology to intelligent control system for a fleet of unmanned aerial vehicles. In: Luo, B., Cheng, L., Wu, Z.-G., Li, H., Li, C. (eds.) Neural Information Processing: 30th International Conference, ICONIP 2023, Changsha, China, November 20–23, 2023, Proceedings, Part IX, pp. 26–37. Springer Nature Singapore, Singapore (2024). https://doi.org/10.1007/978-981-99-8138-0_3

13. Zhang, X., Xie, Y.M., Wang, C., Li, H., Zhou, S.: A non-uniform rational b-splines (NURBS) based optimization method for fiber path design. Comput. Methods Appl. Mech. Eng. **425**, 116963 (2024)

14. Zhao, C., Wang, S., Yang, B., He, Y., Pang, Z., Gao, Y.: A coupling optimization method of production scheduling and logistics planning for product processing-assembly workshops with multi-level job priority constraints. Comput. Ind. Eng. **190**, 110014 (2024)

Cross-Domain Evaluation of CNN-Based and Generative Adversarial Networks Models' Generalisability for (D)DoS Attack Detection in CPS and IoT

Vicky Ngo[1(✉)], Sira Yongchareon[1], Ji Ruan[1], Mahsa Mohaghegh[1], and Roopak Sinha[2]

[1] Auckland University of Technology, Auckland 1010, New Zealand
vicky.ngo@autuni.ac.nz
[2] Deakin University, Burwood, VIC 3125, Australia

Abstract. One of the ongoing concerns for (D)DoS detection models in cyber-physical systems and the Internet-of-Things is the lack of labelled training data, especially with the cost of producing these datasets. A possible solution is to apply an existing detection model in one domain to attack detection in another domain. However, there is currently minimal research into this particular area. In this study, we attempt to generalise detection models in this manner and evaluate their performance. Particularly, we performed a cross-domain evaluation on two (D)DoS attack detection models in Internet-of-Things and cyber-physical systems, respectively. The detection performance of the two models and the computational overhead were evaluated in both domains in a resource-constrained environment to evaluate their detection performance. The results of the cross-domain evaluation suggest that certain elements may have an important role in determining a model's performance after being applied to an unfamiliar domain. We discover that the detection model architecture determines the types of information that these models can learn from, as well as the impact of data distributions on the classification results. Further research is required to determine the specific requirements for generalising other detection models in the Internet-of-Things and cyber-physical systems domains.

Keywords: evaluation · deep learning · distributed denial-of-service attacks · detection · Cyber-Physical Systems · Internet-of-Things

1 Introduction

Denial-of-service (DoS) and distributed-denial-of-service (DDoS) attacks have historically posed great security concerns for both cyber-physical systems (CPS) environments and Internet-of-Things (IoT). CPS and IoT are similar in that they are both used to control physical processes through network communication, where availability requirements are high. Common (D)DoS attack techniques in IoT have been shown to overlap with those in CPS [14]. Thus, a (D)DoS detection model trained for the IoT environment should also be useful for detection in the

CPS environment, and vice versa, compared to knowledge transfer from models trained in unrelated fields.

Various ML/DL-based (D)DoS detection algorithms have been proposed for both the CPS and IoT domains. However, most of these algorithms in CPS consider neither the actual applications of these models in resource-constrained systems nor sufficiently address existing (D)DoS attack techniques targeting. According to [3], the lack of up-to-date, labelled datasets of current (D)DoS attack techniques to train detection models also affects the real-time deployment of models. While it is common for a detection model to achieve high detection performance on a dataset it was trained on, there has been little research in terms of how these models perform on other datasets in the same domain [8]. By extension, the generalisation of detection models across domains has also not been given much consideration, even though this method could expand the detection surface [3].

Previous studies have evaluated machine learning (ML) models' capabilities for generalisation, where the models attempt to achieve high detection performance in other datasets but within the same domain [3,8]. This is called cross-evaluation [3]. However, we have yet to find studies that consider *cross-domain evaluation* of (D)DoS detection models, where the models are generalised for datasets from different domains. We also acknowledge that certain models may not be specifically designed to cover other domains. Furthermore, there is a high degree of asymmetry in which the swapping of source and target domains could greatly affect classification performance [8].

From the literature [3], generalised models have similar sets of features in which they place varying degrees of importance during training and detection. These features are most important when 'transferred' to train a different model, which also implies that models should be able to extract these domain-relevant features to ensure accurate performance. Additionally, it is also necessary to tune the model's parameters appropriately to adapt it to unfamiliar datasets [6]. To evaluate these specifications, we performed a controlled experiment in cross-domain evaluation of two deep learning (D)DoS detection models in CPS and IoT, respectively, which are MAD-GAN (Multivariate Anomaly Detection with Generative Adversarial Networks) [9] and LUCID (Lightweight, Usable CNN in DDoS Detection) [4].

Based on our understanding, the model should be able to achieve benchmark performance by satisfying both specifications. It is possible that not satisfying one of the specifications would lead to reduced detection performance compared to the benchmark. At the same time, there may also be other unseen specifications or other element(s) that have a critical effect on the model's detection performance in an unfamiliar domain. Our experiment aims to evaluate models in this context. The contributions made through this study are as follows:

1. Applying LUCID and MAD-GAN's models toward attack detection in CPS and IoT domains and evaluating their performances across both domains.

2. Identifying LUCID's and MAD-GAN's (D)DoS detection surfaces and whether they could be applied for detection outside of their intended domain.
3. Demonstrating the effect of cross-domain evaluation on a detection model's computational overhead and detection accuracy in a resource-constrained environment. We also discovered that these factors are mostly influenced by the model's architecture and learning capabilities, respectively.

This study is organised as follows: Sect. 2 describes the related work in generalization and evaluation of detection models with different datasets from different domains. Section 3 continues with the design and implementation of the experiment. Section 4 discusses the training results, while Sect. 5 evaluates the detection results in a resource-constrained environment. Section 6 discusses this result. Finally, Sect. 7 concludes the findings and future works.

2   Related Works

In the context of deploying (D)DoS detection models in real-world networks, it is important that the models are able to detect unseen attacks. Existing evaluation methods tend to have detection models trained and tested on the same datasets, which would obviously return high detection performance [3,8]. Up until now, there have been no records as to how these results would transfer to other network scenarios [8]. It is also possible that detection models become overfitted to the chosen datasets [2]. Therefore, questions have been raised about the applicability of models that were evaluated using such methods [2].

As such, a realistic (D)DoS detection model should first and foremost be sufficiently generalised for different datasets within its domain. Extending a model detection surface through generalisation would temporarily address models' need for up-to-date datasets and the lack thereof [3]. There are also additional challenges that come with the costs of creating and labelling new datasets, contributing to the issues [3].

A study in 2018 [2] suggested evaluating different ML models on the three datasets Kyoto2006+, NSL-KDD, and gureKDD. Furthermore, the studies considered alternative evaluation strategies, including using data sets within the same domain but with different traffic distributions. This study argued that high accuracy can be achieved with fine-tuned hyperparameters.

Another study in 2020 [6] performed similar research, testing 12 ML algorithms for detecting attacks such as DoS, port scanning, SQL injection, brute force, etc. These attacks came from the UNSW-NB15 and CICIDS2017 datasets. The study discovered discrepancies in the models' detection performance and concluded that a suitable model for (D)DoS detection should be based on the datasets used. However, the difference in performance can also be marginal in some cases.

A framework named XeNIDS was proposed [3] for cross-evaluation of detection models based on Network Flows (NetFlow) data. XeNIDS's capability for preserving baseline performance and the importance of ensuring networks' compatibility are discussed. Additionally, the same study also demonstrates use cases in which the detection surface is extended by using cross-evaluation. However, it should be noted that the scope is still limited to one domain only, and further research is needed to generalise detection models outside its trained datasets. In the heterogeneous scenario where the data has diverse characteristics, the authors discussed the importance of features that denote malicious behaviour for training models and performing detection. It is also highlighted that different models may place varying importance on such features.

According to [8], seven supervised and unsupervised learning models were extensively evaluated on four benchmark datasets. It was found that none of the models considered were able to generalise over the chosen datasets, especially where the features can have different distributions. There is also a strong impact on classification accuracy when switching sources and target domains.

With regard to the related works above, we found that generalising detection models may affect their detection performance, especially if the source and target domains have different distributions. Additionally, there have not been studies that attempt cross-evaluation of detection models for different domains. However, it may be possible to extend a model's detection surface by ensuring that the models are able to extract, learn, and use domain-relevant features during both the training and detection processes. Our study aims to study this in Sect. 1 in the context of CPS and IoT environments.

3   Methodology

Our study investigates the capabilities of MAD-GAN and LUCID models for generalisation across the CPS and IoT domains, as well as their learning capacities on the datasets of respective domains. Only the most necessary changes to the source code would be made to ensure that the model can execute in our test environments while preserving the model's architecture integrity.

3.1   Model Selection

We have chosen LUCID (Lightweight, Usable CNN in DDoS Detection), designed for detection in network and IoT, and MAD-GAN (Multivariate Anomaly Detection with Generative Adversarial Networks), designed for detection in CPS environments.

LUCID is a CNN-based (D)DoS detection model that can be deployed in online resource-constrained environments, which had been tested in edge computing [4]. In our study, we used LUCID v2.0, which was released in June 2022 [5]. At the time of publication, the LUCID model was trained on the CICIDS2017 dataset and evaluated across another two datasets, namely ISCX2012 and CSE-CIC2018, along with a combination of all three datasets named UNB201X [4].

MAD-GAN is a GAN model that incorporated LSTM-RNN for (D)DoS detection in CPS [9,10]. The MAD-GAN model consists of a generator and a discriminator as two LSTM-RNNs (Long Short Term Memory Recurrent Neural Networks), offering two methods for anomaly detection [9]. The first method uses only the discriminator for direct anomaly detection, while the second method uses both the generator and discriminator for detection [9]. However, there have been reported issues that the second detection method could not function as expected with no solutions found [10]. As such, we have decided to only use the discriminator to evaluate MAD-GAN's performance. This could affect MAD-GAN's detection performance, as mentioned in Sect. 5.

3.2   Data Selection and Balancing

For the test case in an IoT environment, we chose CICIDS2017 [12], CICD-DoS2019 [11,13], which the LUCID model supports. The traffic in both CICIDS2017 and CICDDoS2019 datasets were generated by the University of New Brunswick using profiles representing abstract behaviours of human interactions with the testbed, allowing for the generation of naturalistic benign background traffic [12,13]. Both datasets have been frequently used by deep learning researchers for (D)DoS detection in the IoT domain. In this research, we combine both datasets, CICIDs2017 and CICDDoS2019, into one single dataset to obtain both benign and malicious traffic. We refer to the combined dataset as CICD-DoS throughout this paper. The combination method is similar to the approach taken by the original research for the LUCID model [4].

For the test case in a CPS environment, we use the SWaT dataset with data collected from a real water treatment testbed [1,7]. This dataset was commonly used for training detection models in the CPS environment and relevant domains [7]. We selected the latest network packets collected in December 2019 for the SWaT datasets containing historian exfiltration and sensor disruption attacks targeting SCADA and engineering workstations. Both datasets are sufficiently large for our experiment.

3.3   Experiment Design and Implementation

The experiment design is composed of two phases. The first phase involves training the two models on a CPU. The second phase involves testing the two models' detection performances and computational overhead in resource-constrained environments. A Raspberry Pi with 8 GB RAM and 32 GB of storage space was used to represent such an environment. During training, we set up two separate virtual environments on the PC to account for each model's different package dependencies. Further details on dependencies and hardware used for training are in our GitHub repository[1].

3.3.1   Phase 1 - Training

The purpose of phase 1 is to train MAD-GAN and LUCID, as well as evaluate how well the models would adapt to unfamiliar datasets with different feature

---

[1] **All technical details and relevant resources on our experiment can be found at** https://github.com/vickyngo-code/Domain-Adaptation-of-MAD-GAN-and-LUCID-for-Cyberattack-Detection.

spaces. We first clean the data on the CICDDoS and SWaT datasets before selecting data samples.

We then propose two training cases to apply MAD-GAN and LUCID for (D)DoS detection in CPS and IoT environments. Case 1 trains both models on the CICDDoS dataset to produce LUCID-DDoS and MAD-GAN-DDoS models. Likewise, case 2 trains both models on the SWaT dataset to produce LUCID-SWaT and MAD-GAN-SWaT models.

LUCID-DDoS. We trained LUCID on the CICDDOS dataset with some changes made to the training hyperparameters. These changes are intended to reduce memory consumption while maintaining a high F1 score, based on case studies with LUCID reported in [4]. Further information can be found in our GitHub(See footnote 1).

MAD-GAN-DDoS. The MAD-GAN model's training and testing procedures require two `.txt` files containing the training and testing parameters for any new dataset; therefore, we created similar `.txt` files customised for the CICDDoS dataset to adapt the MAD-GAN model. This suggests that MAD-GAN has sufficient flexibility in its input parameters, which is in line with our original assumption. We trained MAD-GAN on benign samples only in the CICDDoS dataset before evaluating it on malicious samples. During the data preprocessing processes, we removed non-numerical columns. The model was trained over 100 epochs as per the default settings.

LUCID-SWaT. LUCID can adapt to any dataset, as long as the IP addresses of the attacker and victim machines are available [5]. This information is crucial for LUCID's data preprocessing algorithm to label benign and attack traffic accurately before feature extraction and training [5]. The model extracts 11 features that are closely related to the TCP/IP protocols for training and evaluating the model [4]. Note that LUCID does not extract packet payload and is tightly implemented on the TCP/IP protocol [5]. Regardless, LUCID's limited flexibility means that it may be able to be applied to another detection domain, such as CPS.

However, our training results show that LUCID could not learn meaningful patterns from the SWaT dataset because historian exfiltration and sensor disruption attacks require a model to analyse packet payloads. By extension, LUCID would be ineffective against application layer-based attacks in the CPS domain, where packet payloads play an important role in distinguishing benign and malicious traffic. Additionally, a majority of network packets collected from the SWaT dataset utilises the CIP protocols, which LUCID might have excluded during data preprocessing. Overall, this result can be attributed to its feature extraction algorithm. Improvement of the model would require changes to its architecture, which is out of the scope of this study.

MAD-GAN-SWaT. We chose to use the SWaT 2019 dataset as it provides the latest data for training in comparison with the SWaT 2016 dataset. Furthermore, our trial experiments with the SWaT 2016 dataset show that MAD-GAN's performance does not differ greatly from MAD-GAN-SWaT's performance with the

SWaT 2019 dataset. Therefore, it would be more beneficial to use more recent data, as older data might not be representative of current attacks targeting CPS and IoT.

We trained the MAD-GAN model on the 2019 version of the SWaT dataset, which is smaller and has fewer attacks compared to the 2016 version. We first process network packets from the SWaT dataset using CICFlowMeter, a network traffic analyser used by the University of New Brunswick, to develop the CICDDoS2019 and CIC-IDS2017 datasets. This allows network packets from SWaT2019 to be converted into a consistent format as `.csv` for training and detection with MAD-GAN.

3.3.2   Phase 2 - Evaluation on Raspberry Pi

Due to issues with LUCID's ability to train on the SWaT dataset, we only deployed three models, including LUCID-DDoS, MAD-GAN-DDoS, and MAD-GAN-SWaT, for testing. Each model is tested on both CICDDoS and SWaT datasets on a Raspberry Pi representing a resource-constrained environment. Pre-recorded traffic traces from both datasets were used because we did not have suitable testbeds to generate realistic traffic. We also aim to evaluate how generalising detection models for different domains would affect each model's detection performance and computational overhead in a resource-constrained environment. We used the metrics of accuracy (Acc), precision (Pre), recall (Rec), and F1 scores (F1) to compare the models' performances, which are already incorporated in LUCID and MAD-GAN. We made no changes to either model regarding this element.

In this phase, MAD-GAN-SWaT's performance is set as the baseline for comparison When testing all three models on the SWaT dataset. Similarly, when the three models are tested on the CICDDoS dataset, LUCID-DDoS's performance would be the baseline instead. We would also measure the time taken to perform the detection task and the computational overhead of the models, defined as the CPU and Random Access Memory (RAM) usage of each model when performing detection on the Raspberry Pi.

Note that LUCID's only traffic inference output is the DDoS rates in the test samples, meaning that LUCID can mistakenly classify benign traffic as DDoS traffic and vice versa. To account for this, we evaluated LUCID-DDoS on datasets that contain either entirely benign traffic or entirely malicious traffic. We assume a 0% and 100% DDoS rate for the sets of benign and malicious traffic, respectively.

With the above in mind, we propose the following 8 test cases to be carried out on a Raspberry Pi. Note that the CICDDoS dataset includes both benign and attack traffic, combined from CICIDS2017 (benign) and CICDDoS2019 (attack), respectively.

1. LUCID-DDoS on CICIDS2017
2. LUCID-DDoS on CICDDoS2019
3. LUCID-DDoS on SWaT (benign)
4. LUCID-DDoS on SWaT (attack)
5. MAD-GAN-DDoS on CICDDoS
6. MAD-GAN-DDoS on SWaT
7. MAD-GAN-SWaT on CICDDoS
8. MAD-GAN-SWaT on SWaT

Table 1. MAD-GAN Models' Average Discriminator & Generator losses

	MAD-GAN-SWaT	MAD-GAN-DDoS
Discriminator loss	0.046156	0.011935
Generator loss	7.049835	8.493008

4   Training Evaluation

The results of phase 1 are relevant to our assumption that a model must be able to extract features relevant to both domains in order to perform detection more efficiently in an unfamiliar domain during testing.

MAD-GAN Models. We trained MAD-GAN-SWaT on 405,254 fully benign flows from the SWaT2019 dataset, and 529,918 flows from the CICDDoS dataset for MAD-GAN-DDoS. The training results of MAD-GAN models were evaluated in terms of the discriminator's loss and the generator's loss over each epoch they were trained on. Both models were trained over 100 epochs. The GAN model design dictates that as the discriminator's loss decreases, the generator's loss should also decrease, and vice versa, as both the discriminator and generator are training each other [9]. This implies that the two losses should eventually converge toward an optimal number after considerable training time.

Figure 1 shows the discriminators' and generators' losses for both MAD-GAN-DDoS and MAD-GAN-SWaT models during training The MAD GAN-DDoS showed a continuous diverging pattern, which suggests that the MAD-GAN-DDoS model's discriminator was able to discriminate data produced by the generator exceptionally well while not providing sufficient feedback for the generator to improve its performance. This led to increasing generator loss over time, as shown in the average loss over 100 epochs of both MAD-GAN models in Fig. 1. Furthermore, it might also imply that the discriminator did not train very well either, as the generator's worsening performance might not provide helpful feedback for the discriminator to improve. Since training MAD-GAN on both the SWaT and CICDDoS datasets caused a similar issue, the fact that MAD-GAN was not well-trained could be attributed to its architecture rather than the datasets themselves.

Overall, the results have shown that the MAD-GAN model has more balanced training results when trained on the SWaT dataset compared to training on the CICDDoS dataset. Due to the divergence in the generators' losses in both models and the possibility that the discriminators might not be well-trained, we expect that the detection performance of both models will be impacted. Nevertheless, MAD-GAN's training results are in line with our assumptions, as the model was able to train on an unfamiliar dataset from the IoT domain. We expect that MAD-GAN will be able to perform cross-domain detection in phase 2.

LUCID Model. We extracted 872,590 samples, with 436295/436295 benign/D-DoS distribution for training. For training, validation, and testing, 706797, 78534, and 87259 samples were used, respectively. This corresponds to 5297712,

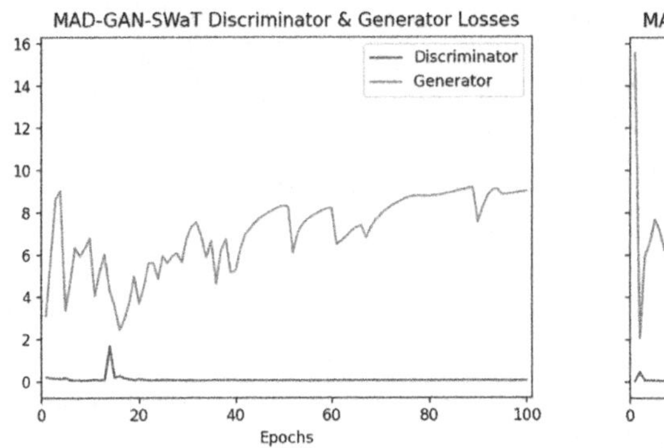

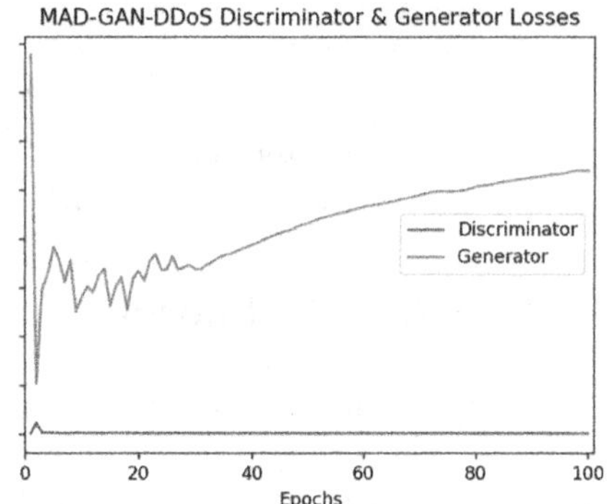

Fig. 1. MAD-GAN Models Training Discriminator & Generator Loss

583791, and 653940 packets, in that order. The LUCID-DDoS model achieved high detection performance on the training and testing sets, as shown in Table 2.

5   Experiment Result

We recorded LUCID-DDoS predicted DDoS percentage when evaluated on CICDDoS and SWaT datasets. Meanwhile, MAD-GAN-DDoS and MAD-GAN-SWaT used the standard metrics, namely accuracy (Acc), precision (Pre), recall (Rec), and F1 scores, to evaluate its detection performance on the labelled data.

It is also suggested that a model's capacity to extract relevant features during training and detection is crucial for the model to be generalised for detection in another domain, given that LUCID was unable to train on the SWaT dataset due to the same reason. This means that LUCID cannot be reused for the detection of attacks in the CPS domain. By extension, we speculate that LUCID-DDoS would not be able to discern between benign and malicious traffic in the SWaT dataset during phase 2.

5.1   MAD-GAN Models

Both MAD-GAN-DDoS and MAD-GAN-SWaT were evaluated with both benign and malicious traffic for a total of 460,623 flows from the SWaT2019 dataset, and 388,768 flows from the CICDDoS datasets. Below, we compare both MAD-GAN-DDoS and MAD-GAN-SWaT performance on each dataset.

Table 2. LUCID-DDoS Training Results

	ACC	PRE	REC	F1
Training	99.42%	99.19%	99.67%	99.43%
Testing	99.42%	99.20%	99.63%	99.42%

We observed that both MAD-GAN-SWaT and MAD-GAN-DDoS achieved near-perfect precision when tested on the CICDDoS dataset. Low accuracy and recall also lowered the F1 score. These results suggest the following: 1) the number of false positives is close to zero, and 2) there might be many false negative predictions. Accuracy being the same as recall implies that the number of true negatives must also be approximately zero. A possible explanation is that the CICDDoS dataset was imbalanced, leading to high numbers of false negatives yet an abysmal number of true negative predictions.

Both models returned higher accuracy on the SWaT dataset but significantly lower precision when compared with previous testing results on the CICDDoS dataset. The recall rates of MAD-GAN-SWaT and MAD-GAN-DDoS for all types of epochs do not exceed 50.35%. The F1 score is also low. Again, it is likely that the high number of false positives caused the low precision shown in Table 3. Interestingly, MAD-GAN-DDoS's average performance on the SWaT dataset was on par with that of MAD-GAN-SWaT's average performance on the same dataset.

The MAD-GAN models require extensive time to perform anomaly detection. Table 3 shows that each model took at least 30 min to classify 10,000 samples. Previous testing has shown a positive correlation between the amount of data and classification time, which could be a concern when deploying MAD-GAN for real-time detection, assuming that similar performance persists.

With regards to MAD-GAN's computational overhead, Fig. 2 shows that both MAD-GAN models's RAM usage ranged between 60% and 70%, with the CPU usage always reaching 100% until they finish and the CPU usage falls to 0% with minimal changes in RAM usage. This suggested that MAD-GAN requires extensive CPU power to perform detection at a reasonable speed, while low RAM usage means that the model is less likely to freeze the machine it was deployed on. Note that deploying the MAD-GAN model results in a resource-constrained environment does not interfere with detection results, based on previous evaluations that we performed in phase 1 after the initial training.

Overall, we have observed that the MAD-GAN model could be applied for attack detection in both CPS and IoT domains. This is demonstrated through the fact that MAD-GAN-DDoS achieved comparable performance with MAD-GAN-SWaT when evaluated on the SWaT dataset. However, MAD-GAN's slow detection time is a concern for real-time detection. The high false negative rate could be attributed to MAD-GAN using only the discriminator for detection

Table 3. MAD-GAN Models Evaluation Results on Raspberry Pi

Test Case	ACC	PRE	REC	F1	Time (s)
MAD-GAN-SWaT on SWaT	49.14%	40.91%	54.93%	45.40%	11847
MAD-GAN-DDoS on SWaT	50.05%	41.25%	51.89%	44.49%	10646
MAD-GAN-SWaT on CICDDoS	54.21%	100.00%	54.21%	68.47%	9722
MAD-GAN-DDoS on CICDDoS	50.68%	100.00%	50.68%	65.57%	9574

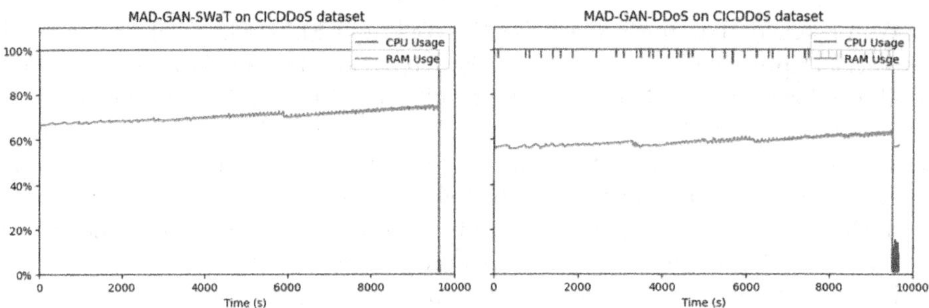

Fig. 2. MAD-GAN Models CPU & RAM Consumption

Table 4. LUCID-DDoS Evaluation Results

Samples	Packets	Time (s)	DDoS%
cicddos_benign_1	44998	605.00	2.04%
cicddos_benign_2	42500	647.65	7.73%
cicddos_benign_3	34215	605.72	11.70%
cicddos_benign_4	25132	378.67	12.63%
cicddos_attack_1	26389	371.32	98.97%
cicddos_attack_2	26535	369.33	99.93%
cicddos_attack_3	26093	361.80	99.83%
cicddos_attack_4	25906	359.36	99.90%
cicddos_attack_5	25943	355.65	100.00%
cicddos_attack_6	25922	364.65	99.97%
swat_benign_1	77746	1007.91	33.22%
swat_benign_2	78190	998.26	34.34%
swat_attack_1	78879	1,010.46	33.79%
swat_attack_2	77137	986.00	30.82%

due to issues described in Sect. 3. MAD-GAN's low RAM consumption is beneficial for deployment in a resource-constrained environment. Utilising both MAD-GAN's discriminator and generator in the future is possible for better detection results.

5.2 LUCID-DDoS

We evaluated LUCID-DDoS with fully benign and fully malicious data from both the SWaT and CICDDoS datasets. For each sample given to LUCID-DDoS, it will then perform detection in multiple instances and return the DDoS rate as a percentage. Table. 4 reported the arithmetic mean of the DDoS rate for each instance. More information on the detected DDoS of each instance can be found in our GitHub repository (See footnote 1).

The (D)DoS rate predicted by LUCID-DDoS on the CICDDoS and SWaT-datasets is shown in Table 4, where it reported near 100% DDoS rate in CICD-DoS attack samples and lower DDoS rate in benign samples. This substantial difference suggested that LUCID-DDoS can effectively discern between benign and malicious traffic in the CICDDoS dataset. We expect that LUCID-DDoS would perform similarly when deployed in a real system for detection in IoT.

In the meantime, We found that LUCID had returned similar (D)DoS rates for both benign and traffic samples in the SWaT datasets, which implies that LUCID-DDoS was not able to distinguish between benign and attack samples in the SWaT dataset. Because both training and testing phases in LUCID use the same feature extraction module, this issue correlates with the fact that LUCID could not be trained on the SWaT datasets or application-layer attacks as described in Sect. 3. Therefore, LUCID may be restricted to attacks based on TCP/IP protocols at the moment, and further improvements are needed to adapt LUCID to a wider range of attack types and protocols.

Figure 3 shows the CPU and RAM consumption from two test cases on the Raspberry Pi. LUCID has a high RAM usage demand for both test cases, while its CPU usage remains mostly stable at 30% except for some initial spikes. From Table 4, we also notice that samples with a higher number of packets tend to have a higher RAM consumption peak. This suggests that while LUCID-DDoS has very low CPU usage, it should only perform anomaly detection on a small number of packets at a time to avoid maxing out the available RAM. In such cases, the machine it was deployed on could freeze and force a shutdown to resume operations. In ICS and CPS, where availability requirements are high, such situations are not acceptable. Further work is needed to address LUCID-DDoS high RAM requirements.

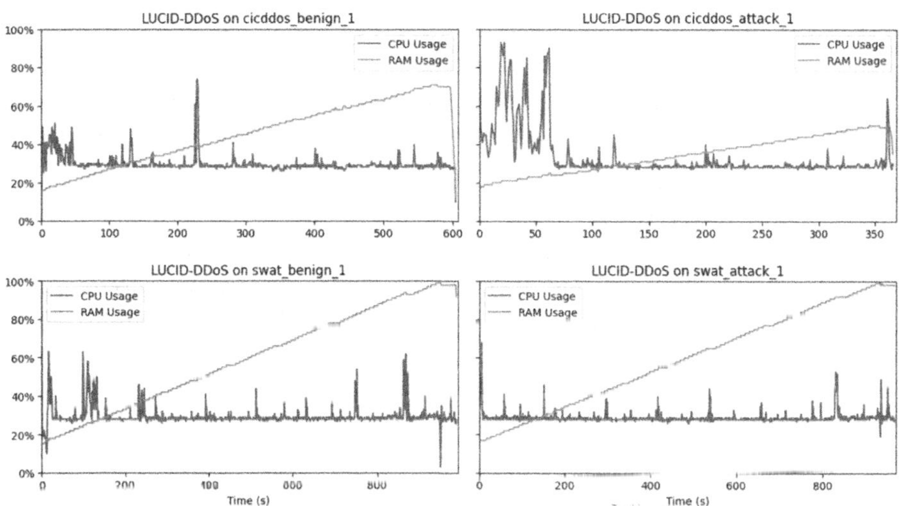

Fig. 3. LUCID-DDoS CPU & RAM Consumption

6   Discussion

In Sect. 1, we introduced 2 specifications that could enable a model to achieve benchmark performance when detecting attacks in an unfamiliar domain. One is the model's ability to extract important, domain-relevant features during training and detection, while the other is how the parameters were tuned.

Our experiment results with MAD-GAN suggested that it met both specifications. Firstly, MAD-GAN requires only minimal changes to its parameters in order to adapt to the unfamiliar CICDDoS dataset. Additionally, we observed that MAD-GAN models' detection results for both the SWaT and CICDDoS datasets are relatively close in Table 3. This further postulates that MAD-GAN was able to extract relevant features during both the learning and testing phases to effectively perform detection. Although MAD-GAN's detection performance was not very good, partly due to the technical difficulties encountered in Sect. 3, the results are promising.

The LUCID experiment result shows that it could not extract domain-relevant features at all despite efforts to tune its input parameters, affecting not only training but also testing processes. This is demonstrated by the fact that LUCID could not learn or detect attacks in the SWaT datasets. Note that LUCID does have a certain level of flexibility when adapting to different datasets. However, our attempts at training LUCID were unsuccessful, as shown in our implementation in Sect. 3. This result is unsurprising, considering that LUCID was implemented tightly around the TCP/IP protocols. However, this design also suggests that LUCID could be generalised in other domains where (D)DoS attack techniques involving TCP/IP protocols have a stronger focus, in contrast to the CPS domain, where application-layer attacks involving industrial protocols such as CIP are more prevalent. A different approach is needed to adapt LUCID's architecture for this purpose.

Additionally, LUCID's detection results in Sect. 5 clearly show that it was unable to distinguish between malicious and benign samples. Combined with our understanding of LUCID's architecture discussed in Sect. 3, it is highly likely that LUCID was unable to extract relevant features from the SWaT dataset for training and performing detection. In the case of LUCID, it may not be possible for it to be generalised for detection in other domains unless major changes are made to its architecture. Generally speaking, the models' fundamental limitations are among the threats to validity in our research.

In terms of detection results, LUCID-DDoS has significantly better performance than both MAD-GAN-SWaT and MAD-GAN-DDoS when evaluated on the CICDDoS dataset. While MAD-GAN-SWaT did not achieve a good detection performance on the SWaT dataset, we still consider it to be more effective than the LUCID-DDoS performance on the same dataset. This is because LUCID-DDoS was not able to distinguish between benign and malicious traffic in the SWaT dataset at all with the data patterns it learned from training on the CICDDoS dataset, which means that there is no basis to compare LUCID-DDoS against MAD-GAN-SWaT. Again, this is a fundamental limitation in LUCID and our research. There are no observable changes to the three detection models' computational overhead when used to detect attacks in a different dataset, as their CPU and RAM usages remain similar to what was observed during training and testing before deployment on the Raspberry Pi.

Furthermore, our experiments have confirmed that a model does not necessarily perform well on testing sets from unfamiliar domains. It remains a prerequisite that the model's design should first allow for the extraction of relevant features for cross-domain training and detection. Ensuring that the networks' are indeed similar and compatible should also be taken into consideration when performing cross-domain evaluation [3]. The transferability of knowledge learned by the detection models remains a crucial part of ensuring high detection accuracy when the models are deployed for cross-domain detection.

It is important to consider model's performance in a resource-constraint environment such as CPS and IoT environments, where there are high availability requirements. MAD-GAN's slow detection performance raises another efficiency issue with how well a model can keep up with the amount of traffic going through the system at any time. Many of these issues can be addressed by optimising the model's architecture, ensuring that it is lightweight and efficient. Furthermore, this approach would enable us to address fundamental limitations in models to better generalise them for cross-domain detection.

7   Conclusion and Future Works

Our cross-domain evaluation on the generalisation of (D)DoS attack detection models for the CPS and IoT shows that detection accuracy is heavily reliant on whether the features and patterns learned could be applicable in another domain. This result was found in previous studies where cross-evaluation was performed within one domain only. In an unfamiliar domain, detection results can be poor due to the high level of asymmetry. Additionally, generalising detection models for different domains does not seem to affect a detection model's computational overhead, which largely depends on the model's architectural design.

Further experiments should be performed across a broader range of models to obtain a more comprehensive evaluation of any unseen elements or specifications that may affect the generalisation of detection models, and their detection performance in different domains. Another research direction would be to explore better model architecture for cross-domain detection.

References

1. iTrust Labs Dataset Info (2022). https://itrust.sutd.edu.sg/itrust-labs_datasets/dataset_info/
2. Al-Riyami, S., Coenen, F., Lisitsa, A.: A re-evaluation of intrusion detection accuracy: Alternative evaluation strategy. In: Proceedings of the 2018 ACM SIGSAC Conference on Computer and Communications Security, pp. 2195–2197. CCS '18, Association for Computing Machinery, New York, NY, USA (2018). https://doi.org/10.1145/3243734.3278490

3. Apruzzese, G., Pajola, L., Conti, M.: The cross-evaluation of machine learning-based network intrusion detection systems. IEEE Trans. Netw. Serv. Manage. **19**(4), 5152–5169 (2022). https://doi.org/10.1109/TNSM.2022.3157344

4. Doriguzzi-Corin, R., Millar, S., Scott-Hayward, S., Martinez-Del-Rincon, J., Siracusa, D.: Lucid: a practical, lightweight deep learning solution for DDoS attack detection. IEEE Trans. Netw. Serv. Manage. **17**, 876–889 (2020). https://doi.org/10.1109/TNSM.2020.2971776

5. Doriguzzi-Corin, R., Millar, S., Scott-Hayward, S., Martínez-del Rincón, J., Siracusa, D.: LUCID: a practical, lightweight deep learning solution for DDoS attack detection. https://github.com/doriguzzi/lucid-ddos (2022)

6. Elmrabit, N., Zhou, F., Li, F., Zhou, H.: Evaluation of machine learning algorithms for anomaly detection. In: 2020 International Conference on Cyber Security and Protection of Digital Services (Cyber Security). pp. 1–8 (2020). https://doi.org/10.1109/CyberSecurity49315.2020.9138871

7. Goh, J., Adepu, S., Junejo, K.N., Mathur, A.: A dataset to support research in the design of secure water treatment systems. In: Havarneanu, G., Setola, R., Nassopoulos, H., Wolthusen, S. (eds.) CRITIS 2016. LNCS, vol. 10242, pp. 88–99. Springer, Cham (2017). https://doi.org/10.1007/978-3-319-71368-7_8

8. Layeghy, S., Portmann, M.: Explainable cross-domain evaluation of ml-based network intrusion detection systems. Comput. Electr. Eng. **108**, 108692 (2023). https://doi.org/10.1016/j.compeleceng.2023.108692, https://www.sciencedirect.com/science/article/pii/S0045790623001167

9. Li, D., Chen, D., Jin, B., Shi, L., Goh, J., Ng, S.-K.: MAD-GAN: multivariate anomaly detection for time series data with generative adversarial networks. In: Tetko, I.V., Kůrková, V., Karpov, P., Theis, F. (eds.) ICANN 2019. LNCS, vol. 11730, pp. 703–716. Springer, Cham (2019). https://doi.org/10.1007/978-3-030-30490-4_56

10. Li, D., Ng., S.K., Chen, D., Goh, J.: Multivariate anomaly detection for time series data with GANs. https://github.com/LiDan456/MAD-GANs (2019)

11. of New Brunswick, U.: DDoS 2019 | Datasets | Research | Canadian Institute for Cybersecurity | UNB. https://www.unb.ca/cic/datasets/ddos-2019.html

12. Sharafaldin., I., Habibi Lashkari., A., Ghorbani., A.A.: Toward generating a new intrusion detection dataset and intrusion traffic characterization. In: Proceedings of the 4th International Conference on Information Systems Security and Privacy - ICISSP, pp. 108–116. INSTICC, SciTePress (2018). https://doi.org/10.5220/0006639801080116

13. Sharafaldin, I., Lashkari, A.H., Hakak, S., Ghorbani, A.A.: Developing realistic distributed denial of service (DDoS) attack dataset and taxonomy. In: 2019 International Carnahan Conference on Security Technology (ICCST), pp. 1–8 (2019). https://doi.org/10.1109/CCST.2019.8888419

14. Zahid, F., Funchal, G., Melo, V., Kuo, M.M.Y., Leitao, P., Sinha, R.: DDoS attacks on smart manufacturing systems: a cross-domain taxonomy and attack vectors. In: 2022 IEEE 20th International Conference on Industrial Informatics (INDIN), pp. 214–219 (2022). https://doi.org/10.1109/INDIN51773.2022.9976172

Robust Design of Echo State Networks for Soft Sensor Applications Based on Risk-Aware Optimization and Stability Testing

Shunsuke Takagaki[1]([envelope]), Ryunosuke Fukuzaki[2], Koki Tateishi[1], Toshikazu Noda[1], and Hiroyasu Ando[3]

[1] Mitsubishi Heavy Industries, Ltd., Tokyo, Japan
{shunsuke.takagaki.qp,koki.tateishi.wd,toshikazu.noda.ew}@mhi.com
[2] University of Tsukuba, Tsukuba, Japan
s2430119@u.tsukuba.ac.jp
[3] Tohoku University, Sendai, Japan
hiroyasu.ando.d1@tohoku.ac.jp

Abstract. "Soft sensors" with time series prediction models can reduce cost for devices and be used for controlling and monitoring systems instead of hardware sensors. They have been actively studied in the field of machine learning. In particular, Echo State Network (ESN) is known to a promising method to realize soft sensors in terms of both maintaining accuracy and reducing training cost. On the other hand, since most of the weights of network are fixed to random values, ESN carry an inherent risk of generating models with unstable prediction result due to variability in weight distribution. In this paper, we introduce several mechanisms, related to hyperparameter optimization and network stability, to eliminate the risk for a typical ESN training procedure and propose a method to automatically and efficiently design a robust ESN. In addition, the proposed method is applied to a soft sensor modeling task in a real-world dataset. As a result of the evaluation, it is clarified that the introduced mechanisms significantly reduce accuracy variation and the risk of divergence of the prediction.

Keywords: Echo State Network · Reservoir Computing · Time-Series Prediction · Soft Sensor · Robust Design

1 Introduction

Recent technological advancements in machine learning have achieved remarkable progress, leading to the development of data-driven problem-solving applications across various fields, from industry to society. In process industries, such as plants, where a large amount of operational data can be obtained continuously by measuring with hardware sensors, research applications using machine learning techniques are particularly popular. A typical example is "soft sensors [1]". This is a technology for real-time prediction of the behavior of an unobserved variable by other observed ones. It can be realized by constructing a model that

M. Mahmud et al. (Eds.): ICONIP 2024, CCIS 2297, pp. 31–45, 2026.
https://doi.org/10.1007/978-981-96-7036-9_3

has been trained on the relationship between observed and accumulated operational data using a time series prediction algorithm.

Applications of soft sensors to power plants [2] and chemical plants [3] are known. Properly trained soft sensors can replace hardware sensors, enabling not only low-cost prediction of process values that are difficult to measure continuously and in real time due to high installation and maintenance costs, as well as long-term reliability problems, but also applications such as anomaly detection [4] and model predictive control [5].

It is desirable that the time-series prediction algorithm used for the soft sensor can adequately represent the dynamics of the prediction object. As a typical example, there are methods based on a structure using the past state, such as classical autoregression (AR) model or Recurrent Neural Network (RNN), and RNN has been especially widely studied due to its high expressiveness. However, RNNs have suffered from problems such as extremely high training cost and the gradient explosion. Among these, the gradient explosion has already been solved by architecture improvement approaches such as Gated Recurrent Unit (GRU) [6] and Long Short-Term Memory (LSTM) [7]. On the other hand, studies have been conducted on simplifying the gate structure [8] and improving the parameter update law [9] to reduce the training cost, but these approaches are limited because they assume the use of gradient descent. Therefore, when training the RNN soft sensor in a realistic time, it is necessary to take measures such as sampling a small number of operation data for training and simplifying the model structure, and the practical accuracy requirement cannot be satisfied, or the experiment cycle for tuning the prediction performance is severely limited.

In recent years, there has been rapid progress in the technological field that can solve this problem in a different context from RNNs. It is called "Reservoir Computing" (RC) and refers to an information processing framework for reading desired information from a given (fixed) mapping to a higher dimensional space. Any mapping in RC can be used as long as it has some causality, and physical RC using physical phenomena directly has become one field of research [10]. On the other hand, the RC concept has also been introduced for RNNs, leading to the development of a time-series model called "Echo State Network" (ESN) [11].

An ESN has a recurrent structure similar to an RNN, where the output layer is defined as a linear layer (called the readout layer) and the weights of the input layer and the hidden layer (called the reservoir layer) are fixed at random values. This weight fixing reduces ESN training to a simple linear regression problem. Therefore, it is not necessary to update the weights many times by the gradient descent method, and extremely fast training is realized. In addition, alternative methods for designing high-precision ESNs have also been developed through previous studies [12,13]. Examples of industrial applications of ESN include calcium carbonate ($CaCO_3$) concentration prediction [13] in flue gas desulfurization (FGD) equipment and overvoltage detection in lithium-ion batteries [14], and each example shows that ESN outperforms RNN-based methods.

However, the fact that most of the weights are fixed at random in an ESN also raises the problem that there is an inherent risk of training models whose predictions are unstable depending on variability in weight distribution. Applying ESN to soft sensors without ensuring robustness can undermines the validity of the prediction and potentially leads to incorrect decision making and inappropriate control in the application. Therefore, eliminating the risk due to the inherent randomness of ESN is a very important issue to be addressed for the practical application of ESN soft sensors.

In this paper, we show that introducing several mechanisms into the ESN design can automatically and efficiently realize the training of ESNs with guaranteed robustness. Section 2 provides an overview of ESN and the challenges associated with robustness. Section 3 presents some ideas for improving robustness and proposes a new ESN design method that combines them. In Sect. 4, we present a numerical experimental procedure for the proposed method, and in Sect. 5, we discuss the results. Finally, in Sect. 6, we conclude the paper.

2   Benchmark ESN Design Methods and Practical Issues

In this section, we describe the basic structure of ESN and its design method with reference to [13]. A typical ESN consists of an input layer, a reservoir layer, and a readout layer. The reservoir layer has a large number of sparsely coupled neurons including self-coupling and corresponds to a recurrent structure in RNNs. Let the input at time t be $\boldsymbol{u}(t) \in \mathbb{R}^{D \times 1}$, the reservoir state be $\boldsymbol{x}(t) \in \mathbb{R}^{N \times 1}$, and the output be $\boldsymbol{y}(t) \in \mathbb{R}^{M \times 1}$, the ESN state transition that outputs the one step ahead prediction $\boldsymbol{y}(t+1)$ is:

$$\boldsymbol{x}(t) = f(\boldsymbol{W}_{\text{in}}[\mathbf{1}; \boldsymbol{u}(t)] + \boldsymbol{W}\boldsymbol{x}(t-1)) \text{ and} \tag{1}$$

$$\boldsymbol{y}(t+1) = \boldsymbol{W}_{\text{out}}[\mathbf{1}; \boldsymbol{u}(t); \boldsymbol{x}(t)], \tag{2}$$

where $\boldsymbol{W}_{\text{in}} \in \mathbb{R}^{N \times (1+D)}, \boldsymbol{W} \in \mathbb{R}^{N \times N}$ and $\boldsymbol{W}_{\text{out}} \in \mathbb{R}^{M \times (1+D+N)}$ are the weights of the input, reservoir, and readout layers. $\mathbf{1} \in \mathbb{R}^{1 \times 1}$ is a bias term, $[\cdot; \cdot]$ represents row-wise concatenation, and $f(\cdot)$ is an activation function.

The ESN is trained by adjusting only $\boldsymbol{W}_{\text{out}}$ so that the desired predictions can be output. Specifically, the parameters other than $\boldsymbol{W}_{\text{out}}$ in these state transition equations, i.e., $f, \boldsymbol{W}_{\text{in}}$, and $\boldsymbol{W}$, are fixed in advance. In general, the weights of $\boldsymbol{W}_{\text{in}}$ and $\boldsymbol{W}$ are randomly generated according to certain rules (e.g., uniform or Gaussian distribution). In Eq. (1), the time-series data of the reservoir state $\boldsymbol{x}(t)$ can be generated sequentially even if $\boldsymbol{W}_{\text{out}}$ is unknown, so we can use it to find $\boldsymbol{W}_{\text{out}}$ such that the error on both sides in Eq. (2) becomes small. A common approach is to apply ridge regression to avoid overfitting due to multicollinearity of $\boldsymbol{r}(t)$, by using matrices $\boldsymbol{X}, \boldsymbol{Y}$ that concatenate the time-series data of $[\mathbf{1}; \boldsymbol{u}(t); \boldsymbol{x}(t)]$ and $\boldsymbol{y}(t+1)$ in the column direction, $\boldsymbol{W}_{\text{out}}$ can be obtained as:

$$\boldsymbol{W}_{\text{out}} = \boldsymbol{Y}\boldsymbol{X}^{\top}(\boldsymbol{X}\boldsymbol{X}^{\top} + \lambda\boldsymbol{I})^{-1}, \tag{3}$$

where $\boldsymbol{I}$ is the $(1+D+N)$-th identity matrix, and λ is a regularization parameter.

The ESN state transition equation (1) has been improved in some previous studies. For example, Leaky-Integrator ESN [11], in which the speed of the dynamics can be adjusted by a parameter called the leak rate (α), and structures

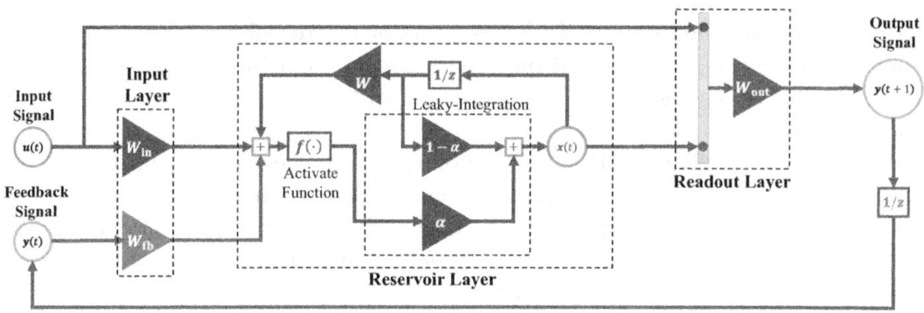

Fig. 1. Block diagram of Leaky-Integrator ESN with feedback connection.

with an output feedback connection are now common research subjects. When these are applied, Eq. (1) is rewritten as:

$$\boldsymbol{x}(t) = (1 - \alpha)\boldsymbol{x}(t - 1) + \alpha f(\boldsymbol{W}_{\text{in}}[\boldsymbol{1}; \boldsymbol{u}(t)] + \boldsymbol{W}\boldsymbol{x}(t - 1) + \boldsymbol{W}_{\text{fb}}\boldsymbol{y}(t)), \quad (4)$$

where $\boldsymbol{W}_{\text{fb}} \in \mathbb{R}^{N \times M}$ and is the weights of the feedback layer, and the sequentially generating of the time-series data of $\boldsymbol{x}(t)$ using this formula gives the last term $\boldsymbol{y}(t)$ the true value (this is called "Teacher Forcing"). The structure of the ESN model according to these equations (2) and (4) is shown in Fig. 1.

Since the performance of ESN largely depends on the parameters related to the model structure fixed in advance (in Eqs. (2) and (4)) and the setting of the regularization parameter λ (in Eq. (3)), it is very important to make adjustments according to the task. Previous studies have shown that using hyperparameter tuning with Bayesian optimization [15] and adjusting λ each time according to randomly generated connection weights [13] are effective for improving accuracy. In addition, although the tanh function is generally used as the activation function f, accuracy improvement with even functions, such as the Sinc function ($y = \sin(x)/x$), has been reported in the specific task [13,16]. The appropriate choice of f is also one of the important factors for high accuracy ESN.

While a benchmark design methodology for high accuracy ESNs has been established, there is still discussion on how to improve robustness to enable industrial applications such as soft sensors. In other words, since the weights are randomly fixed in advance in the training procedure, there is a case in that the accuracy will decrease or the prediction value will diverge (Fig. 2 shows examples of NARMA10 [17]) even if the ESN is trained by the same method. No substantial and practical method has been established that focuses on avoiding this instability.

Previous research has explored ensembling multiple ESNs with different random seeds [13,15] as a naive method to reduce the effects of randomness, but its effectiveness is limited when the robustness of individual ESNs is not guaranteed at all. An alternative approach would be to somehow ensure that "the trained ESN is highly accurate", one of which is the ESN stability testing based on Predictive Generalized Synchronization (PGS) [18] described below.

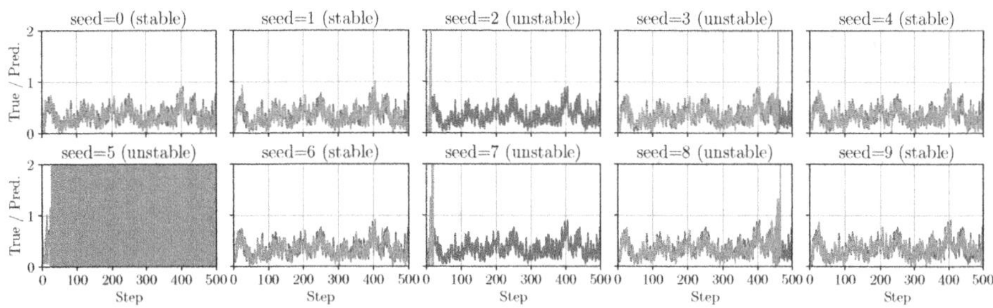

Fig. 2. NARMA10 [17] prediction by ESNs which with the same hyperparameters and different weights (blue line: true, orange line: predicted).

In the field of nonlinear dynamical systems, the existence of a kind of synchronization between different systems is called Generalized Synchronization (GS) [19]. Applying the concept of GS to the dynamics of the ESN, it means that the reservoir state $x(t) \in \mathbb{R}^{N \times 1}$ responds to the input $u(t) \in \mathbb{R}^{D \times 1}$ with a certain relationship, and it is expressed mathematically that there exists a function ψ satisfying $x(t) = \psi(u(t))$. In [18], PGS is defined as ϕ, the inverse function of ψ exists, i.e., $u(t) = \phi(x(t))$. The existence of ϕ implies that the ESN state transition (2) and (4) is independent of the input $u(t)$, i.e., it can be reduced to an autonomous system. In [18], the authors proposed a numerical method to test the validity of PGS by driving ESN with two different initial states $x_A(0), x_B(0)$ by the same input $u(t)$ and evaluating the time evolution of the distances between $x_A(t)$ and $x_B(t)$. That is, the PGS is determined under the definition of $\lim_{t \to \infty} \|x_A(t) - x_B(t)\| \to 0$, and the determination result has been verified to have the same tendency as the accuracy.

This stability testing method focuses on the time evolution of the reservoir state $x(t)$, so that an efficient testing can be realized before training W_{out}. Therefore, the PGS can be used for pre-filtering in hyperparameter tuning. However, the appropriate setting of the threshold value and the number of testing steps depends on the hyperparameters, so that the uniform evaluation is difficult independent of tasks, and it is not practical in the present form.

Based on the above issues, the goal of this paper is to present practical improvements for automatically and efficiently designing highly robust ESN with reduced effect of randomness.

3   Automatic Design Method for Robust ESN

In this section, specific ideas for improving the robustness of ESN are discussed in Sect. 3.1–3.3. The outline is as follows.

Adjusting hyperparameters to account for variation of accuracy (Sect. 3.1).
-   Improved stability metric for hyperparameter evaluation (Sect. 3.2).
-   Controlling randomness of the weights based on Hopfield-type Network and the diversity of the reservoir state (Sect. 3.3).

Then, we propose an automatic design method of robust ESN by integrating these mechanisms into typical methods in Sect. 3.4.

3.1 Risk-Aware Hyperparameter Optimization

In general, the objective function in ESN hyperparameter tuning by Bayesian optimization is the accuracy of the trained model. That is, an ESN is defined for one hyperparameter set (candidate) $\boldsymbol{p}$, the accuracy is evaluated, and the next evaluation candidate is suggested from the evaluation history. The "optimal" hyperparameter $\boldsymbol{p}_{\text{optim}}$ is found by repeating these steps. The ESN used to evaluate prediction accuracy is usually a single model initialized by a single random seed, so the risk of reduced accuracy due to randomness cannot be taken into account. We define the objective function as the maximum value of the RMSE evaluated with several ESNs initialized with the same $\boldsymbol{p}$ and different random seeds. Then, the search for $\boldsymbol{p}_{\text{optim}}$ is formulated as follows:

$$\boldsymbol{p}_{\text{optim}} = \underset{\boldsymbol{p}}{\text{argmin}} \max\{\text{RMSE}_0(\boldsymbol{p}), \text{RMSE}_1(\boldsymbol{p}), \text{RMSE}_2(\boldsymbol{p}), \dots \}, \tag{5}$$

where $\text{RMSE}_i(\boldsymbol{p})$ is the evaluated RMSE of the i-th model generated by $\boldsymbol{p}$.

3.2 Stability Testing Based on Predictive Generalized Synchronization

The open problem with the stability testing method based on PGS proposed in [18] is that it is difficult to set a uniform criterion for the ESN initialized by various candidate of $\boldsymbol{p}$, as described in Sect. 2. For example, the distance threshold of the reservoir state should be set according to its scale, but generally the scale is proportional to the number of dimensions of the reservoir state. In addition, since the number of steps required for an adequate testing depends on the speed of the dynamics expressed by the ESN, the influence of the leakage rate must be taken into account. However, it is not easy to find a method to adaptively set a threshold or a number of steps for any $\boldsymbol{p}$.

In this paper, we propose an improved method based on "smoothness of convergence" as the evaluation metric, instead of reservoir state distance itself. This method focuses on the fact that PGS, i.e., the existence of the GS function ψ, implies that ψ is smooth (differentiable). We model the "smooth" convergence process of the reservoir state distance as:

$$\log \|\boldsymbol{x}_{\text{A}}(t) - \boldsymbol{x}_{\text{B}}(t)\| \approx -a \times t + b, \tag{6}$$

and the coefficients a, b can be easily obtained by linear regression from the data of $\boldsymbol{x}_{\text{A}}(t), \boldsymbol{x}_{\text{B}}(t)$ obtained by driving with $\boldsymbol{u}(t)$. Then, we can obtain the coefficients of determination R^2 as stability metric.

Since the R^2 evaluation does not refer to information about the scale or speed of the dynamics, we can use typical regression analysis criteria, such as $R^2 >$ 0.995. Note that this criteria is deterimined for a sufficient smoothness of the conveging process, but it is not limited to the specific value. In addition, we can fix the number of steps for evaluation without adaptively adjusting, taking care to remove the effect of the initial transient. This allows unified stability testing for any hyperparameter set. In this paper, we consider incorporating this stability metric as a pre-filtering mechanism in hyperparameter tuning. In other words, candidates p judged as "unstable" by this testing are rejected without training, with the goal of improving candidate quality and reducing computational cost. The details are described in Sect. 3.4, including in combination with the ideas described in the next section.

3.3   Controlling the Randomness of the Weights

In this paper, we discuss two methods to control the randomness in the weights of ESN, and discuss their effectiveness through numerical studies.

The first is a constraint on W. In previous research, there have been studies of topology, such as circular or chain [20], as methods for controlling the generation of W, but the theoretical support for these is not always clear. On the other hand, in the context of neural networks, the "Hopfield Network" [21] has been actively studied as a network with stable state transitions. This network is defined as a structure in which there are no self-couplings and the interactions are symmetric, and its stability is theoretically supported by the existence of local minima of the energy function. The stability of the Hopfield Network with synchronous update can be easily introduced in ESN by restricting W to a symmetric matrix with zero diagonal components (called "Hopfield matrix" in this paper).

The other is a constraint on W_{fb}. Since the reservoir state $x(t)$ generally has strong multicollinearity, applying ridge regression to train W_{out} is a common approach, but another approach could be to introduce spatial diversity in $x(t)$ and mitigate multicollinearity. The introduction of W_{fb}, which is a common practice for ESN, is expected to have the effect of providing such diversity, and the variability of weights should be closely related. The problem here is that the distribution of randomly sampled weights from a predefined population distribution may differ from the inherent statistical properties of that population. In other words, if the variance of the sampled weights is less than the variance of the population distribution, there is a concern that the expected diversity-promoting effect will not be achieved, making ESN unstable. In this paper, we propose to impose the constraint: (standard deviation of the elements of W_{fb})$\succ$(expected value of the standard deviation in the population) as a countermeasure.

3.4   Proposed Robust ESN Training Method

The three ideas mentioned above can be introduced naturally for a series of processes ranging from hyperparameter tuning to ESN training. The schematic

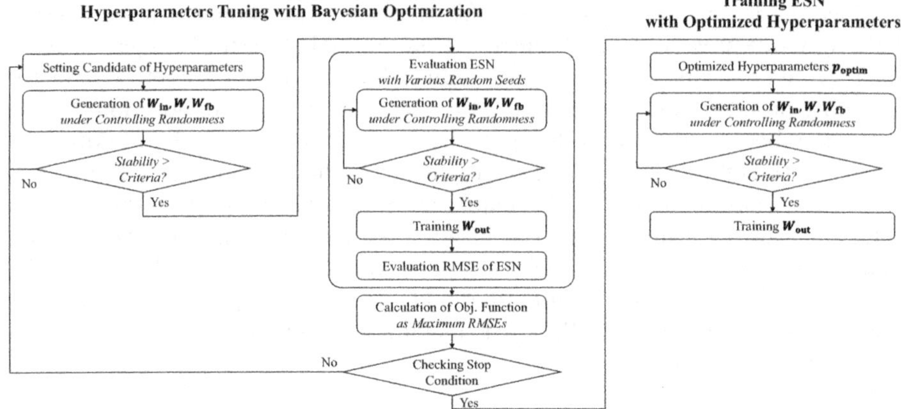

Fig. 3. Schematic diagram of the proposed robust ESN training method.

diagram is shown in Fig. 3 as the proposed method in this paper, and the procedure is described below. Key points for improving robustness are shown in *italics*.

1. Following the Bayesian optimization procedure, a candidate p is obtained.
2. The weights of W_{in}, W, and W_{fb} are randomly generated according to p. W *and W_{fb} are generated to satisfy the randomness constraints.*
3. *The stability index R^2 is evaluated by the input $u(t)$. If R^2 is greater than the specified value, it goes to 4. Otherwise, p is rejected and returns to 1.*
4. *Regenerate the weights with multiple random seeds to satisfy the same constraints as 2. Then, train W_{out} and evaluate the RMSEs for each ESN.*
5. *Evaluate the objective function as the maximum of RMSEs and return to 1.*
6. p_{optim} is obtained by repeating 1.–5. a sufficient number of times.
7. The robust ESN is obtained using p_{optim}. *The constraints of W, W_{fb} are the same as 2., and the random seed is chosen satisfying the condition of R^2.*

4   Experimental Setup

4.1   Datasets

We verified the effectiveness of the stability testing method and the robust ESN training method proposed in Sect. 3 using toy dataset and real-world dataset.

First, NARMA10 [17] was selected as the toy data. This is a time-series that is widely used as a benchmark of ESN, and its time evolution $y(n)$ is shown as:

$$y(n + 1) = 0.3y(n) + 0.5y(n) \sum_{i=0}^{9} y(n - i) + 1.5u(n - 9)u(n) + 0.1, \quad (7)$$

where $u(n)$ is a random input following the uniform distribution $\mathcal{U}(0, 0.5)$. In this paper, time-series data $y(n)$ of 2,500 samples were generated according to

Eq. (7), and the first 2,000 samples were used as training data and the remaining 500 samples were used as test data. This dataset was used to validate the stability testing method described in Sect. 4.2.

In addition, to verify the effectiveness for the soft sensor modeling task, the operational dataset of flue gas desulfurization equipment in actual coal-fired power plant (called FGD dataset) was used as a real-world dataset. This FGD dataset is identical to the one used in [13] and consists of eight dimensional explanatory variables, such as SO_2 concentration at the FGD equipment inlet-outlet, and $CaCO_3$ concentration as the target variable. This consists of 12 batches of continuous time-series data, and each variable and time scales are non-dimensionalized. Note that due to the need to perform a large amount of model training, each batch was undersampled at 15 point intervals and reduced from 35,218 to 2,354 points. This dataset was used to validate the robust ESN training method described in Sect. 4.3.

4.2 Experiment 1: Stability Testing Under Controlled Randomness

The purpose of this experiment is to verify that the stability testing method (Sect. 3.2) works effectively with controlling the randomness (Sect. 3.3). We evaluated the relationship between the stability metric and the accuracy through multiple case studies with various ESN hyperparameters, random seeds of weights, and controlling of randomness, with the NARMA10 dataset.

The model structure is the Leaky-Integrator ESN with feedback connection described in Sect. 2. The weights except $\boldsymbol{W}_{\text{out}}$ were randomly generated by the uniform distribution $\mathcal{U}$ and the constraints described in Sect. 3.3 are imposed. When $\boldsymbol{W}_{\text{fb}} \sim \mathcal{U}(-w_{\text{fb}}, +w_{\text{fb}})$, the constraint can be expressed as:

$$\sigma(\boldsymbol{W}_{\text{fb}}) > \frac{w_{\text{fb}}}{\sqrt{3}}, \tag{8}$$

where and $\sigma(\boldsymbol{W}_{\text{fb}})$ is the (unbiased) standard deviation of the elements of $\boldsymbol{W}_{\text{fb}}$.

The hyperparameters of the ESN were set as shown in Table 1, and in each case where the combination of spectral radius and density was varied, the ESN was initialized and trained with 100 random seed weights. Note that in this experiment, the hyperparameter set were generated exhaustively, and the effectiveness of the risk-aware hyperparameter optimization (Sect. 3.1) was evaluated in the next experiment.

Table 1. Hyperparameter settings in experiments.

Hyperparameter	Description	Exp. 1 (NARMA10)	Exp. 2 (FGD)
N	number of neurons	1000 (fixed)	10 – 200
α	leakage rate	0.3 (fixed)	0.05 – 1.00
d	density of $\boldsymbol{W}$	$0.1, 0.2, \ldots, 1.0$	0.0 1.0
ρ	spectral radius of $\boldsymbol{W}$	$0.1, 0.2, \ldots, 1.5$	0.0 – 1.5
w_{in}	$\boldsymbol{W}_{\text{in}} \sim \mathcal{U}(-w_{\text{in}}, +w_{\text{in}})$	0.5 (fixed)	0.0 – 1.5
w_{fb}	$\boldsymbol{W}_{\text{fb}} \sim \mathcal{U}(-w_{\text{fb}}, +w_{\text{fb}})$	0.5 (fixed)	0.0 – 1.5

The stability metric R^2 was computed by driving the ESN with the training data input and defined as the "Stability" score. In addition, the regularization parameter λ was fixed at 10^{-8}, and the relationship between Stability of the training data and RMSE of the test data was evaluated.

4.3   Experiment 2: Robustness of Proposed Training Method

The purpose of this experiment is to verify that the robustness of the ESN is improved by the training method proposed in Sect. 3.4, which combines risk-aware hyperparameter optimization and stability testing. The FGD dataset was used to compare ESN training methods with and without the implementation of the mechanisms described in Sect. 3.1–3.3.

The model structure and randomness constraints of the model are the same as in Experiment 1, except that each neuron in the reservoir is connected to only one neuron in the input and feedback connections [15]. In this case, the constraint of $\boldsymbol{W}_{\text{fb}}$ can be expressed as:

$$\sigma(\boldsymbol{W}_{\text{fb}}) > \frac{w_{\text{fb}}}{\sqrt{3(1 + D + M)}}, \tag{9}$$

where $D = 8$ and $M = 1$ in this experiment. Another difference is that λ is not fixed, but optimized by grid search in the range of $[10^{-6}, 10^{6}]$, based on the RMSE in the cross-validation.

In a stability testing using a multi-batch dataset such as the FGD dataset, the stability metric R^2 must be calculated for each batch. In this experiment, for given $\boldsymbol{p}$, we computed R^2 by in each batch for ESN initialized by 10 random seeds, and the minimum value of them ($10\times$ number of batches) was defined as the "Stability" score. If the criterion: Stability> 0.995 is not satisfied, the value of the objective function is rejected as $+\infty$ without training.

We used LA-MCTS [22] as the Bayesian optimization algorithm for the six hyperparameters shown in Table 1. The number of iterations in an optimization process was 1,000, and the objective function was defined as the maximum RMSEs in cross-validation of the ESN trained by five random seeds (Sect. 3.1). The hyperparameters are optimized based on the above procedure, and as a result, $\boldsymbol{p}_{\text{optim}}$ and are obtained, the evaluation model was trained, and the maximum error was evaluated by cross-validation.

The above describes a procedure for training only a single model, and randomness must be given to this procedure in order to evaluate the effectiveness of the robust ESN training method. In this context, "randomness" means two things: randomness in the hyperparameter tuning procedure and randomness in the ESN weights, each of which must be taken into account. In this experiment, $n_1(= 100)$ of $\boldsymbol{p}_{\text{optim}}$ were derived with different random seeds, and ESNs with $n_2(= 100)$ of random weights were initialized and trained with each $\boldsymbol{p}_{\text{optim}}$, and the maximum error was evaluated.

Here, $n_1 \times n_2 = 10,000$ evaluation models is obtained, and the maximum error distribution statistic (i.e., the minimum, median, maximum, and unbiased standard deviation of the n_2 maximum errors) can be calculated for each of $\boldsymbol{p}_{\text{optim}}$.

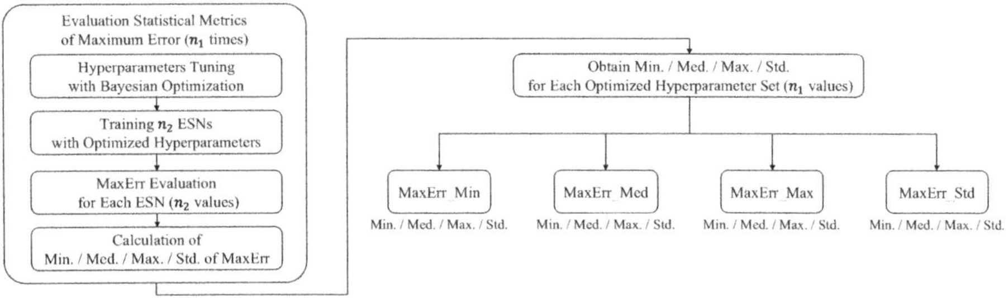

Fig. 4. Robustness evaluation for the proposed method with error distribution.

We defined the set of distribution statistics calculated in this procedure as "MaxErr_Min", "MaxErr_Med", etc., and evaluated the robustness of the proposed method based on their distribution characteristics (Fig. 4). In addition, an ablation study based on error trend analysis was performed for a comprehensive evaluation including the risk of divergence of the prediction.

5 Results and Discussion

5.1 Experiment 1: Stability Testing Under Controlled Randomness

Table 2 shows the results of the evaluation of the relationship between Stability and RMSE using the NARMA10 dataset according to the procedure in Sect. 4.2. In this evaluation, we defined RMSE< 1 as True (T), and Stability> 0.995 as Positive (P), and For the four cases with and without constraints on W and W_{fb}. Then, the proportions of true positives (TP), true negatives (TN), false positivse (FP), and false negatives (FN) for the total number of models in each case: 15,000 were calculated. We focused on the ratio of T ($=$ TP $+$ TN) and Precision ($=$ TP$/($TP $+$ FP$)$), which are related to the performance of the ESN itself and the effectiveness of stability testing, respectively. Note that since we assume that ESNs judged as negative (N) will be rejected without proceeding to the training step, we are not interested in evaluation metrics including FN, such as Accuracy ($=$ TP $+$ FN) and FN$/($FP $+$ FN$)$.

In Table 2, Cases 1 and 2, which do not constrain W, and Cases 3 and 4, which constrain W, show almost the same tendency, and the TP $+$ TN and Precision of Cases 3/4 are larger than those of Cases 1/2. From these results,

Table 2. Evaluation of the relationship between Stability and accuracy.

Case	Const. of W	Const. of W_{fb}	TP	TN	FP	FN	TP+TN	Precision
1 (base)	without	without	0.251	0.268	0.287	0.194	**0.519**	**0.466**
2	without	with	0.245	0.268	0.295	0.192	**0.514**	**0.455**
3	with	without	0.621	0.290	0.059	0.030	**0.911**	**0.913**
4	with	with	0.616	0.290	0.061	0.032	**0.907**	**0.910**

although the contribution of the constraint of W_{fb} was not seen in this experiment, it was confirmed that the stability of the ESN was improved and the stability testing method worked effectively, by using Hopfield matrix as W. We introduced the Hopfield structure of W with the expectation of providing stable state transition, therefore these results were reasonable considering that the stability testing method focuses on the time evolution of the reservoir state $x(t)$.

5.2   Experiment 2: Robustness of Proposed Training Method

Figure 5 and Table 3 show the results of evaluating the distribution of the maximum error statistics in 10,000 evaluation models obtained by the proposed robust training method using the FGD dataset according to the procedure in Sect. 4.3. In these figure and table, the case where the mechanism of Sect. 3.1–3.3 is not introduced at all is reffered to as [base], and the case where it is all introduced is reffered to as [proposed]. As for the distribution of MaxErr_Min, they show almost the same characteristics, but a comparison of MaxErr_Med and MaxErr_Max shows that [proposed] significantly reduces the propotion of models with extreme deterioration of the maximum error, while [base] has a long tail of distribution. In addition, from the comparison of the distribution of MaxErr_Std, it can be seen that the proposed method can significantly suppress the training of the model which results in high accuracy variance.

Next, as an ablation study, we added a case without risk-aware hyperparameter optimization [w/o R-A Optim.], a case without stability testing [w/o Stability Testing], and a case without imposing constraints of W_{fb} [w/o W_{fb} Const.]. The results of visualizing the trend of the prediction error in the evaluation model group of $n_1 \times n_2 =10,000$ are shown in Fig. 6. Note that the constraints of W with Hopfield matrix are treated as prerequisites for stability testing, and these constraints are also excluded in [w/o Stability Testing], considering the results of Experiment 1. [proposed] reduces the variance of the error over the trend and the divergence of the prediction as seen in batches #2 and #7 of [base]. Together with the results described in the previous part, the effectiveness of the proposed method, which combines three mechanisms to improve robustness, was clearly demonstrated.

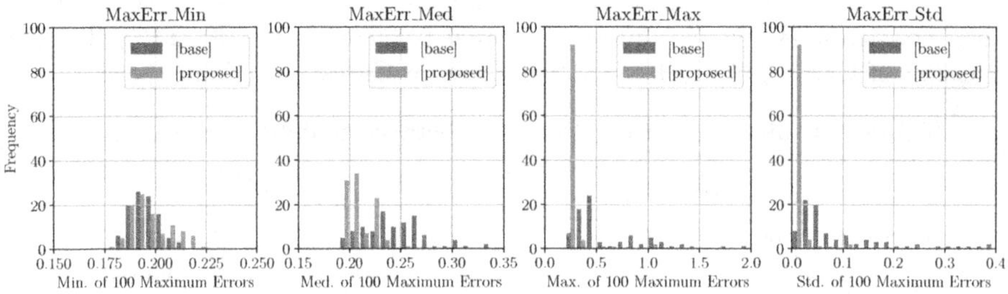

Fig. 5. Distribution Histogram of MaxErr_Min/Med/Max/Std.

Table 3. Statistics of MaxErr_Min/Med/Max/Std (red: [base], blue: [proposed]).

Distribution	Min. Value	Med. Value	Max. Value	Std. Value
MaxErr_Min	0.181 / 0.180	0.195 / 0.195	0.214 / 0.221	0.007 / 0.010
MaxErr_Med	0.198 / 0.194	0.242 / 0.203	0.335 / 0.255	0.030 / 0.013
MaxErr_Max	0.219 / 0.205	0.519 / 0.226	23.054 / 1.058	3.623 / 0.125
MaxErr_Std	0.006 / 0.002	0.060 / 0.004	3.233 / 0.108	0.457 / 0.017

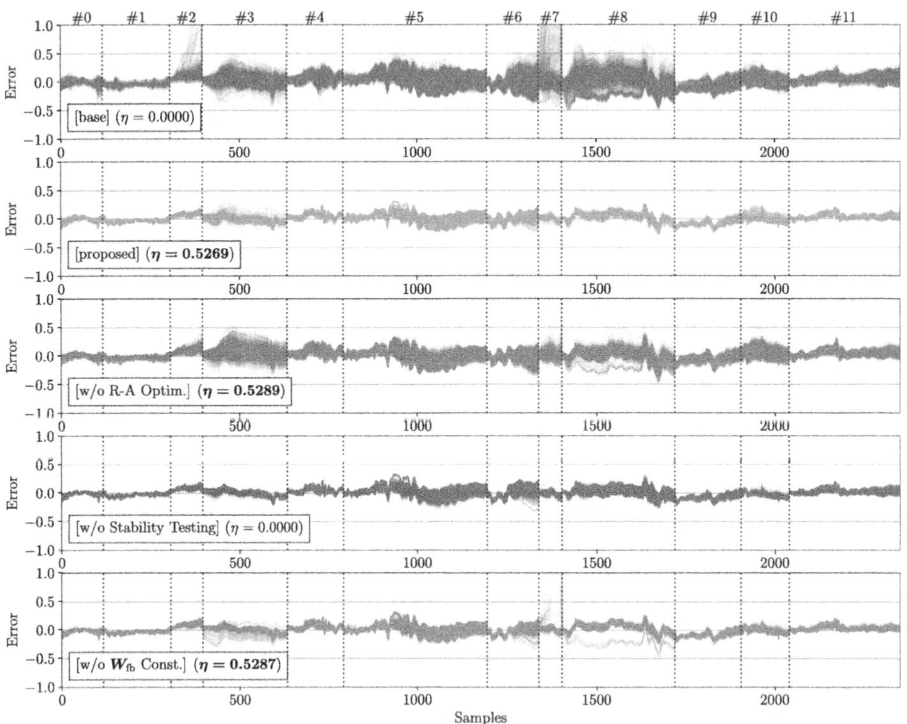

Fig. 6. Prediction error trend and computing efficiency η in each case of ablation study. Note that $0 \le \eta \le 1$, and $\eta = 0$ means that no candidate of p is rejected at all.

We further discussed the effects of each mechanism from the results of three additional cases. First, the comparison of these cases shows that the effect of risk-aware optimization dominates the reduction of error variance. On the other hand, the results of [w/o Stability Testing] show that the introduction of stability testing has a limited impact on the robustness of accuracy. It was considered that the robustness improvement expected from the stability determination was already achieved by the risk-aware optimization and the W_{fb} constraints. However, improving robustness is not the only benefit we expect from introducing stability testing. It is the efficiency improvement of the hyperparameter search. In fact, the Bayesian optimization process in [proposed] actually advanced to the accuracy evaluation step an average of 473.2 times out of a total of 1,000 iterations. That is, about half of the hyperparameter candidates were rejected without W_{out} training. We define the proportion of rejected steps as computing efficiency η, and η for each case is also shown in Fig. 6. In the proposed method, the introduction of stability testing has the effect of reducing the computational cost of hyperparameter search. Also, [w/o W_{fb} Const.] shows an increase in models with divergence of the prediction in batch #7, unlike other additional cases. This shows that the constraint on W_{fb} has the effect of reducing the risk of divergence of the prediction. From the above discussion, the effects of each of the three mechanisms introduced in the proposed method can be clarified.

6   Conclusions

In this paper, we proposed a new method that introduces mechanisms to eliminate the destabilization risk into the ESN training procedure to deal with the robustness degradation caused by the randomness of the fixed weights. Our contributions are as follows: design of the objective function considering risks in ESN hyperparameter optimization, improved stability metric based on PGS, and proposal of a randomness control method.

The verification of the proposed method using the $CaCO_3$ soft sensor modeling as an example task shows that each of these mechanisms works effectively, and that the proposed method automatically and efficiently realizes the training of robust ESN, which can greatly suppress the accuracy variation and the risk of divergence of the prediction. This result allows designers of ESN soft sensors to provide robust models without trial and error in the design process, which greatly improves practicality.

Note that the ideas for training robust ESN presented in this paper is not *ad hoc* with a specific task in mind, and it is considered to be universally applicable to a wide range of problems. However, verification of such generality will be a subject of future investigation. In the future, if it becomes clear that it is necessary to improve the three mechanisms introduced in this paper, or to introduce a new mechanism through verification in various task and strengthening of the theoretical basis, it is expected that the ESN soft sensor will be further improved in practicality under the interpretable methodology.

References

1. Kadlec, S., Gabrys, B., Strandt, S.: Data-driven soft sensors in the process industry. Comput. Chem. Eng. **33**(4), 795–814 (2009)
2. Pan, H., Su, T., Huang, X., Wang, Z.: LSTM-based soft sensor design for oxygen content of flue gas in coal-fired power plant. Trans. Inst. Meas. Control. **43**(1), 78–87 (2021)
3. Kaneko, H., Funatsu, K.: Adaptive soft sensor based on online support vector regression and Bayesian ensemble learning for various states in chemical plants. Chemom. Intell. Lab. Syst. **137**, 57–66 (2014)

4. Cheng, T., Harrou, F., Sun, Y., Leiknes, T.: Monitoring influent measurements at water resource recovery facility using data-driven soft sensor approach. IEEE Sens. J. **19**(1), 342–352 (2018)

5. Shan, B., et al.: Soft sensor model predictive control for azeotropic distillation of the separation of DIPE/IPA/water mixture. J. Taiwan Inst. Chem. Eng. **152**, 105185 (2023)

6. Cho, K., Bahdanau, D., Bengio, Y.: On the properties of neural machine translation: encoder-decoder approaches. arXiv preprint arXiv:1409.1259 (2014)

7. Hochreiter, S., Schmidhuber, J.: Long short-term memory. Neural Comput. **9**(8), 1735–1780 (1997)

8. Zhou, G.-B., Wu, J., Zhang, C.-L., Zhou, Z.-H.: Minimal gated unit for recurrent neural networks. Int. J. Autom. Comput. **13**(3), 226–234 (2016). https://doi.org/10.1007/s11633-016-1006-2

9. Kingma, D. P., Ba, J.: Adam: a method for stochastic optimization. arXiv preprint arXiv:1412.6980 (2014)

10. Tanaka, G., et al.: Recent advances in physical reservoir computing: a review. Neural Netw. **115**, 100–123 (2019)

11. Jaeger, H.: The "echo state" approach to analysing and training recurrent neural networks. German National Res. Center Inf. Technol. Rep., **148** (2001)

12. Lukoševicius, M.: A practical guide to applying echo state networks. Neural Netw. Tricks Trade, 659–686 (2012)

13. Takagaki, S., Tateishi, K., Ando, H.: Time-series prediction of calcium carbonate concentration in flue gas desulfurization equipment by optimized echo state network. In: Proceedings of the 32nd International Conference on Artificial Neural Networks, pp. 281–292 (2023)

14. Sánchez, L., Anseán, D., Otero, J., Couso, I.: Assessing the health of LiFePO4 traction batteries through monotonic echo state networks. Sensors **18**(1), 9 (2017)

15. Racca, A., Magri, L.: Robust optimization and validation of echo state networks for learning chaotic dynamics. Neural Netw. **142**, 252–268 (2021)

16. Chang, H., Nakaoka, S., Ando, H.: Effect of shapes of activation functions on predictability in the echo state network. arXiv preprint arXiv:1905.09419 (2019)

17. Atiya, A.F., Parlos, A.G.: New results on recurrent network training: unifying the algorithms and accelerating convergence. IEEE Trans. Neural Netw. **11**(3), 697–709 (2000)

18. Platt, J. A., Wong, A. S., Clark, R., Penny, S. G., Abarbanel, H. D. I.: Robust forecasting using predictive generalized synchronization in reservoir computing. arXiv preprint arXiv:2103.00362 (2021)

19. Sushchik, M.M., Rul'kov, N.F., Tsimring, L.S., Abarbanel, H.: Generalized synchronization of chaos in directionally coupled chaotic systems. Phys. Rev. E **51**, 980–994 (1995)

20. Strauss, T., Wustlich, W., Labahn, R.: Design strategies for weight matrices of echo state networks. Neural Comput. **24**(12), 3246–3276 (2012)

21. Hopfield, J.J.: Neural networks and physical systems with emergent collective computational abilities. Proc. Natl. Acad. Sci. **79**(8), 2554–2558 (1982)

22. Wang, L., Fonseca, R., Tian, Y.: Learning search space partition for black-box optimization using monte Carlo tree search. Adv. Neural. Inf. Process. Syst. **33**, 19511–19522 (2020)

Logic Error Localization in Student Programming Assignments Using Pseudocode and Graph Neural Networks

Zhenyu Xu[1], Kun Zhang[2], and Victor S. Sheng[1(✉)]

[1] Department of Computer Science, Texas Tech University, Lubbock, TX, USA
{zhenxu,victor.sheng}@ttu.edu
[2] Department of Computer Science, Xavier University of Louisiana,
New Orleans, LA, USA
kzhang@xula.edu

Abstract. Pseudocode is extensively used in introductory programming courses to instruct computer science students in algorithm design, utilizing natural language to define algorithmic behaviors. This learning approach enables students to convert pseudocode into source code and execute it to verify their algorithms' correctness. This process typically introduces two types of errors: syntax errors and logic errors. Syntax errors are often accompanied by compiler feedback, which helps students identify incorrect lines. In contrast, logic errors are more challenging because they do not trigger compiler errors and lack immediate diagnostic feedback, making them harder to detect and correct. To address this challenge, we developed a system designed to localize logic errors within student programming assignments at the line level. Our approach utilizes pseudocode as a scaffold to build a code-pseudocode graph, connecting symbols from the source code to their pseudocode counterparts. We then employ a graph neural network to both localize and suggest corrections for logic errors. Additionally, we have devised a method to efficiently gather logic-error-prone programs during the syntax error correction process and compile these into a dataset that includes single and multiple line logic errors, complete with indices of the erroneous lines. Our experimental results are promising, demonstrating a localization accuracy of 99.2% for logic errors within the top-10 suspected lines, highlighting the effectiveness of our approach in enhancing students' coding proficiency and error correction skills.

1 Introduction

Pseudocode is commonly used in introductory programming courses to teach algorithms to computer science students [6]. It is defined using natural language familiar to the students, conveying the behavior of algorithms. For students with different linguistic backgrounds, pseudocode can be written in their native languages, with the high level of abstraction allowing for translation into English while preserving the original meaning. This approach lowers the language barrier students might face when learning to define algorithms. Pseudocode can

M. Mahmud et al. (Eds.): ICONIP 2024, CCIS 2297, pp. 46–60, 2026.
https://doi.org/10.1007/978-981-96-7036-9_4

include mathematical expressions when they simplify the description of certain behaviors within an algorithm. It allows students to read and understand algorithms by abstracting away the details of programming languages, enabling them to focus on defining the algorithm to solve problems, rather than its technical implementation in code.

During the process where students learn to convert pseudocode into source code, they frequently encounter syntax errors and logic errors, also known as semantic errors. For syntax errors, the compiler can report error messages including suspicious syntax error tokens and syntax error types, which can be used as clues to track and determine syntax errors for students. For example, Microsoft Visual C++ compiler, GCC (the GNU compiler collection), and Clang/LLVM all can catch syntax errors and memory errors in programs [22].

Unlike syntax errors, logic errors do not trigger compiler error reports, are more challenging to localize in student programming assignments. Logic errors can be caused by misunderstanding program specifications or by minor mistakes in the code, such as a wrong iteration number in a loop or a misplaced decimal point. These errors can result in programs to fail test cases and can be challenging for students to locate without any guidance from compilers. Even experienced instructor can spend a considerable amount of time finding these errors, as understanding the program specification and logical structure of the code is an essential step in correcting logic errors. Common types of logic errors are given in Table 1.

Table 1. Common types of logic errors and their descriptions

Type of Logic Error	Description
Loop Condition	Incorrect iteration numbers, inequality, or logical conjunctions in the for/while loops.
Condition Branch	Incorrect logical expressions in the if condition.
Statement Integrity	Statement lacks a self-consistent logical structure after the condition.
Variable Initialization	Incorrect declaration and initialization of variables.
Data Type	Incorrect data type.
Computation	Incorrect basic math symbols or missing mathematical brackets.

The difficulty of identifying logic errors varies by type. Loop condition and data type errors are usually simpler to spot and fix due to their structured and predictable patterns, such as the for keyword in "for (init; condition; increment)" often located on a single line. This clarity and consistency in structure make it easier for automated tools to detect and propose corrections. In contrast, errors in condition branch, such as those involving multiple "if" and "else if" statements over several lines, present more challenges. Their complexity and the variety of conditions and logic across different branches make automated correction harder. This is because accurately fixing such errors often requires an understanding of the program's broader logic, beyond just the immediate context of the errors. Each logic error type has unique characteristics that influence the ease of identification and repair.

Since compilers report error lines for syntax errors, students usually do not face significant challenges in locating them independently. Therefore, our focus is on addressing the challenge of localizing logic errors at the line level. We propose a novel approach to enhance logic error localization in introductory programming courses, drawing inspiration from the DrRepair [27] model. Our method combines the analysis of source code and pseudocode, utilizing bidirectional Long Short-Term Memory (BiLSTM) [8] networks and a graph attention layer [25] to decipher code structure and logic. Furthermore, we employ Code-BERT [7] to evaluate the semantic similarity between code and pseudocode, allowing the model to predict error probabilities in code lines while adjusting its focus based on semantic alignment. Throughout training, we dynamically balance the emphasis between error prediction accuracy and semantic understanding, aiming to improve logic error detection by integrating advanced techniques and adaptive learning strategies.

In our study, we explore a method to collect logic errors during the syntax error repair process of DrRepair, utilizing the SPoC dataset [12] composed of C++ programs from programming competitions. DrRepair, originally designed to fix syntax errors, iteratively repairs programs until they pass all test cases. However, this process often results in programs with logic errors. By analyzing these iterations, we construct a dataset containing various types of logic errors, providing a valuable resource for studying logic error localization and correction. Furthermore, we conducted a comparison of our approach with existing state-of-the-art tools for logic error localization and analyzed the impact of different types of logic errors on localization accuracy.

Our contributions are listed as follows:

1. We introduce a novel technique that leverages pseudocode and employs semantic alignment to assist in localizing logic errors in students' programs.
2. We utilize a graph-based approach in both source code and pseudocode to enhance the localization of logic errors.
3. We create a dataset specifically tailored for logic error analysis, providing a valuable resource for further research in this area.

2   Related Work

In this section, we review the existing literature on automated program repair with deep learning, logic error localization, and graph neural networks, which form the foundation for our approach.

2.1 Automated Program Repair with Deep Learning

Deep learning has significantly advanced the field of program repair, particularly in syntax error correction. Gupta et al. introduced DeepFix, a sequence-to-sequence model that repairs syntax errors in C programs but does not consider the program's structure [10]. To capture this structure, Graph Neural Networks (GNNs) have become popular. Allamanis et al. used Gated Graph Neural Networks to represent both syntactic and semantic aspects of code [2]. Dinella et al. proposed Hoppity, which transforms a buggy program into a graph to predict error locations and their repairs [5]. Yasunaga et al. designed a program-feedback graph, combining source code and compiler feedback to improve syntax error repair using GNNs [27]. Chen et al. introduced PLUR, an algorithm that simplifies program learning and repairing through a program-feedback graph [4]. Li et al. employed context learning and tree transformation to fix syntax errors requiring consecutive changes [14]. These advancements demonstrate the efficacy of combining deep learning with program structures and compiler feedback for syntax error correction.

2.2 Logic Error Localization

In student programming assignments, detecting and fixing logic errors can be achieved using several approaches. Test cases are commonly used to verify if the program's output matches the expected results, helping to identify any discrepancies. Static analysis tools are employed to examine the code without executing it, pinpointing potential logic issues [18]. Additionally, Automated Program Repair tools like Tarantula [11], Ochiai [1], and DStar [26] can suggest corrections at specific lines in the code, using various techniques to locate and address errors effectively. Raana et al. proposed a system to detect logic errors in C++ codes, extract the dependency of methods or functions among source codes, and classify detected logic errors based on a decision tree [19]. Lee et al. presented FixML, a system for automatically generating feedback on logic errors for students' programming assignments [13]. Yoshizawa et al. proposed a logic error detection system based on program structure patterns [28]. Rahman et al. applied a language model to evaluate source codes using a bidirectional long short-term memory (BiLSTM) neural network [3]. Matsumoto et al. provide an iterative trial model to repair multiple logic errors in source codes [15]. Neural-BugLocator (NBL) [9] is a deep learning-based technique developed by Gupta et al. for localizing logic errors in student programs with respect to a failing test without executing the program. Fine-grained Fault Localization [17] combines syntactic and semantic analysis to more accurately localize errors in student programs, outperforming existing techniques on the Prutor and Codeflaws datasets.

2.3 Graph Neural Network

Graph Neural Networks have emerged as a powerful tool for learning representations of graph-structured data, enabling a wide range of applications across various domains. Scarselli et al. introduced the concept of GNNs, proposing a model that extends traditional neural networks to handle graph data by leveraging the graph structure to propagate node features [21]. Veličković et al. introduced the Graph Attention Network (GAT), which incorporates attention mechanisms to weigh the importance of neighboring nodes during feature aggregation, allowing for more flexible and powerful graph representations [24]. Olah and Perez demonstrated the application of GNNs in predicting traffic flow, showcasing their ability to capture spatial and temporal dependencies in complex systems. Monti et al. highlighted the use of GNNs for detecting fake news on social media platforms by modeling the relational information among users and news articles [16]. Stokes et al. showcased the application of deep learning, including GNNs, in discovering new antibiotics, demonstrating the potential of GNNs in drug discovery and biomedical research [23]. Sanchez-Gonzalez et al. presented a graph network-based approach for simulating complex physical systems, illustrating the versatility of GNNs in modeling physical interactions [20]. We employ the attention mechanism from GATs to update the embeddings of nodes in the graph. Specifically, each node updates its representation by considering its connections to other nodes, including those in the source code and pseudo-code, and their corresponding weights.

3   Approach

In this section, we detail our proposed method for logic error localization, including the model architecture, the construction of our logic error dataset, and the experimental setup.

3.1   Model Architecture

The model architecture shown in Fig. 1 is an encoder framework based on DrRepair, to identify logic errors in code. It processes both the source code and corresponding pseudocode to encode the information, then outputs predictions for the indices of erroneous lines.

Model Overview. Initially, lines of source code and pseudocode are each processed by corresponding bidirectional LSTM networks, $LSTM_{source}$ and $LSTM_{pseudo}$, to generate a hidden state h for every line. Following this, a graph attention layer, denoted as $g = Graph(h)$, leverages the structural connections within the code to facilitate information flow and enhance these hidden states. Concurrently, CodeBERT is employed to compute Semantic Alignment Scores, assessing the semantic similarity between the source code and pseudocode across equivalent lines. The model further processes these states with another layer of $LSTM_{source}^{(2)}$ and $LSTM_{pseudo}^{(2)}$, enriching the hidden state for each line. Through a re-contextualization function $x = context(g)$, these states are amalgamated into a cohesive line embedding s_i, transitioning the representation from a token-based to a line-based level, thereby refining the model's predictive capability.

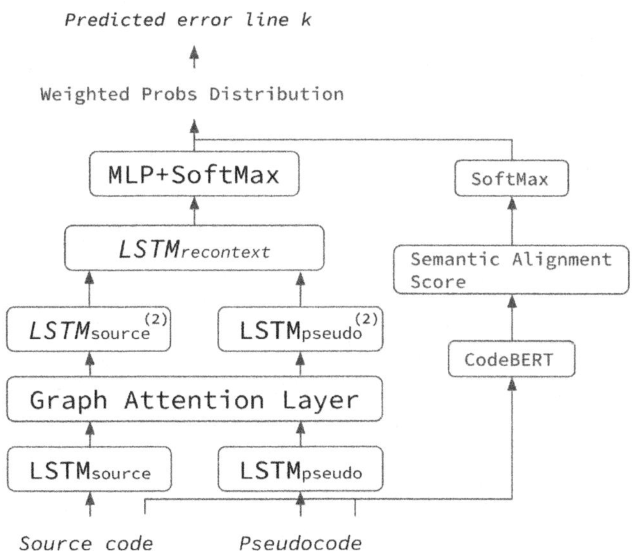

Fig. 1. Model architecture.

In the final step, the model utilizes an MLP followed by a SoftMax layer to deduce the probability of errors across the code lines. This phase incorporates the Semantic Alignment Scores into the model's error prediction, creating a weighted probability distribution. Specifically, if a line is initially deemed likely to contain an error but exhibits a high semantic alignment score, its probability of being the erroneous line is adjusted accordingly.

The training of the model involves dynamic adjustments of the weights assigned to Semantic Alignment Loss. Initially, the training concentrates on reducing the Cross-Entropy Loss to enhance the model's ability to accurately predict logical error lines. As the training progresses and the model's proficiency in identifying error lines increases, the emphasis gradually shifts towards the Semantic Alignment Loss, aiming to foster a deeper semantic understanding and alignment between the source code and pseudocode.

Graph Attention Layer. The graph attention layer forms a key part of our model, creating connections between tokens from the source code and pseudocode that are essential for identifying logic errors. Through a graph $G = (V, E)$, with nodes V representing tokens and edges E connecting matching tokens, the model captures semantic relationships within the code. This approach, advancing beyond DrRepair's framework, leverages pseudocode to enhance logic error localization.

In this layer, token representations are transformed and analyzed to ascertain the importance of each token's connections. Equations (1) through (4) detail this process, which involves linear transformation, attention score computation, normalization through SoftMax, and the aggregation of neighboring node information. This mechanism assigns weights to nodes, guiding the model's focus to the most significant tokens for error correction. The utilization of graph attention, as defined by Velicković et al. [25], empowers the model to trace and emphasize tokens crucial for understanding the logical flow of code.

$$z_i^{(l)} = W^{(l)} h_i^{(l)} \tag{1}$$

$$e_{ij}^{(l)} = \text{LeakyReLU}(a^{(l)^T}(z_i^{(l)} || z_j^{(l)})) \tag{2}$$

$$\alpha_{ij}^{(l)} = \frac{\exp(e_{ij}^{(l)})}{\sum_{k \in \mathcal{N}(i)} \exp(e_{ik}^{(l)})} \tag{3}$$

$$h_i^{(l+1)} = \sigma \left(\sum_{j \in \mathcal{N}(i)} \alpha_{ij}^{(l)} z_j^{(l)} \right) \tag{4}$$

Logic Error Line Prediction with Semantic Alignment. Employing Code-BERT, we derive semantic alignment scores between the source code lines $x_{1:L}$ and their equivalent pseudocode lines $y_{1:L}$ to evaluate their semantic congruence:

$$\alpha_{1:L} = \text{Softmax}(\text{CodeBERT}(x_{1:L}, y_{1:L})) \tag{5}$$

Table 2 showcases a side-by-side comparison of pseudocode and corresponding source code. Throughout the training phase, the model dynamically adjusts the weights assigned to these semantic alignment scores to fine-tune learning priorities. Initially, the emphasis is placed on minimizing Cross-Entropy Loss to expedite the enhancement of error line prediction accuracy. As the model's predictive proficiency evolves, we progressively accentuate the semantic alignment scores to bolster the model's semantic comprehension and correlation:

$$\tilde{p}(k|x_{1:L}, \alpha_{1:L}) = \text{Softmax}(\alpha_{1:L} \odot \text{MLP}(x_{1:L})) \tag{6}$$

where $\alpha_{1:L}$ is adaptively modified over the training period to gradually shift focus towards semantic alignment.

The objective of the training regimen is to minimize a holistic loss that amalgamates the Cross-Entropy Loss, ensuring precise error line detection, with a component that assesses the degree of semantic alignment. This dual-focus strategy endeavors to balance the accurate identification of errors while preserving semantic consistency between the source code and pseudocode. The augmented loss function is delineated as:

$$\mathcal{L}_{CE} = -\sum_{i=1}^{L} y_i \log(p(k = i|x_{1:L})) \tag{7}$$

Here, $\mathcal{L}_{CE}$ represents the Cross-Entropy Loss, with y indicating the true labels of the error lines, ensuring that the model not only identifies error lines with heightened accuracy but also deepens its semantic analysis, fostering an enriched understanding of the logical relationships encoded in the pseudocode.

Table 2. Example of pseudocode and corresponding source code.

i	Pseudocode x_i	Source Code y_i
0	`s = string`	`string s;`
1	`len = integer`	`int len;`
2	`let k, ans be integer`	`int k, ans = 0;`
3	`read a`	`cin >> s;`
4	`set len to size of s`	`len = s.size();`
5	`for i = 0 to len exclusive`	`for (int i = 0; i < len; i++) {`
6	`for j = 1 to len exclusive`	`for (int j = 1; j < len; j++) {`
7	`for k = 0 to infinity`	`for (k = 0; k < len; k++) {`
8	`if i + k is greater than len or j + k is greater than len or s[i + k] is not equal to s[j + k], break`	`if ((i + k > len) || (j + k > len) || (s[i + k] != s[j + k])) break;`
9		`}`
10	`set ans to max of ans, k`	`ans = max(ans, k);`
11		`}`
12	`print ans`	`cout << ans << endl;`
13		`return 0;`
14		`}`

3.2   Logic Error Dataset Construction

SPoC Dataset. The SPoC data consists of 18,356 C++ programs, which are collected from programming competitions. Each program has its own human-written pseudocode, and its public and hidden test cases. Kulal et al. [12] generate a functional correct program from its corresponding pseudocode in SPoC. Each pseudocode line can be translated to a code line, which can have multiple candidate translations. A program can be synthesized by choosing a suitable candidate translation for each pseudocode line. The program is evaluated against both its public and hidden test cases in SPoC. For each program, there are 5 to 10 public test cases to ensure that the program passes its preliminary functional tests.

Construction Procedure. We collect emerging logic errors during DrRepair's syntax error repair process and construct datasets containing programs with logic errors, based on the SPoC dataset. Our observations reveal that DrRepair's repair process on SPoC data generates a significant number of logic errors, encompassing various types of logic errors. DrRepair aims to correct programs with syntax errors, ensuring they pass all test cases. It begins by identifying and fixing syntax errors in an initial program, a process that may require multiple iterations. If the program remains incorrect after an iteration, DrRepair continues to predict and fix errors. Each iteration produces a repaired program by replacing the predicted error line with a modified code line. Successful correction is achieved when a repaired program passes all test cases. However, DrRepair may fail if the repair attempts exceed a set limit (e.g., 100 attempts) or if no suitable code candidates are available to correct the last syntax error in the current corrected program.

DrRepair relies on error messages as hints and cannot directly repair logic errors. During the repair process, if a program is corrected but still contains logic errors, indicating it cannot pass the test cases, DrRepair will backtrack to the previous step and attempt another candidate of correction code lines. This scenario may occur multiple times throughout the repair process, resulting in programs with logic errors being generated. To identify the exact lines with logic errors, we compare programs that still have logic errors after attempted fixes with those that have successfully passed the test cases. By deliberately replacing these identified lines with logic errors, we can artificially create programs that contain these errors for further testing and analysis.

4   Experiments

In this section, we outline the research questions, experimental methodology, and results of our study.

4.1   Research Questions

RQ1: Performance Comparison with State-of-the-Art Tools. How does the performance of our model compare with other state-of-the-art tools for logic error localization?

RQ2: Impact of Logic Error Types. What is the impact of different types of logic errors on the accuracy of error localization? Do certain types of logic errors pose greater challenges?

4.2   Experimental Methodology

Datasets. We have curated two specialized datasets, S-Logic-Err and M-Logic-Err, designed to evaluate programs that pass some but fail other test cases. Programs that fail all tests often need extensive rewriting, which reduces the relevance of fault localization. The S-Logic-Err dataset focuses on single-line code errors, while M-Logic-Err covers multi-line errors, thus addressing a wider spectrum of bug localization complexities. Each dataset includes over 500 unique programming challenges, with S-Logic-Err containing approximately 3800 programs and M-Logic-Err about 1500 programs. On average, each program is about 30 lines long. To rigorously evaluate and fine-tune our models, we utilize a five-fold cross-validation method. Each dataset entry is structured with a program ID, source code, pseudocode, and the index of the logic error line. This format provides a robust framework for analyzing and testing various fault localization strategies, facilitating comprehensive studies on the efficacy of different methods.

Types of Logic Errors and Their Examples. In the Logic-Err dataset we designed, we emphasize the variety of logic errors to reflect real-world coding issues. Table 3 offers examples for each error type, showcasing both the incorrect and the corrected code lines.

Table 3. Examples of common types of logic errors in our dataset, including logic error lines and their correct lines.

Type of Logic Error	Logic Error Line	Correct Line
Loop condition	for (int i = 1; i < i; i++)	for (i = 1; i < 10; i++)
Condition branch	if (n >= 1)	if (n <= 1)
Statement integrity	for (i = 1; i <= 10; i++) { sum = i; printf(sum);}	for (i = 1; i <= 10; i++) { sum += i; printf(sum);}
Variable initialization	int t = red = green = blue = 29;	int t = 29, red, green, blue;
Data type	long long n, m, x	int n, m, x
Computation	int mid = low + high / 2;	int mid = (low + high) / 2;

Baselines. In our study, we compare our method with three established tools for finding errors in code: Tarantula, Ochiai, and DStar. They are spectrum-based fault localization (SBFL) techniques that are primarily used to identify fault locations in software by analyzing the execution traces of passing and failing test cases. They compute suspiciousness scores for each program element (like lines of code or blocks) based on how frequently these elements are executed in passing versus failing test runs. Tarantula checks how each part of the program acts in tests that pass and tests that fail. It uses colors to show how likely it is that each part has an error, helping developers find mistakes more quickly. Ochiai works similarly to Tarantula but uses a specific mathematical formula to decide how suspicious each part of the program is. It considers how often each part appears in passing and failing tests to figure out its connection to errors. DStar also looks for errors in the program but uses a different formula that can be adjusted to better suit different situations.

Metric. We assess our model's performance using localization accuracy, which focuses on accurately identifying the lines where logic errors occur. For evaluation, we measure how effectively the model localizes errors within the top results of our rankings, specifically reporting the accuracy at the top-1, top-5, and top-10 positions.

Implementation Details. In the dataset construction, we increase attempt limit to 300 to generate more candidates containing logic errors. We employ semantic alignment by utilizing the Semantic Textual Similarity (STS) task with CodeBERT. This involves comparing two pieces of text, typically code snippets or a combination of code and its description, to generate a similarity score. For our specific application, we use a version of CodeBERT that has been finetuned on a C++ code corpus, available at https://huggingface.co/neulab/codebert-cpp.

4.3   Experimental Results

In this section, we present the findings from our experiments, focusing on the performance comparison with state-of-the-art tools and the impact of different logic error types on localization accuracy.

RQ1: Performance Comparison. To assess the effectiveness of different methods in localizing logic errors, we conducted comprehensive experiments across two distinct datasets, S-Logic-Err and M-Logic-Err. These datasets are designed to challenge the models with single-line and multi-line logic errors, respectively. The results of these experiments are summarized in Table 4, which presents the localization accuracy at the top-1, top-5, and top-10 ranks.

Table 4. Logic error localization results on the S-Logic-Err and M-Logic-Err datasets, presented in terms of top-n (%) accuracy.

Method	S-Logic-Err Dataset			M-Logic-Err Dataset		
	Top-1	Top-5	Top-10	Top-1	Top-5	Top-10
Tarantula	18.6	41.5	62	11.6	35.1	55.7
Ochiai	20.4	52.2	78.9	18.7	46.2	79.4
DStar	29.7	58.4	78.3	22.6	42.8	71.5
Our Method	36.1	71.2	99.2	28.6	68.3	96.4

The results indicate that our method outperforms traditional fault localization techniques such as Tarantula, Ochiai, and DStar, particularly in higher recall scenarios (Top-5 and Top-10). This suggests that our approach, which integrates advanced semantic analysis, provides more accurate and reliable localizations across different types of logic errors. We observed a consistent improvement in performance across both datasets.

RQ2: Impact of Logic Error Types. To understand the impact of different logic error types on fault localization effectiveness, we analyzed our method's performance using the Top-1 localization ratio. Table 5 presents the distribution

Table 5. Distribution of logic error types and the localization ratio of our method on two datasets (S-Logic-Err and M-Logic-Err)

Type of Logic Error	S-Logic-Err Dataset			M-Logic-Err Dataset		
	Total	Proportion	Loc Ratio	Total	Proportion	Loc Ratio
Loop Condition	844	21.8%	63.2%	1304	37.3%	28.9%
Condition Branch	906	23.4%	26.1%	1231	35.2%	24.9%
Statement Integrity	1212	31.3%	29.7%	591	16.9%	14.1%
Variable Initialization	476	12.3%	27.0%	297	8.5%	10.5%
Data Type	93	2.4%	56.8%	322	9.2%	36.4%
Computation	341	8.8%	24.9%	409	11.7%	21.2%

of logic errors by type, their proportion in our datasets (S-Logic-Err and M-Logic-Err), and the localization success rate for each type.

This table illustrates that certain types of logic errors, such as loop conditions and data type issues, exhibit notably higher localization ratios, particularly in the S-Logic-Err dataset. These types of errors might be more distinctive or have clearer patterns that our model can detect effectively. In contrast, errors related to statement integrity and computation show lower localization ratios, indicating these may be more complex or involve subtler bugs that are harder for the model to pinpoint accurately. The high proportion of errors such as loop conditions and condition branches in both datasets reflects their commonality in programming, emphasizing the need for models like ours that can adeptly handle these frequent issues.

5   Future Work

In the field of programming education, merely localizing logic errors is insufficient. Future efforts should focus on enhancing Large Language Models to not only detect these errors but also generate more actionable feedback on logic errors, including suggested patches. This advancement would significantly improve learning outcomes by providing students with detailed guidance on how to correct their mistakes. Additionally, we recognize that the integration of multiple techniques, such as Graph Neural Networks, LSTM networks, and CodeBERT, increases the complexity and computational cost of our model. To address this issue, future research will explore model compression techniques, such as pruning and quantization, to reduce the computational footprint without sacrificing accuracy.

We also acknowledge the potential limitations of the graph attention mechanism, particularly in cases where the pseudocode-source code mapping is imperfect. To mitigate the impact of these imperfections, we plan to investigate alternative graph structures and incorporate confidence scores into the attention mechanism to handle uncertain mappings more effectively. Improving the accuracy of semantic alignment between pseudocode and source code, especially for abstract pseudocode, is another important direction for future work. We aim to refine the dynamic adjustment of alignment weights during training and further explore techniques to enhance semantic understanding in more complex or abstract cases. Finally, we plan to extend this logic error localization technique to support additional programming languages, such as Python and Java, which will increase the generalizability and applicability of the model in various educational and professional contexts. This extension is a priority in our ongoing research efforts.

6   Conclusion

This paper introduces a novel approach to logic error localization that combines semantic alignment with syntactic analysis, enhancing the identification of logic errors in student programs beyond traditional syntax error detection. Our method, validated through comprehensive experiments, outperforms current state-of-the-art tools in accuracy across multiple datasets. By integrating advanced techniques like bidirectional LSTM networks, graph attention mechanisms, and the semantic capabilities of CodeBERT, our framework not only identifies errors but also provides a foundation for future developments in automated program repair. This work lays the groundwork for further enhancements that could include generating corrective feedback and expanding to more programming languages, thereby improving both educational outcomes and programming proficiency. The code and dataset for independent evaluation are available at: https://github.com/Arrtourz/LogicerrorRepair.

Acknowledgements. This research was supported by the National Institute on Minority Health and Health Disparities (NIMHD) of the National Institutes of Health (NIH) under Award Number U54MD007595.

References

1. Abreu, R., Zoeteweij, P., van Gemund, A.J.: An evaluation of similarity coefficients for software fault localization. In: Proceedings of the 12th Pacific Rim International Symposium on Dependable Computing, pp. 39–46 (2007)
2. Allamanis, M., Brockschmidt, M., Khademi, M.: Learning to represent programs with graphs. arXiv preprint arXiv:1711.00740 (2017)
3. BibitemrefArticle Rahman, M., Mostafizer, Y., Watanobe, K.: A bidirectional LSTM language model for code evaluation and repair. Symmetry **13** (2021)
4. Chen, Z., et al.: PLUR: a unifying, graph-based view of program learning, understanding, and repair. Adv. Neural. Inf. Process. Syst. **34**, 23089–23101 (2021)
5. Dinella, E., Dai, H., Li, Z., Naik, M., Song, L., Wang, K.: Hoppity: learning graph transformations to detect and fix bugs in programs. In: International Conference on Learning Representations (ICLR) (2020)
6. Dirgahayu, T., Huda, S.N., Zukhri, Z., Ratnasari, C.I.: Automatic translation from pseudocode to source code: a conceptual-metamodel approach. In: 2017 IEEE International Conference on Cybernetics and Computational Intelligence (CyberneticsCom), pp. 122–128. IEEE (2017)

7. Feng, Z., et al.: CodeBERT: a pre-trained model for programming and natural languages. arXiv preprint arXiv:2002.08155 (2020)

8. Graves, A., Graves, A.: Long short-term memory. In: Supervised Sequence Labelling with Recurrent Neural Networks, pp. 37–45 (2012)

9. Gupta, R., Kanade, A., Shevade, S.: Neural attribution for semantic bug-localization in student programs. In: Advances in Neural Information Processing Systems, vol. 32 (2019)

10. Gupta, R., Pal, S., Kanade, A., Shevade, S.: DeepFix: fixing common C language errors by deep learning. In: Proceedings of the AAAI Conference on Artificial Intelligence, vol. 31 (2017)

11. Jones, J.A., Harrold, M.J., Stasko, J.: Empirical evaluation of the tarantula automatic fault-localization technique. In: Proceedings of the 20th IEEE/ACM international Conference on Automated Software Engineering, pp. 273–282 (2005)

12. Kulal, S., et al.: SPoC: search-based pseudocode to code (2019)

13. Lee, J., Song, D., So, S., Oh, H.: Automatic diagnosis and correction of logical errors for functional programming assignments. Proc. ACM Program. Lang. 2(OOPSLA), 1–30 (Oct 2018)

14. Li, Y., Wang, S., Nguyen, T.N.: DEAR: a novel deep learning-based approach for automated program repair. In: Proceedings of the 44th International Conference on Software Engineering, pp. 511–523 (2022)

15. Matsumoto, T., Watanobe, Y., Nakamura, K.: A model with iterative trials for correcting logic errors in source code. Appl. Sci. 11(11), 4755 (2021)

16. Monti, F., Frasca, F., Eynard, D., Mannion, D., Bronstein, M.M.: Fake news detection on social media using geometric deep learning. arXiv preprint arXiv:1902.06673 (2019)

17. Nguyen, T.D., et al.: FFL: fine-grained fault localization for student programs via syntactic and semantic reasoning. In: 2022 IEEE International Conference on Software Maintenance and Evolution (ICSME), pp. 151–162. IEEE (2022)

18. Pistoia, M., Chandra, S., Fink, S.J., Yahav, E.: A survey of static analysis methods for identifying security vulnerabilities in software systems. IBM Syst. J. 46(2), 265–288 (2007)

19. Raana, A., Azam, M., Ghazanfar, M., Javed, A., Amin, Y., Naeem, U.: C++ bug cub: logical bug detection for C++ code. The Nucleus 53(1), 56–63 (2016)

20. Sanchez-Gonzalez, A., Godwin, J., Pfaff, T., Ying, R., Leskovec, J., Battaglia, P.W.: Learning to simulate complex physics with graph networks. arXiv preprint arXiv:2002.09405 (2020)

21. Scarselli, F., Gori, M., Tsoi, A.C., Hagenbuchner, M., Monfardini, G.: The graph neural network model. IEEE Trans. Neural Netw. 20(1), 61–80 (2009)

22. Stanier, J., Watson, D.: Intermediate representations in imperative compilers: a survey. ACM Comput. Surv. (CSUR) 45, 1–27 (2013)

23. Stokes, J.M., et al.: A deep learning approach to antibiotic discovery. Cell 181(2), 475–483 (2020)

24. Veličković, P., Cucurull, G., Casanova, A., Romero, A., Liò, P., Bengio, Y.: Graph attention networks. arXiv preprint arXiv:1710.10903 (2017)

25. Velickovic, P., Cucurull, G., Casanova, A., Romero, A., Lio, P., Bengio, Y., et al.: Graph attention networks. Statistics 1050(20), 10–48550 (2017)

26. Wong, W.E., Qi, Y., Zhao, L., Cai, K.Y.: DStar: A d* algorithm for fault localization. In: 2014 IEEE 25th International Symposium on Software Reliability Engineering, pp. 570–581. IEEE (2014)

27. Yasunaga, M., Liang, P.: Graph-based, self-supervised program repair from diagnostic feedback. In: International Conference on Machine Learning, pp. 10799–10808. PMLR (2020)
28. Yoshizawa, Y., Watanobe, Y.: Logic error detection system based on structure pattern and error degree. Adv. Sci. Technol. Eng. Syst. J. 4(5), 1–15 (2019)

FISHER: An Efficient Sim2sim Training Framework Dedicated in Multi-AUV Target Tracking via Learning from Demonstrations

Guanwen Xie[1], Xinqi Wang[2], Yimian Ding[1], Jingzehua Xu[1], Dongfang Ma[3], Jingjing Wang[4(✉)], and Yong Ren[5]

[1] Tsinghua Shenzhen International Graduate School, Tsinghua University, Shenzhen, China
[2] College of Information and Electronic Engineering, Zhejiang University, Hangzhou, China
[3] Ocean College, Zhejiang University, Zhoushan, China
[4] School of Cyber Science and Technology, Beihang University, Beijing, China
`drwangjj@buaa.edu.cn`
[5] Department of Electronic Engineering, Tsinghua University, Beijing, China

Abstract. Multiple autonomous underwater vehicles (AUVs) target tracking problem is a significant challenge for AUV swarm control, which is crucial to the growth of the marine industry. To emphasize the great adaptability while tackling the limitations of reinforcement learning (RL) methods in Multi-AUV target tracking tasks, we propose an efficient two-stage learning from demonstrations (LfD) training framework, FISHER, based on few-shot expert demonstration, featuring imitation learning (IL) and offline reinforcement learning (ORL). In the first stage, we develop a sample-efficient algorithm, multi-agent discriminator actor-critic (MADAC), to facilitate the imitation of expert policy and the generation of offline datasets. In the second stage, based on the decision transformer (DT), the reward function-independent algorithm, multi-agent independent generalized decision transformer (MAIGDT) is utilized for further policy improvement. Simultaneously, we propose a simulation to simulation (sim2sim) method to facilitate the generation of expert trajectories, which is compatible with traditional methods like artificial potential field (APF). Through comparative experiments, we verify the improvement of the proposed MADAC and MAIGDT algorithms. Finally, full target tracking simulation processes show that FISHER can achkmieve performance comparable to expert demonstrations, thereby further demonstrating the strong practicality of FISHER framework. To accelerate relevant research in this direction, the code for simulation will be released as open-source.

Keywords: Autonomous underwater vehicles · Target tracking · Reinforcement learning · Simulation to simulation · Learning from demonstrations

---

G. Xie and X. Wang—These authors contributed equally to this work.

1 Introduction

Due to their powerful maneuverability and wide sensing capabilities, multiple autonomous underwater vehicles (AUVs) have broad application prospects in the construction of the Internet of Underwater Things (IoUT) network, underwater rescue, target tracking etc. Particularly, target tracking is a representative issue, which requires AUVs to keep close to the moving target, while keeping excellent action consistencies and avoiding AUV-target or AUV-AUV collisions simultaneously. The numerous prerequisites make it challenging to use traditional control methods to achieve effective formation control. Fortunately, reinforcement learning (RL) provides an efficient way to solve this problem, due to its strong ability to feature expression and robustness to meet various demands. However, there still exists some challenges when applying RL: (1) The performance of agents considerably relies on the design of the reward function, especially for multiple objectives. A poorly designed reward function may lead to undesirable outcomes, such as sub-optimal policies and reward hacking. (2) RL methods need abundant interactions between agents and the environment, which leads to heavy costs of time and computing resources.

Thanks to the recent booming development of learning from demonstrations (LfD) in RL, these aforementioned issues can be effectively addressed. Imitation learning (IL) and offline reinforcement learning (ORL) are two primary topics in this field. On the one hand, the objective of IL is to learn a policy effectively from limited expert demonstrations. Most current methods are mainly based on generative adversarial imitation learning (GAIL) [3,13], which aligns the policy with expert demonstrations by training a discriminator. However, the original GAIL suffers from the instability of generative adversarial methods. Besides, original GAIL generally utilizes policy obtained via on-policy algorithms for training, such as proximal policy optimization (PPO) [9], which results in low sample efficiency and unsatisfactory performance. Furthermore, IL methods typically have various limitations, such as poor generalization and multitasking performance. On the other hand, ORL is proposed to obtain a generally applicable policy given a dataset with possibly sub-optimal trajectories, without additional online data collection. However traditional ORL methods still rely on the design of the reward function. Besides, ORL usually makes high demands on the scale of the offline dataset, otherwise, bad outcomes may be brought forth [6,8]. These factors mentioned before make it difficult for IL and ORL to be deployed independently in practical LfD scenarios.

To fully exploit the advantages of RL in dealing with complex demands while overcoming its main challenge, we propose a two-stage LfD training network named FISHER, and apply it for the underwater multi-AUV target tracking tasks. Our main contributions can be presented as follows:

- We introduce FISHER, an efficient and reward function irrelevant LfD training framework using few-shot expert demonstrations, which can be easily generated utilizing traditional tracking methods like APF, and transformed by proposed simulation to simulation (sim2sim) procedure. Then IL is used for efficient policy improvement, and ORL is utilized to further enhance both generalization and multi-task performance. The framework is deployed on a high-precision simulation platform for marine target tracking tasks.

- To tackle problems in the GAIL-based IL algorithm, we introduce the discriminator actor-critic (DAC) algorithm and expand it into the multi-agent DAC (MADAC). Leveraging the replay buffer, off-policy RL algorithm, and improvements for generative adversarial networks (GAN) training, MADAC shows a significant boost in training efficiency, while reducing computation loss and demand for environment interaction.

- To tackle the challenges in ORL, we introduce the multi-agent independent generalized decision transformer (MAIGDT), without depending on a reward function. Then we demonstrate through comparative experiments and evaluation of Multi-AUV target tracking processes that MAIGDT significantly outperforms traditional methods, thereby validating the effectiveness of our training framework. To accelerate relevant research in this direction, the code for the simulation will be released as open-source.

2 System Model and Problem Formulation

In this section, we briefly present the AUV dynamic model and underwater detection model for modeling and better simulating the target tracking task. Then, the Markov decision process (MDP) is introduced to lay a foundation for proposed FISHER framework.

Considering that a moving target T, a group of $N(N > 1)$ AUVs are responsible for tracking the target, and both the target and AUVs move on the same plane with a fixed depth d. Target's position vector is denoted as $\boldsymbol{p}_T = [x_T(t), y_T(t)]$. Similarly, the position vectors of tracker AUVs are denoted as $\boldsymbol{p}_i = [x_i(t), y_i(t)], i \in \boldsymbol{N}, \boldsymbol{N} = \{1, ..., N\}$. Besides, there are also M obstacles $\{o_1, ..., o_M\}$ in the environment, and each AUV needs to track the target while avoiding these obstacles as much as possible.

2.1 AUV Dynamics Model

Since AUVs track the target in the horizontal plane, without loss of generality, their dynamic models can be expressed by the three-degree of freedom underdrive model. We denote that AUV i has the body reference frame $\boldsymbol{v}_i = [v_{i,x}(t), v_{i,y}(t), w_i]$, and the world reference frame $\boldsymbol{\eta}_i = [x_i(t), y_i(t), \theta_i]$, where $v_{i,x}(t), v_{i,y}(t), w_i$ and θ_i are surge velocity, sway velocity, angular velocity and yaw angle, respectively. The basic kinematic equation of an AUV is given by

$$\dot{\boldsymbol{\eta}}_i = \boldsymbol{J}(\boldsymbol{\eta}_i)\boldsymbol{v}_i = \begin{bmatrix} \cos\theta_i & -\sin\theta_i & 0 \\ \sin\theta_i & \cos\theta_i & 0 \\ 0 & 0 & 1 \end{bmatrix} \boldsymbol{v}_i. \tag{1}$$

Then, the kinetic equation of AUV can be expressed as

$$\boldsymbol{M}_i\dot{\boldsymbol{v}}_i + \boldsymbol{C}_i(\boldsymbol{v}_i)\boldsymbol{v}_i + \boldsymbol{D}_i(\boldsymbol{v}_i)\boldsymbol{v}_i + \boldsymbol{G}_i(\boldsymbol{\eta}_i) = \boldsymbol{\tau}_i, \tag{2}$$

where M_i represents the inertia matrix including the additional mass of AUV i, C_i denotes the Coriolis centripetal force matrix, while D_i is the damping matrix caused by viscous hydrodynamic. Besides, G_i represents the composite matrix of gravity and buoyancy, and τ_i is the control input of AUV i. Additionally, we discretize the kinematic and kinetic equations above over time, and we obtain

$$\eta_{t+1} = \eta_t + \Delta T \cdot J\left(\eta_t\right) v_t, \tag{3}$$

$$v_{t+1} = v_t + \Delta T \cdot M^{-1} F\left(\eta_t, v_t\right), \tag{4}$$

where $F\left(\eta_t, v_t\right) = \tau_t - C\left(v_t\right) v_t - D\left(v_t\right) v_t - G(\eta_t)$, and ΔT is the time interval.

2.2 Underwater Detection Model

We use the active sonar equation of the underwater environment to model the detection process between the AUV and target, i.e.

$$EM = SL - 2TL + TS - (NL - DI) - DT, \tag{5}$$

where the unit of all parameters is dB, and SL, TL, TS, NL, DI represent the emission sound strength, transmission loss, target strength related to the target reflection area, environmental noise level and directionality index, respectively. DT and EM are the detection threshold and the echo margin of sonar, respectively.

Similarly, we model the communication between AUVs using the passive sonar equation, and we have

$$EM = SL - TL - NL + DI - DT. \tag{6}$$

Furthermore, TL is related to AUV-target distance d and center acoustic frequency f, i.e.

$$TL = 20 \lg(d) + d \times \alpha(f) \times 10^{-3}, \tag{7}$$

$$\alpha(f) = 0.11 \frac{f^2}{1 + f^2} + 44 \frac{f^2}{4100 + f^2} + 2.75 \times 10^{-4} f^2 + 0.003, \tag{8}$$

where $\alpha\left(f\right)$ is an empirical formula for the attenuation of sound waves in water. Since EM and d show a monotonically decreasing relationship, the maximum detection radius r_c of an AUV can be expressed as

$$r_c = \underset{d}{\mathrm{argmax}}\{EM(d) \geq 0\}. \tag{9}$$

2.3 Markov Decision Process

Given the assumption that AUV's behavior only depends on the current state, the target tracking process can be modeled as a Markov decision process (MDP), which includes state space $\mathcal{S}_i$, action space $\mathcal{A}_i$, and reward function $\mathcal{R}_i$.

State Space $\mathcal{S}_i$**:** In MDP, the state of each AUV is observable, and the ith AUV's state $s_i(t) \in \mathbb{R}^{4N+4}$ in the state space $\mathcal{S}_i$ can be expressed as

$$s_i(t) = \{x_{i,t}(t), y_{i,t}(t), v_{x_{i,t}}(t), v_{y_{i,t}}(t), x_{i,j}(t), y_{i,j}(t), v_{x_{i,j}}(t), v_{y_{i,j}}(t),$$
$$EM_k \cos(\theta_{ok_i}), EM_k \sin(\theta_{ok_i})\}_{j \in N, j \neq i, k \in \{1, \ldots, N_o\}}, \tag{10}$$

which consists of three parts: 1) The initial 4 terms denote the target's position and velocity, whose values are defined in the coordinate system of the polar axis in which the direction of the AUV i is facing, namely $x_{i,t}(t) = d_i(t) \cos(\theta_{i,t}(t))$. The same applies hereinafter. 2) The intermediate $4N - 4$ terms are other AUVs' positions and velocities. 3) The final $2N_o$ terms represent the obstacles' position. We assume that an AUV can detect at most N_o of the nearest obstacles, and the echo margin of the obstacle k is EM_k. When less than N_o obstacles are detected, corresponding EM is set to 0dB.

Action Space $\mathcal{A}_i$**:** The action $a_i(t)$ in the action space $\mathcal{A}_i$ can be expressed as two high-level control parameters

$$a_i(t) = [v_i(t), w_i(t)], \tag{11}$$

where $||v_i(t)|| = \sqrt{v_{i,x}(t)^2 + v_{i,y}(t)^2} \in [0, v_{\max}]$ and $||w_i(t)|| \in [0, w_{\max}]$.

Reward Function $\mathcal{R}_i$**:** To some degree, the reward function can reflect the tracking performance of the AUV swarm. It is utilized for traditional RL algorithms to train agents for comparison. **The reward function is not utilized for training the FISHER framework.** There are three parts that are important for the target tracking task

$$r_{ti_i}(t) = \begin{cases} d_i(t) - d_{\min}^t(t), & d_i(t) > d_{\min}^t, \\ 0, & d_i(t) < d_{\min}^t, \end{cases} \tag{12}$$

$$r_{o_i}(t) = \sum_{j=1, j \neq i}^{N} (d_{\text{safe}} - d_{ij}(t)) + \sum_{k=1, k \neq i}^{M} (d_{\text{safe}} - d_{i,o_k}(t)), d_{ij}(t) < d_{\text{safe}} \text{ and } d_{i,o_k}(t) < d_{\text{safe}}, \tag{13}$$

$$r_{l_i}(t) = \begin{cases} d_i^l(t) - d_{\min}^l(t), & d_i^l(t) > d_{\min}^l(t), \\ 0, & d_i^l(t) < d_{\min}^l(t). \end{cases} \tag{14}$$

The definition and meaning of each term in Eq. (12)–(14) are elaborated as follows: 1) The reward term r_{ti_i} is used to encourage a single AUV to track the target, which can be determined by the distance between AUV i and the target. $d_{\min}^t$ denotes the optimal distance from the target. We also introduce the term $r_{tc}(t) = \max_i \{r_{ti_i}(t)\}$ to reflect overall tracking performance. 2) The penalty term r_{o_i} is used to avoid collision with all other AUVs and obstacles. For each AUV or obstacle that is less than the safe distance d_{safe} from the current AUV, a corresponding penalty will be applied and all the penalties will be summed up 3) The reward term r_{l_i} is utilized to encourage each AUV to keep good swarm consistency. To be intuitive, we use a simplified form here, namely an AUV cannot be too far from the nearest AUV in the swarm. where $d_i^l(t) = \min_j \{d_{ij}(t)\}$. Similarly, $d_{\min}^l$ is the optimal distance from other AUVs.

Furthermore, to adjust the positivity of the AUVs tracking target by adjusting the term r_{ti_i} and r_{tc}, we set two weight factors, w_1 and w_2 for r_{ti_i} and r_{tc}, respectively, and we put three settings for signifying them: **Cooperative**: $w_1 = 1$, $w_2 = 0$; **Mixed**: $w_1 = 0.5$, $w_2 = 0.5$; **Split**: $w_1 = 0$, $w_2 = 1$. The cooperative setting only requires that at least one AUV approach the target, while the split setting encourages each AUV to maintain proximity to the target individually. Finally, the overall reward function can be calculated as follows

$$r_i(t) = a \left(w_1 r_{tc}(t) + w_2 r_{ti_i}(t) \right) + w_3 r_{o_i}(t) + w_4 r_{l_i}(t) + r_b, \tag{15}$$

where $W = [aw_1, aw_2, w_3, w_4]$ is the weight vector and r_b is a bias constant.

3   Methodology

In this section, we introduce the training framework FISHER for the multi-AUV target tracking task based on few-shot expert demonstrations. We first introduce our sim2sim method in detail, which can easily generate expert trajectories. Then we present two stages of FISHER: MADAC for sample-efficient imitation learning and MAIGDT for training generalizable policy to complete the target tracking task. The schematic diagram of our proposed training framework is depicted in Fig. 1.

3.1   Sim2sim Expert Demonstration Generation

It is of great difficulty to directly generate expert trajectories through traditional RL methods when the designed reward function is sub-optimal. Therefore, it's necessary to simplify the generation process with the proposed sim2sim method.

To be specific, our sim2sim method consists of the following components: 1) we first simplify the tracking environment, ignoring underwater and other environmental effects, and considering AUVs and the target as particles. This allows us to take advantage of traditional target tracking methods, such as artificial potential field (APF) [4], to obtain AUVs' trajectories. 2) Then we train a simple policy for a single AUV to reach a specific point in the underwater simulation environment, without any obstacle. The state space is composed of positions of the AUV and target point, with the action space being consistent with that adopted by FISHER. The reward function is the negative value of the Euclidean distance to the target point. It's quite simple to optimize the training objective, and the tracking error can be quickly reduced to less than 0.2 m. 3) Finally, we deploy the aforementioned model to each AUV to complete the target tracking tasks in the simulation environment, under the guidance of the AUVs' optimal position obtained previously. We can add some disturbance parameters and repeat this procedure to enhance the diversity of expert trajectories.

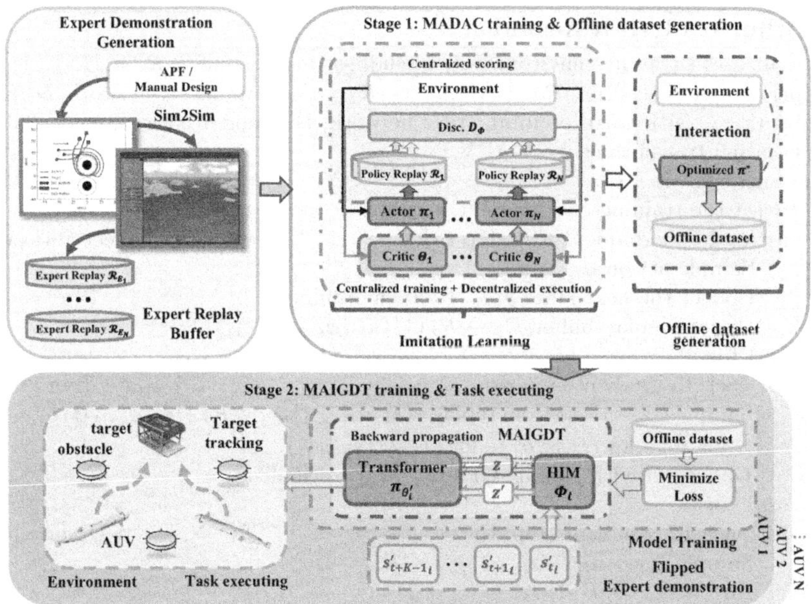

Fig. 1. The schematic diagram of our proposed training framework FISHER.

3.2 Multi-agent Discriminator Actor-Critic

We achieve policy improvement by employing IL using a small number of expert trajectories. Existing methods primarily based on GAIL, which trains a discriminator to distinguish between expert trajectories and policy-generated trajectories, thereby guiding policy improvement and making the generated trajectories approximate the expert trajectories. The primary issue of original GAIL is the need for extensive environmental interaction. To address this, Kostrikov *et al.* [5] introduced the DAC algorithm, which utilized a replay buffer to store previously generated trajectories. Then, similar to Song *et al.* [10] of expanding GAIL to the multi-agent scenario, we can optimize the discriminator network D_i of AUV i as

$$\mathcal{L}_D = \mathbb{E}_{(s,a)\sim\mathcal{R}} \left[\sum_{i=1}^{N} \log\left(D_i(s,a)\right) \right] + \sum_{i=1}^{N} \mathbb{E}_{(s,a)\sim\pi_{E_i}} \left[\log\left(1 - D_i(s,a)\right) \right], \quad (16)$$

where $\mathcal{R}$ denotes the replay buffer, $s = [s_1,\ldots,s_N]$, $a = [a_1,\ldots,a_N]$ and π_{E_i} represents global state and action, and the expert policy of AUV i. The output score D_i of the discriminator can guide policy improvement utilizing off-policy RL algorithms. In addition, our proposed DAC algorithm makes some refinements, such as introducing the absorbing state s_a [11] for termination of episodes, and introducing some improvements for GAN for stabilizing training, including gradient penalty (GP) and spectral normalization (SN) [7,14].

Algorithm 1. FISHER Algorithm

1: Initialize the training environment, including Replay buffer $\mathcal{R} = [\mathcal{R}_1, \ldots, \mathcal{R}_N]$, expert trajectory buffer $\mathcal{R}_E = [\mathcal{R}_{E_1}, \ldots, \mathcal{R}_{E_N}]$, discriminator network D, policy network π_{θ_i} with corresponding critic network, DT model parameters θ_i' with its anti-casual transformer Φ_i of AUV i.

2: **for** each episode k **do**   ▷ Stage 1 : IL with MADAC

3:     Reset the training environment.

4:     **for** each environment timestep t **do**   ▷ Collect trajectories

5:         Sample action $a_{t_i} \sim \pi_{\theta_i} \left(\cdot \mid s_{t_i} \right)$

6:         Collect the next state s_{t+1_i} from environment

7:         Update replay buffer $\mathcal{R}_i \leftarrow \mathcal{R}_i \cup \left\{ (s_{t_i}, a_{t_i}, \cdot, s_{t+1_i}) \right\}$

8:     **end for**

9:     **for** each IL gradient step **do**   ▷ Update discriminator

10:         Sample transitions from replay $\{(s_t, a_t, \cdot, \cdot)\}_{t=1}^{B} \sim \mathcal{R}, \{(s_t', a_t', \cdot, \cdot)\}_{t=1}^{B} \sim \mathcal{R}_E.$

11:         $\mathcal{L}_D = \sum_{b=1}^{B} \log D\left(s_b, a_b\right) - \log\left(1 - D\left(s_b', a_b'\right)\right).$

12:         Update D with Adam+GP+Spectral Normalization

13:     **end for**

14:     **for** each RL gradient step **do**   ▷ Update policy

15:         Sample $\{(s_{t_i}, a_{t_i}, \cdot, s_{t+1_i})\}_{t=1}^{B} \sim \mathcal{R}_i$

16:         **for** $b = 1, \ldots, B$ **do**

17:             $r_i \leftarrow \log D\left(s_{b_i}, a_{b_i}\right) - \log\left(1 - D\left(s_{b_i}, a_{b_i}\right)\right)$

18:             $\left(s_{b_i}, a_{b_i}, \cdot, s_{b+1_i}\right) \leftarrow \left(s_{b_i}, a_{b_i}, r_i, s_{b+1_i}\right)$

19:         **end for**

20:         Update π_{θ_i} with SAC+CTDE

21:     **end for**

22: **end for**

23: Collect trajectories τ_i using optimal policy $\pi_{\theta_i}^*$.   ▷ Make offline datasets

24: Sample n batches of sequence with length K from the offline dataset τ_i.

25: **for** each GDT gradient step **do**   ▷ Stage 2 : ORL with MAIGDT

26:     Flip the state of sequences and get z_i vectors from anti-casual transformer Φ_i.

27:     Update models of GDT by Adam updating on Φ_i and θ_i' by $L_{\text{MSE}}(\theta_i')$ of equation (Eq. (18)).

28: **end for**

29: Get expert demonstration τ_{E_i}' for imitation

30: **while** target tracking task timestep t **do**   ▷ FISHER evaluation loop

31:     Get sequence from timestep t to $t + K - 1$ of τ_{E_i}', flip the state of sequence and get z_{t_i} vector from anti-casual transformer Φ_i

32:     Predict action based on vector z_i, state s_i and a_i of previous K timesteps

33: **end while**

Next, we turn our attention to extending DAC to multi-AUV scenarios. We tested two representative architectures for this extension. The centralized setting sets a discriminator for all AUVs, namely $D_1 = \ldots = D_N = D$, while the policies are trained in the centralized setting. In contrast, The decentralized setting sets a discriminator for each AUV, namely $D_i(s, a) = D_i(s_i, a_i)$. In this paper, we adopt the centralized setting due to its stability in training. We will also compare the performance of the two settings in the subsequent sections.

3.3   Multi-agent Independent Generalized Decision Transformer

We utilize ORL for further policy improvement, effectively enhancing the generalization and multitasking performance. Traditional ORL methods optimize the Bellman objective, therefore the estimation accuracy of the policy gradient is seriously affected by the sufficiency of the dataset. Thus, DT [1] is introduced to abstract ORL problems into seq2seq problems and use sequences to model targets.

DT employs a transformer-based GPT-2 model for autoregressive training and action prediction. Original DT reshapes the trajectory in the offline dataset. A modified trajectory can be denoted as

$$\tau_i' = \left(\hat{r}_{1_i}, s_{1_i}, a_{1_i}, \hat{r}_{2_i}, s_{2_i}, a_{2_i}, \ldots, \hat{r}_{T_i}, s_{T_i}, a_{T_i}\right), \tag{17}$$

where $\hat{r}_{t_i} = \sum_{t'=t}^{T} r_{t_i'}$ denotes the expected total reward of AUV i. When training the model, we sample n batches of sequence with length K from the offline dataset. The prediction head corresponding to the input token $s_i(t)$ is trained to predict $\hat{a}_i(t)$, and the losses for each timestep are averaged. The training objective of the DT model $\pi_{\theta_i'}$ is illustrated as

$$\max_{\pi_{\theta_i'}} J'(\theta_i') = \min_{\pi_{\theta_i'}} \mathcal{L}_{\mathrm{MSE}}(\theta_i') = \min_{\pi_{\theta_i'}} \left[-\frac{1}{N} \sum_{j=1}^{N} (a_j - \hat{a}_j)^2\right]. \tag{18}$$

However, the original DT still relies on the design of the reward function. Furuta *et al.* [2] have demonstrated that DT is doing hindsight information matching, namely using future information to find positive examples with certain contextual parameter values (e.g. returns-to-go for DT). Therefore, we can make DT to match the state transition of expert demonstrations, rather than predicting action using return-to-go. This can be achieved by replacing the return-to-go of the original DT with the information statistics of sequences.

Specifically, we use a second transformer Φ, which takes a reverse-order state sequence as input. The output of transformer Φ is a vector z that contains the information of future states. Since Φ is differentiable to DT's action-prediction loss, Φ can learn sufficient features of states by optimizing equation Eq. (18), and DT is enough to match any distribution to an arbitrary precision. When executing target tracking tasks, we specify an export trajectory τ_E' and use Φ to get features of it, which guides DT to efficiently imitate the demonstration. Figure 1 also shows this process, where z substitutes the return-to-go to facilitate the transformer predicting the action.

Table 1. Simulation Parameters

Parameters	Values
Hydroacoustic parameters SL,TS,DI,DT,NL	100dB, 3dB, 3dB, 20dB, 30dB
Hydroacoustic transmit frequency f	10kHz
Maximum speed parameters $v_{\max}$, $\omega_{\max}$	2.4 m/s, 1.0rad/s
Reward weight factor (a, w_3, w_4, r_b)	-0.125, -0.2, -0.1, 3
Distance parameters $\left(d^t_{\min}, d_{\text{safe}}, d^l_{\min}\right)$	12 m, 8 m, 16 m
Learning rate	0.0003(MADAC),0.0001(MAIGDT)
MADAC gradient penalty factor	1.0
MAIGDT context length K	20

4 Simulation Experiments

In this section, we first introduce the utilized experiment settings. Then we
detail the design of experiment scenarios and corresponding expert trajectories,
followed by the discussion of experiment results and analysis in Sect. 4.3.

4.1 Experiment Settings

We verify the effectiveness of the proposed FISHER by simulating the whole
process of a two-AUV swarm tracking target. In the beginning, the positions
of AUVs are $(-20\,\text{m},\ 8\,\text{m})$ and $(-20\,\text{m},\ -8\,\text{m})$ relative to the target, which is
oriented at the x-axis initially. Then, the policies control AUVs at a frequency
of 12.5 Hz. Other representative parameters of the simulation are provided in
Table 1 for a summary.

4.2 Design of Scenarios and Expert Trajectories

We design several scenarios that feature different target moving trajectories and
obstacle distributions, and all of them have corresponding expert trajectories of
two AUVs. These scenarios are divided into the two parts as follows:

The first part possesses sparse obstacle(s). As the scenarios are not complex,
reward function-based RL methods can achieve acceptable performance. The
obstacle distribution and expert trajectories are shown in Fig. 2, and we label
these scenarios as Scenario 1, and Scenario 2.

However, the second part with dense obstacles makes it hard to correspond
with the reward function, for AUVs must reorganize the formation while passing
through obstacles. Therefore, we introduce some performance indicators to evaluate these scenarios, which will be detailed in Sect. 4.3. Similarly, we introduce
two scenarios and label them as Scenario 3 and Scenario 4.

4.3 Experiment Results and Analysis

We first evaluate the effectiveness of the two stages of FISHER, MADAC and
MAIGDT, by comparative experiments in the scenario(1/2) of sparse obsta-
cle(s) based on the accumulated reward obtained by AUVs. Then, we perform

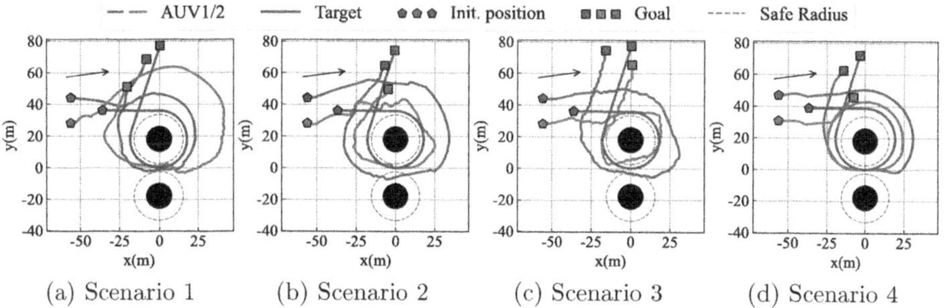

Fig. 2. Trajectories of the target, expert demonstrations of AUVs and obstacle distribution of different scenarios. (a) Scenario 1 (sparse obstacle). (b) Scenario 2 (sparse obstacle). (c) Scenario 3 (dense obstacles). (d) Scenario 4 (dense obstacles).

the target tracking tasks in the scenarios(3/4) of dense obstacles, using some performance indicators to evaluate FISHER and a representative MARL baseline, SAC following the centralized training and decentralized execution(CTDE) manner(denoted as MASAC). Thereby, we can verify the effectiveness and advantages of the proposed FISHER framework.

To start with, we conduct experiments to compare MADAC with the original GAIL implementation (GAIL + PPO)[1] with the centralized multi-agent setting, and the decentralized settings of multi agent DAC (named MAIDAC), given 10 expert trajectories. And the experiment results are shown in Fig. 3.

Observations from Fig. 3(a) illustrate that MAIDAC converges more rapidly and stably than the original GAIL, due to the introduction of replay buffer and off-policy SAC algorithm. And MAIDAC finally achieves expert-level reward after 90 training episodes in Fig. 3(a). Besides, both MADAC and MAIDAC can converge rapidly, but only MADAC can achieve expert-level reward, and MADAC possesses stronger stability. As the number of AUVs increases, MAIDAC shows more distinct disadvantages compared to MADAC, and discussions are deferred to future work.

Moreover, we evaluate the demand of the proposed MADAC algorithm for the number of expert demonstrations, and we conduct comparative experiments in Scenario 1. As Fig. 4 shows, more expert demonstrations can accelerate the training speed and stability. However, generally speaking, our algorithm does not require an extensive number of trajectories, and satisfactory results can be obtained with a limited number of demonstrations. Next, we turn our attention to comparing our proposed MAIGDT with conservative Q-learning (CQL)[2], a typical ORL baseline. The trajectories of the offline dataset adopted here are generally sub-optimal, and there is a significant imbalance in the rewards obtained by two AUVs, making it challenging to obtain satisfactory outcomes. The outcomes of experiments are depicted in Fig. 5. As Fig. 5 shows, MAIGDT outper-

---

[1] https://github.com/ikostrikov/pytorch-a2c-ppo-acktr-gail.
[2] https://github.com/aviralkumar2907/CQL.

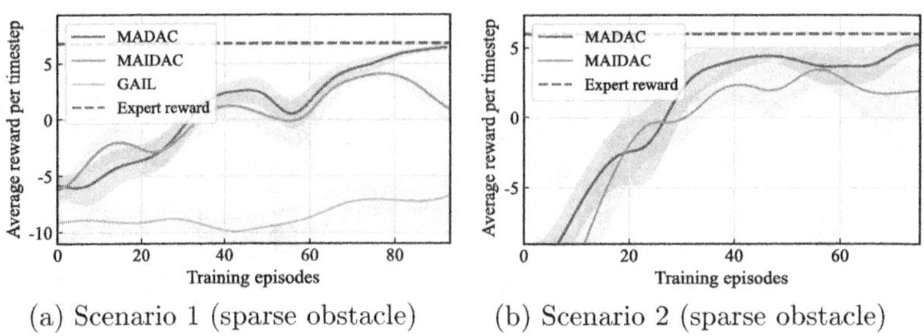

(a) Scenario 1 (sparse obstacle)　　(b) Scenario 2 (sparse obstacle)

Fig. 3. Average total reward curves of all AUVs relying on MADAC, MAIDAC and GAIL for training in different scenarios. (a) Scenario 1 (sparse obstacle). (b) Scenario 2 (sparse obstacle).

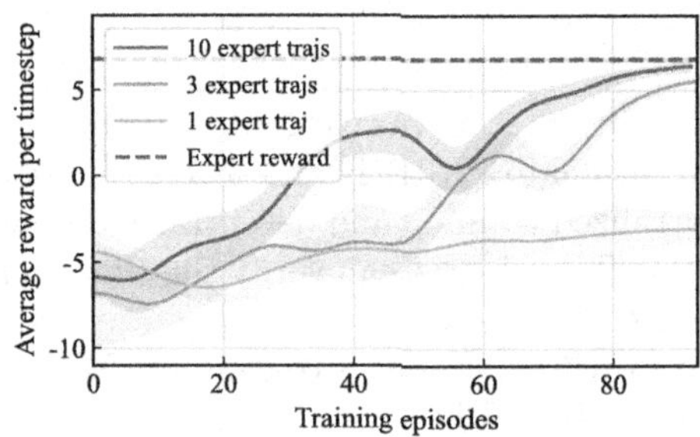

Fig. 4. Average total reward curves of all AUVs utilizing MADAC with different numbers of trajectories for training in Scenario 1 (sparse obstacle).

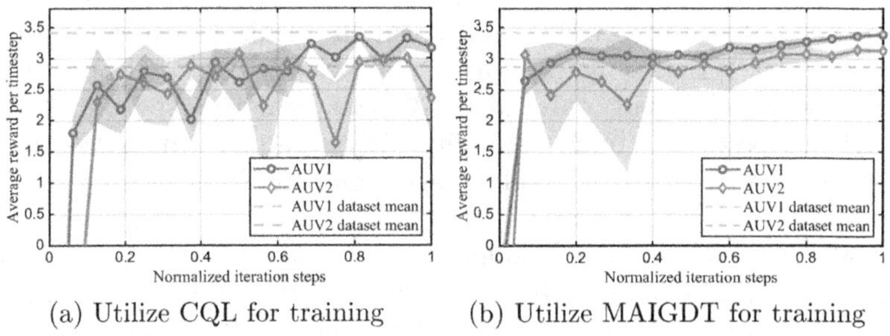

(a) Utilize CQL for training　　(b) Utilize MAIGDT for training

Fig. 5. Average total reward curves of each AUV utilizing different algorithms for training in Scenario 1. (a) Utilize CQL for training. (b) Utilize MAIGDT for training.

forms CQL in terms of training stability and final performance, with MAIGDT's final performance exceeding the dataset's average.

Then we evaluate the multi-task capability of the proposed MAIGDT. To realize this, we design two tasks, both derived from Scenario 2, but with forward directions being clockwise (CW) and counterclockwise (CCW), and conduct experiments to compare the performance between CQL and MAIGDT. As

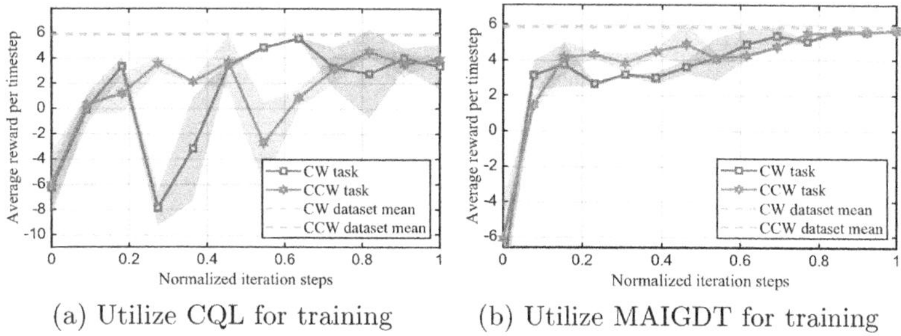

(a) Utilize CQL for training   (b) Utilize MAIGDT for training

Fig. 6. Average total reward curves of all AUVs utilizing different algorithms in CW and CCW tasks taken from Scenario 2. (a) Utilize CQL for training. (b) Utilize MAIGDT for training.

illustrated in Fig. 6, the training of CQL is quite unstable, with the rewards of the two AUVs fluctuating drastically. In contrast, our proposed MAIGDT demonstrates commendable performance and robust stability across both tasks.

Finally, we perform the target tracking tasks in the scenario of dense obstacles (Scenario 4). For comparison, we utilize the MASAC baseline trained with three reward settings respectively, as demonstrated in Sect. 2.3, to reveal the limitations of the reward function design.

For convenience, we introduce six performance indicators similar to Yang et al. [12], i.e., minimum distance mean, minimum distance standard deviation, consistency mean, consistency standard deviation, minimum distance, and danger duration. Minimum distance is the distance between the target and the AUV closest to the target. Minimum obstacle distance represents the minimum distance between the obstacle and the AUVs during the whole process. Consistency refers to the distance between AUVs. While danger duration denotes the time duration during which there is at least one AUV that is less than $d_{\text{safe}} = 8\text{m}$ away from an obstacle. To ensure the validity of the results, we train the policy of AUVs until convergence from scratch 3 times to test the training stability.

As shown in Table 1, the optimal values for minimum distance mean and consistency mean are 12 m and 16 m, respectively. The corresponding results in Scenario 4 are shown in Table 2, while the trajectories of AUVs recorded from the physical simulation environment are shown in Fig. 7, similar to Fig. 2.

It's evident that the benefits of a multi-AUV swarm are scarcely exhibited under the cooperative setting, while AUVs under the split setting tend to disregard the risks of crashing while tracking targets. In addition, the results of MASAC are notably unstable, with severe jiggling while AUVs track the target, reflecting the intrinsic shortcomings of traditional RL methods dependent on reward functions. In contrast, the proposed FISHER effectively acquires knowledge from expert policies, achieving performance that is close to the expert policy, and demonstrating strong stability in both the training process and task execution.

Table 2. Performance of AUVs tracking target in three settings and proposed FISHER framework in Scenario 4. The result is shown through $a \pm b$, where b signifies the standard deviation between policies from multiple training sessions.

Experiments	Cooperative	Mixed	Split	FISHER
$\mathbb{E}$(min-distance)	14.64m $\pm$ 0.27m	14.96m $\pm$ 1.04m	**13.88 m $\pm$ 0.17 m**	14.48m $\pm$ 0.14m
Std(min-distance)	2.60m $\pm$ 0.39m	3.11m $\pm$ 0.43m	1.82m $\pm$ 0.09m	**1.30 m $\pm$ 0.02 m**
$\mathbb{E}$(consistency)	26.50m $\pm$ 1.82m	17.56m $\pm$ 0.70m	**16.20 m $\pm$ 0.66 m**	16.64m $\pm$ 0.36m
Std(consistency)	8.19m $\pm$ 1.60m	3.36m $\pm$ 1.08m	2.79m $\pm$ 0.43m	**1.37m $\pm$ 0.04 m**
Min(obs distance)	6.87m $\pm$ 1.48m	5.57m $\pm$ 1.65m	5.48m $\pm$ 1.23m	**10.41m $\pm$ 0.09 m**
Danger time	9.29s $\pm$ 6.05s	11.59s $\pm$ 6.16s	16.11s $\pm$ 3.53s	**0.00s $\pm$ 0.00s**

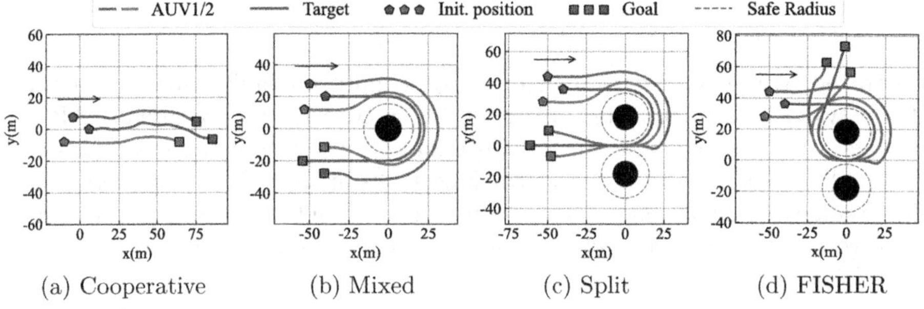

(a) Cooperative   (b) Mixed   (c) Split   (d) FISHER

Fig. 7. Representative tracking trajectories of AUVs utilizing MASAC with three reward settings and FISHER. (a) Cooperative setting. (b) Mixed setting. (c) Split setting. (d) FISHER.

5   Conclusion

In this paper, we propose an efficient training framework FISHER and apply it to train multiple AUVs to complete target tracking tasks via LfD, while under the guidance of expert demonstration transformed by sim2sim. There are two stages in the FISHER: the first stage employs the MADAC algorithm to imitate the expert policy with high sample efficiency and then generates offline datasets. The second stage utilizes MAIGDT, enabling AUVs to make further policy improvements without designing a reward function. Comparative experiments are conducted to compare the performance of MADAC and GAIL, as well as MAIGDT and CQL, to show the superiority of our proposed algorithms. Finally, the target tracking task is evaluated in detail to demonstrate our proposed FISHER framework's remarkable performance and practicality. As a part of future work, we plan to further improve the realism of the simulation, and conduct both simulation and real-world experiments in even more complex tasks.

Acknowledgement. This work of Jingjing Wang was partly supported by the National Natural Science Foundation of China under Grant No. 62071268 and No. 62222101, partly supported by the Young Elite Scientist Sponsorship Program by the China Association for Science and Technology under Grant No. 2020QNRC001, and partly supported by the Fundamental Research Funds for the Central Universities. This work of Yong Ren was partly supported by the National Natural Science Foundation of China under Grant 62127801, partly supported by the National Key Research and Development Program of China under Grant 2020YFD0901000.

References

1. Chen, L., et al.: Decision transformer: reinforcement learning via sequence modeling. In: Advances in Neural Information Processing Systems, pp. 15084–15097 (2021)
2. Furuta, H., Matsuo, Y., Gu, S.S.: Generalized decision transformer for offline hindsight information matching. In: International Conference on Learning Representations, pp. 1–28 (2022)
3. Ho, J., Ermon, S.: Generative adversarial imitation learning. In: Proceedings of the 30th International Conference on Neural Information Processing Systems, pp. 4572–4580 (2016)
4. Khatib, O.: Real-time obstacle avoidance system for manipulators and mobile robots. Int. J. Robot. Res. **5**(1), 90–98 (1986)
5. Kostrikov, I., Agrawal, K.K., Dwibedi, D., Levine, S.: Discriminator-actor-critic: addressing sample inefficiency and reward bias in adversarial imitation learning. In: International Conference on Learning Representations, pp. 1–15 (2019)
6. Macaluso, G., Sestini, A., Bagdanov, A.: Small dataset, big gains: enhancing reinforcement learning by offline pre-training with model-based augmentation. In: The 2nd AAAI Workshop on Artificial Intelligence with Biased or Scarce Data (AIBSD), p. 4 (2024)
7. Miyato, T., Kataoka, T., Koyama, M., Yoshida, Y.: Spectral normalization for generative adversarial networks. arXiv preprint arXiv:1802.05957 (2018)
8. Pan, X., Yao, J., Kou, H., Wu, T., Xiao, C.: HarmonicNeRF: geometry-informed synthetic view augmentation for 3D scene reconstruction in driving scenarios. In: ACM Multimedia 2024 (2023)
9. Schulman, J., Wolski, F., Dhariwal, P., Radford, A., Klimov, O.: Proximal policy optimization algorithms. arXiv preprint arXiv:1707.06347 (2017)
10. Song, J., Ren, H., Sadigh, D., Ermon, S.: Multi-agent generative adversarial imitation learning. In: Advances in Neural Information Processing Systems, pp. 1–12 (2018)
11. Sutton, R.S., Barto, A.G.: Reinforcement Learning: An Introduction. A Bradford Book (2018)
12. Yang, Z., Du, J., Xia, Z., Jiang, C., Benslimane, A., Ren, Y.: Secure and cooperative target tracking via AUV swarm: a reinforcement learning approach. In: IEEE Global Communications Conference, pp. 1–6 (2021)
13. Yao, J., et al.: NDC-Scene: boost monocular 3d semantic scene completion in normalized device coordinates space. In: 2023 IEEE/CVF International Conference on Computer Vision (ICCV), pp. 9421–9431. IEEE Computer Society (2023)
14. Yao, J., Qian, Q., Hu, J.: Multi-modal proxy learning towards personalized visual multiple clustering. In: Proceedings of the IEEE/CVF Conference on Computer Vision and Pattern Recognition, pp. 14066–14075 (2024)

Revisiting Cross-Domain Problem for LiDAR-Based 3D Object Detection

Ruixiao Zhang[1(✉)], Juheon Lee[2], Xiaohao Cai[1], and Adam Prugel-Bennett[1]

[1] ECS, University of Southampton, Southampton, UK
{rz6u20,X.Cai}@soton.ac.uk, apb@ecs.soton.ac.uk
[2] Cambridge, UK

Abstract. Deep learning models such as convolutional neural networks and transformers have been widely applied to solve 3D object detection problems in the domain of autonomous driving. While existing models have achieved outstanding performance on most open benchmarks, the generalization ability of these deep networks is still in doubt. To adapt models to other domains including different cities, countries, and weather, retraining with the target domain data is currently necessary, which hinders the wide application of autonomous driving. In this paper, we deeply analyze the cross-domain performance of the state-of-the-art models. We observe that most models will overfit the training domains and it is challenging to adapt them to other domains directly. Existing domain adaptation methods for 3D object detection problems are actually shifting the models' knowledge domain instead of improving their generalization ability. We then propose additional evaluation metrics – the side-view and front-view AP – to better analyze the core issues of the methods' heavy drops in accuracy levels. By using the proposed metrics and further evaluating the cross-domain performance in each dimension, we conclude that the overfitting problem happens more obviously on the front-view surface and the width dimension which usually faces the sensor and has more 3D points surrounding it. Meanwhile, our experiments indicate that the density of the point cloud data also significantly influences the models' cross-domain performance.

Keywords: 3D object detection · Cross domain · LiDAR point cloud · Deep learning · Generalization

1 Introduction

3D object detection aims to localize and categorize different types of objects in specific 3D space described by 3D sensor data (*e.g.*, LiDAR point clouds). Recently, the application of this technology has achieved significant improvement due to the development of deep neural networks, especially in the field of autonomous driving. Current 3D object detection methods mainly focus on

---

J. Lee—Independent Researcher.

M. Mahmud et al. (Eds.): ICONIP 2024, CCIS 2297, pp. 76–91, 2026.
https://doi.org/10.1007/978-981-96-7036-9_6

Fig. 1. LiDAR point cloud and image data from three datasets: KITTI [7], Waymo [17] and nuScenes [2]. Point cloud density and image shapes are different due to different sensor equipment. For the car objects close to the sensors, a large number of points are collected and most shapes are clearly visible. While for those away from the sensors, only a few points are collected and it is difficult to estimate the dimensions.

specific datasets, *i.e.*, models will be trained and tested independently on a specific dataset. In doing so, a number of models achieved high performances on public benchmarks including nuScenes [2], Waymo [17], and KITTI [7] (see *e.g.*, Fig. 1). However, if the evaluation on a new dataset is needed, in most cases, the training on the new dataset as well as modifications of some training hyperparameters are necessary. In other words, it is hard for models trained on one dataset to adapt directly to another. These domain shifts may arise from different sensor types, weather conditions [21] and object sizes [26] between different datasets or domains. This domain adaptation problem is therefore a big challenge for real-world applications of existing 3D object detection methods, whose retraining steps can be very slow and resource-consuming. It is thus significant to understand the reasons for this cross-domain performance drop and propose efficient methods to raise the cross-domain performance to the same level as within-domain tasks.

The main factors influencing cross-domain performance can be divided into the models and the domains (datasets). How the domains influence cross-domain performances has been investigated in [21]. By comparing the performance of two models [14,25] on several different datasets, the work in [21] has proved that the difference in car size across geographic locations is one of the main challenges for domain adaptation problems. It is then natural to ask: Will there also be any crucial factors influencing the cross-domain performance on the side of models? Therefore, in this paper, our first goal is to investigate the cross-domain performance of existing 3D object detection models with different inputs and structures. We select representative LiDAR-only and multi-modal (*i.e.*, LiDAR + RGB Image) methods including PV-RCNN [13], SECOND [24] and TransFusion [1], some of which are based on 3D convolutional neural networks (CNNs) and the others are based on transformers. Based on our results on three different datasets, *i.e.*, KITTI [7], Waymo [17] and nuScenes [2], we find that all the tested methods similarly fail on the cross-domain tasks, no matter they are based on CNNs or transformers. A more interesting fact is that multi-modal methods achieve poorer performance on some tasks than LiDAR-only methods, even though more data information is taken.

There have been some generic training methods trying to overcome the domain adaptation problem, which can also be considered explorations on the side of models. By reproducing and analyzing one of the state-of-the-art (SOTA) methods, ST3D [26], we surprisingly find that ideas in this field have still limited performance, and more deep ideas are still needed to better solve this problem.

Meanwhile, we notice that current evaluation metrics mainly focus on the average precision (AP) of 3D detection predictions and bird's-eye view (BEV) predictions, which sometimes are not sufficient to evaluate the performance difference between methods especially when different domains are involved. Therefore, we propose two additional evaluation metrics – side-view AP and front-view AP – to make fairer and more comprehensive comparisons. Inspired by the conclusion in [21] that car size difference is the core reason for failure in cross-domain tasks, we evaluate the side-view AP and front-view AP of existing methods, as well as the overlaps between predictions and ground truth in every single dimension.

Our experiments show that the performance under the side-view and front-view metrics is similar, and the prediction accuracy in object length (*i.e.*, depth) is even higher than that in width. By further analyzing the absolute error in each dimension, we prove that the higher overlap accuracy in length is actually because length is usually much higher than width, and the absolute error in length and width are similar. This conclusion might be contradictory to our common sense at first glance, since when we look at the LiDAR point cloud data such as shown in Fig. 1, for most objects, only sides facing or close to the LiDAR sensor can be completely captured. The results also illustrate that the overfitting problem of models makes them not care about the completeness of objects' point cloud data in the target domain, but instead make similar predictions as in the source domain where they are trained.

We, therefore, also suggest more investigations in the evaluation methods for 3D object detection, especially in cross-domain tasks, where the current measurements usually focus on the overall overlap of the entire bounding boxes and ignore the different degrees of influence by different dimensions. To summarize, our main contributions are threefold:

- We analyze the cross-domain performance of representative models of different structures. Our results show that most existing models overfit to the source domain and cannot directly perform well on other domains, no matter whether based on CNNs or transformers. We also suggest that multi-modal methods are harder to be adapted to new domains due to the inconsistency of data and calibrations.
- We analyze one of the SOTA self-training methods ST3D for domain adaptation. Our results reveal a serious problem, *i.e.*, the self-training method ST3D actually shifts the knowledge distribution contained in the model to the new

domain, which cannot improve the models' generalization ability, but instead reduces the models' detection ability on the source domain.
- We propose two additional evaluation metrics – the side-view AP and front-view AP – to evaluate the models' cross-domain performance more comprehensively and locate the errors more accurately. Our results illustrate a surprising phenomenon, *i.e.*, although the incomplete point cloud data occurs more in the length dimension than the width dimension of objects due to occlusion, the cross-domain performance under the side-view and front-view metrics is similar. This suggests that the poor cross-domain detection ability of existing models/methods is not directly related to the objects' incomplete point cloud data, but more related to the overfitting problem caused by the model structures and training strategy.

2   Related Work

3D Object Detection with Point Clouds. The common way of representing real-world 3D space is using LiDAR point cloud data, which uses 3D points to record the 3D environment. Although it benefits from accurate point locations, the main challenge for LiDAR-based 3D object detection methods is finding the best way to process the point cloud data. CNNs have been widely used in 2D object detection problems; however, due to the sparsity and spatial disorder of the point cloud data, they cannot be directly fed into the CNNs. Therefore, current LiDAR-based methods either transform the point clouds into spatial invariant formats to use CNN models or propose new methods that can learn features directly from the 3D points. VoxelNet [28] and SECOND [24] encoded point clouds into voxels so that features can be extracted by CNNs designed for 3D inputs. MV3D [4] projected point clouds into 2D spaces (*i.e.*, the front view and bird's-eye view) and used 2D CNNs to extract features. PointRCNN [14] applied PointNet++ [11] to obtain 3D point-wise features and directly learn the 3D proposals from the points. PV-RCNN [13] voxelized the point cloud data first and then used key-point-wise features to keep more semantic features, in which the combination of point-based and voxel-based methods greatly improves the performance with acceptable computation cost.

3D Object Detection with Images. While using 3D data (*e.g.*, point clouds) to detect 3D objects is a reasonable way, there is still a lot of interest in directly adapting 2D methods to 3D fields, which mainly take 2D images as the input data. The main challenge of applying 2D models to 3D problems is how to estimate the depth information of the relevant 3D scene, which cannot be directly obtained from the image data. In [15], the physical and visual height of objects was used to estimate the depth. The work in [12] estimated the depth of every pixel from the images to get 3D voxel features and projects the voxel features into the bird's-eye view for the final 3D detection. Although there have been many trials on image-only 3D object detection problems, the lack of rich spatial information still results in a large gap in the performance compared with that predicted by models using the point cloud data.

3D Object Detection with Multi-modal Inputs. Since images contain richer semantic information and point clouds contain more spatial information, it is natural to explore the possibility of fusing these two types of data. Based on the position of the fusion step in models, existing works on multi-modal inputs can be divided into three categories. Earlier works [10,16] mainly focused on proposal-level and result-level fusions, in which the models fused the proposals or final predictions obtained from the point cloud channel and the image channel separately. Since the proposals or predictions are based only on one type of data for each channel, both of them suffer from the disadvantages of specific data types and therefore the fusions cannot achieve significantly better results than just using the point cloud data. Afterward, the proposal of Point-Painting [19] proved that fusing these two data types in an early step, *i.e.*, the point-level fusion, can greatly improve performance. TransFusion [1] pointed out the limitations of hard association in previous fusion models and proposed a soft association between image features and point cloud features, which greatly increases the performance to a higher level than point-cloud-only methods. In this paper, we will focus on both the LiDAR-only methods and the multi-modal methods.

Domain Adaptation. Domain adaptation has been widely used in 2D object detection [9,20] and 2D semantic segmentation [5,8]. However, there are only a few approaches specifically designed for 3D object detection. ST3D [26] and ST3D++ [27] used self-training algorithms to generate pseudo labels and train models on the target domain without ground truth. In [22], distillation methods were proposed for LiDAR point clouds to overcome the beam difference between datasets. The work in [21] normalized the object size of different datasets based on additional prior knowledge. Our experiments below show that these existing domain adaptation methods still have limited and unstable performance.

3   Datasets

KITTI. The KITTI object detection dataset [7] is one of the most popular datasets in outdoor 3D object detection tasks. It contains 7,481 training samples and 7,518 test samples. For each sample, KITTI provides its point cloud data with a 64-beam Velodyne LiDAR sensor and its image data with stereo cameras. Following existing works [3,13], the training set is further separated into 3,712 and 3,769 samples as the training and validation sets, respectively.

nuScenes. The nuScenes dataset [2] contains 28,130 training samples and 6,019 validation samples. Following [21], we treat the validation set as the test set and re-split the training set into the training and validation sets. Furthermore, since our experiments mainly focus on car detection, we filter out the samples that do not contain any car objects. Finally, there are 8,614 training samples and 2,395 validation samples. For each sample, nuScenes provides its point cloud data with a 32-beam LiDAR and its image data with five cameras for different angles.

Waymo. The Waymo open dataset [17] contains 122,000 training, 30,407 validation, and 40,077 test samples. It is much larger than the other two above-mentioned datasets. Following existing works, we sub-sample the training and validation sets into 7,905 and 2,000 samples. It should be noted that subsampling the dataset by such ratios will not significantly influence the final performance, since Waymo collects the data as continuous frames and models in our experiments do not consider this time-related information.

Data Integration. Since most existing 3D object detection methods focus on the performance within each specific domain/dataset, they often fine-tune the models for different datasets independently and ignore the influence of gaps between them. However, to investigate the cross-domain performance of these models, we must find a way to merge these datasets. We note that the following differences between datasets have a significant influence on the cross-domain experiments: (i) the point cloud range; (ii) the origin of coordinates; and (iii) the unit for preprocessing the point cloud data, such as voxel sizes in voxel-based methods. Following the ideas in [21,26], some preprocessing methods are therefore adopted. For all datasets, we set the point cloud range to $[-75.2, -75.2, -2, 75.2, 75.2, 4]$ m and shift the whole point cloud space of different datasets so that the X-Y plane always coincides with the horizontal plane; following [26], we set the voxel size of all voxel-based methods to $(0.1, 0.1, 0.15)$ m.

4 Setup and Metrics

4.1 Setup of 3D Object Detection Methods

To better investigate the influence of model structures on cross-domain performance, we train and test the following models on KITTI, Waymo, and nuScenes datasets. We first evaluate two representative LiDAR-only methods, *i.e.*, PV-RCNN [13] and SECOND [24], which are based on 3D CNNs. Afterward, we test the performance of TransFusion [23], a transformer-based method that can take both the LiDAR point clouds and RGB images as the input, to analyze the influence of adding images in cross-domain tasks. We also compare the results of TransFusion-L (*i.e.*, the Transfusion that only takes LiDAR point clouds as the input) with PV-RCNN and SECOND. Last, we apply ST3D [26], a SOTA self-training method for domain adaptation problems in 3D object detection, to PV-RCNN. Following the experiments in ST3D [26], we equip the SECOND model with an extra IoU head for better performance. We first train PV-RCNN, SECOND-IoU, and TransFusion-L using the OpenPCDet [18] toolbox with suggested numbers of epochs and learning rates. Since OpenPCDet only supports LiDAR-only models, we use another toolbox, MMDetection3D [6], for the comparison of TransFusion-L and TransFusion-LC.

We train the LiDAR-only model for 40 epochs with learning rate 5×10^{-5} and batch size 8, and further train with images for 20 epochs with the same learning rate and batch size. We train ST3D with PV-RCNN using the OpenPCDet [18] toolbox. As guided by the original work, we first train the model on the source domain and adapt random object scaling (ROS) to it. Afterward, we train the model with the ST3D method and evaluate the performance on the target domain. Following other works [23,26] based on MMDetection3D and OpenPCDet, we adopt random horizontal flip, rotation and scale transforms during the training process. All the models are trained on RTX 8000.

4.2 Metrics

Most existing methods follow KITTI to evaluate the detection performance in 3D and the BEV by the AP. As concluded in [21], the car size difference is one of the main idiosyncrasies that account for the performance gap between within-domain and cross-domain tasks. We therefore further ask the following question: Does the gap fairly come from the three dimensions or are there specific dimensions

Table 1. Performance of 3D object detection models within and across multiple datasets (evaluated on the validation set). Three representative models are selected for CNN (point-based [14] & voxel-based [24]) and transformer methods [1]. We report the AP (average precision) of the *Car* category objects in the format of BEV/3D with IoU threshold set to 0.7 following the KITTI benchmark. Following [21], we replace the $40, 25, 25$ pixel thresholds on 2D box height with $30, 70, 70$ meters on object depth to better evaluate the performance on Waymo and nuScenes. Results of three within-domain and cross-domain tasks are reported. The results show significant drops when directly adapting models to new domains.

Tasks	Metrics	PV-RCNN	SECOND-IoU	TransFusion-L
KITTI → KITTI	Easy	95.0/91.2	94.0/89.4	90.9/82.6
	Moderate	81.7/70.5	76.5/64.5	73.7/59.6
	Hard	81.4/69.0	76.2/62.7	73.0/57.6
nuScenes → nuScenes	Easy	57.0/41.3	55.0/37.3	54.0/33.0
	Moderate	51.7/37.3	49.9/32.9	48.8/29.3
	Hard	51.7/37.3	49.9/32.9	48.8/29.3
nuScenes → KITTI	Easy	80.9/38.1	56.4/14.7	54.6/14.7
	Moderate	63.0/25.3	37.2/8.5	38.2/9.9
	Hard	62.0/24.6	36.3/7.5	38.4/10.0
Waymo → Waymo	Easy	70.6/62.5	68.2/59.3	68.7/57.1
	Moderate	64.5/53.9	62.3/50.3	62.8/50.3
	Hard	64.5/53.9	62.3/50.3	62.8/50.3
Waymo → nuScenes	Easy	37.0/24.5	32.6/20.7	34.9/19.1
	Moderate	32.7/21.2	29.2/18.4	31.4/16.9
	Hard	32.7/21.2	29.2/18.4	31.4/16.9
Waymo → KITTI	Easy	75.4/25.1	66.3/25.6	73.6/35.3
	Moderate	55.9/18.9	48.1/17.3	55.6/26.9
	Hard	53.2/17.8	45.2/15.0	54.7/26.0

responsible for the majority of the gap? To explore this problem, we propose two additional evaluation metrics, *i.e.*, the side-view AP and the front-view AP, and combine them with the 3D and BEV AP to understand the prediction results better.

As shown in Fig. 2, we project the 3D prediction boxes of the objects into the side-view and front-view planes, *i.e.*, the X-Z plane and the Y-Z plane. Figure 2 shows the calculation methods of these two metrics. Given a 3D bounding box with size $(2l, 2w, h)$, center (x, y, z) and rotation angle θ, then the projected length l_p and width w_p respectively to the front-view plane and side-view plane are given by

$$l_p = 2(w \sin \theta + l \cos \theta), \quad w_p = 2(w \cos \theta + l \sin \theta). \tag{1}$$

Note that there is no need to project the height since the bottom side of the bounding box is parallel to the horizontal plane. We then calculate the related 2D AP with the intersection over union (IoU) thresholds at 0.7. In other words, we mark an object as being correctly detected if the IoU between the prediction box and the ground-truth box is larger than 0.7.

We focus on the performance of the *Car* category – the main focus in most existing works and datasets. KITTI evaluates three cases: Easy, Moderate, and Hard. Following [21], we replace the constraints of box height with object location depth as the criteria for difficulties to better evaluate the performance on other datasets. In detail, we replace the constraints of "larger than $40, 25, 25$ pixels" by "within $30, 70, 70$ meters" for Easy, Moderate, and Hard difficulties.

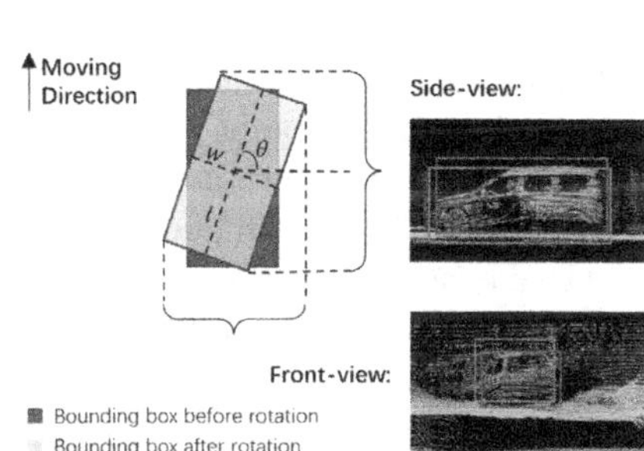

Fig. 2. Definition of the side-view and the front-view AP. The red and blue boxes denote the ground truth and predictions. We project not only the related side to the front/side 2D plane but also consider the other sides that actually can be seen in the related view. For example, the left side is also considered when making a projection into the front view. (Color figure online)

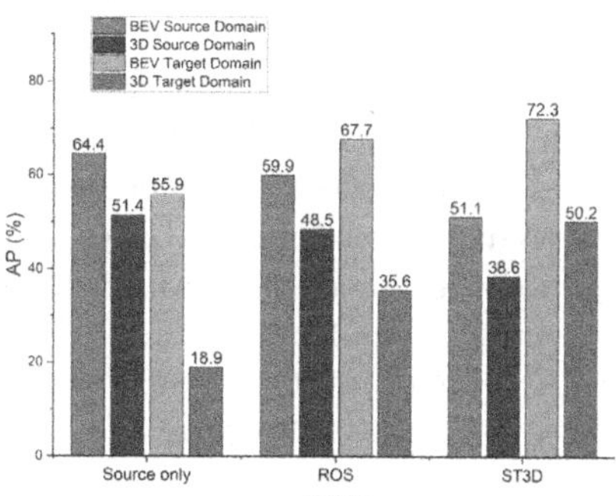

Fig. 3. Performance comparison of the source-only, ROS and the best ST3D (Waymo–KITTI) models on the source domain (Waymo). The results indicate that, with the improvement of the detection ability on the target domain, the performance of ST3D models on the source domain drops significantly.

5   Experiments and Analysis

5.1   Results of LiDAR-Only Methods

We first evaluate the cross-domain performance of existing LiDAR-only methods. We summarize the results in Table 1. Three models are evaluated on three datasets: KITTI, Waymo, and nuScenes (Nusc). We report the results of three within-domain tasks and three cross-domain tasks. Since KITTI only provides the annotations in the front view while Waymo and nuScenes annotate the ring view point clouds, we adapt the models trained on Waymo and nuScenes and only use KITTI as the target domain. Table 1 shows the results of the BEV and 3D AP. For the results of side-view and front-view AP, we defer the discussion in Sect. 5.4.

We see that both PV-RCNN and SECOND-IoU work well on most within-domain tasks. Since nuScenes uses LiDAR sensors with fewer beams and the point cloud data of it is sparser than the others, the Nusc-Nusc task is harder and these results have already been close to the SOTA methods although they are slightly lower than the other two within-domain tasks. Based on the conclusion that both the point-based and voxel-based methods can achieve fairly good results in within-domain tasks, we see heavy drops when evaluating the same models on cross-domain tasks. When trained on nuScenes and evaluated on KITTI, Table 1 shows that the BEV AP of PV-RCNN is even higher than within-domain results (*i.e.*, nuScenes → nuScenes) for the Moderate and Hard difficulties, but the 3D AP drops by 3.2%–12.7%; and the 3D AP of SECOND-IoU drops by 22.6%–25.4%. Similar performance gaps are observed when the models are trained on Waymo. No matter designed with point-based or voxel-based structures, both models fail on all cross-domain tasks with large gaps compared with their within-domain performance if trained on target domains.

Similar to SECOND-IoU, TransFusion is also a voxel-based model. However, it uses a transformer decoder layer to learn from the LiDAR point cloud data and predict the bounding boxes instead of using pre-determined anchors. Furthermore, another transformer decoder is used to combine the image features with predictions from the LiDAR-only channel. We note that the first transformer decoder layer of TransFusion (TransFusion-L) can be independently used as a LiDAR-only detector. We first evaluate and compare it with the above

Table 2. Performance of TransFusion with images. For within-domain tasks, TransFusion-LC (trained with LiDAR point clouds and images) achieves worse results than TransFusion-L (trained with LiDAR only). The results indicate that it is harder to adapt multi-modal methods to new domains without further training.

Models	Nusc-Nusc			Nusc-KITTI		
	Easy	Moderate	Hard	Easy	Moderate	Hard
TransFusion-L	77.9/42.2	45.7/23.5	42.9/22.4	59.6/25.0	48.9/17.4	48.5/17.0
TransFusion-LC	77.4/46.4	42.9/23.8	42.5/22.9	29.6/0.1	24.4/0.5	25.7/0.5

LiDAR-only models. As shown in Table 1, although TransFusion-L replaces the traditional 3D CNNs with transformer decoders, it still achieves similar results as PV-RCNN and SECOND-IoU on different cross-domain tasks.

The results show that most methods fail to obtain acceptable performance when trained and evaluated on different domains without domain transfer training. The cross-domain performance will be poorer when there is a big gap between the distributions of the source and target domains, indicating that the models have overfitted to the source domain. We hypothesize that the poor adaptation ability is not highly related to specific model types such as CNNs or transformers, but results from deeper structural problems.

5.2   Results of LiDAR+RGB Methods

We then analyze the cross-domain performance of TransFusion-LC, *i.e.*, the multi-modal version of TransFusion taking both the LiDAR point clouds and RGB images as the input. The second decoder layer is the key point of TransFusion-LC, which helps build a soft association between the features of the point clouds and images instead of heavily relying on the calibration files. Therefore, the influence of images can be better observed without the interference of calibration files. Since OpenPCDet only supports LiDAR-only models, we use another code base, MMDetection3D [6], to compare the performance of TransFusion-L and TransFusion-LC.

Table 3. Performance of ST3D from Waymo to KITTI in the format of BEV/3D AP. Different proportions of KITTI data are used for the self-training step of ST3D for comparison. When using 50% or fewer data, we randomly sub-sample the dataset twice and report the worst and the best results of each for ST3D. In particular, as an example, 'A - Epoch 13 & 17' means that this is the first randomization with *e.g.* 50% of KITTI, where the worst and the best results obtained at epochs 13 and 17 are shown in the middle three columns and the last three columns, respectively. The results of the source only and ROS are only shown once (in the middle three columns) since they are unique. The best results of ST3D for all three difficulties are indicated in bold.

Data	Worst & Best	Easy	Moderate	Hard	Easy	Moderate	Hard
100%	Source only	75.4/25.1	55.9/18.9	53.2/17.8	–	–	–
	ROS	88.5/46.6	67.7/35.6	68.3/36.8	–	–	–
	ST3D (w/ ROS)	88.4/50.9	66.0/38.2	66.4/39.0	91.0/67.3	**72.3**/50.2	72.6/50.8
50%	A - Epoch 13 & 17	88.8/57.7	80.6/52.8	78.9/52.2	90.6/74.5	81.6/64.1	79.7/62.3
	B - Epoch 01 & 05	90.0/59.7	80.4/53.6	78.7/52.9	91.5/75.3	82.5/65.0	80.5/63.3
25%	A - Epoch 21 & 07	77.7/27.9	69.6/28.3	70.0/28.5	89.7/60.9	79.1/54.8	77.1/53.8
	B - Epoch 10 & 20	85.3/46.6	77.3/42.6	75.7/43.6	85.4/66.2	76.4/59.6	76.3/59.7
10%	A - Epoch 17 & 15	84.2/44.5	74.7/41.7	74.9/41.6	89.9/64.1	79.6/56.7	79.5/55.3
	B - Epoch 02 & 21	86.8/52.4	77.8/48.3	76.1/48.0	**91.8**/**77.4**	81.1/**68.2**	**81.0**/**66.5**

We must note that it is much harder to adapt multi-modal models to different domains than LiDAR-only models. Besides ensuring the preprocessing settings are the same or similar, we also need to make the images consistent. Unfortunately, existing datasets use different RGB cameras to collect the image data, which raises a challenge in making the models able to take different sizes of images as the input, during the training and evaluation stages. For example, nuScenes collects images of size 1600×900, while KITTI collects images of size around 1280×384. When adapting the models trained on nuScenes to KITTI, we can either simply pad the KITTI images to the nuScenes size during evaluation or downsample the nuScenes image size during training to keep the image size consistent. We also try to use an equally small image size of 400×224 or 192×640 to focus on the center contents of the images. Surprisingly, all methods failed to obtain reasonable results on nuScenes to KITTI tasks.

We report the best results in Table 2, with the third strategy using an image size of 400×224. TransFusion-LC performs similarly to TransFusion-L when trained and evaluated on nuScenes, but the performance drops heavily when evaluated on KITTI. We hypothesize that this drop comes from the large gap of images in nuScenes and KITTI. Although we have tried different methods to overcome the size difference, the pixel distribution and semantic knowledge of images are still very different in these two datasets, which makes it hard to extract useful features from KITTI images using a model trained on nuScenes. As a result, the image features become noise in the second decoder layer and lead to a heavy drop in the final detection performance. Approaches to combining the two modal data consistently and better camera-only 3D object detection methods are both necessary to reduce the noise.

5.3 Results of Self-training Methods

The performance of a self-training algorithm, ST3D, is analyzed. We focus on applying ST3D on PVRCNN, from Waymo to KITTI. As a self-training method, ST3D learns from the target domain without requiring ground truth annotations. Following the original work, it takes three steps in the model training process: (i) training the model normally in the source domain; (ii) using the ROS method to improve the generalization ability of the model; and (iii) training the model with the ST3D self-training algorithm by using the target data without annotations. Since self-training models are hard to converge, selecting the epoch with the best evaluation results is required [26]. We argue that this operation should not be used; otherwise this implies the information from the annotations is used and ST3D will no longer be a self-training model. Therefore, we select the worst and the best epochs for a fairer comparison. Note that a random epoch selection may be more reasonable when applying ST3D in practice.

Table 3 (first row) shows that the cross-domain performance improves by a surprising degree benefiting from the additional prior knowledge about the target domain, compared with the source-only (*i.e.*, directly adapt models from the source domain to the target domain) results. ROS improves the results by over 37% and ST3D improves the results by over 48% on average for the 3D

AP. However, we notice two problems with ST3D. The first is that the ROS method has already boosted the performance, which means further self-training is not necessary. The second is that ST3D requires all the training data from the target domain although the annotations are not needed. This leads to the following question: Is the performance of ST3D related to the number of data samples from the target domain? We, therefore, train ST3D with fewer samples from KITTI and evaluate the models on the full KITTI dataset. We sample 50%, 25% and 10% of the KITTI training data and randomize the procedure twice. We observe huge fluctuations in the results of ST3D as shown in Table 3. It indicates that when trained with the full KITTI dataset, the result of the worst epoch is even lower than that of the ROS model, while the best result is around 12% better than ROS. When trained with 25% of KITTI, the results of the first randomization fluctuate between 27.9% and 60.9%, while the results of the second randomization fluctuate between 46.6% and 66.2%, for the 3D AP of the Easy difficulty case. When trained with just 10% of KITTI, the results become even better with a smaller fluctuation, showing the instability of ST3D regarding different data sizes from the target domain.

It is also worth investigating whether the final ST3D model can still work well on the *source* domain. In Fig. 3, we compare the results of the source-only, ROS and the best ST3D models on Waymo. We see that with better performance on the target domain, the detection ability on the source domain becomes worse. In

Table 4. Performance of 3D object detection models within and across multiple datasets in side-view, front-view and BEV AP with the IoU threshold set to 0.7. Results are reported in Easy/Moderate/Hard difficulties. Results show that the side-view AP is much lower than the front-view and BEV AP for most tasks, which attributes the problem to length (depth) and height errors.

Tasks	Metrics	PV-RCNN	SECOND-IoU	TransFusion-L
KITTI → KITTI	Side-view	95.3/84.0/82.6	94.8/81.1/79.6	91.2/77.4/76.0
	Front-view	98.1/85.0/84.7	97.7/82.5/80.5	95.0/77.7/76.6
	BEV	95.0/81.7/81.4	94.0/76.5/76.2	90.9/73.7/73.0
nuScenes → nuScenes	Side-view	54.9/49.5/49.5	53.5/48.2/48.2	48.5/43.6/43.6
	Front-view	56.1/50.4/50.4	53.9/48.4/48.4	49.3/45.0/45.0
	BEV	57.0/51.7/51.7	55.0/49.9/49.9	54.0/48.8/48.8
nuScenes → KITTI	Side-view	82.0/60.9/61.1	61.9/39.4/38.1	57.9/35.9/36.7
	Front-view	80.9/58.3/57.6	49.3/30.0/29.6	42.2/27.4/27.9
	BEV	80.9/63.0/62.0	56.4/37.2/36.3	54.6/38.2/38.4
Waymo → nuScenes	Side-view	33.2/29.6/29.6	31.0/27.2/27.2	31.4/28.0/28.0
	Front-view	34.7/30.5/30.5	32.6/28.9/28.9	34.3/30.2/30.2
	BEV	37.0/32.7/32.7	32.6/29.2/29.2	34.9/31.4/31.4
Waymo → KITTI	Side-view	89.7/72.8/72.4	77.9/62.5/59.4	86.4/70.5/69.7
	Front-view	72.1/59.4/60.1	57.2/46.0/44.5	78.3/61.8/63.0
	BEV	75.4/55.9/53.2	66.3/48.1/45.2	73.6/55.6/54.7

other words, ST3D actually shifts the knowledge domain learned by the model, rather than preserving the generalization ability. We thus reach three below arguments.

(I) ST3D learns the data distribution from the target domain and its performance highly relies on the quality of the available target domain data. In other words, if the sampled 10% of KITTI has a similar distribution to the full KITTI, ST3D can achieve good performance (*e.g.*, 77.4% of 3D AP in the Easy difficulty); otherwise, ST3D's performance may degrade (*e.g.*, 27.9% of 3D AP in the Easy difficulty).

(II) The performance of ST3D is unstable due to its self-supervised algorithm. Even if the sampled data has a close distribution to the full target domain, it may still achieve poor results if models from bad epochs are selected.

(III) Models trained with ST3D can only work well on the target domain but cannot work well on the source domain anymore, which means the generalization ability of the models is still at a low level and the cost of generalizing the models would still be an open problem.

5.4   Analysis with Additional Evaluation Metrics

To analyze the cross-domain problem more deeply as discussed in Sect. 4.2, we use the two proposed AP metrics, *i.e.*, the side-view and the front-view AP, to evaluate the models' cross-domain performance more comprehensively.

Table 5. Performance of 3D object detection models within and across multiple datasets in different dimensions with the IoU threshold set to 0.85. We report the overlaps of the ground truth and the predictions in length (depth), width, and height in the format of Easy/Moderate/Hard difficulties.

Tasks	Metrics	PV-RCNN	SECOND-IoU	TransFusion-L
KITTI → KITTI	Length	91.0/77.6/75.9	89.0/71.8/70.1	83.2/69.9/66.9
	Width	92.8/75.5/75.1	92.1/72.2/70.2	84.3/65.8/64.1
	Height	94.2/78.5/78.6	93.2/77.1/75.7	88.0/72.0/70.6
nuScenes → nuScenes	Length	50.9/46.9/46.9	46.8/42.9/42.9	43.6/40.2/40.2
	Width	52.7/47.4/47.4	49.3/44.3/44.3	47.1/42.2/42.2
	Height	45.9/40.9/40.9	43.7/38.9/38.9	38.2/33.9/33.9
nuScenes → KITTI	Length	57.8/45.7/45.2	40.1/28.0/26.5	28.9/22.8/23.7
	Width	53.4/42.0/41.5	18.7/13.7/14.1	15.6/12.4/14.0
	Height	61.8/44.8/46.3	45.8/28.6/28.7	41.3/25.8/27.3
Waymo → nuScenes	Length	26.8/23.9/23.9	22.5/19.7/19.7	22.3/20.0/20.0
	Width	32.2/28.9/28.9	29.6/26.4/26.4	31.1/27.4/27.4
	Height	27.1/24.2/24.2	26.5/23.6/23.6	26.0/22.9/22.9
Waymo → KITTI	Length	73.9/54.2/50.5	58.9/43.0/39.4	65.3/50.0/47.8
	Width	12.1/13.3/14.3	10.5/10.9/11.4	15.7/15.7/17.4
	Height	89.0/73.2/74.1	66.9/55.3/54.2	86.8/70.7/72.2

Table 6. The average size (meters) of 3D ground-truth bounding boxes of the five datasets and percentage differences between selected datasets.

Metric	KITTI	Argoverse	nuScenes	Lyft	Waymo	W → K	N → K	W → N
Width (m)	1.62	1.96	1.96	1.91	2.11	+30.2%	+21.0%	+7.7%
Height (m)	1.53	1.69	1.73	1.71	1.79	+17.0%	+13.1%	+3.5%
Length (m)	3.89	4.51	4.64	4.73	4.80	+23.4%	+19.3%	+3.4%

We evaluate the previous models on the same tasks and report the results of the side-view, front-view, and BEV AP in Table 4. We see that the AP under these three metrics are similar in within-domain tasks and the cross-domain Waymo → nuScenes task; however, in the cross-domain Waymo → KITTI and nuScenes → KITTI tasks, the front-view AP is obviously lower than the side-view and BEV AP. We thus further calculate the AP of the single-dimension IoU of the length (depth), width, and height below.

Since it is easier for the single-dimension IoU to reach 0.7, we set the threshold as 0.85 instead of 0.7, and the threshold 0.85 is a more equivalent threshold against the threshold 0.7 of 2D IoUs (*i.e.*, side-view, front-view, and BEV) for a single dimension. As shown in Table 5, the length AP is much higher than the width AP for the cross-domain tasks where KITTI is the target domain, which contradicts our common sense that width might be easier to predict since it will face us in most cases and therefore has more points surrounding it. By analyzing the difference in average object size between different datasets as shown in Table 6, we find that the surprising results are actually due to the degree of size difference. Specifically, for Waymo → KITTI and nuScenes → KITTI tasks, the object size differences in width (*i.e.*, 30.2% and 21.0%) are bigger than that in length (*i.e.*, 23.4% and 19.3%). Meanwhile, since more points surround the front-view surfaces consisting of width and height, it is easier for models to get sufficient information from the point clouds to predict the width based on the knowledge obtained from their training domains, which, together with the larger gap in width, results in bigger errors in predicting the width.

We also notice that for the cross-domain Waymo → nuScenes task in Table 5, the width AP is higher than the length AP and height AP. Since the point cloud data in nuScenes is much sparser than that in Waymo, the results indicate that the sparsity of the dataset has a larger influence on the correctness of size prediction. In detail, since there are only a few points surrounding the front-view surface of the object, and the number of points in the Z-axis is limited by the LiDAR beams, the models can thus predict the width a bit better than the length and height.

6   Conclusion

Deep investigations on domain adaptation for 3D object detection are undertaken in this paper. Since researchers are currently focusing on achieving higher performance on a specific dataset, it is unsurprising that existing models actually overfit the training domain and cannot be directly adapted to other domains with different data distributions. It is however worth pointing out that better domain adaptation approaches are still waiting to be explored to improve the generalization ability of models instead of shifting the knowledge domain. Meanwhile, we propose two new evaluation metrics – the side-view and the front-view AP – to provide a more comprehensive measurement of models' cross-domain performance. By using the proposed metrics and further analyzing the performance in each dimension, we notice that the poor cross-domain performance mainly results from the width dimension when the source and target domain have similar point cloud densities, which further indicates the severe overfitting problem of existing model structures and training strategies. Our results also show that the original evaluation metrics are sometimes insufficient to analyze and guide the learning situation of models. We hope that the new side-view and front-view metrics proposed in this paper can be widely applied in the design of new 3D object detection models and the evaluation on different datasets for within-domain and cross-domain tasks.

Disclosure of Interests. The authors have no competing interests to declare that are relevant to the content of this article.

References

1. Bai, X., Hu, Z., et al.: Transfusion: robust lidar-camera fusion for 3d object detection with transformers. In: CVPR, pp. 1090–1099 (June 2022)
2. Caesar, H., Bankiti, V., et al.: nuscenes: a multimodal dataset for autonomous driving. In: CVPR, pp. 11618–11628 (2020)
3. Chen, X., Kundu, K., et al.: 3d object proposals for accurate object class detection. In: NIPS (2015)
4. Chen, X., Ma, H., et al.: Multi-view 3d object detection network for autonomous driving. In: CVPR, pp. 6526–6534 (2017)
5. Chen, Y., Li, W., Gool, L.V.: Road: Reality oriented adaptation for semantic segmentation of urban scenes. In: CVPR, pp. 7892–7901 (2018)
6. Contributors, M.: MMDetection3D: OpenMMLab next-generation platform for general 3D object detection (2020). https://github.com/open-mmlab/mmdetection3d
7. Geiger, A., Lenz, P., Urtasun, R.: Are we ready for autonomous driving? the kitti vision benchmark suite. In: CVPR, pp. 3354–3361 (2012)
8. Huang, H., Huang, Q., Krahenbuhl, P.: Domain transfer through deep activation matching. In: ECCV (September 2018)
9. Khodabandeh, M., Vahdat, A., Ranjbar, M., Macready, W.: A robust learning approach to domain adaptive object detection. In: ICCV, pp. 480–490 (2019)
10. Qi, C.R., Liu, W., Wu, C., Su, H., Guibas, L.J.: Frustum pointnets for 3d object detection from rgb-d data. In: CVPR, pp. 918–927 (2018)
11. Qi, C.R., Yi, L., Su, H., Guibas, L.J.: Pointnet++: deep hierarchical feature learning on point sets in a metric space. In: Advances in Neural Information Processing Systems., vol. 30. Curran Associates, Inc. (2017)

12. Reading, C., Harakeh, A., Chae, J., Waslander, S.L.: Categorical depth distribution network for monocular 3d object detection. In: CVPR (2021)
13. Shi, S., et al.: Pv-rcnn: point-voxel feature set abstraction for 3d object detection. In: CVPR (June 2020)
14. Shi, S., Wang, X., Li, H.: Pointrcnn: 3d object proposal generation and detection from point cloud. In: CVPR, pp. 770–779 (2019)
15. Shi, X., Ye, Q., Chen, X., Chen, C., Chen, Z., Kim, T.K.: Geometry-based distance decomposition for monocular 3d object detection. In: ICCV (2021)
16. Shin, K., Kwon, Y.P., Tomizuka, M.: Roarnet: a robust 3d object detection based on region approximation refinement. In: IV, pp. 2510–2515 (2019)
17. Sun, P., Kretzschmar, H., et al.: Scalability in perception for autonomous driving: Waymo open dataset. In: CVPR (June 2020)
18. Team, O.D.: Openpcdet: An open-source toolbox for 3d object detection from point clouds (2020). https://github.com/open-mmlab/OpenPCDet
19. Vora, S., Lang, A.H., Helou, B., Beijbom, O.: Pointpainting: sequential fusion for 3d object detection. In: CVPR, pp. 4603–4611 (2020)
20. Wang, T., Zhang, X., Yuan, L., Feng, J.: Few-shot adaptive faster r-cnn. In: CVPR, pp. 7166–7175 (2019)
21. Wang, Y., Chen, X., et al.: Train in germany, test in the usa: making 3d object detectors generalize. In: CVPR, pp. 11710–11720 (2020)
22. Wei, Y., Wei, Z., et al.: Lidar distillation: bridging the beam-induced domain gap for 3d object detection. arXiv preprint arXiv:2203.14956 (2022)
23. Xuyang, B., Zeyu, H., et al.: TransFusion: robust lidar-camera fusion for 3d object detection with transformers. In: CVPR (2022)
24. Yan, Y., Mao, Y., Li, B.: Second: sparsely embedded convolutional detection. Sensors **18**(10) (2018)
25. Yang, B., Luo, W., Urtasun, R.: Pixor: real-time 3d object detection from point clouds. In: CVPR, pp. 7652–7660 (2018)
26. Yang, J., Shi, S., et al.: St3d: self-training for unsupervised domain adaptation on 3d object detection. In: CVPR, pp. 10363–10373 (2021)
27. Yang, J., Shi, S., et al.: St3d++: Denoised self-training for unsupervised domain adaptation on 3d object detection. IEEE Trans. Pattern Anal. Mach. Intell. (2022)
28. Zhou, Y., Tuzel, O.: Voxelnet: end-to-end learning for point cloud based 3d object detection. In: CVPR, pp. 4490–4499 (2018)

Self-supervised Pretraining-Enhanced Intelligent Quality Control for Ocean Observations with Limited Historical Data

Shunfang Wu[1,2], Xiang Li[1,2(✉)], and Zhigang Zhao[1,2(✉)]

[1] Key Laboratory of Computing Power Network and Information Security, Ministry of Education, Shandong Computer Science Center (National Supercomputer Center in Jinan), Qilu University of Technology (Shandong Academy of Sciences), Jinan, China
{xiangli,zhaozg}@sdas.org
[2] Shandong Provincial Key Laboratory of Computing Power Internet and Service Computing, Shandong Fundamental Research Center for Computer Science, Jinan, China

Abstract. The integrity of ocean observation data, gathered through various sources, is frequently marred by issues such as instrumental biases, extraneous interferences, and encoding errors. For the establishment of a centralized oceanographic big data hub, the rigorous quality control (QC) of these observations is both essential and critical to the reliability of subsequent oceanographic analyses and smart oceanographic applications. Current AI-driven QC methods, while promising, are often hindered by the need for substantial datasets, a requirement that is not met in many observational locations due to the paucity of historical data. This challenge necessitates the innovation of an intelligent QC model framework capable of reasoning in low- or no-sample contexts. Addressing this need, we have crafted a framework that leverages unsupervised pre-training on vast datasets, allowing for a data-driven discernment of the underlying patterns in ocean observation data. This framework is designed to provide high-precision predictive capabilities and robust QC for a diverse array of marine data types. This study marks the first instance of applying pre-trained modeling methodologies to the domain of marine data QC.

Keywords: Ocean observation data · Quality control · Prediction · Anomaly detection · Pre-trained model

1 Introduction

Ocean observation is crucial for marine science and technology. Understanding oceanic processes and managing marine resources depend on ocean observation data from buoys, research vessels, and satellites. For example, the Argo program is a global initiative providing high-resolution, real-time measurements of seawater temperature and salinity worldwide.

X. Li—Co-first author.

To integrate ocean observations into marine science databases, data quality control is essential [1]. Initially, data quality control relied on traditional techniques like range checking, which ensures measured parameters fall within plausible values, flagging any that don't as suspect. While these methods have been somewhat successful, automatic quality control of ocean observations still depends heavily on expert-defined rules and basic statistical techniques. Recently, machine learning and deep learning methodologies have gained traction in quality control. There is a growing international push to adopt AI-based quality control for oceanographic data. However, research on machine learning-aided quality control is still in its early stages and has yet to be widely implemented in real-time operational settings like ocean data exchange centers.

Conventional AI-based quality control methods generally demand an adequate dataset from the target domain to construct domain-specific models that exclusively capture the intrinsic data distribution of that domain, as depicted in the left subplot of Fig. 1. The middle subplot represents the transitional paradigm that we are currently engaged in, wherein the inference model is developed by fine-tuning a general model that has been pre-trained on a wide range of data sources. The right subplot illustrates the ultimate paradigm, where the pre-trained model is directly applied for few-shot or zero-shot inference, necessitating only a minimal dataset from the target domain to facilitate effective inference.

In the past two years of AI research, models underpinned by the Transformer architecture, including GPT and Llama, have gained prominence and have seen remarkable success in both textual and visual domains. We postulate that a model imbued with knowledge from a variety of sources can enhance inference within the target domain. Consequently, in this study, we endeavor to construct an ocean-centric pre-trained model, nurtured on extensive ocean observation data, which can be leveraged for numerous downstream tasks, with a particular focus on aiding prediction and quality control tasks. We will elaborate on our methodology in subsequent sections and conduct comprehensive experiments to substantiate our hypotheses. The main points of this work's contributions can be outlined as follows:

- This study marks the pioneering effort in leveraging pre-trained AI models for enhancing the quality control processes in ocean observation data management.
- Our model has demonstrated superior performance in both prediction and anomaly detection tasks through comparative experiments, affirming our hypothesis that the integration of knowledge from diverse data sources has enabled our model to surpass the baseline models, which are trained exclusively on data from the target domain.
- Our study highlights the pre-trained models' remarkable ability to infer in data-sparse target domains. The proposed framework enables two inference modes: fine-tuning on ample target data or leveraging zero-shot/few-shot learning on limited target data.
- The model demonstrates robust generalization in two key aspects. Firstly, it shows strong cross-data-type generalization, effectively performing not only

on marine profile data but also on other types of marine data, such as moored buoy data. Secondly, it exhibits strong temporal adaptability, effectively performing across different time periods and marine areas.

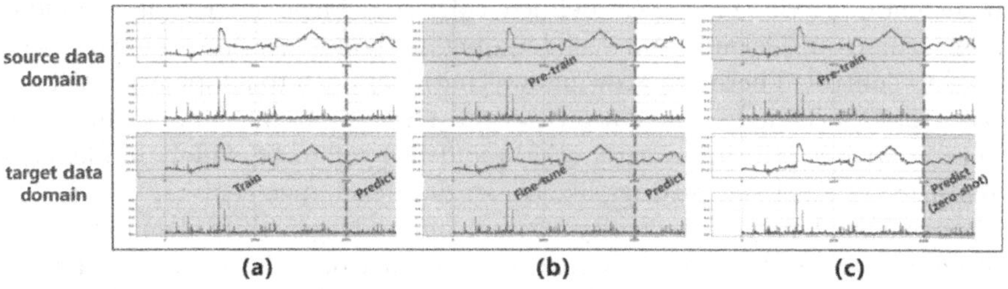

Fig. 1. The progression of three inference paradigms for ocean data, from traditional to advanced methodologies.

2   Related Work

2.1   Quality Control of Ocean Observation Data Based on Traditional Methods

Generally speaking, the quality control of ocean observation data refers to identifying and filtering out spurious observation data generated by various errors to improve the accuracy and reliability of the data.

Conventional quality control methods for observational data predominantly rely on predefined rules and statistical models. Prior to the 1990s, quality control was largely dependent on expert judgment [2]. In the 21st century, there has been a growing preference for automated QC techniques, which are more efficient and widely applicable [3]. Quality control can be categorized into two types based on the timing of implementation: near real-time QC (NRQC) and delayed mode QC (DMQC). NRQC involves immediate, straightforward quality checks performed at data exchange centers shortly after the acquisition of ocean observation data [4], typically with a delay of between 1 h and 1 week. In contrast, DMQC employs more comprehensive and rigorous quality control procedures to evaluate the integrity of ocean observation data, usually with a delay ranging from 6 to 12 months [5].

Quality control can further be classified into automated quality control (AutoQC) and expert-reviewed quality control (ExpertQC). AutoQC encompasses a range of methods including range checks, spike checks, equivalence checks, gradient checks, and various calibration processes. For instance, a spike check may be used to identify unreasonable fluctuations in data at certain depths, which could be due to instrument malfunction or improper handling. This involves calculating the difference in data between three consecutive observations; if the difference exceeds a set threshold, it is flagged as an outlier. However, ExpertQC is resource-intensive, requiring significant human, material, and financial investment, as well as considerable time expenditure.

2.2 Using Machine Learning Methods to Assist Quality Control

The profile shapes across various regions. For instance, the signature method interprets the profile shape as a distinct nonlinear function for the corresponding supervised learning process [6]. This method transforms each vertical sequence of Argo observations into a signature, thereby representing it as a nonlinear function of the profile shape, and subsequently utilizes this mathematical characteristic to assign quality control signatures. Smith et al. conducted an online evaluation of marine sensor data quality through the use of Dynamic Bayesian Networks (DBN) [7], a framework that models the causal relationships between quality assessment variables without being bound by restrictive assumptions about outlier detection or classification methods that may arise in practical applications. Furthermore, Mieruh et al. have integrated artificial neural network techniques with traditional quality control algorithms [8], employing a fully connected MLP network structure that aids conventional automated quality control procedures need time-consuming manual verification of anomalous data by experts.

2.3 Unsupervised Training Methods for Time Series

Unsupervised training methods do not require labeling of time series samples and are therefore more widely applicable than supervised training methods. Unsupervised training is typically performed using self-encoders and reconstruction loss, both traditional self-encoders and more recent model architectures such as Transformer Encoder. For example, Denoising self-encoders can improve the robustness of time series representations and are widely used in the construction of pre-trained models for time series data. For example, in the field of time series, Ma et al. [9] introduced a framework for learning incomplete temporal data representations based on joint interpolation and clustering denoising self encoders. The Transformer model combined with the "mask+prediction" training paradigm is also seen as a reconstruction-based training method. Shi et al. [10] further introduced an unsupervised pre-training approach utilizing the Transformer's self-attention mechanism, which utilises the aforementioned denoising prerequisite task to capture the local dependencies and trends. In contrast to the above studies, Hou et al. [11] introduced a token-based approach for training on traffic flow time series data. Zhang et al. [12] presented a cross-reconstruction Transformer, pre-trained via a cross-domain descent reconstruction process, to capture the time-frequency domain correlations in time series. In addition, Zhao and colleagues [13] developed a bidirectional Transformer-based encoder to comprehend the correlation between two distinct temporal spans at various magnitudes, thereby capturing the temporal interdependencies within traffic flow datasets.

3   Methodology

3.1   Model Structure

We propose a pre-training framework for ocean observation data based on the PatchTST model, which uses the original Transformer encoder to map the observed signal to a latent representation, and employs the Mean Squared Error (MSE) as the loss function, which is normalized to zero mean and unit standard deviation for each time series instance [14]. The model possesses two characteristics: Patching and Channel-independence, as shown in the Fig. 2.

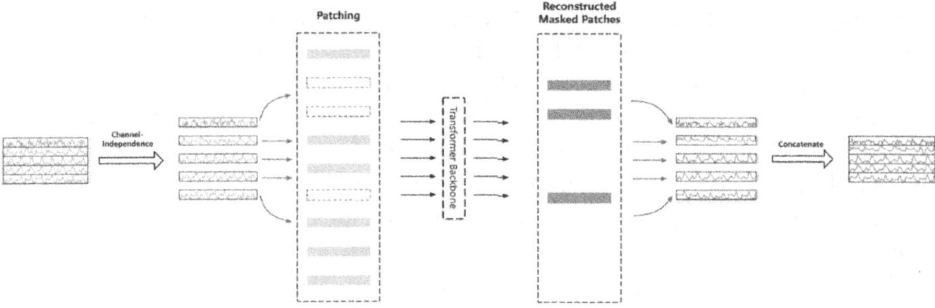

Fig. 2.   Two characteristics of the model for pretraining: Patching and Channel-independence.

Patching: For an univariate sequence, let the length of each patch be P. The sequence is divided into N patches with or without overlap, where $N = \left\lfloor \frac{(L-P)}{S} \right\rfloor + 2$, and S denotes the non-overlapping region between consecutive patches. Each patch is then treated as a token, followed by embedding and positional encoding, and inputted into the Transformer Encoder. Finally, the vectors are flattened and fed into a prediction head (Linear Head) to obtain the prediction of the univariate sequence.

Channel-Independence: Each dimension of the multivariate time series is processed separately, i.e., each dimension is individually inputted into the Transformer Backbone. This implies that each input token contains information from a single channel only. Subsequently, the prediction results obtained from each dimension are concatenated along the dimension axis. This approach treats different dimensions as independent, but the embedding and Transformer weights are shared across all dimensions. Applying this characteristic to marine observation data prediction allows the model to capture the unique characteristics of each dimension, ensuring that the prediction process is not influenced by other factors. This is crucial for analyzing changes in the marine environment.

Initially, in the pre-training stage, the model is trained on a vast collection of unlabeled data from various source domains. These domains are different from the one we're ultimately interested in for our task, but they're still related. Next, in the fine-tuning stage, we adjust the pre-trained model to make it more suitable for a specific task in the target domain. This adjustment improves the model's ability to match the target domain's data patterns and task needs.

Our pre-training model follows the "Mask and Predict" strategy, a popular self-supervised learning method. With this method, input sequences are broken down into non-overlapping segments. A random selection of these segments is then uniformly chosen, and we apply zero-value masks to them. The model is tasked with reconstructing these masked segments, guided by the Mean Squared Error (MSE) loss. The pre-training phase leverages the rich information within the source domain data. Later, fine-tuning on other domains enhances the model's effectiveness and adaptability for particular time series tasks.

3.2 Anomaly Probability Sliding Window (APSW) Based Detection Methods

The proposed method for detecting anomalous data is a prediction-based approach that harnesses the forecasts produced by the pre-trained model. The procedure involves subtracting the predicted value of ocean observation data at time t from the actual measured value at that time to yield the prediction bias. This bias is then used to gauge the degree of anomaly.

Detecting anomalies directly from prediction bias is a widely used method. However, setting an appropriate threshold for bias to determine anomalies is not straightforward. To tackle this issue, we have developed a sliding window-based framework for anomaly probability computation. This framework archives historical prediction biases and uses statistical modeling to calculate the anomaly probability. This method quantifies the anomalousness of a data point with a score that ranges from 0 to 1. As a result, determining the threshold becomes a simpler process. The detection method based on Anomaly Probability Sliding Window (APSW) is shown in Fig. 3.

Specifically, the sliding window is denoted as W, and within this window, we further define a smaller window W'. The larger window W captures a broader context of prior prediction biases (with the sequence of biases in this window denoted as S), while the smaller window W' stores the most recent prediction biases from a smaller, immediate context. By analyzing the distributional change between the larger window W and the smaller window W', we can identify anomalous data. This detection process is carried out sequentially, starting from the initial observation and continuing to the end of the time series. Assuming that the distribution adheres to a normal distribution, the key parameters are as follows:

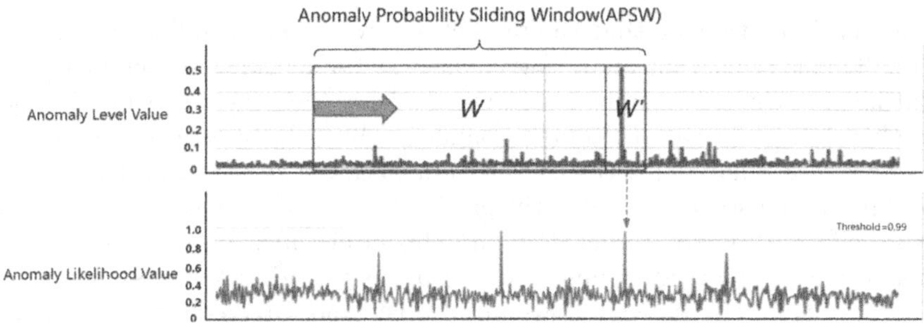

Fig. 3. Anomaly Probability Sliding Window (APSW) based detection methods.

$$\mu_t = \frac{\sum_{i=0}^{i=W-1} S_{t-i}}{W} \tag{1}$$

$$\sigma_t^2 = \frac{\sum_{i=0}^{i=W-1} (S_{t-i} - \mu_t)^2}{W-1} \tag{2}$$

In addition, we need to calculate the mean of the bias sequence within the shorter window W':

$$\tilde{\mu}_t = \frac{\sum_{i=0}^{i=W'-1} S_{t-i}}{W'} \tag{3}$$

Finally, the anomalous probability can be calculated by the complementary probability of the right tail function of the standard Gaussian distribution (Q function) to obtain the anomaly likelihood value of the ocean observation data at time t: $1 - Q((\tilde{\mu}_t - \mu_t)/\sigma_t)$, denoted as L_t. Set the threshold probability for anomaly determination as ϵ. If the output data $L_t \geq \epsilon$, then the target data x at time t is considered as anomalous data.

4   Experiments

4.1   Experiment Setting

During pre-training and fine-tuning, the hyperparameters of our framework are set as follows: the batch size is 128, the learning rate is 0.0001, the dropout rate is 0.2, and the number of epochs is 10. Additionally, the patch length is set to 10, and the stride between patches is set to 5.

For anomaly detection experiments, the large historical window W for storing anomalies is set to 10,000, and the small historical window W' is set to 100. Based on experience from multiple experiments, the threshold for determining anomalies is set to 0.99.

The proposed framework has been developed using Python and the deep learning library PyTorch, with Python version 3.7 and PyTorch version 1.11.0. The model underwent pre-training on the NVIDIA A100 Tensor Core GPU platform and was subsequently fine-tuned and inferred on the NVIDIA GeForce GTX 1650 platform.

This experiment plans to pre-train on global Argo temperature and salinity profile data and test on the SeaDataNet Mediterranean dataset and the Copernicus Global Ocean near-real-time in-situ quality-controlled observation dataset. The experimental framework is shown in Fig. 4. Both test datasets include quality control labels, which will help us verify the prediction performance and further validate the quality control effectiveness. Our hypothesis is that a well-pre-trained model should achieve good performance on similar datasets after pre-training on Argo data.

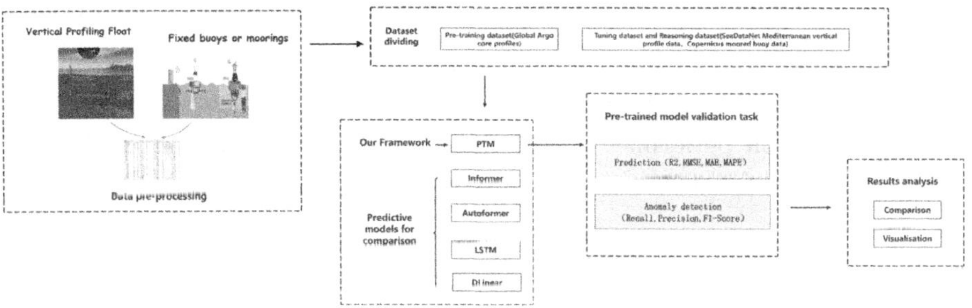

Fig. 4. Framework of experiments.

4.2 Datasets

The training data were derived from the Argo dataset[1], which encompasses over 2.3 million observation profiles. Each profile represents the data collected by an individual buoy during a single cycle. The Argo profile data includes key parameters such as seawater temperature, salinity, and pressure. Additionally, a subset of these profiles measures other biogeochemical elements, such as dissolved oxygen, chlorophyll, and pH.

For the testing phase, our research employed the SDC MED DATA TS V2 dataset[2], which houses a wealth of historical seawater temperature and salinity data from the Mediterranean region [15]. Furthermore, validation experiments were conducted using the Copernicus Global Ocean Near Real-time In Situ Observational Dataset[3], which encompasses a wide range of parameters

---

[1] ftp://ftp.argo.org.cn/pub/ARGO/global/core/.

[2] https://files.seadatanet.org/aggregated_datasets/Med_Sea/SDC_MED_DATA_ TS_V2.

[3] https://data.marine.copernicus.eu/product/INSITU_GLO_PHYBGCWAV_DISC RETE_MYNRT_013_030/description.

including meteorological, biogeochemical and oceanographic data. The dataset gathers observational data from moored buoys, which includes 2,800 locations with over 7 billion records. Our study leveraged elements such as dissolved oxygen, chlorophyll, pH, seawater temperature as well as salinity.

4.3   Evaluation Metrics

To assess the efficacy of ocean observation data predictions, this paper employs a suite of evaluation metrics: R-squared (R^2), Root Mean Squared Error (RMSE), Mean Absolute Error (MAE), and Mean Absolute Percentage Error (MAPE). These metrics are used to compare the performance of different models in forecasting ocean observation data. Where the closer R^2 is to 1, the better the model fits the data, and the closer RMSE, MAE, and MAPE are to 0, the closer the predicted values are to the true values.

To assess the efficacy of anomaly detection within the context of data quality control, this paper employs recall, precision, and the F1 score as evaluation metrics. These metrics are used to comprehensively evaluate the effectiveness of anomaly detection methods. Where higher recall and precision indicate better anomaly detection performance, and the F1 score represents a good balance between recall and precision.

4.4   Results

Given that anomalies are identified through model predictions, the predictive accuracy is a critical determinant of subsequent anomaly detection performance. Thus, we begin by evaluating the predictive capabilities of ocean observation data using pre-trained models, employing the four metrics: R^2, RMSE, MAE, and MAPE.

We benchmarked our model against cutting-edge (SOTA) time series prediction models, which included Informer (Zhou et al., 2021) [16], Autoformer (Wu et al., 2021) [17], LSTM (Hochreiter and Schmidhuber, 1997) [18], and DLinear (Zeng et al., 2022) [19]. Unlike these conventional approaches that require extensive historical data within the target domain for training a model from the ground up, our proposed framework is designed to initiate efficient inference with just a minimal amount of historical data from the target domain. Specifically, the comparison results are shown in the Fig. 5, we note a significant decline in the prediction capabilities of traditional SOTA models when trained on a limited dataset of target domain. Conversely, our model exhibits robustness against variations in the number of samples from the target domain, ensuring a stable level of predictive performance. Moreover, our model surpasses other SOTA models in terms of performance even when they are trained with a large number of samples. As illustrated in Table 1, we note a significant decline in the prediction capabilities of traditional SOTA models when trained on a limited dataset of target domain.

To comprehensively validate the few-shot inference capability of the proposed framework, we compared the prediction performance of our model against other

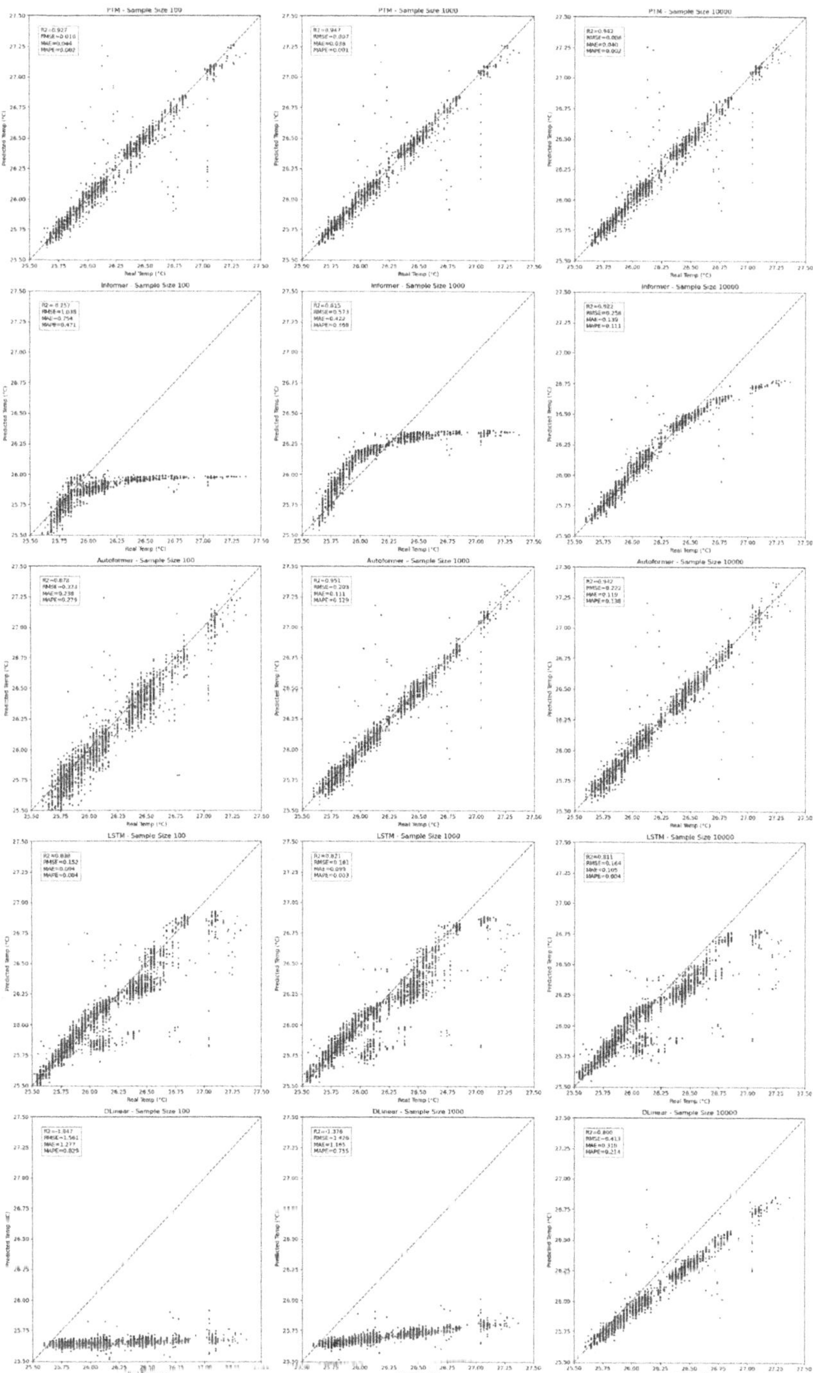

Fig. 5. Comparison of the predictive performance of the pre-trained model (PTM) and other models for seawater temperature with different sample sizes.

Table 1. Comparison of pre-trained models (PTMs) and traditional models with different historical sample sizes of seawater temperature in the target domain.

Models	Number of historical samples for fine-tuning PTM or training SOTAs	R2	RMSE	MAE	MAPE
PTM	100	**0.927**	**0.010**	**0.044**	**0.002**
	1000	0.947	**0.007**	**0.038**	**0.001**
	10000	**0.942**	**0.008**	**0.040**	**0.002**
Informer	100	-0.257	1.038	0.754	0.471
	1000	0.615	0.573	0.422	0.366
	10000	0.922	0.258	0.139	0.111
Autoformer	100	0.878	0.323	0.238	0.279
	1000	**0.951**	0.203	0.111	0.129
	10000	**0.942**	0.222	0.119	0.138
LSTM	100	0.838	0.152	0.094	0.004
	1000	0.821	0.161	0.099	0.003
	10000	0.811	0.164	0.105	0.004
DLinear	100	-1.847	1.561	1.277	0.829
	1000	-1.376	1.426	1.165	0.755
	10000	0.800	0.413	0.318	0.214

representative oceanic observations with a sample size of 1000, including seawater temperature (Temp), salinity (Sal), dissolved oxygen (DO), chlorophyll (Chl-a), and pH. The outcomes of these predictive experiments are detailed in Table 2, where our framework consistently delivers the highest predictive accuracy for all five elements.

Table 2. Comparison of the inference effectiveness for five ocean elements using our proposed PTM and four other SOTA models with limited historical samples (sample size = 1000)

Models	Elements																		
	Temp				Sal				Chl-a				DO				pH			
	R2	RMSE	MAE	MAPE	R2	RMSE	MAE	MAPE	R2	RMSE	MAE	MAPE	R2	RMSE	MAE	MAPE	R2	RMSE	MAE	MAPE
PTM	**0.925**	**0.004**	**0.038**	**0.001**	**0.929**	**0.007**	**0.050**	**0.002**	**0.897**	**0.305**	**0.354**	**0.409**	**0.926**	**0.010**	**0.057**	**0.011**	**0.895**	**0.001**	**0.008**	**0.001**
Informer	0.460	0.388	0.314	0.250	0.353	0.495	0.367	0.615	0.361	2.174	1.364	1.200	0.686	1.580	0.985	0.931	0.493	1.857	1.390	0.799
Autoformer	0.812	0.212	0.141	0.922	0.775	0.119	0.166	0.971	0.831	0.938	0.646	2.969	0.852	0.322	0.080	0.309	0.786	0.313	0.196	0.272
LSTM	0.339	0.171	0.138	0.007	0.369	0.226	0.156	0.050	0.457	1.194	0.635	0.428	0.709	0.795	0.506	0.246	0.686	0.047	0.031	0.003
DLinear	-3.730	1.310	1.026	1.511	0.427	0.775	0.600	0.899	-1.668	1.674	1.278	1.470	0.428	0.776	0.600	0.900	0.417	0.715	0.526	0.897

We further validated the generalizability of our pre-trained model through two experiments. First, Fig. 6 illustrates the testing results within the Argo dataset. Specifically, Fig. 6(a) presents the inference performance when pre-training and fine-tuning on data from different years and months, showcasing the model's temporal adaptability. In contrast, Fig. 6(b) demonstrates the inference performance when pre-training and fine-tuning on data from different sea areas, highlighting the model's spatial adaptability.

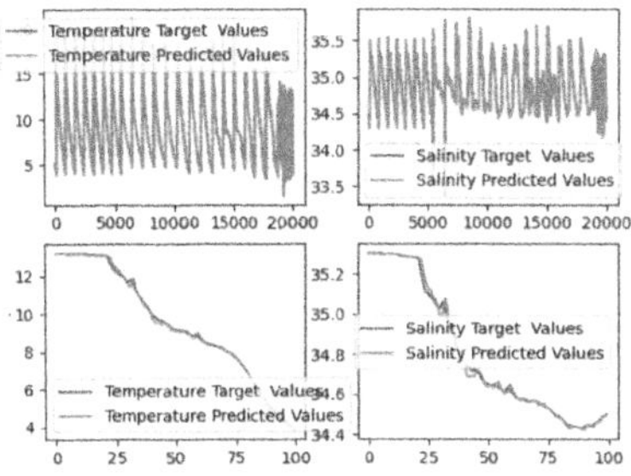

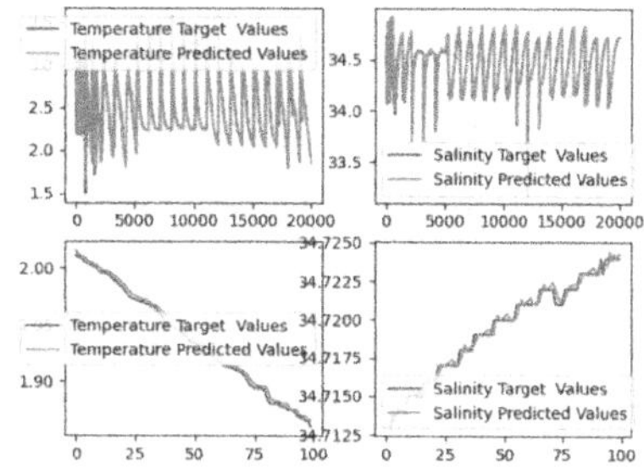

(a) Pre-training data: September 2018, fine-tuning and inference data: June 2023

(b) Pre-training data: Pacific Ocean, fine-tuning and inference data: Atlantic Ocean

Fig. 6. Fine-tuning and inference within the Argo dataset.

Second, we aimed to assess whether the model pre-trained on the Argo profile dataset can perform well on other datasets. Figure 7 illustrates the model's cross-dataset generalization capability. Specifically, Fig. 7(a) shows the effect of fine-tuning inference on SeaDataNet Mediterranean Sea profile data, demonstrating the model's ability to generalize to a different marine profile dataset. Figure 7(b) depicts the effect of fine-tuning inference on Copernicus moored buoy data, highlighting the model's capability to adapt to stationary buoy data.

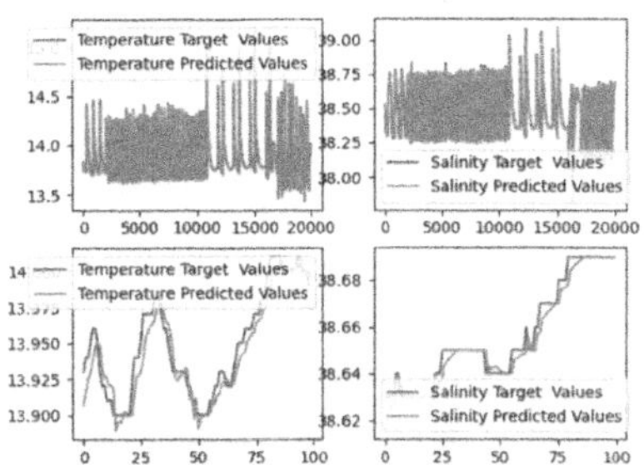

(a) Fine-tuning inference on SeaDataNet Mediterranean vertical profile data

(b) Fine-tuning inference from Copernicus moored buoy data

Fig. 7. Fine-tuning and inference across dataset.

To validate the quality control performance, we conducted experiments using both traditional methods and the Anomaly Probability Sliding Window (APSW) method. The APSW method calculates anomaly probabilities based on the prediction results of pre-trained models. The traditional methods used were the 3-Sigama criterion-based detection method and the spike detection method [20], which is commonly employed in the quality control of oceanic observations to identify significant and unreasonable abrupt changes. The comparative experimental results are shown in Table 3. As seen in the table, compared to traditional quality control methods, our anomaly detection method improves the Recall by 15%–20% and the Precision by more than 50%.

Table 3. Comparison between traditional quality control methods and our APSW method on data of seawater temperature.

Method	3-Sigama	Spike test	APSW
Recall	0.527	0.669	0.748
Precision	0.108	0.318	0.881
F1	0.179	0.431	0.809

Since our Recall and Precision metrics are derived from comparing the predicted quality labels with the original quality labels in the dataset, it is important to note that these original quality labels can only serve as a reference. The original quality labels often contain errors, and their quality can negatively impact the evaluation of our model. To visually demonstrate the effectiveness of our model in anomaly detection, as evident from the fourth subplot of Fig. 8, the proposed 'Pretrained Model + APSW' based anomaly detection approach successfully identifies the data that were clearly anomalous (indicated by the points in red).

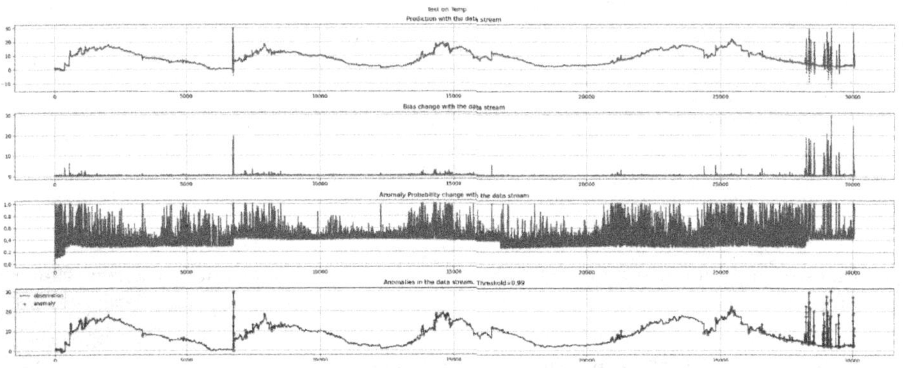

Fig. 8. APSW based method for detecting and labelling anomalies. (Color figure online)

5   Conclusions

This paper constructs an ocean pre-trained model capable of supporting few-shot inference on ocean observation data in the absence of sufficient historical data. The model is applied to various downstream tasks, particularly prediction and quality control. Our pre-training framework aims to enhance the model's generalization ability by learning general features from large-scale data. Compared to traditional models that require extensive training from scratch, this approach can achieve effective inference with minimal historical data from the target domain. Experimental results demonstrate that our pre-trained model performs robustly even with limited data from the target domain. In various element predictions and few-shot inference, our approach significantly outperforms SOTA methods in all metrics, including R^2, RMSE, MAE, and MAPE.

In downstream anomaly detection tasks, our method achieves an impressive F1 Score of approximately 80%, validated against original quality control labels, underscoring its effectiveness in anomaly identification with high accuracy. Moving forward, our research will prioritize refining APSW-based detection methods to further enhance anomaly detection performance. At the same time, we will add ablation experiments to prove the superiority. In addition, the model has not done enough in modeling and mining the intrinsic correlation of multi-source data, which is the part we will focus on breaking through to solve in the next step.

Acknowledgement. This work was supported by the Major Innovation Project for the Integration of Science, Education, and Industry of Qilu University of Technology (Shandong Academy of Sciences) (2024GH24, 2023HYZX01, 2023JBZ02), the Talent Research Projects of Qilu University of Technology (Shandong Academy of Sciences) (2023RCKY136), the Technology and Innovation Major Project of the Ministry of Science and Technology of China (2022ZD0118600) and the Jinan '20 New Colleges and Universities' Funded Project (202333043). Research on the Development Strategy of Marine Big Data Application Services in Shandong Province(2022-DFZD-35-03).

References

1. Taleb, I., Serhani, M.A., Bouhaddioui, C., Dssouli, R.: Big data quality framework: a holistic approach to continuous quality management. J. Big Data **8**(1), 1–41 (2021). https://doi.org/10.1186/s40537-021-00468-0
2. Thadathil, P., Ghosh, A.K., Sarupria, J., Gopalakrishna, V.: An interactive graphical system for xbt data quality control and visualization. Comput. Geosci. **27**(7), 867–876 (2001)
3. Cummings, J.A.: Ocean data quality control. Operational Oceanography i 21st Century, 91–121 (2011)
4. Goni, G.J., et al.: More than 50 years of successful continuous temperature section measurements by the global expendable bathythermograph network, its integrability, societal benefits, and future. Front. Mar. Sci. **6**, 452 (2019)
5. Roemmich, D., et al.: On the future of argo: a global, full-depth, multi-disciplinary array. Front. Mar. Sci. **6**, 439 (2019)
6. Castelao, G.P.: A framework to quality control oceanographic data. J. Open Source Softw. **5**(48), 2063 (2020)
7. Smith, D., Timms, G., De Souza, P., D'Este, C.: A bayesian framework for the automated online assessment of sensor data quality. Sensors **12**(7), 9476–9501 (2012)

8. Mieruch, S., Demirel, S., Simoncelli, S., Schlitzer, R., Seitz, S.: Salaciaml: a deep learning approach for supporting ocean data quality control. Front. Mar. Sci. **8**, 611742 (2021)
9. Ma, Q., Zheng, J., Li, S., Cottrell, G.W.: Learning representations for time series clustering. Adv. Neural Inform. Process. Syst. **32** (2019)
10. Shi, P., Ye, W., Qin, Z.: Self-supervised pre-training for time series classification. In: 2021 International Joint Conference on Neural Networks (IJCNN), pp. 1–8. IEEE (2021)
11. Hou, L., Geng, Y., Han, L., Yang, H., Zheng, K., Wang, X.: Masked token enabled pre-training: a task-agnostic approach for understanding complex traffic flow. IEEE Trans. Mobile Comput. (2024)
12. Zhang, W., Yang, L., Geng, S., Hong, S.: Self-supervised time series representation learning via cross reconstruction transformer. IEEE Trans. Neural Netw. Learn. Syst. (2023)
13. Zhao, L., Gao, M., Wang, Z.: St-gsp: spatial-temporal global semantic representation learning for urban flow prediction. In: Proceedings of the Fifteenth ACM International Conference on Web Search and Data Mining, pp. 1443–1451 (2022)
14. Nie, Y., Nguyen, N.H., Sinthong, P.,Kalagnanam, J.: A time series is worth 64 words: Long-term forecasting with transformers, arXiv preprint arXiv:2211.14730 (2022)
15. Simoncelli, S., Oliveri, P., Mattia, G., Myroshnychenko, V.: Seadatacloud temperature and salinity historical data collection for the mediterranean sea (version 2) (2020)
16. Zhou, H., et al.: Informer: Beyond efficient transformer for long sequence time-series forecasting. In: Proceedings of the AAAI Conference on Artificial Intelligence, vol. 35, pp. 11106–11115 (2021)
17. Wu, H., Xu, J., Wang, J., Long, M.: Autoformer: decomposition transformers with auto-correlation for long-term series forecasting. Adv. Neural. Inf. Process. Syst. **34**, 22419–22430 (2021)
18. Hochreiter, S., Schmidhuber, J.: Long short-term memory. Neural Comput. **9**(8), 1735–1780 (1997)
19. Zeng, A., Chen, M., Zhang, L., Xu, Q.: Are transformers effective for time series forecasting? In: Proceedings of the AAAI Conference on Artificial Intelligence, vol. 37, pp. 11121–11128 (2023)
20. Wong, A., Keeley, R., Carval, T., et al.: Argo quality control manual for ctd and trajectory data (2023)

SHAPE: Smart Shaping with Adaptation Physically Excited Networks

Zhaoqing Leng[1,2(✉)], Zhengang Zhao[1,2(✉)], Xiaoyu Zhou[1,2], Yican Zhang[1,2], and Xu Dong[1,2]

[1] School of Software Engineering, University of Science and Technology of China, Suzhou 215123, China

[2] Suzhou Institute for Advanced Research, University of Science and Technology of China, Suzhou 215123, China

`lengzhaoqing@mail.ustc.edu.cn`, `gavin@ustc.edu.cn`

Abstract. In the field of material shaping, traditional manual methods and physical modeling can no longer meet the precision and efficiency requirements of modern manufacturing. To address these challenges, this paper introduces a novel neural network model that leverages an advanced attention mechanism derived from Transformer architectures. Specifically, the model—termed Smart Shaping with Adaptation Physically Excited Networks (SHAPE)—is designed to predict changes in material straightness and height at multiple shaping points for various materials to ensure high-quality shaping outcomes with minimal material waste. Considering the changes in the physical properties of materials during the processing, the complex implicit deformation relationships between multiple points of material data, and the long-distance feature dependencies, we have specifically modified the multi-head attention mechanism. Combined with the adaptive attention mechanism, this allows for identifying changes in straightness and height at multiple points during shaping. Convolutional layers are added to simulate physical diffusion deformation. Considering the inherent physical constraints on materials, we have also embedded Swift's law, which describes material deformation, into the loss function. This ensures that the intelligent shaping network aligns more closely with the actual physical laws of materials during the prediction process. Potential rebound effects are also taken into account in the network.

Experimental results show that the proposed model performs excellently across various evaluation metrics, significantly outperforming existing TG-DNN and LSTM networks. Ablation experiments demonstrate that the multi-head attention mechanism and integrating Swift's law are critical to improving the model's performance.

Keywords: Material Smart Shaping · Neural Network · Physical Constraint

© The Author(s), under exclusive license to Springer Nature Singapore Pte Ltd. 2026
M. Mahmud et al. (Eds.): ICONIP 2024, CCIS 2297, pp. 107–121, 2026.
https://doi.org/10.1007/978-981-96-7036-9_8

1   Introduction

Material forming is a cornerstone of modern manufacturing, applied across industries such as aerospace, automotive, and electronics. Traditional forming methods, including stamping, forging, and extrusion, though prevalent, are increasingly unable to meet the high precision and efficiency demands of contemporary manufacturing processes. Particularly, stamping technology, essential for shaping high-strength steel and aluminum alloys into complex parts, faces challenges in consistency and cost-efficiency when relying on manual adjustments and outdated physical modeling techniques.

Modeling methods have been introduced into the forming field to address these issues. Traditional physical modeling methods, such as partial differential equations (PDEs), struggle with nonlinear material properties and high-dimensional data. They require extensive computational resources and complex boundary conditions, which limit their ability to simulate nonlinear behaviors.

This paper introduces the Smart Shaping with Adaptation Physically Excited Networks (SHAPE), a novel neural network model designed to enhance the precision of material shaping processes in manufacturing. Specifically, SHAPE aims to predict the optimal deformation parameters for various materials, ensuring high-quality shaping outcomes with minimal material waste. By leveraging a modified multi-head attention mechanism integrated with Swift's law, the model dynamically adjusts to the physical properties of materials during processing, predicting changes in material straightness and height at multiple shaping points.

The main contributions of this paper include:

- **Integration of Multi-head and Adaptive Attention Mechanisms:** The SHAPE model innovatively combines multi-head attention and adaptive attention layers to extract and utilize complex features in material processing data effectively. This dual attention mechanism enhances the model's ability to focus on relevant features dynamically, improving feature representation and prediction accuracy.
- **Simulation of Complex Shaping Scenarios:** Considering the complex situations that may arise during actual material processing, such as non-uniform deformation under multi-point loading and rebound behavior during actual deformation, SHAPE integrates a physical diffusion model and a rebound compensation mechanism. This ensures the model can accurately simulate complex material behaviors and closely align with actual processing conditions.
- **Physics-based Loss Function:** We introduce a loss function based on physical principles, incorporating the dynamic changes of material properties during the processing stage into the model training. This integrated approach respects the physical characteristics of materials, allowing the system to comprehensively consider and handle material rebound effects, ensuring predictions are accurate while preventing data overfitting.
- **Empirical Validation Across Various Material Processing Scenarios:** Extensive data demonstrates the effectiveness of SHAPE in various material

forming environments. Compared with traditional methods, our approach significantly improves processing precision and efficiency, showcasing its practical benefits and adaptability to different industrial applications.

2   Related Work

In recent years, methods and tools capable of modeling and simulating the manufacturing, processing, and forming of the structure and properties of steel and metal alloys have developed dynamically. Computer-aided modeling is used in scientific and industrial research [12].

Advancements in AI for Material Forming. AI technology has been applied across various industries, including intelligent forming processes. Specifically, Shun Wang [13] employed the Sparrow Search Algorithm-Based Backpropagation (SSA-BP) to successfully predict the deformation effects of titanium alloy after line heating, outperforming traditional Backpropagation (BP) and Genetic Algorithm-Based Backpropagation (GA-BP) methods. Liu S [9] employed a deep-Q-network (DQN) to approximate action value functions and, through Reinforcement Learning (RL) model convergence evaluation, predicted optimal forming routes, enabling operators to achieve desired material shapes without expert guidance. Sherwan Mohammed compared different training and transfer functions in predicting aluminum alloy sheet Single Point Incremental Forming (SPIF) precision [11], while Daniel J. Cruz used Long Short-Term Memory (LSTM) to predict hardening parameters of processed materials [2]. These studies demonstrate AI's strong potential in intelligent forming but have drawbacks: These simple network architectures are unsuitable for capturing high-dimensional data features and lack adjustments based on the physical properties of materials.

Moreover, the rebound effect of materials is a crucial factor when applying AI to forming processes. Existing studies [3] have used AI to predict rebound angles but only applied them to shallow neural networks without integrating them into deep neural networks, leading to feature extraction omissions. Shiming Liu specifically considered the rebound effect [10], integrating a CNN architecture with a multi-task processing mechanism proposing the novel TG-DNN neural network.

Physics-Informed Neural Networks (PINNs). PINNs have become powerful tools for directly encoding PDEs within neural network architectures. These models leverage the structure of PDEs to guide the training process, ensuring that predictions conform to known physical laws. PINNs are particularly effective when traditional data-driven methods fail to comply with conservation laws or when training data is scarce [4]. Currently, PINNs are widely applied in fields with well-defined physical laws, such as fluid dynamics and material science. However, solving PDEs typically incurs high computational costs, and when dealing with practical engineering problems, there is no universal PDE-like governing equation that can adequately describe the primary physical characteristics of different materials. Moreover, solving PDEs introduces high computational costs and low computational efficiency.

Swift Model. The Swift model is a classical model used in material processing to describe and predict material deformation behavior. Joun M-S pointed out in his paper [8] that the Ludwik series models (including the Ludwik and Swift models) are very suitable for describing the classical strain hardening behavior of metallic materials. Compared to the Ludwik model, the Swift model is more suitable for stamping shaping scenarios. The introduced virtual initial strain ϵ_0 helps the model simulate plastic behavior before the initial hardening stage, making the model more accurate in strain applications. Furthermore, the Swift model has been widely applied in actual engineering fields. Early literature described the changes in strain values of materials upon reloading based on Swift's law [5]. Recently, researchers have successfully described the deformation behavior of Ti-6Al-4V alloy using the Swift model combined with a fourth-order polynomial [1]. However, the Swift model has parameter dependencies, and its model parameters (such as K, ϵ_0, n) vary significantly under different materials and processing conditions. How to adapt this model to different materials is also a problem that this paper aims to address.

Transformers in Physical Systems. While Transformers are primarily known for their success in natural language processing, their applications in the physical sciences are growing. The self-attention mechanism in Transformers excels at capturing long-range dependencies and can adapt to recognize patterns in spatial and temporal data streams in material science. Recent studies have shown that projecting dynamic systems into vector representations allows Transformer models to predict physical phenomena in dynamic systems, with prediction results surpassing those of classical methods commonly used in the machine learning literature. However, these applications are still in their early stages [6].

3   Methodology

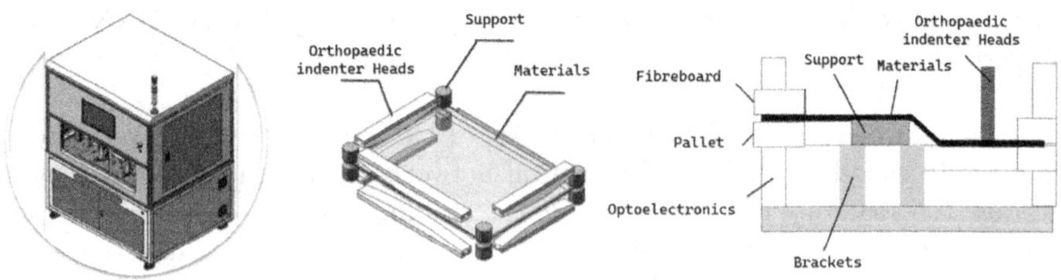

Fig. 1. Material Shaping Instrument: Detailed Operation of Each Component in the Material Shaping Process

The material processing scenario for this experiment is shown in Fig. 1. This figure illustrates that the material is first placed on the machine, which collects data from various points. The collected data, combined with expert prior knowledge, is transmitted to the processor. The machine's processing range includes the left, center, and right sides of the material. After the processor predicts the processing depths for these three sides, the shaping machine uses mechanical heads to apply multi-point presses to the material, completing the shape change.

The overall structure of the SHAPE model is depicted in Fig. 2. This figure describes our proposed network based on the material shaping architecture. The SHAPE intelligently participates in and controls the shaping process, addressing the issue of manual judgment in determining the shaping amount. The figure details each step and process of our network, illustrating the data flow and how each component contributes to the smart material shaping process.

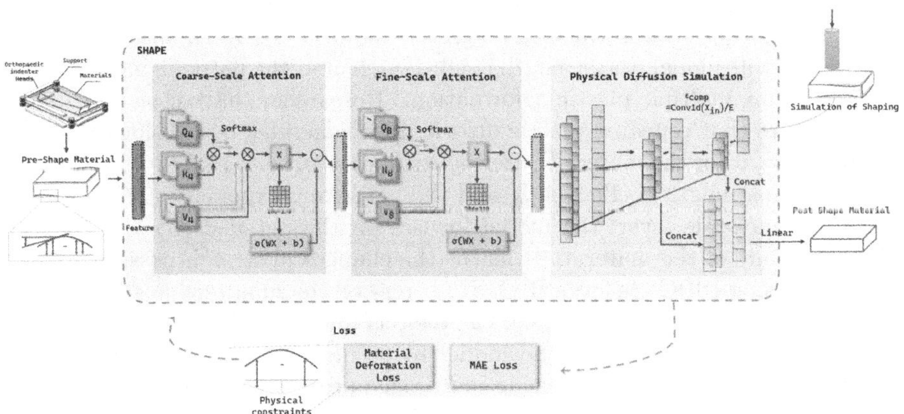

Fig. 2. SHAPE: Detailed network architecture and data flow. The coarse-scale attention layer learns material types, processing counts, and historical records. The fine-scale attention layer captures implicit relationships between pointwise straightness and height values. Convolutional layers simulate physical deformation, incorporating physical rebound characteristics. Finally, physical constraints in the loss function, including Swift's law and rebound mechanisms, feedback into the model, optimizing the entire smart shaping process.

3.1   Multi-scale Multi-head and Adaptive Attention Mechanism

Multi-scale Multi-head Attention Mechanism. For a given material, deformation at one point usually causes different degrees of deformation at other points, indicating a contextual relationship between different points. By utilizing the multi-head attention mechanism of the Transformer, we can discover the contextual relationships between different points. Additionally, we have hierarchically modified the multi-scale multi-head attention mechanism to include a coarse-scale structure with 4 heads and 64 keys and a fine-scale structure with 8 heads and 32 keys. This mechanism enhances the network's ability to focus on deformation features at different positions in the material data sequence, accommodating the complexity and diversity of material properties.

In the coarse-scale layer, the model processes the broad characteristics of the input material, identifying patterns in the entire processing condition sequence and learning the effects of material type, prior processing history, and the number of processing times on material shaping. Since subsequent physical laws are parameter-dependent, the effectiveness of learning material characteristics in this layer directly impacts the estimation of subsequent physical parameters and shaping outcomes. Specifically, the material type determines the material's basic physical and chemical properties, such as hardness, which significantly influence the final processing outcome. Additionally, prior processing history and the number of processing times reflect information about the material being reused and processed multiple times, which can alter the material's performance. For example, in the Swift model, the strength coefficient K and the hardening exponent n are affected by multiple plastic deformations. Researchers have clearly stated in the literature that repeated plastic deformation generates new dislocations and interactions within the material, altering its microstructure and thus specifically affecting the strength coefficient K and the hardening exponent n [7].

In the fine-scale layer, the model focuses on the detailed changes between different points of the material. It learns the changes in straightness before and after processing (SPC_before, SPC, SPC_res) or the minute changes in point-wise height changes (Ci_before, Ci, Ci_res), as well as the interactions between different points. This enables the model to accurately predict the mechanical behavior of peripheral points during multi-edge processing.

The multi-head attention mechanism splits the input material's various point data into multiple heads, independently learning the weight of each point concerning different points and then recombining these weights into a single output. This captures the multidimensional features of material deformation from different subspace perspectives simultaneously, revealing internal latent deformation relationships, thereby more comprehensively capturing the dynamic changes of the material during the processing. The formula is as follows:

$$\text{Attention}(Q, K, V) = \text{softmax}\left(\frac{QK^T}{\sqrt{d_k}}\right) V \tag{1}$$

where Q, K, and V represent the query, key, and value vectors respectively, and d_k is the dimension of the key vectors. By scaling the dot-product attention, the model's focus and differentiation on different points are enhanced.

Adaptive Attention Mechanism. When shaping materials, there is not only a connection between points, but changes at different positions also affect the overall deformation of the material to varying degrees. Adaptive attention layers are utilized to learn the relative importance of each point within the overall material. This adjusts the contribution weights of changes in straightness and height at different points to the overall deformation, allowing the model to focus on important features while suppressing unimportant information. This mechanism allows the model to adaptively adjust the importance given to different features, improving the network's interpretability and prediction accuracy for different points of the material.

$$\text{AdaptiveAttention}(X) = \sigma(WX + b) \odot X \tag{2}$$

where σ is the sigmoid activation function, W and b are the learned weights and biases, and $\odot$ denotes element-wise multiplication. X represents the feature tensor input to the adaptive attention layer.

Adaptive attention layers are applied in multiple key steps within SHAPE to enhance the focus on important features. This layer is especially useful after the fine-scale multi-head attention layer, as changes in straightness and point height are considered critical indicators during the shaping process. Focusing on these detailed features can improve the impact on the final processing quality.

The combination of these two attention mechanisms provides the SHAPE model with powerful data processing capabilities. The multi-scale multi-head attention mechanism allows the model to interpret data at multiple levels, discovering contextual relationships between different points, while the adaptive attention mechanism ensures the model focuses on the most informative features. This enables SHAPE to handle and analyze high-dimensional and multivariate material processing data effectively.

3.2 Simulating Physical Diffusion Deformation and Rebound Effect with Convolutional Layers

In the SHAPE, The convolutional layers are specifically designed to simulate materials' physical diffusion process and compensate for the rebound effect after physical processing. This layer primarily includes two convolutional layers that capture dependencies within local regions through convolution operations, thereby simulating how materials are affected by various physical forces during processing and predicting deformations.

Convolution operations can capture spatial and temporal characteristics of material shaping through filters. When processing data on straightness or point height changes, convolutional layers can identify local patterns and trends in these features. In the SHAPE model, we apply this concept to material processing data, using multiple filters to extract specific physical information from the data simultaneously. The parameters of these filters are learned automatically during training, enabling the network to adaptively recognize and utilize the most important physical characteristics in the processing data.

The mathematical model of the physical diffusion process, describing how the convolutional layer computes changes in material state, is as follows:

$$X_{\text{out}} = \text{ReLU}(\text{BN}(W * X_{\text{in}} + b)) \tag{3}$$

where X_{in} is the input material state, X_{out} represents the material state after simulating physical diffusion, W and b are the convolution kernel parameters and bias, $*$ denotes the convolution operation, BN represents batch normalization, and ReLU is the activation function.

During the elastic deformation phase, materials generally follow Hooke's Law, which is expressed as follows:

$$\epsilon = \frac{\sigma}{E} \tag{4}$$

where ϵ is the elastic strain, σ is the stress, and E is the material's elastic modulus.

Combining Hooke's Law with convolution operations, the mathematical model is as follows: Assuming the convolution operation outputs the stress after plastic deformation, the combination with Hooke's Law is modeled mathematically as:

$$\epsilon_{\text{comp}} = \frac{\text{Conv1d}(X_{\text{in}})}{E} \tag{5}$$

where Conv1d represents the convolution operation, and the tensor X_{in} represents the material's state, i.e., the output after simulating the physical phenomenon diffusion. $\text{Conv1d}(X)$ is the output of the convolution operation, representing the stress. E is the material's elastic modulus. This material's parameter is obtained from an expert prior knowledge base, based on the type of material.

Upon completion of the processing, ϵ_{comp} and the output after the physical diffusion processing are passed to the network's linear layer for dimensional expansion.

3.3   Physical Deformation Loss and MAE Loss

After simulating physical diffusion deformation using convolutional layers, the SHAPE model needs to add physical constraints to achieve network convergence. We define the loss function by combining physical laws and data-driven methods, ensuring that predictions are not only accurate but also comply with physical principles, making the network suitable for material science and engineering applications.

Physical Deformation Property Loss. The physical deformation loss layer is a core component of the model, guiding the learning process by integrating actual physical laws to ensure that predictions are statistically accurate and physically feasible. Unlike traditional loss functions, the physical deformation loss layer directly incorporates the mechanical properties of materials, introducing stronger physical constraints during prediction.

In designing the physical deformation loss, we use a well-known theory in material mechanics—Swift's law—to replace the PDE control equation and establish the relationship between stress and strain. Swift's law describes the nonlinear relationship between stress and strain during the plastic deformation phase of materials. Its formula is as follows:

$$\sigma = K(\epsilon_0 + \epsilon)^n \tag{6}$$

where σ is the stress, ϵ is the plastic strain, ϵ_0 is the initial strain, K is the material's strength coefficient, and n is the material's hardening exponent.

For improved model interpretability, we use von Mises stress and equivalent plastic deformation to define the plastic model. The simplified Swift formula is as follows:

$$\sigma = K\epsilon^n \tag{7}$$

The loss function calculates the difference between the model's predicted strain and the strain calculated from actual measurement data, mainly constraining straightness or point height changes.

We transform the actual machining depths in three directions into a deformation representation. The theoretical actual strain tensor ϵ_i can be expressed as

$$\epsilon_i = \left(\frac{\sigma}{K}\right)^{\frac{1}{n}} \tag{8}$$

where K is the material's strength coefficient and n is the material's hardening coefficient. To increase the model's applicability, K and n can be adaptively adjusted in the neural network due to the material parameter feature extraction by the multi-head attention layer. The initial values are set based on the most commonly used material in shaping, aluminum alloy, with $K_0 = 500\,\mathrm{MPa}$ and $n_0 = 0.2$.

Assuming the predicted strain value after processing by the simulated physical diffusion deformation convolutional layer is $\epsilon_{\mathrm{predicted},i}$, the physical deformation loss can be expressed as the mean squared difference between the predicted strain value and the actual strain value calculated by Swift's law:

$$L_{\mathrm{physical}} = \frac{1}{N}\sum_{i=1}^{N}\left(\epsilon_{\mathrm{predicted},i} - \epsilon_i\right)^2 \tag{9}$$

where $\epsilon_{\mathrm{predicted},i}$ is the model's predicted strain value for the i-th sample, and ϵ_i is the corresponding actual strain tensor.

Combined Loss Function. The final loss function is a combination of the physical deformation loss and the traditional mean absolute error (MAE) loss, defined as follows:

$$L_{\mathrm{MAE}} = \frac{1}{N}\sum_{i=1}^{N}|y_i - \hat{y}_i| \tag{10}$$

where y_i is the true value of the i-th sample and $\hat{y}_i$ is the model's predicted value for the i-th sample.

$$L_{\mathrm{total}} = \alpha L_{\mathrm{physical}} + (1 - \alpha)L_{\mathrm{MAE}} \tag{11}$$

where α is the weight coefficient used to adjust the contribution of the losses, L_{physical} is the physical deformation loss, and L_{MAE} is the mean absolute error loss.

The final loss function combines the physically based deformation property loss with the traditional mean absolute error loss, enabling the model to learn from data while ensuring that the learning process complies with physical laws. This ensures that the network understands deformation variables in the smart shaping process, demonstrating excellent adaptability in applications requiring high precision and efficiency in material processing.

4   Experiments

4.1   Experimental Setup

To verify the effectiveness and feasibility of the proposed SHAPE, we designed a series of experiments where the machine shaping force is fixed at 100 N.

Data Collection. The experimental dataset was constructed using camera and infrared scanner data of material edges, combined with an expert experience database. The final dataset includes 30,000 detailed records of the material processing process, each containing raw material characteristics, processing history, and straightness and point height changes before and after processing.

Data Preprocessing. The categorical features in the raw data (e.g., material type and state before and after processing) were digitized to meet the input requirements of the neural network. Specifically, material types were converted to numerical features using one-hot encoding, while the number of processing times and changes in straightness were used directly as numerical inputs.

Input vector:

- **Raw Material (IM, KK, LX)**: Represents the material type. One-hot encoded to a 3-dimensional feature vector.
- **FR (new, old)**: Indicates whether it has been processed before. Directly digitized, with 1 for new and 2 for old.
- **Times**: Indicates the number of processing times, occupying 1 dimension.
- **Status (material status)**: Indicates the material status. Occupies 1 dimension, including RS-1 and post. Digitized, with 0 for RS-1 and 1 for post.
- **SPC_before, SPC, SPC_res**: Records the straightness before and after the previous processing, as well as the differences in straightness changes on the left, center, and right sides, each occupying 3 dimensions, totaling 9 feature dimensions.
- **Ci_before, Ci, Ci_res**: Records the heights before and after the previous processing, as well as the differences in height changes, at 71 points, totaling 213 feature dimensions.

Output Vector:

- **PressDepth**: Represents the predicted machining depths for the left, center, and right sides of the material. It is a key indicator for evaluating the prediction accuracy of the model. It includes 3 dimensions (PressDepth_L, PressDepth_R, PressDepth_F).

4.2   Comparative Experiments

Comparison with Other Network Models. This section compares our proposed SHAPE model with BP-SSA, Deep-Q-Learning, TG-DNN, and LSTM network structures proposed by previous researchers. All models are trained and tested using the same dataset. The experimental parameters are as follows:

- SSA-BP (Shun Wang's): The population size of the sparrow search algorithm is set to 30, with 100 training cycles.
- Deep-Q-Learning (Shiming Liu's): The learning rate is set to 2.5×10^{-5}, and the discount factor is 0.9.
- LSTM (Daniel J. Cruz's): The number of hidden layers is set to 5, with 100 training cycles.
- TG-DNN (Shiing Liu's): The Adam optimizer is used, with a learning rate of 10^{-3}, and 100 training cycles.
- SHAPE (OURS): The Adam optimizer is used, with a learning rate of 10^{-3}, a batch size of 32, and 100 training cycles.

Performance evaluation metrics: Mean Squared Error (MSE), Mean Absolute Error (MAE), and Adjusted R-squared (Adjusted R^2) are used to evaluate model performance.

After training, 40 coordinate points are selected for display. The three graphs below show the prediction results of different network models in the three output dimensions (PressDepth_L et al.). The x-axis represents the point number, and the y-axis represents the predicted and actual values. When the y-value is negative, the material shaping is excessive and needs to be reversed for correction.

As shown in the Fig. 3, both SHAPE and LSTM exhibit good fitting effects, with predictions closely following actual values most of the time. This is likely because both SHAPE and LSTM are good at capturing long-distance dependencies—SHAPE effectively focuses on information and dependencies between different points through the multi-head attention mechanism. In contrast, LSTM captures them through its recurrent structure. Additionally, the multi-layer structures of both SHAPE and LSTM can extract and process complex features layer by layer, enhancing model performance.

However, in cases where PressDepth > 0, LSTM performs better overall. SHAPE may over-focus on certain features due to attention allocation issues, leading to larger prediction errors at certain test points. The LSTM model, however, has a fatal flaw: when the true value < 0, LSTM cannot make effective predictions, resulting in an output value of 0. This is likely due to LSTM units

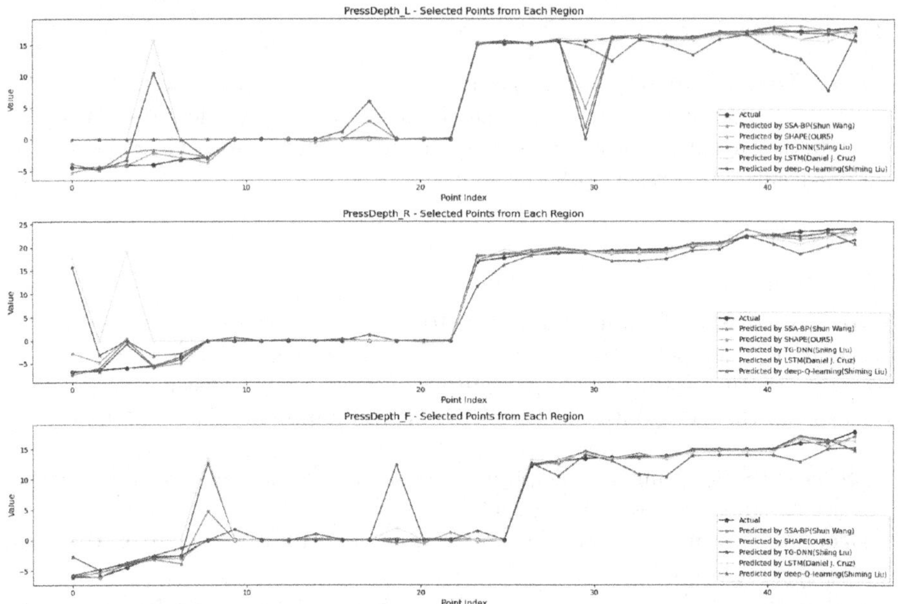

Fig. 3. Prediction results of different network models in three output dimensions.

often using Sigmoid and Tanh activation functions, which do not perform well with negative outputs.

TG-DNN, which has a slightly worse fitting effect, incorporates the rebound effect through a multi-task processing mechanism in its CNN architecture but fails to capture dependencies and global features between multiple points. With its ability to adaptively adjust attention weights during training, SHAPE better fits data transformation and feature distribution, whereas TG-DNN shows instability with noisy data.

The models with the worst fitting effects are SSA-BP and deep-Q-learning. SSA-BP combines the sparrow search algorithm for global optimization, and deep-Q-learning trains through a reinforcement learning framework, focusing more on global optimal solutions or state-action pair optimization. Compared to SHAPE and LSTM, SSA-BP and deep-Q-learning may lack the ability to extract features and handle complex dependencies, resulting in poor performance in predicting complex tasks.

The precise fitting indicators R^2, MAE, and MSE of multiple models are shown in Table 1. Each indicator is better than other network models.

Compared to LSTM, SHAPE can handle cases with the predicted value < 0. Compared to TG-DNN, although both consider rebound and physical constraints, SHAPE's multi-head attention mechanism captures dependent features better and adapts to noise. Compared to SSA-BP and deep-Q-learning, SHAPE better captures complex features, demonstrating its superiority.

Table 1. Performance evaluation metrics of different models.

MODEL	$R^2\uparrow$	MAE$\downarrow$	MSE$\downarrow$	RMSE$\downarrow$
SSA-BP	0.9510	0.5595	3.2175	1.9979
SHAPE	**0.9897**	**0.1924**	**1.1197**	**0.7685**
TG-DNN	0.9613	0.3937	2.7980	1.2976
LSTM	0.9649	0.3527	3.0080	1.3786
deep-Q-learning	0.9425	1.2890	4.7591	2.4166

Ablation Experiments. To verify whether the multi-head attention mechanism and the modified loss function considering Swift's law indeed improve data fitting accuracy, we conducted new experiments on the same dataset with the following network architecture modifications while keeping other settings unchanged:

- Change the head_nums in the multi-head attention layer to 1.
- Set the weight α of the material deformation loss in Eq. 11 to 0.

After these modifications, the fitting effect of the model is shown in Table 2. The evaluation metrics of the complete SHAPE model are superior to those of the comparative models. Specifically, when the material deformation loss is not considered, SHAPE's R^2, MAE, and MSE performance are inferior to the complete SHAPE, indicating better learning effects after introducing physical characteristics. Additionally, removing multi-head attention significantly reduces the model's effectiveness, demonstrating the necessity of the multi-head attention mechanism for SHAPE.

Table 2. Performance evaluation metrics of SHAPE and its ablated versions.

MODEL	$R^2\uparrow$	MAE$\downarrow$	MSE$\downarrow$	RMSE$\downarrow$
SHAPE	**0.9897**	**0.1924**	**1.1197**	**0.7685**
SHAPE without Multi-Head Attention	0.9710	0.2395	1.9877	1.0413
SHAPE without Material Deformation Loss	0.9632	0.2922	2.3916	1.1741

In conclusion, through the above experiments, SHAPE demonstrates its application potential in smart material shaping, especially in scenarios with long-range dependencies and complex features in input data. Using SHAPE will improve prediction accuracy and provide good data support for the next operation of the machine.

5   Conclusion

This paper proposes an innovative deep learning architecture, the SHAPE, specifically designed for smart material shaping. SHAPE integrates multi-head attention mechanisms and physical deformation loss. By combining deep learning techniques with physical laws through a carefully designed neural network structure and loss function, SHAPE significantly improves the prediction accuracy of the material processing process and validates the feasibility of directly incorporating physical properties into the loss function.

In summary, the following conclusions can be drawn:

- The combination of multi-head attention and adaptive attention layers effectively captures the complexity and diversity of material data, enhancing the model's understanding and predictive ability. Networks without these architectures (e.g., TG-DNN) struggle with noisy data, whereas SHAPE remains stable. Additionally, fitting the current material's physical properties based on previous processing data further enhances the prediction accuracy of SHAPE.
- Incorporating physical laws into the loss function ensures statistical accuracy and physical reliability of the model's predictions. Ablation experiments show that adding physical characteristics is crucial for improving prediction accuracy, especially in high-precision processing scenarios.
- Experiments on a large dataset of 30,000 entries demonstrate SHAPE's superior performance across various material processing scenarios. Compared to other network architectures, SHAPE provides more accurate predictions and shows higher computational efficiency and better generalization capabilities.

References

1. Cao, J., et al.: Constitutive equation for describing true stress-strain curves over a large range of strains. Philos. Mag. Lett. **100**(10), 476–485 (2020)
2. Cruz, D.J., Barbosa, M.R., Santos, A.D., Amaral, R.L., de Sa, J.C., Fernandes, J.V.: Recurrent neural networks and three-point bending test on the identification of material hardening parameters. Metals **14**(1), 84 (2024)
3. Cruz, D.J., Barbosa, M.R., Santos, A.D., Miranda, S.S., Amaral, R.L.: Application of machine learning to bending processes and material identification. Metals **11**(9), 1418 (2021)
4. Cuomo, S., Di Cola, V.S., Giampaolo, F., Rozza, G., Raissi, M., Piccialli, F.: Scientific machine learning through physics-informed neural networks: where we are and what's next. J. Sci. Comput. **92**(3), 88 (2022)
5. Fernandes, J.V., Rodrigues, D.M., Menezes, L.F., Vieira, M.F.: A modified swift law for prestrained materials. Int. J. Plast **14**(6), 537–550 (1998)
6. Geneva, N., Zabaras, N.: Transformers for modeling physical systems. Neural Netw. **146**, 272–289 (2022)
7. Hosford, W., Caddell, R.: Metal forming. mechanikcs and metallurgy. Cambridgy University Press (2007)
8. Joun, M.S., Razali, M.K., Jee, C.W., Byun, J.B., Kim, M.C., Kim, K.M.: A review of flow characterization of metallic materials in the cold forming temperature range and its major issues. Materials **15**(8), 2751 (2022)
9. Liu, S., Shi, Z., Lin, J., Li, Z.: Reinforcement learning in free-form stamping of sheet-metals. Proc. Manufact. **50**, 444–449 (2020)

10. Liu, S., Xia, Y., Shi, Z., Yu, H., Li, Z., Lin, J.: Deep learning in sheet metal bending with a novel theory-guided deep neural network. IEEE/CAA J. Automatica Sinica **8**(3), 565–581 (2021)
11. Najm, S.M., Paniti, I.: Artificial neural network for modeling and investigating the effects of forming tool characteristics on the accuracy and formability of thin aluminum alloy blanks when using spif. Inter. J. Adv. Manufact. Technol. **114**, 2591–2615 (2021)
12. Sha, W., et al.: Artificial intelligence to power the future of materials science and engineering. Adv. Intell. Syst. **2**(4), 1900143 (2020)
13. Wang, S., Wang, J., Xu, Z., Wang, J., Li, R., Dai, J.: Deformation intelligent prediction of titanium alloy plate forming based on bp neural network and sparrow search algorithm. J. Marine Sci. Eng. **12**(2), 255 (2024)

Accelerating Attentional Generative Adversarial Networks with Sampling Blocks

Chong Zhang[1], Mingyu Jin[1], Qinkai Yu[2], Haochen Xue[1], Xi Yang[1], and Xiaobo Jin[1(✉)]

[1] School of Advanced Technology, Xi'an Jiaotong-Liverpool University, Suzhou, China
`{Chong.zhang19,Mingyu.jin19,Haochen.Xue20}@student.xjtlu.edu.cn,`
`{Xi.Yang01,Xiaobo.Jin}@xjtlu.edu.cn`
[2] Department of Mathematics Science, University of Liverpool, Liverpool, UK
`sgqyu9@liverpool.ac.uk`

Abstract. Synthesizing text-to-image models for high-quality images by guiding generative models through text descriptions is an innovative and challenging task. In recent years, AttnGAN has been proposed based on the Attention mechanism to guide GAN training, which improves the details and quality of images by stacking multiple generators and discriminators. However, the combination of multiple enhancements in GAN architecture introduces redundancy, hindering the practical application of the model. These redundancies adversely affect its performance, increasing inference time and space complexity. In this paper, we propose an Accelerated AttnGAN (AccAttnGAN) to optimize the structure and training efficiency of AttnGAN by (1) removing redundant structures and improving the backbone network of AttnGAN; (2) integrating and reconstructing multiple losses for the training of deep attention model. Experimental results show that AccAttnGAN significantly reduces the model's space complexity and time complexity during inference while maintaining performance. Code is available at https://github.com/jmyissb/SEAttnGAN.

Keywords: Text-to-image · AttnGAN · Sampling Blocks · Alignment of Semantics and Images · Efficiency

1 Introduction

Generative Adversarial Networks (GAN) [8] are a sophisticated type of deep learning model that can generate new sample data that statistically match the training data. Generative Adversarial Networks (GANs) comprise two networks: a generator network and a discriminator network. The generator network generates new data, while the discriminator network strives to differentiate between

This work was partially supported by Research Development Fund with No. RDF-22-01-020, the top talent award project RDF-TP-0019, and the "Qing Lan Project" in Jiangsu universities and National Natural Science Foundation of China under Grant U1804159.

this generated data and authentic data. With ongoing training, the generator network can generate data that is practically identical to real data. GANs are widely utilized in numerous industries, including image manipulation and creation, music and text generation, and video generation.

GAN technology has developed rapidly in recent years, resulting in many innovative variants that employ different techniques to create data with specific properties. InfoGAN [3] proposes an information encoder network that encodes input data into vectors that can control the characteristics of the generated data. Conditional GAN [18] generates data based on given conditions such as images belonging to a specific category or text of a specific style. CycleGAN [40] uses two generator and two discriminator networks to convert the input image into a target image and back and compare the results they generate. The Integral Probability Metric (IPM) [29] evaluates probabilistic prediction models by comparing the Probabilistic Integral Transform (PIT) to a uniform distribution. Finally, the loss-sensitive GAN [22] trains the generator by minimizing the discriminator loss, which can improve the quality of the generated images.

In the field of image generation, the current influential framework is Generative Adversarial Network (GAN). By adopting adversarial learning, the generator and discriminator are trained simultaneously so that the quality of generated images gradually approaches that of real images. Since the introduction of text description as a supervisory signal, GAN-based text-to-image models have developed rapidly. GAN-INT-CLS [24] demonstrated the power of text as supervisory information, where the main work includes Generative Adversarial What-Where Network (GAWWN) with its scene layout estimator [25], a pixel recurrent neural network (PixelCNN) using CNN to model the distribution of image pixels [20] and stacked generative adversarial networks (StackGAN and StackGAN-v2) [38,39] enhancing image quality by multi-stage model or multi-stage generator and discriminator. In addition, there is Semantic Disentanglement GAN (SD-GAN) [36], which solves the potential spatial factor confusion through a self-distillation mechanism. Unfortunately, current approaches do not provide an efficient way to train models, either by estimating and optimizing the scene or by improving image quality with multiple models.

To improve the quality of text-generated images, Attention Generative Adversarial Network (AttnGAN) [33] introduces a multi-modal attention mechanism of text and images in the generation work, allowing AttnGAN to capture fine-grained word-level information. Thus, it can focus on relevant words in natural language descriptions, which can be used to synthesize fine-grained details in different sub-regions of the image. However, AttnGAN has multiple attention models and multiple sets of baseline networks, where the existence of multiple attention models increases the complexity of the architecture, where each attention module requires additional computation to capture relevant visual information to generate realistic and meaningful images, resulting in higher computational overhead.

To reduce the adverse impact of these redundancies on its performance and reduce the time and space complexity during model running, we propose a new

AttnGAN with multiple sampling blocks, remove the redundant structure and improve the backbone of the AttnGAN network to optimize the structure and training efficiency of AttnGAN. Furthermore, our model integrates multiple losses to reconstruct DAMSM (Deep Attention Multimodal Similarity Model) to generate images that are more consistent with text. Experimental results on public datasets show that our AttnGAN significantly improves model size and training efficiency while maintaining performance.

Overall, our contributions are as follows:

- We greatly accelerate the training and inference performance of the model through the up-and-down sampling module called Accelerated Attention Generative Adversarial Network (AccAttnGAN), which simplifies the structure of AttnGAN and reduces its number of parameters but achieves comparable performance to the baseline network.
- To generate images that better match the semantics of the text, we introduce semantics and the alignment loss of semantics and images into the model as auxiliary information, which can guide the GAN generator to generate images that more accurately match text descriptions.
- The results on two public data sets show that our method greatly reduces the size and running time of the model while improving the generation quality, which strongly reflects the superiority of our algorithm.

2   Related Work

2.1   Text-to-Image Model

AttnGAN [33] uses symbolic attention to enhance image generation quality by focusing on significant text description parts. This mechanism guides the generator to link words and phrases, creating corresponding images. Multimodal GAN [1] generates multiple high-quality images for a single text description, offering varied interpretations. VAE (variational autoencoder) [13], another generative model, learns to create images from a latent representation of text descriptions. Its main advantage is interpretability, though generating realistic images remains challenging. PixelCNN [20] generates images pixel by pixel, each conditioned on the previous one, using a convolutional neural network (CNN). It effectively generates high-quality images and can be applied to text-to-image synthesis.

2.2   GAN and GANs' Variants

GAN [8] can generate realistic images that are indistinguishable from real ones. Dong et al. [6] proposed generating real images from natural language descriptions by altering the GAN structure with a style encoder network to invert the generator. Recent years have seen significant GAN advancements. PGGANs [10] by Tero Karras et al. generate high-resolution images by progressively increasing the generator and discriminator resolution. StyleGAN [11] excels in high-resolution image generation, widely used in face generation and art. CycleGAN [40] performs domain image translation without paired training data. BigGAN [2] by Andrew Brock et al. achieved a breakthrough in realistic image generation through large-scale training. StyleGAN2 [11] improves StyleGAN with a new generator architecture and path length regularization for greater image variability.

2.3 Text-to-Image Task with GAN

Current methods generate images with novel properties but struggle to retain text-independent details. To address this, Nam et al. [19] introduced TAGAN (Text Adaptive Generative Adversarial Network) for generating semantically manipulated images while preserving these details. PPGN (Plug and Play Generative Network) inputs random noise vectors and uses optimization to guide generation. LAPGAN (Laplacian Generative Adversarial Network) [4] combines Laplacian pyramid decomposition with GAN for multi-scale progressive high-resolution image generation. Mirror GAN [23] achieves text-to-image conversion by learning fine-grained correspondences, ensuring text-image consistency through reverse information propagation. Conditional GAN [18] uses text descriptions to generate matching images, with recent adjustments improving image quality and diversity.

GAN-INT-CLS [24] was the first to use textual descriptions as supervisory signals for image generation. GAWWN [38] aims to estimate scene layout during image generation to enhance text-to-image performance. StackGAN [38] generates high-resolution images step-by-step to address traditional GAN challenges. StackGAN-v2 [39] improves this by adding multiple generators, discriminators, super-resolution networks, and triangular loss functions. SD-GAN [17] tackles traditional GAN latent space confounding issues with a self-disentangled learning mechanism, enhancing the distiller for "self-distillation" and auxiliary generator training [41].

3 Baseline Method

AttnGAN [33] achieves fine-grained text-to-image generation through attention-driven multistage refinement, which focuses on related words to synthesize some details in different regions of the image.

The AttnGAN model consists of $m-1$ generators ($\{G_i\}_{i=1}^{m-1}$), and each generator will generate an image as follows ($i = 1, 2, \cdots, m-1$)

$$h_0 = F_0(z, F^{ca}(T)), \tag{1}$$

$$h_i = F_i(h_{i-1}, F^{attn}(T, h_{i-1})), \tag{2}$$

$$\hat{x}_i = G_i(h_i), \tag{3}$$

where z is a noise variable following a Gaussian distribution, T is a matrix of word vector representations, and F^{ca} uses conditional augmentation to encode each word vector in T. The attention network F^{attn} will generate a weighted representation of the word vector according to the relationship between the image vectors and the text vectors.

The attn network F^{attn} takes word features and image features as input: 1) maps word features to image feature space; 2) generates a new representation of word features according to the correlation between word features and image features, which is a linear weighting of image features combination. The specific description is as follows

$$\hat{t} = Ut, \quad \alpha_{i,j} = \frac{\exp(h_j^T \hat{t}_i)}{\sum_{k=1}^{m-1} \exp(h_k^T \hat{t}_i)}, \quad c_i = \sum_{j=1}^{m-1} \alpha_{ij}\hat{t}, \tag{4}$$

where U is the transformation matrix, which transforms word vectors $\hat{t}$ into the semantic space of image features h_i.

AttnGAN model jointly optimizes generative adversarial loss and deep attention multimodal similarity loss. The generative adversarial loss contains the likelihood function loss of conditional distribution and unconditional distribution, where the unconditional distribution likelihood loss determines whether an image is real or fake, and the conditional distribution likelihood loss determines whether an image and a sentence match. DAMSM learns two neural networks to map image subregions and sentence words into a common semantic space, respectively, to compute fine-grained loss for image generation.

3.1 Our Proposed Method

In AttnGAN, multiple generators generate images of increasing size, each of which uses an attention mechanism. To reduce the model complexity, we replace multiple generators with upsampling modules to output feature maps of increasing size. At the same time, to align with the representation vector of the word, we insert multiple downsampling modules and only use attention to the feature map of one scale. The overall framework is shown in Fig. 1.

In our work, we simplify the network structure of AttnGAN. The text encoder is a bidirectional LSTM (long-short-term memory), which is obtained from the pre-trained model provided by AttnGAN. For the input to the generator, we mix sentence features with noise. After data augmentation, the input is first reshaped through a fully connected layer. Then a series of ResDF-Block layers and upsampling layers are applied to extract image features, where the ResDF-block consists of Resnet blocks and DF-Blocks to fuse text and image features. Image features and text features are used together as the input of the attention layer. Finally, image features are re-extracted through a layer of Resnet Block and convolutional layers. The image encoder is a convolutional neural network (CNN) that maps images to semantic vectors generated by the Inception-v3 model pre-trained on ImageNet. The detailed architecture is shown in Fig. 1. In order to better integrate image features and semantic features in the training process, we simplify the existing DFBlock structure, as shown in Fig. 2, which uses a new DFBlock structure consisting of a two-layer linear map structure and an additional hidden layer (with ReLU activation) to extract sentence features,

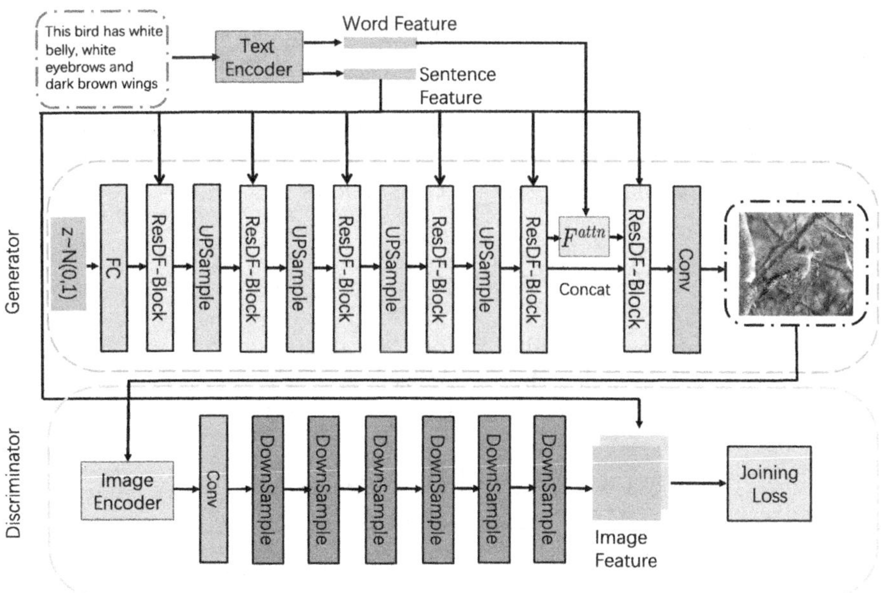

Fig. 1. Framework of AccAttnGAN, which continuously incorporates sentence information into feature maps at multiple scales through the upsampling and downsampling blocks, and uses word-image attention to generate image-text consistent images.

where sentence features are remapped from a 256-dimensional space into a 64-dimensional feature representation. We double the original structure into the ReLU layer, which can double the dimension and provide more information. The lightweight structural model has the advantages of fewer parameters and faster training speed. The discriminator reduces the feature map to the same size

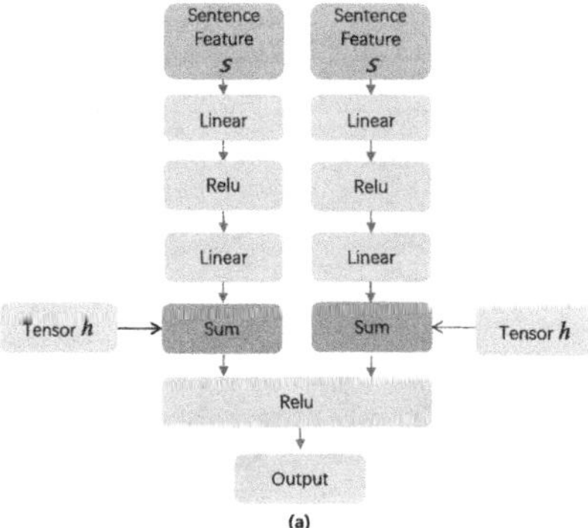

Fig. 2. Structure of ResDF Block module: two parallel branches map the sentence features, respectively.

128   C. Zhang et al.

as the original image through a series of downsampling blocks. The discriminator
will distinguish generated images from real samples. The discriminator, on the
one hand, prompts the generator to generate higher-quality images and, on the
other hand, ensures that the semantics of the text and the semantics of the image
are consistent. First, we reformulate each word t_i into a weighted linear form c_i
($i = 1, 2, \cdots, l$) of n images $x_1, x_2, \cdots, x_n$ as follows

$$c_i = \sum_{j=1}^{n} \alpha_{ij} x_j, \quad \alpha_{ij} = \frac{\exp\left(\omega \cdot t_i^T x_j\right)}{\sum_{k=1}^{n} \exp\left(\omega \cdot t_i^T x_k\right)}, \tag{5}$$

where ω is a probability smoothing factor.

To this end, we minimize the normalized similarity loss function between
each t_i word and its new representation c_i (μ is a probability smoothing factor)

$$\mathcal{L}_W = -\sum_{i=1}^{l} \log \frac{\exp\left(\mu \cdot c_i^T t_i / (\|c_i\| \|t_i\|)\right)}{\sum_{k=1}^{l} \exp\left(\mu \cdot c_i^T t_k / (\|c_i\| \|t_k\|)\right)}. \tag{6}$$

As for adversarial training, we introduce generative loss and hinge loss (discriminative loss) as follows

$$\mathcal{L}_G = -E_{G(z)\sim p_g}[D(G(z), s)], \tag{7}$$

$$\begin{aligned}
\mathcal{L}_D = &-E_{x\sim P_r}[\min(0, -1 + D(x, s)] - (1/2)E_{G(z)\sim P_g}[\min(0, -1 - D(G(z), s))] \\
&- (1/2)E_{x\sim P_{mis}}[\min(0, -1 - D(x, s))] \\
&+ kE_{x\sim P_r}\left[(\|\nabla_x D(x, s)\| + \|\nabla_t D(x, s)\|)^p\right],
\end{aligned} \tag{8}$$

where z is the noise vector, s is the sentence vector generated by the text encoder,
k and p are two hyperparameters for balancing the effectiveness of the gradient
penalty; and P_g, P_r, P_{mis} represent the synthetic data distribution, real data
distribution, and mismatched data distribution with real images and mismatched
text, respectively.

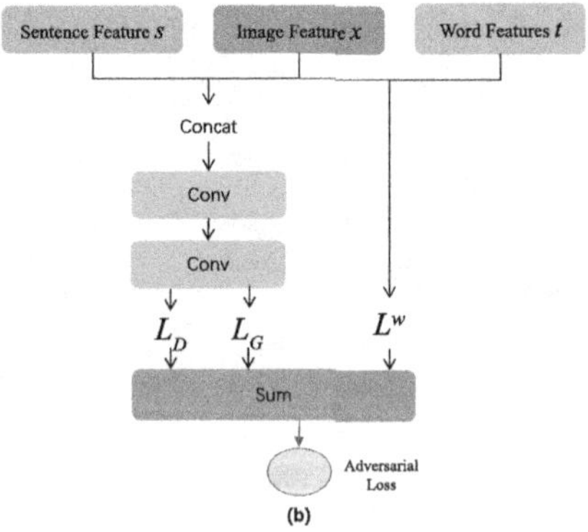

Fig. 3. Calculation of the loss function with the sentence features, image features and
word features as input.

Ultimately our algorithm will optimize the following weighted loss function in the way shown in Fig. 3

$$\mathcal{L} = \mathcal{L}_G + \gamma(\lambda\mathcal{L}_D + (1 - \lambda)\mathcal{L}_W) \tag{9}$$

where γ and λ are hyperparameters for balancing multiple tasks.

4 Experiments

Below we compare our method with previous methods on two publicly available datasets. First, we describe the composition of the two datasets, followed by the execution details and evaluation criteria of our experiments. Next, we present qualitative and quantitative comparisons among various algorithms. Finally, the effectiveness of each component and loss is discussed through the ablation study.

4.1 Datasets and Implementation of Algorithms

For our research, we have carefully selected two datasets, namely CUB Bird [32] and COCO [15]. The CUB Bird dataset comprises 200 bird species categories, totaling 11,788 images, each with 10 distinct descriptions. Additionally, we have included the MS COCO dataset to demonstrate the robustness of our approach. This more challenging dataset features multiple objects and diverse backgrounds, making it more complex. It comprises a training set of 80k images and a validation set of 40k images, each with 5 diverse descriptions.

Both our AccAttnGAN generator and discriminator are optimized using the Adam optimizer. In the generator, Adam's parameters [12] are set to $\beta_1 = 0$, $\beta_2 = 0.9$, and the learning rate is 0.0001. For the discriminator, Adam's parameters are set to $\beta_1 = 0.0$, $\beta_2 = 0.9$, and the learning rate is 0.0004. As for other hyperparameters, they will be set to $\gamma = 2$, $\lambda = 0.2$, $\mu = 5$, $\omega = 10$.

We chose IS (Inception Score) and FID (Fréchet Inception Distance) to evaluate the quality of the generated images. IS is measured by the KL divergence between the generated image category distribution obtained by the pre-trained Inception V3 model and the real image categories. Given a recognizable image x, the probability that it belongs to a certain class y should be as large as possible. That is, the entropy of $p(y|x)$ is as small as possible, and for all images at the same time, they should have a certain diversity, so the bigger $p(y)$ is, the better, so we define the following measurement criteria

$$IS(g) = \exp\left(E_{p_g} D_{KL}(p(y \mid x \| p(y)))\right), \tag{10}$$

where x denotes a generated sample, and y is the label predicted by the Inception model. However, this measurement criterion is very sensitive to the internal weight of the neural network and cannot determine whether the network is overfitting. More importantly, it cannot reflect the distance between real data and generated samples. FID can measure the distance between generated samples and real-world samples.

FID (Fréchet Inception Distance) [9] is a metric used to measure the quality of image generation. It is based on the statistical feature differences between generated and real images: it first uses a pre-trained image classifier (such as the Inception V3 model) to extract feature representations of real and generated images. Then, it uses the Fréchet distance to measure the difference between the two feature distributions. For any two multidimensional Gaussian distributions, including the distribution of real images $x_1 \sim N(\mu_1, \Sigma_1)$ and the distribution of generated images $x_2 \sim N(\mu_2, \Sigma_2)$, the Fréchet distance is computed as

$$d_F \left(x_1, x_2\right)^2 = \|\mu_1 - \mu_2\|_2^2 + \text{tr}\left(\Sigma_1 + \Sigma_2 - 2\left(\Sigma_1 \Sigma_2\right)^{1/2}\right). \tag{11}$$

The lower the difference between the sample mean and the covariance matrix, the better the quality of the generated image and the higher the consistency with the real image distribution. FID's advantage over other indicators is that it can capture more subtle differences in image quality and has a higher correlation with human subjective evaluation results. However, FID also has some limitations, such as the need for pre-trained classifiers and many image samples. We refer to the way FID and IS are evaluated in similar works [34,35].

4.2   Comparison Between Algorithms

Below we will give a quantitative comparison of the results between our method and previous popular methods, as well as a visual comparison of the results of some methods.

Quantitative Comparison. Table 1 shows the comparison of results between our method and other methods on the datasets CUB [32] and COCO [15], where larger IS values and smaller FID indicate better generation results. At the same time, we also provide the memory and time consumed by these methods during the test process as shown in Table 2. During the model training process, some models use additional pre-trained models, as shown in brackets: SD-GAN [36] (pre-trained COCO model [26]), CPGAN [14] (pre-trained YOLO v3 [7]), XMC-GAN [37] (pre-trained VGG-19 [28] and Bert [5]) and DAE-GAN (NLTK POS markers and 2D positions encoding). As shown in the results in Tables 1 and 2, our method achieves comparable performance to other methods on IS and FID metrics for both datasets. In particular, our method achieves better performance compared to the baseline model AttnGAN. For example, on CUB, our method achieves an FID value of 15.03, which is better than AttnGAN's 23.98. Regarding inference time, our method runs about 45 times faster than the baseline method (from 4.05 s to 0.11 s). Considering the memory footprint of the model, our method is much lower than other methods. In particular, compared with the baseline method AttnGAN, it reduces the memory footprint to about 1/9 (26.37M vs. 230M). Our AccAttnGAN does not achieve an absolute advantage in COCO due to the large variety of items in the COCO dataset and the difficulty of training a generative model in a short time. Comparing the FID values on the two datasets CUB and COCO, we also find that the image generation task

Table 1. Results of our method and other methods on the datasets CUB and COCO, where larger IS values and smaller FID values indicate better generation results.

Model	CUB		COCO	
	IS ↑	FID ↓	IS ↑	FID ↓
StackGAN [38]	3.70	–	8.45	–
StackGAN++ [39]	3.84	35.30	8.4	81.59
MirrorGAN [23]	4.56	18.34	26.47	34.71
SD-GAN [36]	4.67	–	**35.19**	–
DM-GAN [41]	4.75	16.09	30.49	32.64
CPGAN [14]	–	–		55.80
TIME [16]	**4.91**	**14.30**	27.85	31.14
SAM-GAN [21]	4.61	20.49	27.31	33.41
XMC-GAN [37]	–	–	30.45	**9.30**
DF-GAN [30]	4.46	18.23	–	41.83
DAE-GAN [27]	4.42	15.19	35.08	28.12
AttnGAN [33]	4.36	23.98	25.89	35.49
AccAttnGAN	4.32	15.03	27.74	34.78

Table 2. Comparison of memory (M) and time (seconds) during the testing stage on the CUB dataset between our method and previous methods.

Model	Cost in the test stage	
	Memory (M) ↓	Time (seconds) ↓
StackGAN [38]	–	–
StackGAN++ [39]	–	–
MirrorGAN [23]	–	–
SD-GAN [36]	335.00	6.18
DM-GAN [41]	46.00	7.06
CPGAN [14]	318.00	–
TIME [16]	120.00	–
SAM-GAN [21]	–	–
XMC-GAN [37]	166.00	–
DF-GAN [31]	46.79	3.71
DAE-GAN [27]	98.00	–
AttnGAN [33]	230.00	4.05
AccAttnGAN	**26.37**	**0.11**

on COCO is more difficult than on CUB. DF-GAN [31] uses the same simple and effective idea to also significantly reduce the size of the model. Nonetheless, our AccAttnGAN outperforms DF-GAN regarding generated image quality (IS, FID) and generation time.

Qualitative Comparison. We compared the visualization results of our model with three different models in Fig. 4: AttnGAN [33] (baseline model), DF-GAN and DM-GAN [41]. As shown from Fig. 4, the results obtained by AttnGAN and DF-GAN are not as realistic as those obtained by DM-GAN and our model. The

Fig. 4. Comparison of qualitative results between our method and several classical methods on multiple sets of text-image data

difference between the bird and the background is less obvious, especially in the leg and tail areas. Additionally, the feathers on the bird's back lack some clarity. The image generated by DM-GAN also loses part of the image. On the contrary, the images generated by our model show a remarkable feature: the birds and the background are well distinguished, making the foreground parts of the birds in our model easy to distinguish. Furthermore, the background in the images of our model has stronger contrast than the backgrounds generated by the other three models.

Finally, we also observe from Fig. 4, that the images in the four columns are distorted on all three models on the COCO dataset, which means that some images are missing, duplicated, or folded. However, compared to images generated by the other three models, the background in our model's images shows stronger brightness, increasing the visual appeal and overall aesthetics of the output.

4.3   Ablation Study

In this section, we conduct an ablation study on the number of ResDF blocks and the loss function. Specifically, we selected the CUB bird dataset, which showed promising results in our previous comparison experiments.

ResDF-Blocks/DF-Blocks. As one of our proposed structures, we replace our ResDF-Block with UP-Block and test it on the CUB dataset. In this ablation study, to verify the necessity of our proposed ResDF block, the UP block was replaced while other external conditions remained unchanged. The experimental results are shown in Table 3. The ID and FID measurements of UP-Block show

Table 3. Ablation results of block type on CUB datasets

Type of blocks	CUB dataset	
	IS ↑	**FID ↓**
UP-blocks	4.18	18.56
ResDF-blocks	**4.36**	**15.17**

that IS and FID in DF-GAN are successfully improved compared with DF-Blocks, which shows the importance of ResDF-Block.

Number of ResDF-Blocks. Our network structure consists of 6 ResDF blocks. We further explored whether the model's performance can be improved by increasing or decreasing the number of ResDF-Blocks. Comparative experiments on the number of ResDF-Blocks are shown in Table 4. The IS and FID indicators of UP-Block show that the IS of UP-Block has been successfully improved without changing the main structure. In addition, regarding the structural design of AccAttnGAN, when the number of ResDF-Block is 6, it has higher image generation performance than when the variable value is 4. It is worth noting that when this variable takes 8 or 10, the performance does not improve significantly, indicating that AccAttnGAN is still effective.

Table 4. Comparison results of the number of ResDF-blocks on CUB datasets

Number of ResDF blocks	CUB dataset	
	IS ↑	FID ↓
4× ResDF-blocks	2.87	59.84
8× ResDF-blocks	4.23	18.14
6× ResDF-blocks	4.29	**15.11**
10× ResDF-blocks	**4.38**	17.42

Loss Functions. Our algorithm for the discriminator uses the loss function $\mathcal{L}D + \mathcal{L}W + \mathcal{L}G$. Furthermore, we provide the optimization results of the algorithm driven by the loss functions $\mathcal{L}D$ and $\mathcal{L}D + \mathcal{L}W + \mathcal{L}DAMSM$. The results, as shown in the table, indicate that $\mathcal{L}W$ plays a key role, but the impact of $\mathcal{L}_{DAMSM}$ on our algorithm is not very pronounced.

Table 5. Ablation Study of Loss Functions in the Discriminator on the CUB Dataset

Loss Functions	CUB Dataset	
	IS ↑	FID ↓
$\mathcal{L}_D$	3.89	59.84
$\mathcal{L}_D + \mathcal{L}_W + \mathcal{L}_{DAMSM}$	4.53	32.12
$\mathcal{L}_D + \mathcal{L}_W + \mathcal{L}_G$ **(Ours)**	**4.67**	**16.73**

Hyperparameters. λs and γs. Since we improved the structure of the baseline model AttnGAN, we redefined the loss function, as shown in Eq. (9). In this section, we study the impact of two parameters λ and γ on the algorithm's performance. Specifically, we first fixed $\lambda = 0.2$ and then took different scaling coefficients $\gamma : \lambda$ to explore the role of the parameter γ, shown in Table 6. Subsequently, we kept $\gamma : \lambda = 1 : 10$ unchanged, and set different values of parameter λ to obtain the impact of λ on the algorithm (Table 7).

Table 6. Comparison results of hyperparameter γ on the CUB dataset with $\lambda = 0.2$.

Ratio $\lambda : \gamma$	CUB dataset	
	IS ↑	FID ↓
1 : 1	2.88	26.51
1 : 5	3.69	23.84
1 : 10	**4.67**	16.73
1 : 15	4.54	**16.46**

From Table 6 we find that the relatively the best result is a hyperparameter ratio of 1:10. It is worth noting that there is no significant improvement in performance between 1:10 and 1:15. Based on the approximate ratio derived from our experiments, our final result uses $\lambda : \gamma = 1 : 10$.

To determine the value of λ, we keep the ratio $\lambda : \gamma = 1 : 10$ fixed and list the algorithm's performance under different λ values, as shown in Table 7. According to experimental data, when γ is equal to 0.2 and 0.4, there is no obvious change in the IS and FID measurements. But when λ is equal to 0.6 and 0.8, there is still a significant gap in the quality of the generated images. To choose the best of the two sets of values, we set $\lambda = 0.2$, $\gamma = 2$ as the optimal hyperparameters.

Table 7. Ablation results of hyperparameter λ on the CUB dataset with the ratio $\lambda : \gamma = 1 : 10$.

Hyperparameter λ	CUB dataset	
	IS ↑	FID ↓
0.2	**4.67**	**16.73**
0.4	4.56	19.31
0.6	3.76	27.74
0.8	2.68	38.81

5   Conclusion

AttnGAN is a classic fine-grained image generation method, however, it has high time and space complexity. To this end, we propose a new acceleration algorithm called AccAttnGAN to reduce its complexity: by proposing a new up-and-down sampling module, the size of the original model is greatly reduced, and at the same time, semantic alignment and hinge discriminant loss are introduced to guide the GAN generator to produce an image that better matches the text.

Experimental results show that our algorithm performs equivalent or better than the baseline algorithm on two public datasets. In particular, on the CUB dataset, our algorithm reduces the model size to one-tenth of AttnGAN, and the running time is approximately one-fourth of AttnGAN. Our improvement eliminates a set of attention models and adds upsampling and residual blocks, resulting in better image quality.

References

1. Amirian, J., Hayet, J.B., Pettré, J.: Social ways: learning multi-modal distributions of pedestrian trajectories with gans. In: CVPR (2019)
2. Brock, A., Donahue, J., Simonyan, K.: Large scale GAN training for high fidelity natural image synthesis. In: 7th ICLR (2019)
3. Chen, X., Duan, Y., Houthooft, R., Schulman, J., Sutskever, I., Abbeel, P.: Info-gan: interpretable representation learning by information maximizing generative adversarial nets. NeurIPS **29** (2016)
4. Denton, E.L., Chintala, S., Fergus, R., et al.: Deep generative image models using a laplacian pyramid of adversarial networks. NeurIPS **28** (2015)
5. Devlin, J., Chang, M.W., Lee, K., Toutanova, K.: Bert: pre-training of deep bidirectional transformers for language understanding. arXiv preprint arXiv:1810.04805 (2018)
6. Dong, H., Yu, S., Wu, C., Guo, Y.: Semantic image synthesis via adversarial learning. In: ICCV, pp. 5706–5714 (2017)
7. Farhadi, A., Redmon, J.: Yolov3: an incremental improvement. In: CVPR, vol. 1804, pp. 1–6. Springer Berlin/Heidelberg, Germany (2018)
8. Goodfellow, I., et al.: Generative adversarial networks. Commun. ACM (2020)
9. Heusel, M., Ramsauer, H., Unterthiner, T., Nessler: Gans trained by a two time-scale update rule converge to a local nash equilibrium. NeurIPS (2017)
10. Karras, T., Aila, T., Laine, S., Lehtinen, J.: Progressive growing of gans for improved quality, stability, and variation. arXiv preprint arXiv:1710.10196 (2017)
11. Karras, T., Laine, S., Aittala, M., Hellsten, J., Lehtinen, J., Aila, T.: Analyzing and improving the image quality of stylegan. In: CVPR, pp. 8110–8119 (2020)
12. Kingma, D.P., Ba, J.: Adam: A method for stochastic optimization. arXiv preprint arXiv:1412.6980 (2014)
13. Kingma, D.P., Welling, M.: Auto-encoding variational bayes. arXiv preprint arXiv:1312.6114 (2013)
14. Liang, J., Pei, W., Lu, F.: CPGAN: content-parsing generative adversarial networks for text-to-image synthesis. In: Vedaldi, A., Bischof, H., Brox, T., Frahm, J.-M. (eds.) ECCV 2020. LNCS, vol. 12349, pp. 491–508. Springer, Cham (2020). https://doi.org/10.1007/978-3-030-58548-8_29
15. Lin, T.-Y., et al.: Microsoft COCO: common objects in context. In: Fleet, D., Pajdla, T., Schiele, B., Tuytelaars, T. (eds.) ECCV 2014. LNCS, vol. 8693, pp. 740–755. Springer, Cham (2014). https://doi.org/10.1007/978-3-319-10602-1_48
16. Liu, B., Song, K., Zhu, Y., de Melo, G., Elgammal, A.: Time: text and image mutual-translation adversarial networks. In: AAAI, vol. 35, pp. 2082–2090 (2021)
17. Ma, J., Zhang, L., Zhang, J.: Sd-gan: saliency-discriminated gan for remote sensing image superresolution. IEEE Geosci. Remote Sens. Lett. **17**(11), 1973–1977 (2019)
18. Mirza, M., Osindero, S.: Conditional generative adversarial nets. arXiv preprint arXiv:1411.1784 (2014)

19. Nam, S., Kim, Y., Kim, S.J.: Text-adaptive generative adversarial networks: manipulating images with natural language. NeurIPS (2018)
20. Van den Oord, A., Kalchbrenner, N., Espeholt, L., Vinyals, O., Graves, A., et al.: Conditional image generation with pixelcnn decoders. NeurIPS **29** (2016)
21. Peng, D., Yang, W., Liu, C., Lü, S.: Sam-gan: self-attention supporting multi-stage generative adversarial networks for text-to-image synthesis. Neural Netw. **138**, 57–67 (2021)
22. Qi, G.J.: Loss-sensitive generative adversarial networks on lipschitz densities. IJCV **128**(5), 1118–1140 (2020)
23. Qiao, T., Zhang, J., Xu, D., Tao, D.: Mirrorgan: learning text-to-image generation by redescription. In: CVPR, pp. 1505–1514 (2019)
24. Reed, S., Akata, Z., Yan, X., Logeswaran, L., Schiele, B., Lee, H.: Generative adversarial text to image synthesis. In: International Conference on Machine Learning (2016)
25. Reed, S.E., Akata, Z., Mohan, S., Tenka, S., Schiele, B., Lee, H.: Learning what and where to draw. NeurIPS **29** (2016)
26. Rombach, R., Blattmann, A., Lorenz, D., Esser, P., Ommer, B.: High-resolution image synthesis with latent diffusion models (2021)
27. Ruan, S., Zhang, Y., Zhang, K., Fan, Y., Tang, F., Liu, Q.: Dae-gan: dynamic aspect-aware gan for text-to-image synthesis. In: CVPR, pp. 13960–13969 (2021)
28. Simonyan, K., Zisserman, A.: Very deep convolutional networks for large-scale image recognition. arXiv preprint arXiv:1409.1556 (2014)
29. Sriperumbudur, B.K., Gretton, A., Fukumizu, K., Schölkopf, B., Lanckriet, G.R.: Hilbert space embeddings and metrics on probability measures. J. Mach. Learn. Res. **11**, 1517–1561 (2010)
30. Tao, M., Tang, H., Wu, F., Jing, X.Y., Bao, B.K., Xu, C.: Df-gan: a simple and effective baseline for text-to-image synthesis. In: CVPR, pp. 16515–16525 (2022)
31. Tao, M., et al.: Df-gan: deep fusion generative adversarial networks for text-to-image synthesis. arXiv preprint arXiv:2008.05865 (2020)
32. Wah, C., Branson, S., Welinder, P., Perona, P., Belongie, S.: Cub. Tech. Rep. CNS-TR-2011-001, California Institute of Technology (2011)
33. Xu, T., et al.: Attngan: fine-grained text to image generation with attentional generative adversarial networks. In: ICCV, pp. 1316–1324 (2018)
34. Xue, H., Jin, M., Zhang, C., Huang, Y., Weng, Q., Jin, X.: Image blending algorithm with automatic mask generation. In: International Conference on Neural Information Processing. pp. 234–248. Springer (2023). https://doi.org/10.1007/978-981-99-8132-8_18
35. Xue, H., Zhang, C., Liu, C., Wu, F., Jin, X.: Multi-task prompt words learning for social media content generation. arXiv preprint arXiv:2407.07771 (2024)
36. Yin, G., Liu, B., Sheng, L., Yu, N., Wang, X., Shao, J.: Semantics disentangling for text-to-image generation. In: CVPR, pp. 2327–2336 (2019)
37. Zhang, H., Koh, J.Y., Baldridge, J., Lee, H., Yang, Y.: Cross-modal contrastive learning for text-to-image generation. In: CVPR, pp. 833–842 (2021)
38. Zhang, H., et al.: Stackgan: text to photo-realistic image synthesis with stacked generative adversarial networks. In: ICCV, pp. 5907–5915 (2017)
39. Zhang, H., et al.: Stackgan++: realistic image synthesis with stacked generative adversarial networks. IEEE TPAMI (2018)

40. Zhu, J.Y., Park, T., Isola, P., Efros, A.A.: Unpaired image-to-image translation using cycle-consistent adversarial networks. In: ICCV, pp. 2223–2232 (2017)
41. Zhu, M., Pan, P., Chen, W., Yang, Y.: Dm-gan: dynamic memory generative adversarial networks for text-to-image synthesis. In: CVPR, pp. 5802–5810 (2019)

Weak Supervision Techniques Towards Enhanced ASR Models in Industry-Level CRM Systems

Zhongsheng Wang[1], Sijie Wang[1], Jia Wang[2]([✉]), Yung-I Liang[2], Yuxi Zhang[2], and Jiamou Liu[1]([✉])

[1] The University of Auckland, Auckland 1010, New Zealand
{zwan516,swan387}@aucklanduni.ac.nz, jiamou.liu@auckland.ac.nz
[2] Atom Intelligence, Hong Kong SAR, China
{johnny.wang,yoyo.liang,sharon.zhang}@atom-intelligence.com

Abstract. In the design of customer relationship management (CRM) systems, accurately identifying customer types and offering personalized services are key to enhancing customer satisfaction and loyalty. However, this process faces the challenge of discerning customer voices and intentions, and general pre-trained automatic speech recognition (ASR) models make it difficult to effectively address industry-specific speech recognition tasks. To address this issue, we innovatively proposed a solution for fine-tuning industry-specific ASR models, which significantly improved the performance of the fine-tuned ASR models in industry applications. Experimental results show that our method substantially improves the crucial auxiliary role of the ASR model in industry CRM systems, and this approach has also been adopted in actual industrial applications.

Keywords: ASR in CRM · Data Augmentation · Model Fine-tuning · Industrial Application

1 Introduction

A *Customer Relationship Management* (CRM) system is essential for managing customer interactions and data across various communication channels [1]. By centralizing customer information, CRMs improve communication, personalize service, and enhance customer satisfaction. They automate routine tasks, boosting productivity, and provide analytics that help in making data-driven decisions. An emerging trend in CRM is to incorporate voice technology, which is becoming increasingly vital for enhancing user experience and operational efficiency. For example, SuiteCRM's mobile application uses voice-to-text processing to allow users to navigate the app, search records, and record notes verbally, improving multitasking and speed [11]. Similarly, the iSpeak system automates customer service interactions traditionally handled by human agents, thereby enhancing efficiency and reducing the need for direct human contact [3].

Z. Wang and S. Wang—These authors contributed equally to this work.

M. Mahmud et al. (Eds.): ICONIP 2024, CCIS 2297, pp. 138–152, 2026.
https://doi.org/10.1007/978-981-96-7036-9_10

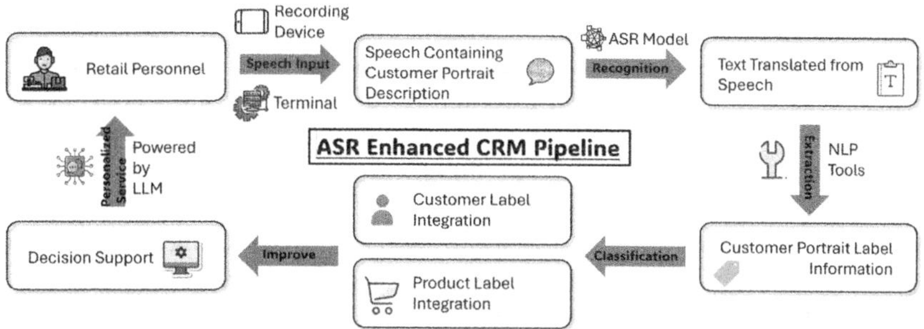

Fig. 1. A CRM Pipeline that is enhanced by voice technology and is designed for industrial deployment.

Integrating voice technologies into CRM systems presents several challenges that can complicate their adoption and functionality [11]. Ensuring compatibility with existing infrastructure is crucial, as CRM systems must interact seamlessly with various tools like customer databases and analytics platforms. Additionally, the technical complexity of implementing accurate *voice-to-text* processing is significant. These systems must handle various accents, dialects, languages, and speech patterns with high precision. Moreover, managing errors from voice recognition, which can occur due to background noise or unclear speech, without diminishing the user experience, remains a substantial challenge. Figure 2 presents our CRM pipeline where voice-to-text processing plays a pivotal role (See Sect. 3). The design leverages *Automatic Speech Recognition* (ASR) to transform spoken interactions from retail personnel into actionable text data. This pipeline not only streamlines the data collection process but also enhances the efficiency and personalization of customer service.

Although ASR models have made significant advancements in recent years, they often struggle to meet the unique needs of specific industries, hindering their effectiveness in the CRM pipeline where precise customer portrait labels are crucial. Fine-tuning ASR models to address these industry-specific requirements is essential but complicated by the difficulty of obtaining large volumes of accurately labeled data in real-world scenarios. The voice data recorded by sales representatives is typically unlabeled, features complex regional accents, and includes numerous proprietary brand names and colloquial terms. This low-quality data is not directly usable or easily cleaned for effective model fine-tuning, presenting a significant obstacle to improving ASR accuracy in CRM applications.

The main problem addressed in this paper is the accurate extraction of customer key features and intentions from voice inputs recorded by sales personnel, especially when general pre-trained ASR models fall short in recognizing industry-specific speech patterns. To address this problem, we designed a weak-supervision framework for fine-tuning pre-trained ASR models to meet CRM application needs. This approach leverages large language models (LLMs) and text-to-speech (TTS) models to make use of existing small, high-quality labeled datasets. With minimal human and financial costs, it generates large, high-quality datasets that can be directly used for ASR model fine-tuning. The ASR models fine-tuned with these datasets are capable of handling complex speech transcription tasks and have demonstrated superior performance. The results of experiments on various baseline models show that compared with the native model, the highest performance improvement is 63%, and the average improvement is 51.5%. We apply our method in real-world industry scenarios and has received positive feedback. Our main contributions are summarized as follows:

- We invented an efficient weak-supervision solution for fine-tuning ASR models in industrial CRM systems, which includes high-quality fine-tuning data generation. See Sect. 5.1.
- We proposed an indicator for evaluating the reasoning performance of ASR models, the **Integrated Error Rate (IER)**, which has a more objective and comprehensive evaluation significance for hybrid speech recognition tasks. See Sect. 5.2.
- We deployed the ASR model fine-tuning method proposed in this paper in relevant industrial fields and verified its superiority in high-quality work performance through verification. See Sect. 6.

2   Related Work

Voice Technologies in CRM Systems. The applications of voice and AI technologies in CRM systems are documented in several studies. [3] introduce iSpeak, a voice-activated relationship management system that automates customer care services, reducing human interaction and improving efficiency. [11] details the development of an Android application for SuiteCRM, incorporating speech-to-text processing for user interface navigation and database interactions, thereby enhancing usability. [1] discussed the importance of e-CRM technologies while highlighting the benefits of chatbots, cloud-based solutions, and Interactive Voice Response (IVR) systems in enhancing customer service and operational efficiency. More recently, [21] introduced a system integrating ASR and text classification to automate transcription, track issues, and detect customer emotions in real time. [10] examines the integration of voice recognition and Natural Language Processing (NLP) in Salesforce Einstein; this work showcases AI's role in understanding customer context and sentiment to improve sales and marketing strategies. Overall, employing ASR in CRM systems enhances efficiency, accuracy, and personalization by automating the transcription of customer interactions, enriching customer profiles, and allowing sales personnel to focus on relationship building.

ASR Models. Recent ASR models such as Whisper, wav2vec, wav2vec 2.0, and SpeechT5 have significantly advanced the field. Whisper [15], developed by OpenAI, achieves near-human transcription accuracy with minimal fine-tuning using a large dataset [14]. Wav2vec, by Facebook AI Research, uses unsupervised pre-training to achieve state-of-the-art performance [17]. Wav2vec 2.0 improves on this with a self-supervised learning framework, showing impressive results with minimal annotated data [4]. Microsoft's SpeechT5 [2] features a unified-modal architecture for various tasks, including ASR and TTS, improving performance across these areas.

Weakly Supervised ASR Model Fine-Tuning. Advancements in weakly supervised learning have enhanced ASR model fine-tuning with limited labeled data. Noisy Student Training (NST) combines labeled and unlabeled data, achieving state-of-the-art results with techniques like SpecAugment [13]. Almost unsupervised approaches using denoising auto-encoders also show substantial gains with minimal paired data [16]. Self-training methods generate and refine pseudo-labels, effectively improving ASR models in low-resource settings [18]. Data augmentation strategies, including TTS-generated synthetic data, enhance ASR performance for minority languages [5]. Leveraging synthetic data from ASR and machine translation improves speech-to-text translation, and lightweight models are crucial for noise-resistant systems in low-resource environments [9]. The LRSpeech system integrates pre-training and knowledge distillation, showing notable improvements for low-resource languages [20]. These advancements highlight the potential of weakly supervised learning to improve ASR models with minimal labeled data.

3   ASR-Enhanced CRM Pipeline

Our desired CRM system should be suited for businesses engaged in complex customer interactions and reliant on detailed customer data for personalized service delivery. For example, the system will be valuable in the luxury goods retail sector, where personalized marketing strategies and tailored product recommendations can significantly enhance the shopping experience and customer satisfaction. This system is also suitable for sectors such as financial services, healthcare, telecommunication, and technology retailers, etc., which offer customized solutions and continuously integrate customer feedback into product development.

To obtain insights into customer behavior, preferences, and trends, the system's priority is in obtaining accurate *customer portrait label* from sales personnel. Therefore, the focus of our design should be on ease of data collection, ensuring that customer information is captured accurately and in real time, without the need for manual entry. The ASR-Enhanced CRM Pipeline is illustrated in Fig. 2. After an interaction between a customer and retail personnel, the system cycles over the following steps:

1. **Voice Input Capture:** Retail personnel use a mobile app equipped with a voice input device to describe customer information.

2. **Speech-to-Text Conversion:** The system processes the captured speech through ASR technology, converting it into text.
3. **Data Extraction and Classification:** Natural Language Processing (NLP) techniques are used to extract key customer portrait labels from the text, which are then categorized within the CRM system.
4. **Data Integration and Analytics:** The organized customer data is integrated with product data.
5. **Decision Support:** Enabled by the integrated data, the CRM system offers personalized pricing, recommendations, and promotions tailored to each customer.
6. **Iterative Learning and Improvement:** This continuous cycle of data collection, processing, categorization, and analytics facilitates iterative learning and improvement.

In the rest of the paper, we shift our focus on the second step above, which is key to the CRM pipeline.

4 ASR Problem Definition

Our goal is to provide an ASR model suitable for the CRM pipeline above, addressing practical challenges to provide accurate transcription into text data. The key challenge here is the lack of real domain-specific labeled data, in the form of voice-text pairs, for training an accurate ASR model.

More formally, given a small amount of real audio data $\mathcal{D}_r = \{(\mathbf{x}_i^r, \mathbf{y}_i^r)\}_{i=1}^{N_r}$, where $\mathbf{x}_i^r$ is the i-th real audio sample and $\mathbf{y}_i^r$ is its corresponding text label, with N_r being the number of real samples, and a domain-specific keyword list $\mathcal{K} = \{k_1, k_2, \ldots, k_m\}$, where m is the number of keywords, we leverage a pre-trained large language model M_{LLM} to generate synthetic text labels $\hat{\mathbf{T}} = \{\hat{\mathbf{y}}_j^s\}_{j=1}^{N_s}$ based on $\mathcal{K}$ and $\{\mathbf{y}_i^r\}_{i=1}^{N_r}$, where N_s is the number of synthetic samples.

We then utilize a pre-trained Text-to-Speech (TTS) model $M_{\text{TTS}}(\hat{\mathbf{T}})$ to generate synthetic audio $\hat{\mathbf{X}}_s = \{\hat{\mathbf{x}}_j^s\}_{j=1}^{N_s}$ from the synthetic text $\hat{\mathbf{T}}$. An optional step involves applying a filtering function $\mathcal{F}$ to select high-quality synthetic data, resulting in $\mathcal{D}_s' = \mathcal{F}(\mathcal{D}_s) = \{(\hat{\mathbf{x}}_j^s, \hat{\mathbf{y}}_j^s) \mid \text{quality}(\hat{\mathbf{x}}_j^s, \hat{\mathbf{y}}_j^s) \geq \tau\}_{j=1}^{N_s'}$, where τ is a predefined quality threshold and N_s' is the number of filtered synthetic samples.

Finally, we fine-tune the ASR model $M_{\text{fine}}(\mathbf{x}; \theta)$, which maps input audio $\mathbf{x}$ to a predicted label $\hat{\mathbf{y}}$ with parameters θ, using $\mathcal{D}_s'$ by minimizing the loss function $L(\mathbf{x}, \mathbf{y}; \theta) = \text{CrossEntropy}(M_{\text{fine}}(\mathbf{x}; \theta), \mathbf{y})$. The fine-tuned optimal ASR model M will be deployed in industrial applications.

5 Methodology

5.1 Weak Supervision ASR

Figure 2 details the specific process by which we build a dedicated ASR model for a specific retail company and apply it to the corresponding CRM system. It consists of two main parts: *data expansion* and *data filtering*.

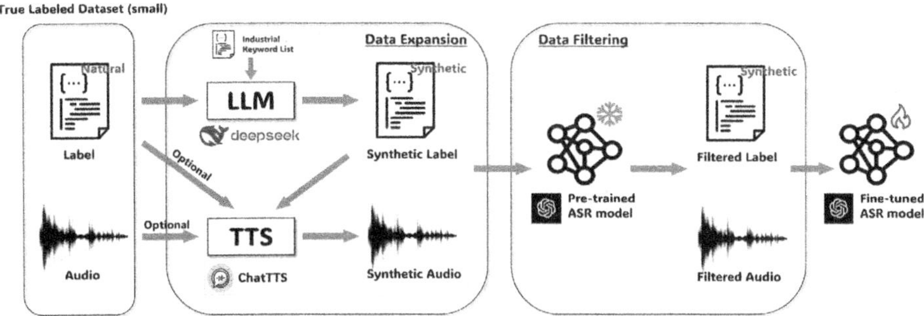

Fig. 2. The detailed main framework of the ASR fine-tuning method, which combines the LLM (DeepSeek) and TTS model (ChatTTS), includes two main parts: data expansion and data screening.

We use the labeled speech data of real business scenarios the industry provides as the input framework, aiming to generate a large amount of high-quality labeled synthetic data. First, we use LLM (DeepSeek V2 [6]) to generate labels for all synthetic data. The way to generate labels is to imitate the expression in the original data labels. In this process, we provide the language model with a keyword list containing most of the industry's professional terms. This keyword list is mainly obtained by crawling relevant industry data on social media and manually cleaning and filtering. After generating labels for all synthetic data, we use an advanced TTS model (ChatTTS[1]) to complete the speech synthesis task, taking the relevant source data and the interference components in the original data (such as dialects, accents, etc.) as one of the input parameters, simulating various complex situations encountered in real scenarios, and thus obtaining a large amount of synthetic speech data and its corresponding real labels.

After expanding the data to obtain a large dataset, we need to further conduct an additional data filtering process to ensure the high quality of the synthetic data. First, we use a pre-trained ASR model without fine-tuning (here we use whisper-large-v2) to infer the corresponding tags for all synthetic data. We then compare these inferred tags with the originally generated tags to calculate the CER metric. We set a threshold for this metric and exclude all data that do not meet this criterion. The remaining synthetic speech data and tags will form the final synthetic dataset for fine-tuning the ASR model. Also, in the fine-tuning stage of the ASR model, we consider using LoRA fine-tuning instead of SFT fine-tuning [7] to reduce computing resource overhead while minimizing performance loss.

5.2   Integrated Error Rate

Traditional speech recognition model indicators include word error rate (WER) [12] and character error rate (CER) [19], which are respectively applicable

---

[1] https://github.com/2noise/ChatTTS.

to monolingual speech recognition tasks in different languages. For example, English recognition is more suitable for WER, while Chinese recognition is more suitable for CER. However, in some industrial scenarios, such as the sales process of luxury brands, the communication between customers and sales staff usually retains the original pronunciation of the brand without translating it into a common language. Therefore, the above two indicators cannot be used as a good evaluation criterion for such multilingual mixed speech recognition tasks.

To address this problem, we propose the Integrated Error Rate (IER). IER combines the advantages of WER and CER and evaluates the accuracy of Chinese phrases and single Chinese characters in detail. In addition, we introduce the concept of keyword recognition and integrate it into IER to evaluate the accuracy of ASR models in recognizing unknown keywords.

The label of speech data X can be considered to consist of three parts: words $\mathcal{W}$, characters $\mathcal{C}$ and keywords $\mathcal{S}$, that is, $X = \bigcup_{i=1}^{n}\{W_i, C_i, S_i\}$. To understand the composition of the sentence in detail and calculate the score comprehensively, we used the Chinese word segmentation tools. Since the potential disagreement in the semantics of Chinese word segmentation may lead to deviations in the word segmentation results, we combined the current mainstream Chinese word segmentation tools jieba[2] and HanLP[3] [8] to divide the text into the above three parts and calculate the IER results.

We define the integrated error rate (IER) as follows:

$$
\begin{aligned}
\mathrm{IER} = \frac{1}{M} \sum_{m=1}^{M} \Bigg[&\left(\frac{|\mathcal{W}_m|}{|\mathcal{W}_m| + |\mathcal{C}_m| + |\mathcal{S}_m|} \right) \cdot \left(\frac{1}{|\mathcal{W}_m|} \sum_{i=1}^{|\mathcal{W}_m|} \delta_i \right) \\
+ &\left(\frac{|\mathcal{C}_m|}{|\mathcal{W}_m| + |\mathcal{C}_m| + |\mathcal{S}_m|} \right) \cdot \left(\frac{1}{|\mathcal{C}_m|} \sum_{j=1}^{|\mathcal{C}_m|} \epsilon_j \right) \\
+ &\left(\frac{|\mathcal{S}_m|}{|\mathcal{W}_m| + |\mathcal{C}_m| + |\mathcal{S}_m|} \right) \cdot \left(\frac{1}{|\mathcal{S}_m|} \sum_{k=1}^{|\mathcal{S}_m|} \zeta_k \right) \Bigg]
\end{aligned}
$$

Here, M is the total number of segmentation methods, m is the current segmentation method, and $|\mathcal{W}_m|$, $|\mathcal{C}_m|$, and $|\mathcal{S}_m|$ represent the number of words, characters, and special terms in the current segmentation method, respectively. The indicator functions δ_i, ϵ_j, and ζ_k represent the errors for corresponding words, characters, and special terms. Specifically, $\delta_i = 1$ if the i-th word is incorrect, otherwise $\delta_i = 0$; $\epsilon_j = 1$ if the j-th character is incorrect, otherwise $\epsilon_j = 0$; and $\zeta_k = 1$ if the k-th keyword is incorrect, otherwise $\zeta_k = 0$.

Due to the potential overlap between WER and CER, we implemented a de-duplication process during calculation. Specifically, for each incorrect word, we count it only once in the word error rate and do not double-count it in the character error rate. The following is an analysis of a real case:

The inference result in Table 1 shows 1 character error: "喜", 1 word error: "老化", and 1 keyword error: "PT" and "二十". "Speedy 20" is an inseparable part in the keyword lists but was split into two incorrect words in the inference results.

---

[2] https://pypi.org/project/jieba/.

[3] https://github.com/hankcs/HanLP.

Table 1. Original true label content of a fictitious label and the version inferred by the ASR model. The second row of data is the result through HanLP word segmentation.

Original Label	Inference Result by ASR Model
上海人，喜欢老花，偏好speedy 20系列，喜欢健身。	上海人，喜老化，偏好PT二十系列，喜欢健身。
上海人，喜欢 老花，偏好 speedy 20 系列，喜欢 健身 。	上海人，喜 老化，偏好 PT 二十 系列，喜欢 健身 。

However, by considering the positions of the correct words before and after, we summarize this as containing only one error. Since this method requires a list of industry-specific keywords and involves complex differences in word segmentation strategies, we manually verified all automatically calculated indicators to ensure their theoretical correctness. This approach allows us to more accurately evaluate the overall performance of ASR models in multilingual environments and specific industries.

6 Experimental Settings

6.1 Datasets

Three original datasets are mentioned in this paper, all provided by Atom Intelligence Group, namely **GUCCI100**, **LV100** and **Test Set**. Our well-trained sales staff record these datasets in real luxury business scenarios. They are entered according to some fixed language expressions and manually annotated to ensure the details and accuracy of the labels. Each voice tag may contain a rough user portrait obtained through observations of sales staff communicating with customers, including information such as age and preferences. The following is an approximate example:

Table 2. The true labels and English translations. These examples are fictitious, they just mimic the label format and are not real data from actual scenarios.

Original Label	English Translation Version
中年男性，上海人，偏好 speedy 系列，喜欢健身。	Middle-aged man, Shanghainese, prefers the Speedy series and likes fitness.
留学生，喜欢手链，喜欢带 logo 的。	International student, like bracelets, especially those with logos.
被香水瓶子吸引，想要购买 gucci 的炼金术士的花园，喜欢高雅。	Attracted by the perfume bottle and want to buy Gucci's The Alchemist's Garden. Like elegant products.

GUCCI100 and LV100 each contain 100 real data information from two luxury brands, while Test Set is a labeled dataset combining different luxury brands. In the experimental part of this paper, the Test Set is mainly used for testing the fine-tuned ASR models and does not participate in the construction of any fine-tuning datasets. Detailed statistics for these datasets are presented in Table 3. In addition to these datasets, the keywords list $\mathcal{K}$, detailed in Table 4, features seven categories used to generate synthetic training samples. Note that the individual audio recordings for each keyword is absent. Relying on these two small datasets and keywords list $\mathcal{K}$, we used the paradigm framework mentioned in the paper to generate 10,000 synthetic data for each brand. These two datasets will be used as independent datasets in the model fine-tuning process and will be merged to form a final version containing 20,000 data for the model fine-tuning process.

6.2   Evaluation Metrics

We mainly used CER and WER as indicators in the experiment. In addition, since this task involves multi-language translation (classified as Chinese and other languages), we calculate the recognition scores of the two languages separately, including CER_cn, CER_oth, WER_cn, and WER_oth. At the same time, we use the integrated error rate (IER) mentioned above as an additional evaluation indicator to evaluate the model performance independently.

6.3   Pseudo Label Generation

We use a small labeled dataset and a keywords list $\mathcal{K}$ to generate pseudo-labels for training. The objective is to ensure diversity in the synthetic dataset by covering as many keywords in $\mathcal{K}$, and as many styles of the reference sentences as possible, thereby producing varied audio samples. To achieve this, we randomly sample s sentences from real data and i keywords from each category in $\mathcal{K}$. If a category contains fewer than i keywords, we select all available keywords in that category. These sentences and keywords are then used as input prompts for DeepSeek V2 to generate the pseudo-labels. In this study, s is set to 5 and i is set to 8.

We use the following prompt to generate pseudo-labels, where the content inside curly braces represents variables. The prompt includes SENTENCE, which is a concatenation of real labels separated by newline character, and other variables separated by commaspattern:

Table 3. Data Statistics

Dataset	Samples	Duration (seconds, avg $\pm$ std)	Audio length (minutes)
GUCCI100	100	8.58$\pm$4.07	14.87
LV100	100	8.92$\pm$3.39	14.31
Test Set	1000	8.74$\pm$3.34	145.46
LVChatTTS	10000	8.81$\pm$3.21	1468.98
GUCCIChatTTS	10000	8.98$\pm$3.12	1496.14

Table 4. Keywords Statistics

Category	Samples	Example
SERIES	408	objets nomades
TYPE	273	装饰品 (decorations)
BRAND	92	balenciaga
MATERIAL	42	empreinte
NICKNAME	42	水桶包 (bucket bag)
LINES	19	monogram
SOCIAL	3	保值 (value preservation)

```
System prompt:

Please refer to the following examples and consider the existing
    ↪ product information. Take a deep breath and think carefully
    ↪ ---can you provide additional examples? These examples should
    ↪ closely align with the original samples. For product-related
    ↪ information, please use the product list we have provided.

User prompt:


Below is the product information for your reference:
Attribute: {SOCIAL}; Brand: {BRAND}; Pattern: [LINES};
Material: {MATERIAL}; Product: {NICKNAME}; Series: {SERIES};
Type: {TYPE};
Here are some examples: {SENTENCE}
Based on the template in the examples, please generate {s} new
    ↪ sentences. The content and style should be aligned with the
    ↪ examples.
```

6.4   Experimental Environment/Startup

We use the power of LLMs by calling the API of DeepSeek V2. The speech data
was generated using the current mainstream open-source TTS model ChatTTS.
The fine-tuning of the ASR model refers to the open-source LoRA fine-tuning
solution on GitHub, which can be viewed here. The important parameters
include the number of fine-tuning epochs is 5, the batch size is 4, the learning rate is 1e-3, and the maximum audio length is 30 s (all label data does not
exceed this limit). We use a NVIDIA RTX A6000 GPU with 40G VRAM.

We will use each of the above fine-tuning dedicated datasets to fine-tune three
different versions of the Whisper model and perform result testing and metric
calculation on the test set. We refer to the GitHub open-source solution[4] for
LoRA fine-tuning methods and codes. Due to the access restrictions of HanLP
and the uncertainty of the word segmentation strategy mentioned in the method,
we invested a lot of human resources to verify the calculation results of IER. In
the experiment, we only calculated the IER indicator for the best-performing

---

[4] https://github.com/yeyupiaoling/Whisper-Finetune.

models after fine-tuning among the three versions. We reduced the amount of data in the test set to 200.

7 Experimental Results

Table 5 shows the inference results of three different versions of the **Whisper** model (whisper-medium, whisper-large-v2, whisper-large-v3) after fine-tuning on different datasets mentioned in this paper, where some of the best results are shown in bold.

Table 5. The performance of the three base models after fine-tuning on different training sets. Model names without any special suffixes are the native model capability results. GUCCI100/LV100 are the results of fine-tuning the models using the real labeled datasets of two brands respectively. GUCCIChatTTS/LVChatTTS are the results of using the high-quality datasets generated by our proposed framework. GUCCCI&LV are the results of merging two synthetic datasets.

Model	CER	CER_cn	CER_oth	WER	WER_cn	WER_oth
whisper-medium	0.54125	0.50068	0.70524	1.48446	0.98699	0.94388
medium-GUCCI100	0.38658	0.30772	0.45721	0.85850	0.69	0.36364
medium-GUCCIChatTTS	0.21169	0.11968	0.38572	1.30250	0.72773	0.73488
medium-LV100	0.42269	0.34472	0.67785	0.69223	0.47222	0.58127
medium-LVChatTTS	0.21234	0.14897	0.34662	1.13762	0.70	0.88423
medium-GUCCCI&LV	0.20007	0.19243	0.22237	0.77381	0.68225	0.73338
whisper-large-v2	0.11958	0.06540	0.24480	1.56604	0.58	0.36471
v2-GUCCI100	0.12333	0.06677	0.18500	0.70142	0.35	0.35498
v2-GUCCIChatTTS	0.08795	0.07739	0.14031	0.71664	**0.33921**	0.44834
v2-LV100	0.07996	0.05031	**0.12779**	0.99662	0.44989	0.57324
v2-LVChatTTS	0.07743	0.05552	0.17980	0.83076	0.42877	**0.29445**
v2-GUCCCI&LV	**0.07390**	**0.04593**	0.13383	**0.62264**	0.44	0.31461
whisper-large-v3	0.18793	0.09832	0.22471	1.17422	0.82	0.99314
v3-GUCCI100	0.14223	0.07774	0.16648	1.22439	0.897	0.88442
v3-GUCCIChatTTS	0.08732	0.07002	0.09956	1.04436	0.66793	0.98092
v3-LV100	0.11339	0.11042	0.13398	1.11147	0.87	0.93732
v3-LVChatTTS	0.11271	0.10887	0.16643	0.90742	0.72	0.80452
v3-GUCCCI&LV	0.09332	0.07741	0.13346	0.80201	0.69	0.78147

Since the main language used in our fine-tuning and test datasets is Chinese, in theory, CER-related indicators are more suitable for evaluating Chinese speech recognition tasks. From this, we can see that the whisper-large-v2 basic model achieved the best CER performance after fine-tuning the final dataset after merging the two synthetic datasets of GUCCI and LV, and the latest whisper-large-v3 model did not have the best performance as expected. In addition, for indicators of other languages (CER_oth and WER_oth), whisper-large-v2 did not achieve the best performance indicators. This may be because the false-labeled speech generated by ChatTTS deviates from the pronunciation of other languages in real scenarios, resulting in a large bias in the synthetic dataset itself.

Similarly, we calculated the IER of the best fine-tuned versions of the three models and manually verified the indicators. The results are shown in Table 6. As expected, the best fine-tuned version of whisper-large-v2 achieved the best results on this indicator. However, the performance of whisper-large-v3 was unexpected again, and it did not even exceed the model fine-tuned by whisper-medium. We will analyze the reasons for this kind of accident in detail later.

Table 6. The values of the IER indicator for the models with the best fine-tuning results for the three versions (based on CER).

Model	IER
whisper-medium-GUCCCI&LV	0.3640
whisper-large-v2-GUCCCI&LV	0.2227
whisper-large-v3-GUCCIChatTTS	0.3962

8 Detailed Analysis

8.1 Case Study

Why Whisper-Large-V3 Performs Worse?. We checked the inference results of the whisper-large-v3 fine-tuned version of the model. As shown in Table 7, in the second half of the inference task, the results of whisper-large-v3 showed abnormal repetition of single Chinese characters/phrases, resulting in poor results in various indicators. However, whisper-large-v2 has achieved almost zero error translation performance after fine-tuning.

Table 7. The original labels and inference results by different models in the test set. (Data anonymization has been completed)

Original Label	Finetuned Model	Inference Result
苏州人，休闲年轻，时尚，喜欢老花，喜欢小包，可推荐男装，喜欢宽松款式。	whisper-medium	苏州，休闲年轻时尚，喜欢老化西湖小包，可推荐男装
	whisper-large-v2	苏州人，休闲，年轻时尚，喜欢老化喜欢小包，可推荐男装。
	whisper-large-v3	苏州人，休闲年轻时尚，喜欢老花喜欢小包，可推荐男装装装装装装装装装装装...
无锡人，喜欢男款西装裤，偏爱复古一点的风格。	whisper-medium	巫溪人，喜欢男款西装，偏复古一点的风格
	whisper-large-v2	无锡人，喜欢男款，西装，裤，偏爱复古一点的风格
	whisper-large-v3	无锡人喜欢阿阿阿阿阿阿阿阿阿阿阿阿阿i阿i阿i阿i阿i阿i阿阿...

The performance of the whipser-large-v3 baseline model after fine-tuning is not as stable as whisper-large-v2 in most cases. This, in fact, has been a well-recognised problem with whisper-large-v3 (see OpenAI official forum). We suspect that the increase in the number of parameters of the model may cause bias in the synthetic data to be learned as data features by the model after fine-tuning, which affects the model performance.

8.2   Selection of Num_epoch and CER Threshold

Our experiments determined that each model's optimal number of fine-tuning epochs is 5. To further validate this conclusion, we conducted additional experiments. Sub-fig. 3a of Fig. 3 shows the results of further fine-tuning the three best-performing models while keeping other parameters constant and calculating their CER metrics. As the number of epochs increased, whisper-large-v3 was significantly affected, with its performance deteriorating substantially. The other two models showed fluctuations but generally trended downward, possibly due to overfitting. Therefore, we chose 5 epochs as one of the parameters for fine-tuning the ASR model.

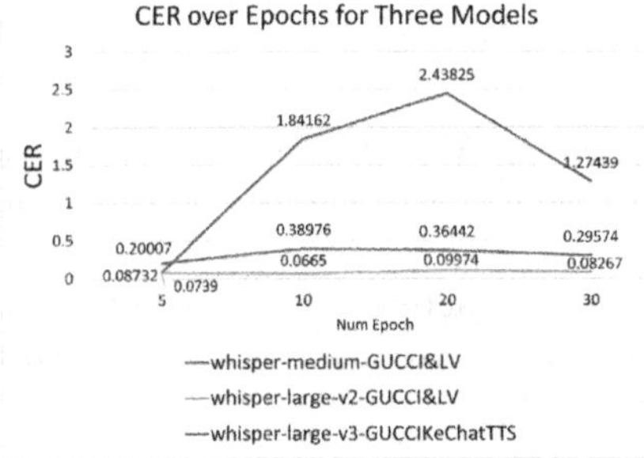

(a) CER indicator results after further fine-tuning the three models with 10, 15, 20, 30 epoch.

(b) Number of remaining data in the two synthetic datasets after filtering with different CER indicators.

Fig. 3. Results of additional experiments.

In addition, a CER threshold is determined to screen the dataset. In Sub-fig. 3b, we show the remaining data in the two synthetic datasets under different CER boundaries. By weighing the amount and quality of data, we choose 0.15 as the filtering boundary. By manually randomly checking the synthetic data, we found that their quality is within an acceptable range.

9   Conclusion and Limitation

We proposed a weak supervision ASR model fine-tuning framework suitable for the industry-specific CRM system, including detailed data expansion and filtering processes. The ASR model fine-tuned by this framework is competent for its task and performs well.

While this paradigm framework can generate results from a small amount of industry-specific labeled data, it is sometimes limited by the training data, causing it to overlook parts of the original prompt or include illegal text in the output (depending on the text review capabilities of the large model platform). In rare cases, due to language models' autoregressive and generative nature, the "fake labels" generated by the LLM may be semantically inconsistent with the original prompt, and the TTS model may produce partially unusable audio. These issues will be addressed in our future work. Additionally, using multiple models during data construction and fine-tuning can introduce some bias and error accumulation.

Acknowledgements. This project is supported by Atom Intelligence Group, which provides the necessary data information and computing resources. The method described in the paper has been tested in Atom Intelligence Group.

References

1. Adlin, F.N., Ferdiana, R., Fauziati, S.: Current trend and literature on electronic crm adoption review. J. Physics: Conf. Ser. **1201**, 012058 (2019)
2. Ao, J., et al.: Speecht5: unified-modal encoder-decoder pre-training for spoken language processing (2022)
3. Atayero, A.A., Alatishe, A.S., Iruemi, J.O.: Development of ispeak: a voice activated relationship management system. Inter. J. Emerging Technol. Adv. Eng. **1**(2) (2011)
4. Baevski, A., Zhou, H., Mohamed, A., Auli, M.: wav2vec 2.0: a framework for self-supervised learning of speech representations (2020)
5. Bartelds, M., San, N., McDonnell, B., Jurafsky, D., Wieling, M.: Making more of little data: improving low-resource automatic speech recognition using data augmentation (2023)
6. DeepSeek-AI: Deepseek-v2: A strong, economical, and efficient mixture-of-experts language model (2024)
7. Han, Z., Gao, C., Liu, J., Zhang, S.Q., et al.: Parameter-efficient fine-tuning for large models: A comprehensive survey. arXiv preprint arXiv:2403.14608 (2024)
8. He, H., Choi, J.D.: The stem cell hypothesis: Dilemma behind multi-task learning with transformer encoders. In: Proc. EMNLP, pp. 5555–5577. Online and Punta Cana, Dominican Republic (Nov 2021)
9. Fendji, J.L.K.E., Metalom, D.C.C.M.T., Yenke, B.O., Atemkeng, M.: Automatic speech recognition using limited vocabulary: a survey. Appli. Artifi. Intell. **36**(1), 2095039 (2022)
10. Kaliuta, K.: Implementing voice recognition and natural language processing in salesforce. IJMELR (2023)
11. Mustapha, M.: Implementing speech-to-text technologies for on-the-go crm. University of Stirling (2016)
12. Park, C., Chen, M., Hain, T.: Automatic speech recognition system-independent word error rate estimatio. arXiv preprint arXiv:2404.16743 (2024)
13. Park, D.S., et al.: Improved noisy student training for automatic speech recognition. In: Interspeech 2020 (2020)

14. Radford, A., Kim, J.W., Xu, T., Brockman, G., McLeavey, C., Sutskever, I.: Robust speech recognition via large-scale weak supervision (2022). https://arxiv.org/abs/2212.04356
15. Radford, A., Kim, J.W., Xu, T., Brockman, G., McLeavey, C., Sutskever, I.: Robust speech recognition via large-scale weak supervision. In: International Conference on Machine Learning, pp. 28492–28518. PMLR (2023)
16. Ren, Y., Tan, X., Qin, T., Zhao, S., Zhao, Z., Liu, T.Y.: Almost unsupervised text to speech and automatic speech recognition (2020)
17. Schneider, S., Baevski, A., Collobert, R., Auli, M.: wav2vec: unsupervised pre-training for speech recognition. arXiv preprint arXiv:1904.05862 (2019)
18. Singh, S., Hou, F., Wang, R.: A novel self-training approach for low-resource speech recognition (2023)
19. Wigington, C., Stewart, S., Davis, B., Barrett, B., Price, B., Cohen, S.: Data augmentation for recognition of handwritten words and lines using a cnn-lstm network. In: Proceedings of ICDAR, vol. 1, pp. 639–645. IEEE (2017)
20. Xu, J., Tan, X., Ren, Y., Qin, T., Li, J., Zhao, S., Liu, T.Y.: Lrspeech: extremely low-resource speech synthesis and recognition. In: SIGKDD, KDD 2020, pp. 2802–2812. Association for Computing Machinery, New York (2020)
21. Yunlong, Z., Xiangyu, L., Hongyan, X., Yingzhe, H., Jialiang, Q.: Design of intelligent customer service report system based on automatic speech recognition and text classification. E3S Web Conf. (2021). https://doi.org/10.1051/e3sconf/202129501064

Guided Safe Diffusion: Prohibiting Diffusion Models from Generating Inappropriate Content

Sidong Jiang[1], Rui Zhang[1], Xi Yang[1], Bin Dong[2], and Kaizhu Huang[3]($\boxtimes$)

[1] Xi'an Jiaotong - Liverpool University, Suzhou, Jiangsu, People's Republic of China
Sidong.Jiang20@student.xjtlu.edu.cn,
{Rui.Zhang02,Xi.Yang01}@xjtlu.edu.cn
[2] Ricoh Software Research Center, Beijing, People's Republic of China
Bin.Dong@cn.ricoh.com
[3] Duke Kunshan University, Kunshan, Jiangsu, People's Republic of China
Kaizhu.Huang@dukekunshan.edu.cn

Abstract. The increasing deployment of large generative models has heightened concerns over security and privacy, particularly regarding the generation of inappropriate content such as violent, explicit, or sensitive images, as well as the potential for creating fake images that spread misinformation and cause social problems. In this work, we propose Guided Safe Diffusion (GSD), an inference-time method specifically designed for diffusion models to prevent the generation of images with undesirable content as defined in a prohibited content list. Our method integrates safety guidance during the denoising steps of the model's inference process, modifying the predicted noise to steer the generation process away from unwanted content. This approach allows the model to accept both an input image and a text description, facilitating controlled image generation. Unlike previous methods, our technique does not necessitate retraining or fine-tuning of the model. We conduct qualitative and quantitative experiments to assess the effectiveness of our method, demonstrating that GSD can remove the unwanted content while preserving unrelated content. The results validate our method's ability to mitigate risks while maintaining the generative utility of diffusion models.

Keywords: Image protection · Diffusion models · AI ethics

1 Introduction

The recent surge in the field of Artificial Intelligence for Generative Content (AIGC), particularly within the last year, is largely attributable to advancements in large language models like GPT [2,21,22], and BERT [4]. These models have demonstrated an unprecedented ability to understand, interpret, and generate language with human-like proficiency. The advent of CLIP [20], a multi-modal model that maps text embeddings to image embeddings, has opened new frontiers in text-to-image tasks. The field has been further revolutionized by powerful image generative models such as DALL-E [15,16,23] and Stable Diffusion [25],

M. Mahmud et al. (Eds.): ICONIP 2024, CCIS 2297, pp. 153–164, 2026.
https://doi.org/10.1007/978-981-96-7036-9_11

which create visually compelling images from textual descriptions. The transformative power of these models has had a profound impact on our lives, opening up a world of new possibilities and applications.

Despite the remarkable capabilities of large generative models, their widespread application raises a series of ethical and societal challenges. People can easily misuse these powerful models to generate images with inappropriate content such as violence, harassment, explicit material, humiliation, and illegal activities. In response to these issues, current efforts have focused on several approaches. Pre-processing techniques involve filtering harmful content from training datasets, such as filtering out images containing people to avoid potential privacy violations [13], and removing unwanted classes of images from training data [15,28]. However, these techniques are computationally expensive and primarily effective for training new models or fine-tuning with large datasets; their efficacy diminishes with smaller datasets on downstream tasks. Another technique involves adversarial example filtering, which uses adversarial inputs to identify and mitigate vulnerabilities, enhancing model robustness and ensuring safe image generation [7,19,26,33]. Despite their effectiveness, adversarial example filtering may miss certain vulnerabilities and could be bypassed by sophisticated attacks. Model modification approaches include retraining or fine-tuning models with adversarial objectives, curated datasets, and regularization terms to guide the models to generate safe content [1]. These methods require significant computational resources and time for retraining and fine-tuning. Post-processing techniques, such as non-local means smoothing [17], block matching [10], and concept erasure [18], are employed to filter or modify generated outputs, ensuring compliance with ethical standards. However, post-processing can degrade image quality and is limited to addressing the symptoms rather than the root cause of inappropriate content generation.

In this work, we propose a novel inference-time guidance method specifically designed for diffusion models, such as Stable Diffusion, to address these concerns. Our approach integrates safety guidance during the denoising steps of diffusion model's inference process, modifying the predicted noise to control the generation process and avoid unwanted content. The model accepts two inputs: a text description and an image reference, similar to ControlNet [37], enabling controlled image generation. This technique ensures that the model generates images devoid of undesirable content without requiring retraining or fine-tuning of the generative model. To evaluate the effectiveness of our method, we conduct comprehensive qualitative and quantitative experiments, measuring the protection failure rate and collateral impact rate. Experimental results demonstrate the efficacy of our method in mitigating risks while preserving the utility of diffusion-based image generation models.

We summarize our contributions as follows:

- We propose Guided Safe Diffusion, an inference-time method that prevents diffusion models from generating images with user defined prohibited content.

- We introduce an algorithm that utilizes BLIP and SAM to identify regions likely containing prohibited content, allowing the safety guidance to accurately edit the target area without affecting other parts of the image.
- Our experiments demonstrate the effectiveness of Guided Safe Diffusion across five different scenes, outperforming the baseline Safe Latent Diffusion [27] in both prohibited content removal rate and prompt fidelity rate.

2   Related Works

Previous efforts to prevent the generation of ethically inappropriate images in generative models have primarily focused on three main approaches: (1) pre-processing techniques such as data filtering and adversarial example filtering; (2) model modification including retraining or fine-tuning the model; and (3) post-processing techniques to filter model outputs.

Pre-processing Techniques. One common approach is to remove potentially harmful or undesirable content from the training dataset. For example, Glide [13] filters out images containing people to avoid potential privacy violations and other inappropriate uses; removing unwanted classes of image or carefully selecting training data [15,24,28,32] is also a common practice. Another technique involves adversarial example filtering, which uses adversarial inputs to identify and filter out potential vulnerabilities. Researches [3,14,31] utilize adversarial training and purification methods to enhance model robustness and ensure safe image generation [34]. These methods ensure that generative models produce safer and more appropriate content.

Model Modification involves retraining or fine-tuning generative models to enhance their robustness and ensure they do not produce undesirable content. Retraining with adversarial objectives helps models learn to avoid generating harmful outputs. For example, iteratively retraining with adversarial inputs can reduce the generation of inappropriate images [1,29]. Fine-tuning models on curated datasets that exclude undesirable content ensures safe image generation. Studies [5,35,36] have shown that fine-tuning generative models on carefully selected data effectively prevents the creation of harmful content. Introducing regularization terms into the model's loss function penalizes the generation of undesirable images. Research [11] demonstrates the effectiveness of regularization techniques in guiding models to generate safe content while maintaining high-quality outputs. These model modification techniques enhance the ethical deployment of generative models, ensuring their outputs are safe and appropriate for various applications.

Post-processing Techniques involve filtering or modifying the outputs of generative models after the generation process to remove images that contain undesirable content. These techniques provide an additional layer of safety and refinement by targeting specific prohibited content related to safety, such as violence, sexual content, and politically sensitive material. One approach is to use image-based filters that apply transformations to remove or blur unwanted content. Techniques like non-local means smoothing and block matching methods help enhance the quality and safety of the final images by reducing unwanted elements [7,10]. Another method involves using adversarial filters that detect and modify generated content based on predefined criteria, ensuring compliance with ethical standards. For instance, studies have used adversarial filtering to remove or alter images containing prohibited elements, thus preventing the dissemination of harmful content [17,18,33]. These post-processing techniques provide a flexible and effective means of ensuring the ethical deployment of generative models, allowing for the refinement of outputs to meet safety and quality standards.

3   Methodology

The goal of our method is to ensure that generative models produce images free from prohibited content. Modern diffusion models, trained on large datasets, require significant time and computational resources for retraining or fine-tuning. Therefore, we developed an inference-time method that allows users to specify objects or concepts they wish to prevent from being generated. Starting from the DDIM [30] latents, GSD locates potentially risky area of image, then applies safety guidance and self-attention guidance to the identified regions. This approach accurately removes the prohibited content while keeping the background unchanged and prevents distortions around the modified areas, ensuring the images appear realistic. Figure 1, 2 and 3 illustrates the three main steps of our approach, which we in detail in the following sections.

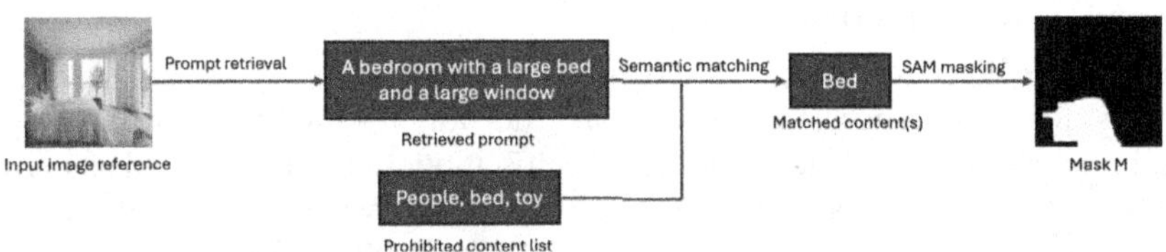

Fig. 1. Step 1: Extract edit region.

Step 1: Extract Edit Region. We use BLIP (Bootstrapping Language-Image Pre-training) [9] to extract text description of the input image. Then, we apply a semantic match between the predefined prohibited content and the prompt retrieved by BLIP. The retrieved prompt is segmented into words or phrases of 1–3 words in length. We calculate the semantic distance between these segments and the terms in the prohibited content list, recording any segment with a cosine distance below the threshold of 0.2 as a matched prompt segment. Next, we use SAM (Segment Anything Model) [8] to extract masks of the matched prompt segment from the input image. Finally, we merge the extract masks into a single mask M, representing the risky regions that potentially contain the prohibited content.

Step 2: Prohibited Content Removal. Classifier-free guidance [6] is a technique widely used in diffusion models to improve the quality and relevance of

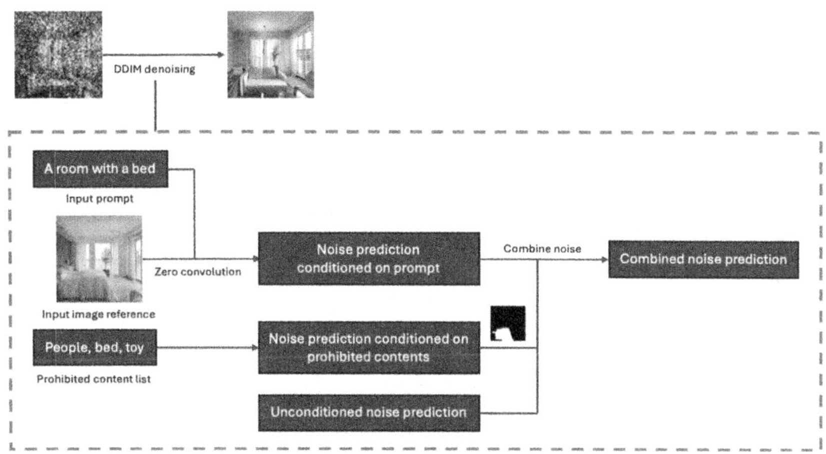

Fig. 2. Step 2: Prohibited content removal.

the generated images without relying on explicit class labels. It achieves this by training the model to handle both conditional and unconditional generation tasks, allowing for a smooth interpolation between the two during inference. The guidance is controlled by a parameter w that adjusts the influence of the conditioning information. The formula for classifier-free guidance is given by:

$$\tilde{\epsilon}_\theta(\mathbf{z}_t, \mathbf{c}_p) = (1 + w)\epsilon_\theta(\mathbf{z}_t, \mathbf{c}_p) - w\epsilon_\theta(\mathbf{z}_t), \tag{1}$$

where $\epsilon_\theta(\mathbf{z}_t)$ represents the unconditioned predicted noise, $\epsilon_\theta(\mathbf{z}_t, \mathbf{c}_p)$ represents the predicted noise conditioned on the prompt p. Safety guidance follow the similar concept to classifier-free guidance. We subtract a safety guidance term to shift the predicted noise away from the risky region in semantic space. The noise conditioned on the prohibited content $\epsilon_\theta(\mathbf{z}_t, \mathbf{c}_s)$ is defined as the safety guidance. We compare difference between the predicted noise conditioned on the prompt and the prohibited content within the masked risky region element-wisely. Values under the threshold δ indicate that the predicted noise is semantically close to the prohibited content, identifying unsafe elements that should be adjusted using safety guidance. This is recorded as:

$$D(\mathbf{c}_p, \mathbf{c}_s, \delta) = M \cdot (\epsilon_\theta(\mathbf{z}_t, \mathbf{c}_p) - \epsilon_\theta(\mathbf{z}_t, \mathbf{c}_s)). \tag{2}$$

Following classifier-free guidance, the full safety guidance term is:

$$S(\mathbf{z}_t, \mathbf{c}_p, \mathbf{c}_s) = D(\mathbf{c}_p, \mathbf{c}_s, \delta)(\epsilon_\theta(\mathbf{z}_t, \mathbf{c}_s) - \epsilon_\theta(\mathbf{z}_t)). \tag{3}$$

With s_q representing the safety guidance scale, the noise prediction incorporating safety guidance is:

$$\bar{\epsilon}_\theta = \epsilon_\theta(\mathbf{z}_t) + s_g(\epsilon_\theta(\mathbf{z}_t, \mathbf{c}_p) - \epsilon_\theta(\mathbf{z}_t) - S(\mathbf{z}_t, \mathbf{c}_p, \mathbf{c}_s)). \tag{4}$$

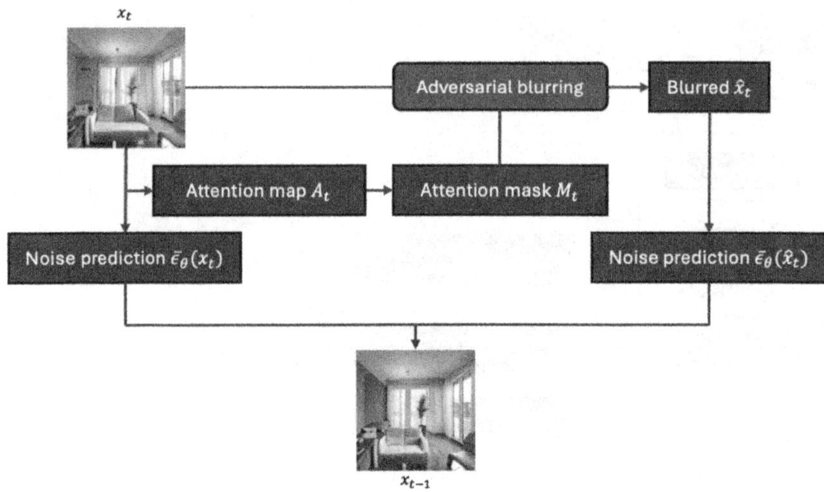

Fig. 3. Step 3: Mitigate edited distortions.

Step 3: Mitigate Edited Distortions. While safety guidance effectively removes prohibited content, it can also distort the region around the edited area, making the image look unrealistic. To address this issue, we apply self-attention guidance to the edited region. The attention map A_t is derived by performing global average pooling (GAP) on the stacked self-attention maps $A_t^S \in \mathbb{R}^{N \times (HW) \times (HW)}$ along the dimension $\mathbb{R}^{HW}$, followed by reshaping it to $\mathbb{R}^{H \times W}$ and applying nearest-neighbor up-sampling to match the size of $\mathbf{x}_t$:

$$A_t = Upsample(Reshape(GAP(A_t^S))). \tag{5}$$

The attention mask M_t is then obtained by applying a masking threshold τ, which is defined as the mean value of the attention map A_t:

$$M_t = (A_t > \tau), \tag{6}$$

$$\hat{\mathbf{x}}_t = (1 - M_t) \cdot \mathbf{x}_t + M_t \cdot \tilde{\mathbf{x}}_t, \tag{7}$$

where $\tilde{\mathbf{x}}_t$ is obtained by defusing the unconditioned noise $\epsilon_\theta(\mathbf{z}_t)$. The final noise prediction is then:

$$\bar{\epsilon}_{final} = \epsilon_\theta(\hat{\mathbf{x}}_t) + (1 + s)(\bar{\epsilon}_\theta - \epsilon_\theta(\hat{\mathbf{x}}_t)). \tag{8}$$

4   Experiment

4.1   Experimental Setup

For edit region extraction, we utilize the BLIP model from the Transformers module for prompt retrieval and the Language Segment-Anything Model [12] to extract masks of risky regions from input images. Our backbone models include Stable Diffusion v1.4 [25] and ControlNet v1.0 [37], both sourced from Hugging Face's Diffusers module.

For the stable diffusion pipeline, the encoder hidden states input for both ControlNet and the denoise UNet are constructed by concatenating the prompt embedding with embeddings conditioned on the prompt input, unconditioned, and conditioned on the prohibited content list. During each denoising step, the denoise UNet predicts the noise for these three conditions. The predicted noises are then concatenated and used in the guidance operation to derive the final noise, which is subsequently applied in the denoising loop.

Our experimental setup is similar to the baseline model, Safe Latent Diffusion (SLD) [27], with the primary difference on the task. While SLD focuses on the text-to-image generation task, our method supports both text and image inputs, similar to ControlNet. This allows for more comprehensive experiments, including scenarios where the input is solely an image (image-to-image). This extension of functionality demonstrates the versatility and robustness of our approach compared to traditional methods that only handle text-to-image tasks.

4.2 Evaluation Setup

Our evaluation framework employs two key metrics: the prohibited content removal rate and the prompt fidelity rate. To comprehensively assess our guidance method, we test it across five distinct scenes, each with four different prompts. For each prompt, we generate 50 images for evaluation. Our results are compared against the baseline model, Safe Latent Diffusion (SLD) [27]. Due to the limited number of comparable existing works that specifically address the prevention of generating prohibited content in diffusion models at inference time, our evaluation primarily focuses on the comparison with SLD.

Prohibited content removal rate is calculated by dividing the number of successful cases where unwanted content is removed by the total number of cases. This metric evaluates the effectiveness of our guidance method in eliminating prohibited content from the generated images.

Prompt fidelity rate measures the percentage of generated images that accurately match the input prompts. In cases where there are conflicts between the prompt and the prohibited content list, the conflicting content is disregarded in the fidelity calculation. This metric is crucial for assessing not only how well the generated images adhere to the specified prompts but also ensuring that the guidance method does not inadvertently alter content that does not need to be changed.

4.3 Quantitative Result

Table 1 illustrates the prohibited content removal rate and the prompt fidelity rate of our method compared to the baseline model Safe Latent Diffusion (SLD). We evaluate three distinct scenarios: (1) the prohibited content is present in the original input image; (2) the prohibited content is introduced by the prompt;

and (3) both the input image and the prompt contain prohibited content. The third scenario is particularly challenging, as the unwanted content is further emphasized by ControlNet's prompt and image conditions. The quantitative results demonstrate that our method outperforms SLD in all three scenarios.

Table 1. Comparison of prohibited content removal rate (CRR) and prompt fidelity rate (PFR) between SLD and our guidance method. The scenarios evaluated are: (1) "Image" - prohibited content is present in the original input image; (2) "Prompt" - prohibited content is introduced by the prompt; (3) "Both" - prohibited content is present in both the input image and the prompt.

	Image	Prompt	Both
Ours	0.955	0.915	0.650
SLD	0.923	0.903	0.485

(a) Prohibited content removal rate

	Image	Prompt	Both
Ours	0.938	0.840	0.620
SLD	0.913	0.810	0.440

(a) Prohibited content removal rate

4.4   Qualitative Result

In this section, we demonstrate the effectiveness of our proposed method by implementing guidance using prohibited content "weapon", "blood", "animal abuse", "bed", and "fruit" in five different scenes. We present the original image without guidance and compare the results using our method and the baseline model, SLD, in Fig. 4. For each scene, we provide two pairs of results generated by SLD and our guidance method, each pair using the same random seed.

4.5   Ablation Study

In this section, we conduct an ablation study to investigate the effectiveness of adding masks and applying self-attention guidance. According to the results shown in Table 2, when the mask is not applied, the prohibited content removal rate decreases significantly. This indicates that the mask allows the safety guidance to accurately edit the unsafe region. By applying the mask, we can increase the safety guidance scale without notably affecting the remaining parts.

Compared to the results without self-attention guidance, there is no significant improvement in the content removal rate. In fact, in scenarios where the prohibited content appears in both the input image and the prompt, the content removal rate without self-attention guidance is slightly higher. This is because, in our model, self-attention guidance is primarily used to enhance the realism of the edited image. For more challenging tasks, where safety guidance struggles to fully remove the prohibited content, self-attention guidance may cause the region near the edited area to revert to match the attention map. While this improves the overall realism of the image, it can reduce the effectiveness of prohibited content removal.

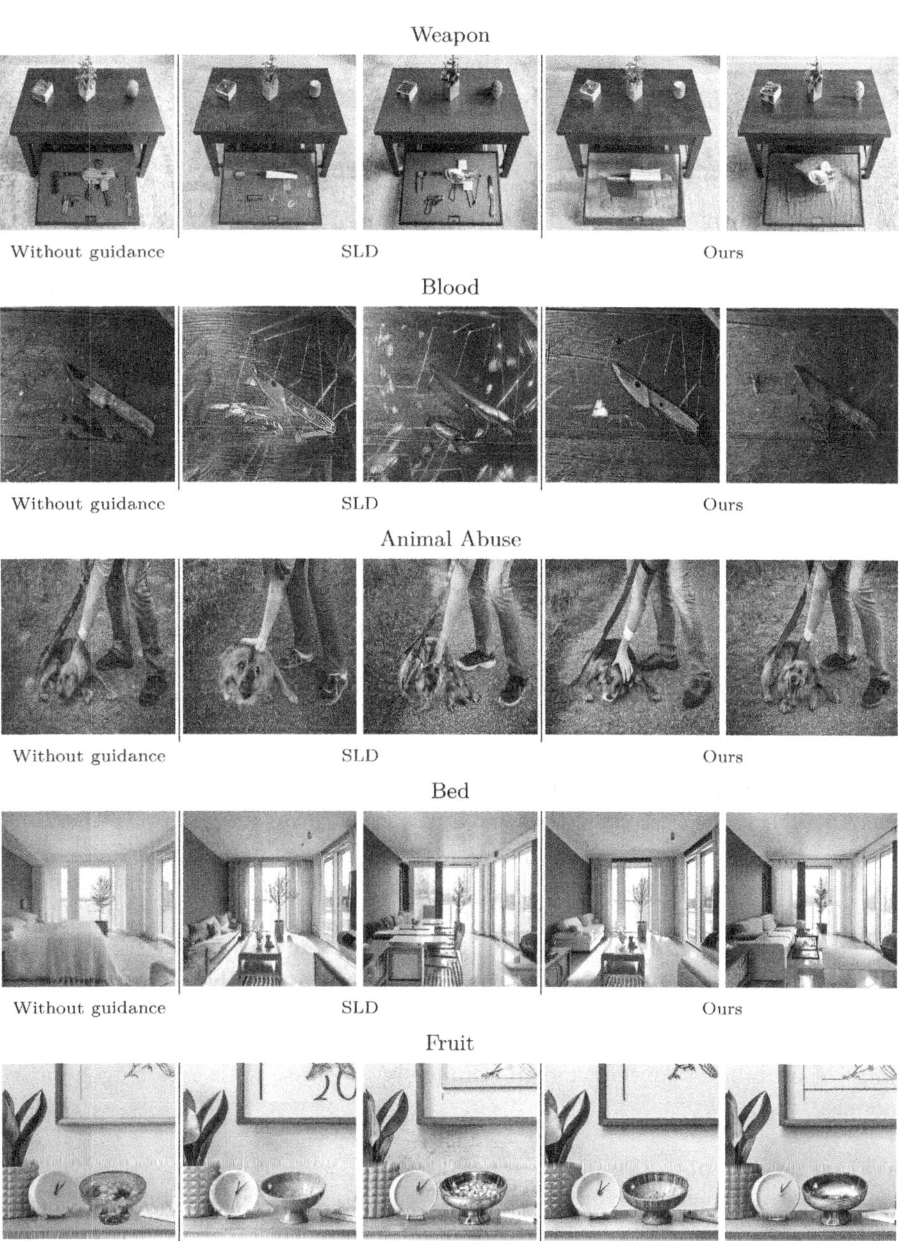

Fig. 4. Sample images generated using our guidance method compared to those guided by the SLD method. The prohibited contents are "weapon", "blood", "animal abuse", "bed", and "fruit", respectively.

Table 2. Ablation Study: Comparison of prohibited content removal rate (CRR) and prompt fidelity rate (PFR) between our guidance method, Safe Latent Diffusion (SLD), our method without the application of a mask on safety guidance, and our method without self-attention guidance (SAG). The scenarios evaluated are: (1) "Image" - prohibited content is present in the original input image; (2) "Prompt" - prohibited content is introduced by the prompt; (3) "Both" - prohibited content is present in both the input image and the prompt.

	Image	Prompt	Both		Image	Prompt	Both
Ours	0.955	0.915	0.650	**Ours**	0.938	0.840	0.620
No mask	0.885	0.838	0.460	**No mask**	0.808	0.793	0.455
No SAG	0.950	0.863	0.665	**No SAG**	0.925	0.808	0.630
SLD	0.923	0.903	0.485	**SLD**	0.913	0.810	0.440

(a) Prohibited content removal rate   (b) Prompt fidelity rate

5   Conclusion

The immense potential of large generative models underscores the need for mechanisms to prevent the creation of inappropriate images. This paper introduces a novel inference-time method to control the output of diffusion models, thereby avoiding the generation of undesirable content. By utilizing guidance on noise in the denoising loop, our approach effectively suppresses unwanted elements during the image generation process, addressing critical ethical concerns. Our method demonstrates significant improvements in both the prohibited content removal rate and prompt fidelity rate compared to the baseline Safe Latent Diffusion (SLD) model, ensuring that generated images remain faithful to the input prompts while excluding inappropriate content.

Future work will focus on enhancing the precision of our method to more accurately detect and target risky regions. By refining the localization of these regions, we can further improve guidance accuracy, allowing for increased guidance scale with minimal impact on unrelated areas. This refinement will strengthen the balance between content removal and image fidelity. Additionally, we aim to extend the flexibility of these techniques to manage a wider range of prohibited content and to seamlessly integrate with various generative models.

Acknowledgements. The work was partially supported by the following: National Natural Science Foundation of China under No. 92370119, 62376113, and 62206225.

References

1. Bertrand, Q., Bose, A.J., Duplessis, A., Jiralerspong, M., Gidel, G.: On the stability of iterative retraining of generative models on their own data. arXiv preprint arXiv:2310.00429 (2023)

2. Brown, T.B., et al.: Language models are few-shot learners. arXiv preprint arXiv:2005.14165 (2020)
3. Chen, X., Gao, X., Zhao, J., Ye, K., Xu, C.Z.: AdvDiffuser: natural adversarial example synthesis with diffusion models. In: Proceedings of the IEEE/CVF International Conference on Computer Vision, pp. 4562–4572 (2023)
4. Devlin, J., Chang, M.W., Lee, K., Toutanova, K.: BERT: pre-training of deep bidirectional transformers for language understanding. In: Proceedings of the 2019 Conference of the North American Chapter of the Association for Computational Linguistics: Human Language Technologies, Volume 1 (Long and Short Papers), pp. 4171–4186 (2019)
5. Gandikota, R., Materzynska, J., Fiotto-Kaufman, J., Bau, D.: Erasing concepts from diffusion models. In: Proceedings of the IEEE/CVF International Conference on Computer Vision, pp. 2426–2436 (2023)
6. Ho, J., Salimans, T.: Classifier-free diffusion guidance. arXiv preprint arXiv:2207.12598 (2022)
7. Kim, S., Jung, S., Kim, B., Choi, M., Shin, J., Lee, J.: Towards safe self-distillation of internet-scale text-to-image diffusion models. arXiv preprint arXiv:2307.05977 (2023)
8. Kirillov, A., et al.: Segment anything. In: Proceedings of the IEEE/CVF International Conference on Computer Vision, pp. 4015–4026 (2023)
9. Li, J., Li, D., Xiong, C., Hoi, S.: BLIP: bootstrapping language-image pre-training for unified vision-language understanding and generation. In: International Conference on Machine Learning, pp. 12888–12900. PMLR (2022)
10. Li, X., et al.: SafeGen: mitigating unsafe content generation in text-to-image models. arXiv preprint arXiv:2404.06666 (2024)
11. Li, Z., Hong, J., Li, B., Wang, Z.: Shake to leak: fine-tuning diffusion models can amplify the generative privacy risk. In: 2024 IEEE Conference on Secure and Trustworthy Machine Learning (SaTML), pp. 18–32. IEEE (2024)
12. Medeiros, L.: Language segment-anything (2023). https://github.com/luca-medeiros/lang-segment-anything
13. Nichol, A., et al.: GLIDE: towards photorealistic image generation and editing with text-guided diffusion models. arXiv preprint arXiv:2112.10741 (2021)
14. Nie, W., Guo, B., Huang, Y., Xiao, C., Vahdat, A., Anandkumar, A.: Diffusion models for adversarial purification. arXiv preprint arXiv:2205.07460 (2022)
15. OpenAI: Dall-e 2 (2022). https://openai.com/dall-e-2/
16. OpenAI: Dall-e 3 (2023). https://openai.com/dall-e-3/
17. Pain, C.D., Egan, G.F., Chen, Z.: Deep learning-based image reconstruction and post-processing methods in positron emission tomography for low-dose imaging and resolution enhancement. Eur. J. Nuclear Med. Molecular Imaging **49**(9), 3098–3118 (2022)
18. Pham, M., Marshall, K.O., Cohen, N., Mittal, G., Hegde, C.: Circumventing concept erasure methods for text-to-image generative models. In: The Twelfth International Conference on Learning Representations (2023)
19. Qian, Z., et al.: Perturbation diversity certificates robust generalization. Neural Netw. **172**, 106117 (2024)
20. Radford, A., et al.: Learning transferable visual models from natural language supervision. arXiv preprint arXiv:2103.00020 (2021)
21. Radford, A., Narasimhan, K., Salimans, T., Sutskever, I.: Improving language understanding by generative pre-training. https://s3-us-west-2.amazonaws.com/openai-assets/research-covers/language-unsupervised/languageunderstandingpaper.pdf (2018)

22. Radford, A., Wu, J., Child, R., Luan, D., Amodei, D., Sutskever, I.: Language models are unsupervised multitask learners. OpenAI Blog **1**(8), 9 (2019)
23. Ramesh, A., et al.: Zero-shot text-to-image generation. arXiv preprint arXiv:2102.12092 (2021)
24. Rombach, R.: Stable diffusion 2.0 release (November 2022). https://stability.ai/news/stable-diffusion-v2-release
25. Rombach, R., Blattmann, A., Lorenz, D., Esser, P., Ommer, B.: High-resolution image synthesis with latent diffusion models. arXiv preprint arXiv:2112.10752 (2021)
26. Sauer, A., Lorenz, D., Blattmann, A., Rombach, R.: Adversarial diffusion distillation. arXiv preprint arXiv:2311.17042 (2023)
27. Schramowski, P., Brack, M., Deiseroth, B., Kersting, K.: Safe latent diffusion: mitigating inappropriate degeneration in diffusion models. In: Proceedings of the IEEE/CVF Conference on Computer Vision and Pattern Recognition, pp. 22522–22531 (2023)
28. Schuhmann, C., et al.: LAION-5B: an open large-scale dataset for training next generation image-text models. Adv. Neural. Inf. Process. Syst. **35**, 25278–25294 (2022)
29. Sharma, R., Zhou, S., Ji, K., Chen, C.: Discriminative adversarial unlearning. arXiv preprint arXiv:2402.06864 (2024)
30. Song, J., Meng, C., Ermon, S.: Denoising diffusion implicit models. arXiv preprint arXiv:2010.02502 (2020)
31. Tan, Z., Wang, S., Yang, X., Huang, K.: PAG: protecting artworks from personalizing image generative models. In: Luo, B., Cheng, L., Wu, ZG., Li, H., Li, C. (eds.) International Conference on Neural Information Processing, pp. 433–444. Springer, Singapore (2023). https://doi.org/10.1007/978-981-99-8070-3_33
32. Tan, Z., Yang, X., Huang, K.: Semantic-aware data augmentation for text-to-image synthesis. In: Proceedings of the AAAI Conference on Artificial Intelligence, vol. 38, pp. 5098–5107 (2024)
33. Wu, Z., Gao, H., Wang, Y., Zhang, X., Wang, S.: Universal prompt optimizer for safe text-to-image generation. arXiv preprint arXiv:2402.10882 (2024)
34. Xiao, C., et al.: DensePure: understanding diffusion models for adversarial robustness. In: The Eleventh International Conference on Learning Representations (2022)
35. Yang, Y., Gao, R., Yang, X., Zhong, J., Xu, Q.: GuardT2I: defending text-to-image models from adversarial prompts. arXiv preprint arXiv:2403.01446 (2024)
36. Zeng, X., Gao, Z., Ye, Y., Zeng, W.: IntentTuner: an interactive framework for integrating human intents in fine-tuning text-to-image generative models. arXiv preprint arXiv:2401.15559 (2024)
37. Zhang, L., Rao, A., Agrawala, M.: Adding conditional control to text-to-image diffusion models. In: Proceedings of the IEEE/CVF International Conference on Computer Vision, pp. 3836–3847 (2023)

A Fine-Tuned Multi-classifier Optimization Framework Towards Safety-Critical Classes

Jiarui Wu[1(✉)] and Baowen Zhang[1,2]

[1] Institute of Cyber Science and Technology, Shanghai Jiao Tong University, Shanghai 200240, China
`{sznswjr,zhangbw}@sjtu.edu.cn`
[2] Shanghai Key Laboratory of Integrated Administration Technologies for Information Security, Shanghai 200240, China

Abstract. It is critically important to minimize the risk of harmful decisions possibly made by neural networks in Safety-Critical Systems (SCS). In this paper, a relative safety-criticality metric is introduced to quantitatively characterize the importance of different classes for a multi-classifier, according to the probability of harm failures induced by its misclassifications in SCS. With this metric, Safety-Critical Classes (SCCs), which require a higher level of protection than other classes, can be identified for classification tasks. A fine-tuning optimization framework is proposed for multi-classifiers to minimize the possibility of losses potentially caused by their wrong operations on SCCs. By constraining the cross-entropy loss function with a tunable penalty term, our proposed training process can effectively improve the recalls of SCCs. Comparative experiments demonstrate the effectiveness of our work. And we believe our work can be potentially used as an optimization scheme to migrate existing multi-classifiers into SCS.

Keywords: Safety-Critical Systems · Classification Tasks · Safety-Critical Classes · Multi-classifiers · Cross-Entropy Loss

1 Introduction

System failures in SCS can result in loss of life, significant property damage, or damage to the environment [11]. ANN classifiers are applied in safety-critical domains such as autonomous driving. Therefore, it's important for researchers to develop more reliable and suitable classifiers that can be extensively employed in SCS for decision-making [1].

The Protection Rings mechanism has been used as the kernel security pattern in the protection of operating systems and CPUs, where important processes run in an internal ring for better security protection [21]. Similarly, in SCS, objects of some classes can exhibit greater safety criticality than others. For instance, humans deserve more protection than animals in autonomous driving fields, and

This work is supported by the National Key Research and Development Program of China under Grant No. 2020YFB1807504 and No. 2020YFB1807500.

the misidentification of humans as animals by the classifiers can cause more severe safety hazards than the reverse.

In fact, previous research work have noticed the phenomenon that misclassifications of different objects can result in different costs by classifiers. Cost-sensitive learning is such an approach in machine learning fields that aims to reduce the differing costs associated with various types of errors during the training stages of model development. Cost-sensitive learning methods are originally employed to tackle the difficulties posed by imbalanced datasets, and have been extended to include weighted strategies for various sample types in specialized areas. However, almost all classifications are associated with cost in these penalty terms, which makes it difficult for the training process to fully capture the differences of safe-criticality between classes [7].

Harmful failure, as one type of severe system failures, is used to describe an unacceptable system outcome that violate safety constraints [15]. In this paper, we introduce a relative safety-criticality metric to quantitatively define the different importance of classes for a multi-classifier operating under safety-critical systems, according to the probability of harm failures that it generate. And then we propose a fine-tuning optimization framework for multi-classifiers to minimize the possibility of losses potentially caused by their wrong operations on SCCs. In general, the contributions of this paper are as follows:

1. A relative safety-criticality metric $Y \prec_{critical} X$ is designed to probabilistically distinguish the damaging effects of different classes when they are misclassified by the classifier in the safety-critical domains.
2. A new optimized loss function for multi-classifiers, $L_{SCCPT}(y, p)$, is designed by combining the penalty term with cross-entropy loss. Through penalizing the misclassification of hierarchical SCC objects, the loss function aims to optimize the models in the growth direction of the recalls of SCCs.
3. Based on the loss function $L_{SCCPT}(y, p)$, a fine-tuning optimization framework is proposed for multi-classifiers to minimize the possibility of harmful losses potentially caused by their wrong operations on SCCs. The framework can be potentially used as an optimization scheme to migrate existing multi-classifiers into SCS.

2   Related Work

2.1   Related Research of ANNs in Safety-Critical Systems

Operational errors in SCS such as autonomous driving can lead to catastrophic outcomes. The use of neural networks in such applications has garnered significant attention from researchers [2,18,24]. The implementation of machine learning in SCS typically involves four stages: Data Management, Model Learning, Model Verification, and Model Deployment [17]. Optimizing ANNs to prevent failures during the ML stage is challenging.

The work in [19] highlights the propensity of neural network systems for unexpected biases in decision-making processes. For instance, [13] emphasizes the challenges in pedestrian detection under low-visibility conditions, such as dimly lit environments. Consequently, interpretability is an increasing concern in neural network applications tailored for SCS, with some scholars even arguing against the use of black-box models in SCS [20,23].

Many machine learning problems are formulated through the optimization of objective functions [22]. The choice of hyperparameters for loss functions plays a crucial role in the ML stage [8,17], significantly impacting the overall performance of the model. Automated hyperparameter optimization strategies such as grid search have been proposed [12], but the complexity of datasets and models in SCS remains a challenge for hyperparameter configuration.

2.2 Cost-Sensitive Neural Networks

[14] initially put forward the foundational concepts and framework of cost-sensitive learning for ANN classifiers, and introduced the concept of the cost matrix. As a significant advancement, [4] introduced the first cost-sensitive algorithm for deep learning models. They modified the softmax layer of neural network outputs to propose a cost-sensitive loss function. In their context, the class represented by the lowest softmax value becomes the predicted class. Specifically, the method is as follows:

$$\delta_{n,k} = \ln\left(1 + \exp\left(z_{n,k} \cdot \left(r_k\left(\mathbf{x}_n\right) - \mathbf{c}_n[k]\right)\right)\right) \tag{1}$$

Equation 1 calculates $\delta_{n,k}$, a smooth approximation of the cost-sensitive loss for each instance n and class k. It utilizes the indicator function $z_{n,k}$ and the predicted probability $r_k(\mathbf{x}_n)$ to compute the loss based on the cost vector $\mathbf{c}_n$. $z_{n,k}$ is a binary indicator function, which takes the value 1 if the predicted class k for instance n matches the actual class and -1 otherwise.

Further, [3] introduced CSADA framework endowing over-parameterized models with cost sensitivity. In the medical domain, [16] proposed a cost-sensitive multi-task learning framework SMILE to address the problem of data heterogeneity, achieving great performance on CoNSeP and MoNuSAC 2020 datasets.

For multi-classifiers some classes can show greater safety criticality than others in SCS due to their high likelihood of harmful failures. Costs cannot adequately reflect the safety priority relationships between classes, making most existing cost sensitive learning methods insufficient for SCS.

3 Method

3.1 SCCs of Multi-Classifiers

Firstly, this section defines a relative safety-criticality metric for a multi-classifier to identify SCCs. Then, a learning framework is proposed for multi-classifiers potentially used in safety-critical scenarios, optimizing the models for those SCCs.

Referring to the description in [15], we provide a series of definitions in SCS:

Definition 1 (Harmful Failure, HF). *In safety-critical domains, harm failure denotes failures that caused by a neural network model that leads to outcomes which can cause physical harm to people, significant damage to property, or detriment to the environment. Such failures are characterized by their potential to significantly compromise the safety and reliability of the domain in question, whether or not a task is completed. Specifically, in classification tasks, we call such harmful failures as Harmful Classification Failures(HCF).*

Definition 2 (Classification Action Set). *Given an input space X, let $f : X \to L$ be a multi-classifier defined on it. Assume that an instance $x \in X$ has a ground truth label l_i, and f predicts x as l_j, i.e., $f(x) = l_j$, we call this prediction a classification action $ca(x, l_i, l_j)$. The set of classification actions performed by the multi-classifier f on the input space X is denoted as CA.*

Obviously, for an action $ca(x, l_i, l_j)$, if $i = j$, the action is a correct classification action performed by f on x; otherwise, if $i \neq j$, it's an incorrect one. We use $MISC(M, N)$ to denote the set of all classification actions $ca(x, M, N)$:

$$MISC(M, N) = \{ca(x, M, N) \mid \forall ca(x, M, N) \in CA, \ M, N \in L\} \tag{2}$$

Definition 3 (Expectations of HCFs). *Let f be a multi-classifier deployed in SCS, with the set of classification actions CA. We define $h_f(ca) = 1$ to indicate that the occurrence of classification action ca that results in a HCF within the application; otherwise ca results no HCF. Let $P_f^H(CA)$ represent the probability distribution of HCFs over the set of classification actions CA by classifier f. The HCF expectation resulting from instances with label M being classified as N is denoted as*

$$H_f^{M,N}(h) = \mathbb{E}[h(ca)], \quad where$$

$$ca \sim P_f^H(CA)$$

$$ca \in MISC(M, N).$$

Definition 4 (Expectations Matrix of HCFs). *Let $f : X \to L$ be a multi-classifier deployed in SCS, expectations matrix of HCFs (HM_f^X) is defined as:*

$$HM_f^X = \begin{bmatrix} HM_f^{L_1,L_1} & \cdots & HM_f^{L_1,L_n} \\ \vdots & \ddots & \vdots \\ HM_f^{L_n,L_1} & \cdots & HM_f^{L_n,L_n} \end{bmatrix}$$

where the element $HM_f^{L_i,L_j}$ within HM_f^X represents the expectation of HCFs when a multi-classifier f classifies instances with original label L_i as L_j.

Next, we give a definition of relative safety-criticality to identify the SCCs in classification tasks as bellow:

Definition 5 (Relative Safety-Criticality). *Given* $X, Y \in L$, *and* $H_f^{(Y,X)}(h) \neq 0$, *if the relative criticality* $rc = \dfrac{H_f^{(X,Y)}(h)}{H_f^{(Y,X)}(h)} > 1$, *then label* X *is considered relatively more safety-critical than* Y *in the SCD, denoted as* $Y \prec_{critical} X$.

According to the definition of relative safety-criticality, if $Y \prec_{critical} X$, then misclassifying samples with original label X as samples of Y results in a greater damage than the reverse case in the probabilistic sense. $\prec_{critical}$ is an ordinal relationship defined on L that reflects the safety criticality of different classes. In this paper we assume that $\prec_{critical}$ satisfies the transitivity of relation. The assumption means that $\prec_{critical}$ is a partial order relationship which can eliminate the existence of possible loops between the relationships.

Definition 6 (Safety-Critical Level of Labels). *Given* $f : X \to L$ *be a multi-classifier with labels* L, *and* SC *is a hierarchical partition* $SC = \{SC_1, SC_2, \ldots, SC_m\}$ *on* L, *where* $(\forall_{1 \leq i < j \leq m}) Y \in SC_i$ *and* $X \in SC_j$ *implies* $Y \prec_{critical} X$, *the safety-critical level of labels in* SC_i *is defined as* i.

Hierarchical classification of labels may not be suitable for all SCS. However, fault classification theory has already been extensively applied in SCS. In addition to the previously mentioned protection ring mechanisms in operating systems, fault tree analysis (FTA) and failure modes and effects analysis (FMEA) [6] have been applied in aerospace [26] and medical [10] domains. Based on this, we believe that it is also necessary to introduce hierarchical classification of labels in classification tasks for SCS.

We use a quantitative threshold to distinguish SCCs from other classes. For classification tasks in SCS, if the safety-critical protection level is set to a positive integer σ, the classes SC_i with $i \geq \sigma$ are SCCs, and the others are non-SCCs.

3.2 Optimizing Cross-Entropy Loss Towards SCCs

Cross-entropy loss functions are commonly used in supervised classification tasks to maximize overall accuracy across all classes. For a classification problem with N classes, the cross-entropy loss for an instance in class i is calculated as follows:

$$L(y, p) = - \sum_{i=0}^{N-1} y_i \log(p_i) \tag{3}$$

To better protect SCCs from harmful risks, we aim to improve their recognition accuracy during multi-classifier training. Our approach introduces a penalty term to focus optimization on SCCs, while non-SCCs follow the standard loss $L(y, p)$. The penalty term applies in two cases: 1) an SCC instance is misclassified as a non-SCC, and 2) a higher-level SCC is misclassified as a lower-level one.

Given a multi-classifier $f : X \to L$ with parameters θ, a hierarchical partition $SC = \{SC_1, SC_2, \ldots, SC_m\}$ on L, and a safety-critical protection level $1 < \sigma \leq m$, a penalty mask is generated as follows.

$$M_{\text{penalty}} = \begin{bmatrix} m_{11} & \cdots & m_{1n} \\ \vdots & \ddots & \vdots \\ m_{n1} & \cdots & m_{nn} \end{bmatrix}, \quad m_{ij} = \begin{cases} 1 & \text{if } i \in \text{SCC} \\ 0 & \text{otherwise} \end{cases}$$

Subsequently, the penalty expectation matrix of HCFs HM_f^X is obtained by:

$$HM_f^X = HM_{f\,\text{origin}}^X \circ M_{\text{penalty}}$$

With the penalty expectation matrix of HCFs, we construct the penalty goal for f as follows:

$$\text{argmin}_\theta \sum_{\substack{\sigma \le i \le m \\ 1 \le j < i \\ X \in SC_i \\ Y \in SC_j}} HM_f^{X,Y} P(h) \tag{4}$$

The penalty objective penalizes the model only when a high-level SCC is misclassified as a lower-level one or when an SCC is misclassified as a non-SCC. Other misclassifications are handled by the cross-entropy loss. Since SCCs are usually a minority in SCS, the penalty expectation matrix of HCFs is typically sparse. Given the penalty matrix HM_f^X, a weight vector w can be derived, where each element w_i represents the expected harm from misclassifying class i:

$$\hat{w}_i = \sum_{j=0, j \neq i}^{N-1} HM_f^{L_i, L_j} \tag{5}$$

The normalization of w can be achieved by dividing each weight by the sum of all weights of classes:

$$w_i = \frac{\hat{w}_i}{\sum_{k=0}^{N-1} \hat{w}_i} \tag{6}$$

Finally, we design the optimized loss function for a multi-classifier by combining the penalty term with cross-entropy loss:

$$L_{SCCPT}(y, p) = -\sum_{i=0}^{N-1} y_i \left[\log(p_i) - \alpha w_i \log(\delta) \right] \tag{7}$$

In the penalty term, w_i quantifies the severity of misclassifying classes in safety-critical contexts. The variable δ denotes the probability that the current sample being misclassified, and α is used as an adjustment coefficient to balance the cross entropy and penalty term in the loss function. The usage of these variables is elaborated as the fine-tuning mechanism below.

3.3 A Fine-Tuning Multiclass Classification Framework for SCCs

In this subsection, we first explain the fine-tuning mechanism provided by the loss function learning, and then propose an SCC-priority training method.

Fine-tuning Mechanism. According to the structure of L_{SCCPT}, the cross-entropy item optimizes the overall accuracy of the multi-classifier, and the penalty item makes the recall of the SCCs preferentially optimized. Thus, optimization directions induced by the two items are semantically different during the training process. In fact, the accuracy of the SCCs can also be fundamentally achieved by cross-entropy. Therefore, the use of L_{SCCPT} can be seen as a fine-tuning process that optimize the model towards the improvement direction of the SCCs' recall.

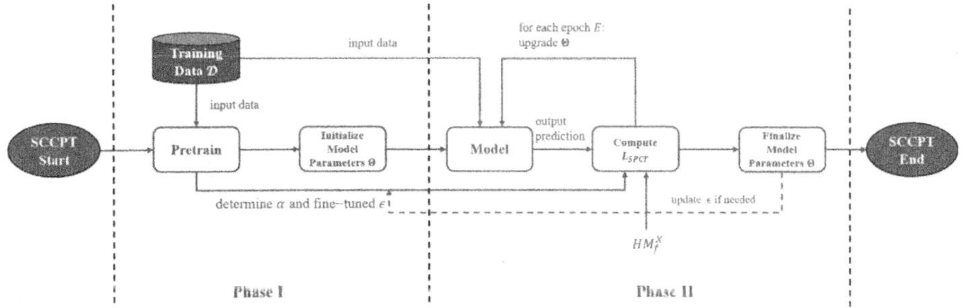

Fig. 1. A Two-phase Optimization Framework for SCCs

In most cases setting $\alpha = 1$ is sufficient. When the overall accuracy of the model trained solely by cross-entropy exceeds 90%, α needs to be increased to ensure the penalty item works. For instance, in the subsequent experiments with the MNIST dataset, when the original model's overall recognition rate surpassed 99%, we set $\alpha = 10$ to generate meaningful training results.

In order to avoid over-optimizing the SCCs, we set a threshold ϵ for δ in the penalty item to work. ϵ filters out the impact of extremely low probability of misclassifications.

$$\delta = \begin{cases} \epsilon & \text{if } 1 - p_i < \epsilon, \\ 1 - p_i & \text{otherwise.} \end{cases} \tag{8}$$

The value of α can be estimated based on the overall accuracy of the model pre trained with cross-entropy. And while for the value of ϵ, it can be set according to the accuracy's magnitude of the SCCs.

Algorithmic Formulation. We propose a fine-tuning multiclass classification framework for SCS with a two-phase training process. The first phase pretrains the model using cross-entropy, while the second phase fine-tunes it for SCCs. The output from the first phase guides the optimization and sets parameters α and ϵ. Algorithm 1 shows the SCCs priority training process for the second phase.

Algorithm 1. SCCs Priority Training Process, SCCPT

Input: Training data $\mathcal{D}$, Expectations matrix of HCFs HM_f^X, Learning rate η, Epochs E, Expected loss threshold ϵ and Safety-critical protection level σ

Output: Trained model parameters Θ

1: Perform pre-training process
2: Initialize model parameters Θ from pre-trained model
3: Compute relative safety-criticality between labels using HM_f^X
4: Filter out the SCCs using relative safety-criticality and σ
5: Compute and normalize SCCs weights w using Equation (5) and Equation (6)
6: Determine α and ϵ with the accuracy and recall of the pre-trained model
7: **for** $epoch = 1$ to E **do**
8:     **for all** instances (x, y) in $\mathcal{D}$ **do**
9:         Compute prediction $p = \text{Model}(x; \Theta)$
10:         Compute δ using Equation (8)
11:         Compute loss L_{SCCPT} using Equation (7)
12:     **end for**
13:     Compute current average loss over $\mathcal{D}$: $L_{SCCPT_{sum}} = \frac{1}{|\mathcal{D}|} \sum_{(x,y) \in \mathcal{D}} L_{SCCPT}$
14:     Update Θ using gradient descent: $\Theta = \Theta - \eta \nabla_\Theta L_{SCCPT_{sum}}$
15: **end for**
16: **return** Θ

The SCCPT algorithm iteratively refines model parameters over multiple epochs. It incorporates a weight vector and the rectification term δ to address high misclassification costs in SCS. If the model's performance is unsatisfactory, SCCPT updates δ and initiates another training round for further fine-tuning.

4   Experiments

This section use experiments to validate the SCCPT method and compare it with the classic cost-sensitive learning method proposed by [4]. All experiments were conducted using Python 3.9.5 and PyTorch 1.10.2+cu113. The main goal is to test if SCCPT improves SCC recall across different datasets and to evaluate its overall accuracy and convergence.

4.1   Setup

Datasets. As far as we know, there are still no open, real-world datasets that incorporate safety-critical damage due to misclassification. Therefore, we used the widely recognized MNIST and CIFAR-10 datasets for our experiments. These benchmarks offer diverse challenges to evaluate our approach. We simulated the expectations of HCF by assigning each class a random integer within $[0,10]$ as its expectation sum w_i. Due to the normalization process of w_i in Eq. 6, the simulation does not affect the plausibility of the experiments.

Models. Kaggle, as a popular data science competition platform, hosts numerous simple yet effective ANN models for solving the classification task on the

MNIST dataset. For our experiments, we selected one of the top six CNN models from Kaggle, based on their high accuracy in MNIST data recognition [5].

Methods. Both MNIST and CIFAR-10 datasets comprise samples of 10 distinct labels. In our experiment, we use a hierarchical partition of these labels for tests, $SC = \{SC_1, SC_2, SC_3\}$, with ascending levels of safety criticality $SC_1 \prec_{critical} SC_2 \prec_{critical} SC_3$, as shown in Table 1. Specifically, there is one label in SC_3 randomly picked from the labels of datasets, two other labels are randomly put in SC_2, and SC_1 collects the remaining 7 labels considered less critical.

Table 1. Safety-Critical Levels in Experiments

Partitions	SC_1	SC_2	SC_3
Numbers of Labels	7	2	1

Quite a few experiments have been carried out in our work. However, due to space limitations, we only demonstrate two groups of comparative experiments here: I and II.

In the first group of experiments (Experiment I), the safety-critical protection level σ is set to 3. Thus, SC_3 is treated as the only set of SCCs, the expectation sum of HCFs of whose element is randomly set as 5. And SC_1 and SC_2 are sets of non-SCCs with 0 expectation sum of HCFs as of their elements. Actually each class in datasets is iteratively chosen as the element of SC_3 to test our methods.

In the second group of experiments (Experiment II), the safety-critical protection level σ is set to 2, thereby considering $SC_2 \bigcup SC_3$ as the set of SCCs, and SC_1 as the set of non-SCCs. The expectation sums of HCFs are randomization as 5 for labels in SC_3, and 3 for labels in SC_2.

Additionally, for the model that inherently has high classification accuracy on the dataset MNIST, we set the adjustment coefficient α as 5 for the penalty term; for the model trained on the dataset CIFAR-10, α is set to 1.

We use the cost-sensitive loss function L_{SOSR} proposed by [4] as a baseline for comparison, whose computation is introduced in Sect. 2.2. To make the results more comparable, the one-dimensional **weight** vector previously set is used for computing the cost vector $\mathbf{c}_n$. The calculation of $\mathbf{c}_n$ is as follows: For a given sample n belonging to the actual class k, the cost vector is determined by[1]:

$$\mathbf{c}_n[i] = \begin{cases} 0 & \text{if } i = k, \\ \max\left(\frac{w_k+1}{w_i+1}, 1\right) & \text{if } i \neq k. \end{cases} \tag{9}$$

---

[1] Adding 1 to w_i before computation is to avoid the error of division by zero. By this approach, the cost vector reflects the relative importance of misclassification between different classes, and prevent excessively small misclassification costs.

4.2 Results

Experiment I. In experiment I, three training methods are used respectively for comparison: ORIGIN, SOSR, SCCPT. The ORIGIN method employs the original cross-entropy loss function in [5]. The SOSR method uses cost-sensitive loss function L_{SOSR} mentioned in Eq. 1. And SCCPT is our process previously proposed in Algorithm 1.

Recall of SCCs. Table 2 displays the recall of class i as SCC generated in the experiment. The formal training on the MNIST dataset was conducted for 100 epochs and for 500 epochs on the CIFAR dataset, so as subsequent experiments. The column corresponding to "class-i" indicates that the ith class set as SCC. On the MNIST dataset, while the ORIGIN method has demonstrated a high recall rate, our SCCPT method is still able to improve the recall rate of the SCC class noticeably. In contrast, the optimization effect of SOSR method is relatively small, which is significantly weaker than that of SCCPT. On the CIFAR dataset, SCCPT significantly enhances the recall rate for SCC classes: the recall rate for most SCCs has increased by more than 10%. However, SOSR, as a cost-sensitive learning method, does not demonstrate a marked advantage on this dataset.

Table 2. Recall of Single SCC Trained in Experiment I

Method	MNIST-SCCPT	MNIST-SOSR	MNIST-ORIGIN	CIFAR-SCCPT	CIFAR-SOSR	CIFAR-ORIGIN
CLASS-0	0.999	0.997	0.997	0.813	0.758	0.758
CLASS-1	1.000	0.995	0.997	0.920	0.810	0.815
CLASS-2	1.000	0.993	0.996	0.714	0.587	0.579
CLASS-3	1.000	0.992	0.992	0.765	0.522	0.582
CLASS-4	1.000	0.986	0.994	0.931	0.789	0.793
CLASS-5	0.997	0.990	0.990	0.834	0.649	0.666
CLASS-6	0.996	0.990	0.990	0.959	0.906	0.894
CLASS-7	0.998	0.985	0.985	0.867	0.811	0.803
CLASS-8	0.999	0.991	0.992	0.957	0.912	0.916
CLASS-9	0.999	0.984	0.991	0.950	0.887	0.896

Learning Rate: 2e−06, α in MNIST-SCCPT: 5, α in CIFAR-SCCPT: 1

Precision and F1-Score of SCCs. We also calculated the precision and F1-score for each SCC in Experiment I, as shown in Tables 3 and 4. On both datasets, the precision for SCCs shows a varying degree of decline compared to the original method. As for the F1-score, SCCs show a slight decline on the MNIST dataset, but on the CIFAR dataset, the performance is mixed when compared to the original method. Overall, this is expected because the model tends to bias the prediction results towards SCC.

Table 3. Pricision of Single SCC Trained in Experiment I

Method	MNIST-SCCPT	MNIST-SOSR	MNIST-ORIGIN	CIFAR-SCCPT	CIFAR-SOSR	CIFAR-ORIGIN
CLASS-0	0.960	0.986	0.984	0.812	0.842	0.849
CLASS-1	0.981	0.996	0.996	0.853	0.918	0.903
CLASS-2	0.955	0.989	0.987	0.712	0.798	0.751
CLASS-3	0.966	0.991	0.993	0.565	0.630	0.610
CLASS-4	0.955	0.995	0.998	0.492	0.679	0.667
CLASS-5	0.963	0.990	0.990	0.581	0.678	0.687
CLASS-6	0.968	0.994	0.998	0.593	0.694	0.732
CLASS-7	0.962	0.989	0.992	0.816	0.842	0.841
CLASS-8	0.968	0.987	0.990	0.657	0.804	0.823
CLASS-9	0.940	0.990	0.985	0.683	0.789	0.797

Learning Rate: 2e-06, α in MNIST-SCCPT: 5, α in CIFAR-SCCPT: 1

Overall Accuracy. The overall accuracy of experiments is also illustrated in Table 5. The table shows that both the cost-sensitive approach and ours will result in a slight decrease in overall accuracy of models. We believe that this can be explained as the cost of the introduction of the penalty item. Optimization of the model in the direction of SCCs may lower the accuracy of non-SCCs and thus weakens the overall accuracy. However, judging from the return on SCC's improved recall rate, a certain amount of decline in overall accuracy is completely acceptable.

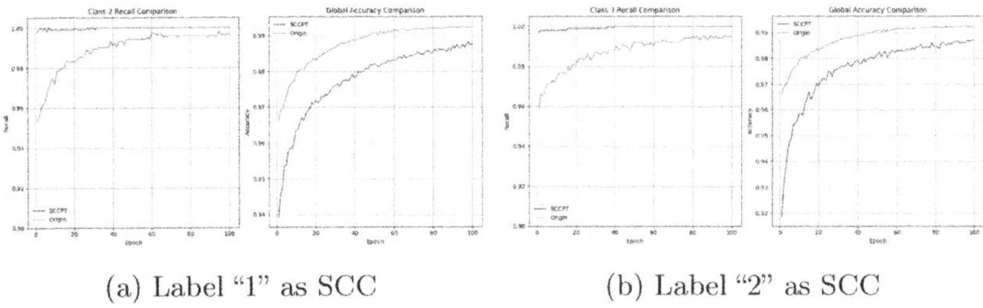

(a) Label "1" as SCC   (b) Label "2" as SCC

Fig. 2. Convergence Comparison of SCCPT and ORIGIN on MINST

Convergence. We also present the variation in recall for each classification when considered as SCCs across training epochs, and we choose two sets of experiments illustrated in Figs. 2 and 3. Overall, our proposed SCCPT method demonstrates quicker convergence in recall for SCCs, maintaining higher levels than those achieved with the original cross-entropy method.

Table 4. F1-score of Single SCC Trained in Experiment I

Method	MNIST-SCCPT	MNIST-SOSR	MNIST-ORIGIN	CIFAR-SCCPT	CIFAR-SOSR	CIFAR-ORIGIN
CLASS-0	0.979	0.991	0.991	0.813	0.798	0.793
CLASS-1	0.990	0.995	0.996	0.885	0.861	0.853
CLASS-2	0.977	0.991	0.991	0.713	0.676	0.662
CLASS-3	0.982	0.992	0.992	0.650	0.571	0.586
CLASS-4	0.977	0.990	0.990	0.644	0.730	0.726
CLASS-5	0.980	0.990	0.992	0.685	0.663	0.659
CLASS-6	0.981	0.992	0.991	0.733	0.786	0.808
CLASS-7	0.979	0.987	0.988	0.841	0.826	0.825
CLASS-8	0.983	0.989	0.991	0.779	0.855	0.862
CLASS-9	0.969	0.987	0.986	0.795	0.831	0.838

Learning Rate: 2e-06, α in MNIST-SCCPT: 5, α in CIFAR-SCCPT: 1

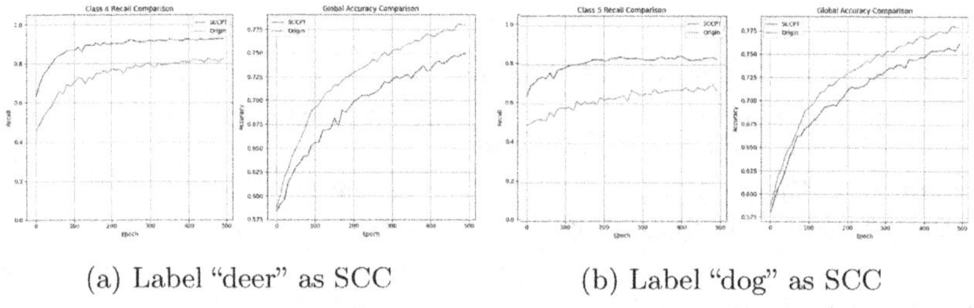

(a) Label "deer" as SCC   (b) Label "dog" as SCC

Fig. 3. Convergence Comparison of SCCPT and ORIGIN on CIFAR-10

In our experiments on the MNIST dataset, it was observed that the SCC classification recalls of the SCCPT method starts at a higher value than original method, and sometimes jitters before slowly rises and converges. We hypothesize that this phenomenon is attributable to the penalty term in the SCCPT method, which initially biases the model towards classifying other categories as SCC. However, as the number of epochs increases, the model corrects the recognition of non-SCC categories, leading to a temporary jitter in the recall rate of the SCC. This is then followed by a gradual stabilization due to the overall improvement in the model's training performance. The steadily rising curve of the model's overall performance over epochs on the right side corroborates this observation.

Experiment II. In this experiment, the safety-critical protection level is set as 2. This means that labels in both SC_3 and SC_2 are optimized in the SCCPT process. Table 6 presents the statistical comparison of overall average accuracy, average recall for SC_3 and SC_2 against the original method. While there is a slight decrease in overall accuracy, the recall for SC_3 class shows a notable

Table 5. Overall Accuracy of Models Trained in Experiment I

Method	MNIST-SCCPT	MNIST-SOSR	CIFAR-SCCPT	CIFAR-SOSR
ORIGIN	0.991	0.991	0.764	0.764
CLASS-0	0.987	0.991	0.759	0.760
CLASS-1	0.989	0.991	0.760	0.763
CLASS-2	0.986	0.992	0.767	0.764
CLASS-3	0.986	0.990	0.768	0.768
CLASS-4	0.986	0.990	0.727	0.762
CLASS-5	0.987	0.990	0.756	0.765
CLASS-6	0.988	0.992	0.743	0.761
CLASS-7	0.987	0.990	0.758	0.769
CLASS-8	0.988	0.991	0.737	0.761
CLASS-9	0.986	0.991	0.725	0.761

Learning Rate: 2e-06, α in MNIST-SCCPT: 5, α in CIFAR-SCCPT: 1

Table 6. Comparison of SCCPT with Origin on MNIST and CIFAR-10 in Experiment II

	MNIST			CIFAR-10		
	Overall Accuracy	SC_3 Recall	$SC2$ Recall	Overall Accuracy	SC_3 Recall	$SC2$ Recall
SCCPT	0.982	0.998	0.996	0.762	0.811	0.739
ORIGIN	0.993	0.992	0.993	0.774	0.745	0.726
Change (%)	−1.12%	+0.59%	+0.35%	−1.54%	+8.95%	+1.73%

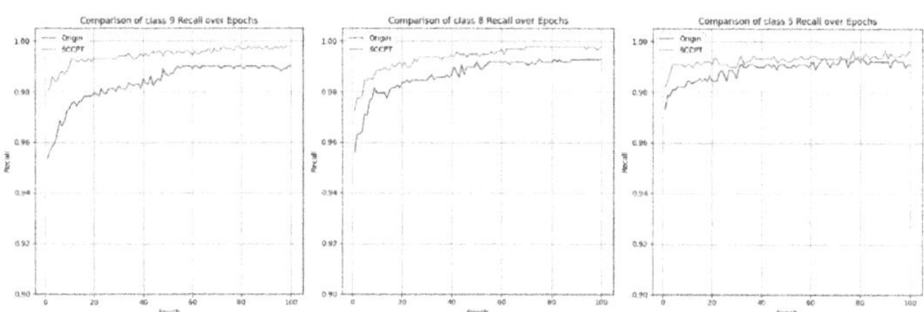

Fig. 4. Label "9" as SC_3, label "8" and "5" as SC_2 in MNIST

increase, and there is a modest improvement in recall for SC_2 classes, aligning with our expectations.

Figures 4 and 5 show the convergence behavior of recall rates for both SC_3 and SC_2 classes in comparison to the original method on both MNIST and CIFAR-10. The SC_3 class demonstrates significant improvements in convergence speed and final recall rate. However, the improvement in SC_2 classifications is less pronounced. Overall, the results meet expectations.

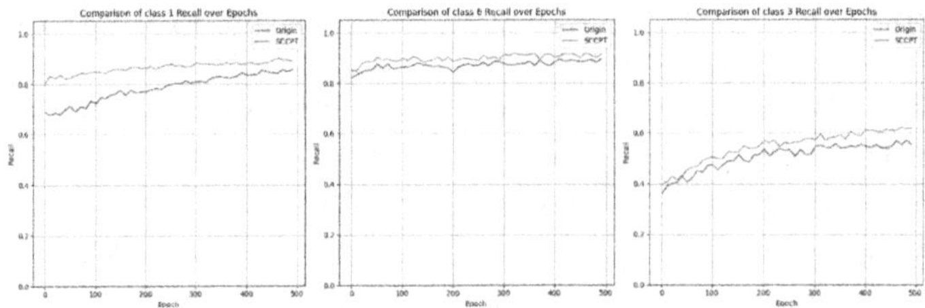

Fig. 5. Label "airplane" as SC_3, label "dog" and "bird" as SC_2 in CIFAR-10

5   Discussion

In this paper, we treat the learning process in SCS as a constrained optimization problem: the safety risks caused by the insufficient accuracy of the model are introduced and penalized, along with the cross-entropy loss function in the training process. This approach is similar to the objective function used in cost-sensitive learning methods. However, the differences between our method and existing cost-sensitive learning approaches are as follows:

- In SCS, objects with higher safety-critical levels need to be protected with higher priority. The safety risks caused by the misclassification of these objects should be avoided as much as possible. Therefore, our method focuses the penalty term only on risks of SCCs while ignoring other trivial ones. Existing cost-sensitive learning methods treat all costs equally in their training processes, which makes them less targeted for SCS. Experiments show that our method can achieve a better improvement on the recognition accuracy of SCCs than that of cost-sensitive ones.
- In some cases, there is limited room for model optimization. For example, multi-classifiers on MNIST datasets can achieve over 97% accuracy, making it hard for cost-sensitive methods to improve certain critical classes. We propose a fine-tuning optimization mechanism by introducing an adjustment coefficient α into the penalty term. If the model to be optimized already has a high recognition rate, then higher orders of magnitude of α can be set such as tenfold, hundredfold, etc. Thus, the impact of the penalty effects is amplified so as to achieve the best possible optimization. For the above limited optimization cases, comparative experiments have verified the advantages of our approach over mainstream cost-sensitive learning methods.

Our experiments indicate that there are potential trade-offs between improving SCC recall and maintaining overall accuracy. To address this dilemma, our current approach involves adjusting the parameter α in Eq. (7). Setting $\alpha = 0$ corresponds to reverting to the standard training method entirely; larger values of α imply a stronger inclination towards enhancing SCC recall. Researchers need to adjust α based on specific contexts and requirements.

Our experiments rely on simulated safety-critical damages. One issue is that the core of our method is based on modifications to the cross-entropy loss function. Therefore, any multi-class classification model that is trained based on the cross-entropy loss function can apply our method, although certain parameter metrics may vary. Conversely, training frameworks that do not utilize cross-entropy cannot apply our method. Another issue that needs to be explored is the training datasets for SCS. To the best of our knowledge, there is still no open, real-world training datasets that incorporate safety-critical damages. Therefore, researchers have to conduct simulation tests to verify their methods. We use classic classification datasets, MNIST or CIFAR-10, to demonstrate our approach. It should be pointed out that collecting real data can be time-consuming and costly, impacting the application of safety-critical algorithms, including ours. In fact, different levels of protected objects are conventionally set according to the opinions of domain experts in many ring protection mechanisms. For initial applications, a possible alternative is to turn to domain experts to set the safety criticality of the classes based on their experiences. Then once a real domain dataset is formed, models can be calibrated or even retrained with the real data to fit more realistic application environments.

Interpretability of models is one of the hot topics of concern in neural network research, which denotes their ability to provide explanations in understandable terms to a human [25]. As illustrated in [9], most ML models behave as black-box models and generate unexplainable, unjustifiable, and unaccountable outputs. Our research also follows the principle of interpretability. The motivation of the proposed method, which aims to minimize the risk of harmful decisions possibly made by neural networks in SCS, is intuitively understandable for human. The Protection Rings mechanism underlying hierarchical safety-critical levels has been widely used and justified in security-related research. The semantic of the penalty item used in SCCPT is also accountable in that it penalizes the model only when a high-level SCC is misclassified as a lower-level one or when an SCC is misclassified as a non-SCC.

6   Conclusion

Misclassifying objects of safety-critical classes as non-SCCs can pose harmful risks to multi-classifier systems operating in SCS. In this paper, we propose a fine-tuned multi-classifier optimization framework towards SCCs according to their hierarchical safety-critical levels. Experiments demonstrate that our approach significantly enhances the recall rates of SCCs for the original models while showing negligible impact on their overall performance. Higher recall rates of SCCs generated by our optimization framework help to mitigate the harmful risk posed possibly by multi-classifiers in SCS.

We believe our work can be potentially used as an optimization scheme to migrate existing multi-classifiers into SCS. If at the initial stage there is not enough posterior harmful risk data to be utilized, an alternative is to resort to the prior experience of domain experts to identify the SCCs and their safety-critical levels. Next, as posterior risk data accumulates, the model can be continuously optimized by our method until it meets the requirements of safety-critical applications.

References

1. Alves, E.E., Bhatt, D., Hall, B., Driscoll, K., Murugesan, A., Rushby, J.: Considerations in assuring safety of increasingly autonomous systems. Contractor Report (CR) NASA/CR-2018-220080, NF1676L-30426, NASA (2018), document ID: 20180006312
2. Azad, S., Mahmud, M., Zamli, K.Z., Kaiser, M.S., Jahan, S., Razzaque, M.A.: iBUST: an intelligent behavioural trust model for securing industrial cyber-physical systems. Expert Syst. Appl. **238**, 121676 (2024)
3. Chen, Q., Kontar, R.A., Nouiehed, M., Yang, J., Lester, C.: Rethinking cost-sensitive classification in deep learning via adversarial data augmentation. arXiv preprint arXiv:2208.11739 (2022)
4. Chung, Y.A., Lin, H.T., Yang, S.W.: Cost-aware pre-training for multiclass cost-sensitive deep learning. arXiv preprint arXiv:1511.09337 (2015)
5. Deotte, C.: 25 million images! 0.99757 mnist. https://www.kaggle.com/code/cdeotte/25-million-images-0-99757-mnist (2019)
6. Duphily, R.J., Command, A.F.S.: Space vehicle failure modes, effects, and criticality analysis (FMECA) guide. Space Missile Syst. Center, El Segundo, CA, USA, Aerosp. Rep. No. TOR-2009 (8591)-13 (2009)
7. Elkan, C.: The foundations of cost-sensitive learning. In: International Joint Conference on Artificial Intelligence, vol. 17, pp. 973–978. Lawrence Erlbaum Associates Ltd (2001)
8. Géron, A.: Hands-on machine learning with Scikit-Learn, Keras, and TensorFlow. " O'Reilly Media, Inc." (2022)
9. Hassija, V., et al.: Interpreting black-box models: a review on explainable artificial intelligence. Cogn. Comput. **16**(1), 45–74 (2024)
10. Hegde, V.: Case study—risk management for medical devices (based on ISO 14971). In: 2011 Proceedings-Annual Reliability and Maintainability Symposium, pp. 1–6. IEEE (2011)
11. Knight, J.C.: Safety critical systems: challenges and directions. In: Proceedings of the 24th International Conference on Software Engineering, pp. 547–550 (2002)
12. Koch, P., Wujek, B., Golovidov, O., Gardner, S.: Automated hyperparameter tuning for effective machine learning. In: Proceedings of the SAS Global Forum 2017 Conference, pp. 1–23. SAS Institute Inc. Cary, NC (2017)
13. Kohli, P., Chadha, A.: Enabling pedestrian safety using computer vision techniques: a case study of the 2018 Uber Inc. self-driving car crash. In: Arai, K., Bhatia, R. (eds.) Advances in Information and Communication: Proceedings of the 2019 Future of Information and Communication Conference (FICC), Volume 1, pp. 261–279. Springer International Publishing, Cham (2020). https://doi.org/10.1007/978-3-030-12388-8_19
14. Kukar, M., Kononenko, I., et al.: Cost-sensitive learning with neural networks. In: ECAI, vol. 15, pp. 88–94. Citeseer (1998)
15. O'Brien, M., Goble, W., Hager, G., Bukowski, J.: Dependable neural networks for safety critical tasks. In: Shehory, O., Farchi, E., Barash, G. (eds.) Engineering Dependable and Secure Machine Learning Systems: Third International Workshop, EDSMLS 2020, New York City, NY, USA, February 7, 2020, Revised Selected Papers, pp. 126–140. Springer International Publishing, Cham (2020). https://doi.org/10.1007/978-3-030-62144-5_10
16. Pan, X., et al.: Smile: cost-sensitive multi-task learning for nuclear segmentation and classification with imbalanced annotations. Med. Image Anal. **88**, 102867 (2023)

17. Paterson, C., Calinescu, R., Ashmore, R.: Assuring the machine learning lifecycle: desiderata, methods, and challenges. ACM Comput. Surv. (2021)
18. Perez-Cerrolaza, J., et al.: Artificial intelligence for safety-critical systems in industrial and transportation domains: a survey. ACM Comput. Surv. **56**(7), 1–40 (2024)
19. Righetti, L., Madhavan, R., Chatila, R.: Unintended consequences of biased robotic and artificial intelligence systems [ethical, legal, and societal issues]. IEEE Robot. Autom. Mag. **26**(3), 11–13 (2019)
20. Rudin, C.: Stop explaining black box machine learning models for high stakes decisions and use interpretable models instead. Nature Mach. Intell. **1**(5), 206–215 (2019)
21. Saltzer, J.H., Schroeder, M.D.: The protection of information in computer systems. Proc. IEEE **63**(9), 1278–1308 (1975)
22. Wagstaff, K.: Machine learning that matters. arXiv preprint arXiv:1206.4656 (2012)
23. Wang, Y., Chung, S.H.: Artificial intelligence in safety-critical systems: a systematic review. Ind. Manage. Data Syst. **122**(2), 442–470 (2022)
24. Xu, Z., Saleh, J.H.: Machine learning for reliability engineering and safety applications: review of current status and future opportunities. Reliab. Eng. Syst. Saf. **211**, 107530 (2021)
25. Zhang, Y., Tiňo, P., Leonardis, A., Tang, K.: A survey on neural network interpretability. IEEE Trans. Emerging Top. Comput. Intell. **5**(5), 726–742 (2021)
26. Zolghadri, A., Henry, D., Cieslak, J., Efimov, D., Goupil, P.: Fault Diagnosis and Fult-Tolerant Control and Guidance for Aerospace Vehicles, vol. 236. Springer (2014). https://doi.org/10.1007/978-1-4471-5313-9

Behavior-Driven Data Augmentation
for Non-Intrusive Load Monitoring

Wassim Nijaoui and Manar Amayri

Concordia University, Montreal, QC, Canada
`manar.amayri@concordia.ca`

Abstract. Non-Intrusive Load Monitoring (NILM), or energy disaggregation, involves predicting the use of individual appliances based solely on the total energy consumption of a household. This device-level data, obtained without installing multiple meters, allows users to monitor their consumption, detect faulty components and make better decisions based on the new insights. Deep learning methods have shown their superiority compared to other methods in recent literature. However, they often lack the ability to capture consumer behavior to increase their robustness. One way to achieve this is to use a generic data augmentation technique based on predefined thresholds. In this article, we propose a statistical approach capable of dynamically generating synthetic data and improving the performance of a given model. The results obtained on the datasets used show that this method achieves higher performance by generating synthetic data outperforming current state-of-the-art data augmentation techniques.

Keywords: NILM · Data Augmentation · Energy Disaggregation

1   Introduction

Non-Intrusive Load Monitoring (NILM) is the task of predicting the energy consumption of individual appliances based on the total energy consumption of a household. This approach, first introduced by Hart in 1992 [1], simplifies the collection of energy consumption data by avoiding the need for sensors on each appliance. Instead, NILM analyzes the aggregate current and voltage signals to identify the operational states of individual devices in a less intrusive manner. This method has numerous applications, including energy auditing, fault detection, and enhancing consumer energy awareness, making it a vital tool for both consumers and utility providers [2,9].

Over the years, researchers have focused on improving NILM methods to address challenges such as accurately predicting the energy consumption of specific appliances and detecting anomalies within their usage patterns [2,10]. Despite significant advancements, NILM remains an active area of research due to the dynamic nature of household appliances, which can become faulty or be

M. Mahmud et al. (Eds.): ICONIP 2024, CCIS 2297, pp. 182–198, 2026.
https://doi.org/10.1007/978-981-96-7036-9_13

replaced, thus altering their consumption patterns. This variability can quickly render pre-existing models obsolete if they assume constant usage patterns [3,11].

Deep learning techniques have demonstrated superior performance over traditional methods in recent NILM studies [5]. However, these models require extensive training data to achieve high accuracy, especially for complex problems. A common approach to mitigate the scarcity of training data is through data augmentation. Traditional data augmentation methods, which involve adding or subtracting small random biases, are widely used but offer limited flexibility and improvement [3,8]. This limitation is particularly evident in tasks like anomaly detection and energy disaggregation, where the performance of the models can be significantly affected by seasonal variations and consumer behavior [12].

One of the primary challenges in NILM is the imbalance in the availability of data across different appliances [11]. Certain appliances are used seasonally, and training models with off-season data can lead to poor predictions. Additionally, the variability in consumer behavior from one household to another further complicates the training process. Current NILM models do not effectively incorporate such dynamic information within their training loops [1,11].

Traditional methods for anomaly detection in NILM often rely on statistical techniques such as thresholding, where an appliance is considered anomalous if its power consumption exceeds a predefined threshold [11]. More advanced methods like the Three-Points method use the gradient of the last three points of the signal to identify anomalies. However, these methods can fail if an appliance's consumption pattern changes due to aging or replacement, leading to frequent misclassifications and necessitating model retraining [11].

To address these challenges, we propose a robust and dynamic approach that enhances both anomaly detection and energy disaggregation in NILM. Our method involves dynamically computing the distribution of each appliance's consumption and using this information for data augmentation. This approach not only generates more realistic synthetic data but also adapts to changes in appliance behavior, thereby improving model robustness and accuracy.

In summary, this work contributes the following:

1. A dynamic method for performing data augmentation and generating synthetic data.
2. Improved accuracy for appliance level on-off state classification.

Our approach leverages statistical techniques to dynamically generate synthetic data and enhance model performance. The results achieved on both datasets used for appliance state classification task and energy disaggregation task demonstrate that our method outperforms existing data augmentation techniques and provides higher accuracy in NILM tasks.

The rest of this paper is organized as follows: Sect. 2 reviews the related work, including deep learning models, post-processing techniques, sliding window approach, and data augmentation techniques used in NILM. Section 3 presents our proposed method, detailing the model architecture and dynamic data augmentation technique. Section 4 describes the experiments and results, including datasets used, methodology, and performance comparison. Section 5 discusses the limitations of our approach. Finally, Sect. 6 concludes the paper and outlines directions for future research.

2   Related Work

Non-intrusive load Monitoring (NILM) has evolved significantly since its inception, with various methods proposed to improve the accuracy and efficiency of disaggregating total energy consumption into individual appliance-level data. Recent advancements have focused heavily on the application of deep learning models and the development of data augmentation techniques to enhance model performance.

2.1   Deep Learning Models in NILM

Deep learning has shown considerable promise in NILM, offering robust solutions for pattern recognition and time series analysis. Various architectures have been explored, including Convolutional Neural Networks (CNNs) [23], Recurrent Neural Networks (RNNs) [22], and Long Short-Term Memory (LSTM) networks [21, 22].

Nguyen and Arboleya [22] applied RNNs on the REDD dataset, demonstrating that RNNs can outperform traditional optimization approaches by accurately recognizing unseen appliances. Kelly and Knottenbelt [5] utilized deep neural networks for energy disaggregation, achieving high accuracy in predicting the usage patterns of household appliances.

Buddhahai et al. [6] introduced a multitarget classification framework using Random k-labelsets (RAkEL) with decision trees. This approach improved the identification of appliance power states by employing a multistate model, yielding better predictive performance for high-power appliances such as air conditioners and water heaters.

More recent work by Massidda et al. [13] demonstrated the effectiveness of convolutional and recurrent neural networks for NILM, showcasing significant improvements in disaggregation accuracy over traditional methods. Similarly, Zhou et al. [14] explored LSTM networks combined with attention mechanisms to capture long-term dependencies in energy consumption patterns, further enhancing model performance.

2.2   Post-processing Techniques and Sliding Window Approach

Post-processing techniques and the sliding window approach are essential for enhancing the accuracy and efficiency of NILM systems. These methods help in refining the raw disaggregation outputs and improving the overall model performance.

Post-processing Techniques. Post-processing techniques are employed to refine the initial predictions made by NILM models, ensuring higher accuracy and reliability. These techniques include smoothing, filtering, and correcting the disaggregated signals based on known appliance characteristics and consumption patterns.

Basu et al. [8] introduced a classification framework for power events in NILM that leverages both steady-state and transient-state features. Their approach effectively detects and classifies appliance events, providing a more accurate disaggregation output. The use of high-frequency data in their method enhances the precision of event detection, making post-processing a critical step in the NILM pipeline.

Similarly, Linh and Arboleya [19] discussed various post-processing strategies, including the application of filtering techniques to remove noise from the disaggregated signals. These methods improve the clarity and accuracy of the output, ensuring that the predicted appliance usage closely matches the actual consumption patterns.

Sliding Window Approach. The sliding window approach is a widely used technique in NILM to handle long sequences of energy consumption data. This method involves segmenting the input data into overlapping windows, allowing the model to focus on smaller, more manageable portions of the signal.

Zhang et al. [20] introduced sequence-to-point (seq2point) learning with a sliding window approach, demonstrating significant improvements in prediction accuracy. By predicting only the midpoint of each window, the seq2point method allows the neural network to concentrate its representational power on a specific segment, reducing the complexity of the learning task.

Kolter and Jaakkola [21] applied sequence-to-sequence (seq2seq) learning using LSTM networks with a sliding window approach. Their experiments showed that this method effectively captures long-term dependencies in energy consumption data, leading to better disaggregation results. The sliding window approach also helps mitigate the computational challenges associated with processing long sequences.

Passing a sequence through the model, as opposed to individual data points, allows for the capture of temporal dependencies and patterns in the data. This approach is particularly beneficial in NILM, where the energy usage of appliances often follows specific temporal patterns. By considering the context provided by the sequence, the model can make more informed predictions.

Linh and Arboleya [15] emphasized the importance of combining data filtering with power state clustering techniques in a sliding window framework. This combination allows for more accurate identification of appliance states and reduces the impact of noise on the disaggregation process.

Benefits of Passing a Sequence. Passing a sequence rather than isolated data points enables NILM models to leverage the temporal structure of energy consumption data. This approach improves the model's ability to recognize patterns and make accurate predictions about appliance usage.

Nguyen and Arboleya [22] highlighted that RNNs and LSTM networks are particularly well-suited for sequence-based learning tasks in NILM. These models can effectively handle the temporal dependencies present in energy consumption data, resulting in more accurate disaggregation.

Moreover, the seq2seq and seq2point methods discussed by Zhang et al. [20] and Kolter and Jaakkola [21] demonstrate the practical advantages of passing sequences through NILM models. These methods have been shown to outperform traditional approaches by capturing the temporal dynamics of appliance usage, leading to more reliable and precise disaggregation results.

2.3 Data Augmentation Techniques in NILM

Data augmentation plays a crucial role in enhancing the robustness and generalization capability of NILM models, especially when dealing with limited datasets. Traditional methods often involve adding random noise or biases to the aggregate energy signal to create synthetic data points.

One common approach involves adding a random value between 0 and the mean of the aggregate signal or a fixed value based on domain knowledge, helping capture variability in appliance usage and improving model training. Nour et al. [3,29] proposed advanced strategies for high-frequency NILM datasets, including threshold-based augmentation and clustering techniques to generate realistic synthetic data. These methods significantly enhanced the performance of NILM models in detecting on/off states and disaggregating energy consumption.

Linh and Arboleya [15] focused on RNN applications in NILM, employing data filtering and power state clustering techniques to preprocess raw datasets. This combination of deep learning and effective preprocessing demonstrated superior performance over traditional methods.

Additionally, Pereira and Nunes [16] compared various data augmentation strategies for NILM, emphasizing the importance of realistic synthetic data generation to improve model robustness. Their findings suggest that augmenting data using appliance-specific characteristics can lead to better generalization and performance in unseen scenarios.

3 Proposed Method

This section elaborates on the model architecture and the underlying methodologies employed for tasks such as appliance state classification and energy disaggregation. Our approach utilizes a combination of Recurrent Neural Network (RNN) models [18] and lasso regression [26], both of which are prevalent in the field of energy disaggregation for their robustness and predictive accuracy.

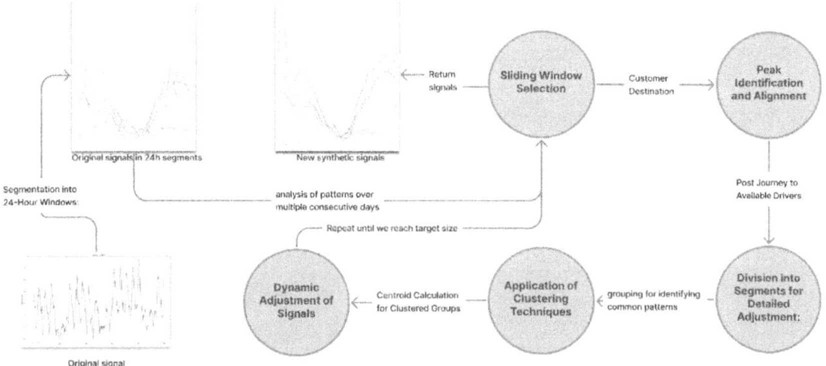

Fig. 1. Workflow of our proposed method

3.1   Signal Segmentation and Alignment

The process begins by segmenting the energy consumption data into discrete 24-hour windows. This segmentation is critical as it captures the daily usage patterns which are often consistent due to habitual consumer behavior, albeit with some variations. We apply a sliding window approach with a width of d days to meticulously select subsets of data, ensuring a dynamic but consistent representation of daily activities as time progresses.

3.2   Peak Matching and Signal Adjustment

Within each selected subset, we identify and align the peaks of the energy signals. This alignment is essential for overlaying the signals to maximize their congruence, which is crucial for subsequent analysis and modeling. Figure 1 illustrates this alignment, which standardizes the dataset and ensures that the energy consumption patterns are accurately represented.

After alignment, the signal is divided into two segments for each 24-hour period. Adjustments are made separately for the left and right segments based on predefined centroids and weights. This segmentation and targeted adjustment allow for the creation of synthetic data points that maintain behavioral similarities to the original data while introducing necessary variations as shown in Fig. 2.

3.3   Data Augmentation and Synthetic Signal Generation

The newly formed signals are then amalgamated to create an augmented dataset. This dataset not only balances but also increases the number of data points, which is vital for enhancing the robustness of our machine learning models. Figure 2 demonstrates the transformation from the original signal to the augmented one. Here, dynamic adjustments are made to smooth out extreme values,

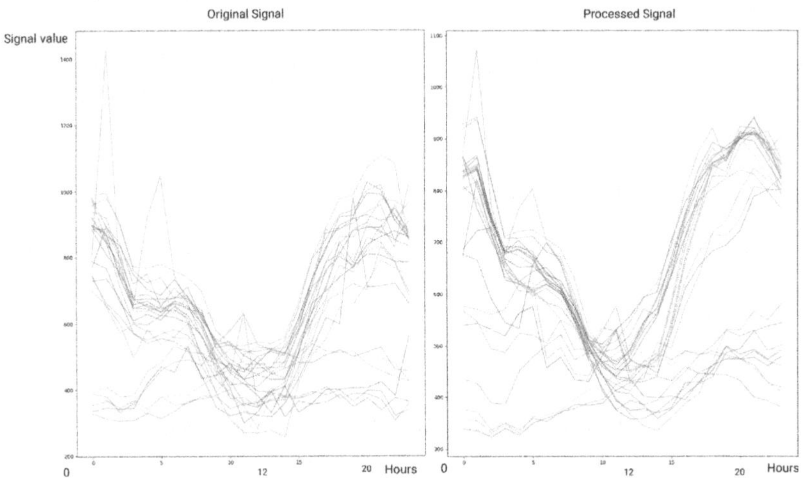

Fig. 2. Comparison of original and adjusted aggregate energy signals

ensuring uniformity and consistency across the dataset. These transformations include scaling, shifting, and clipping operations as detailed in Algorithm 1 in the Appendix, which describes the specific methods used to adjust outliers and overshoots. Finally, we append and form a new augmented signal that should result in more robust preformance·as shown in Algorithm 4.

3.4   Handling Day-to-Day Variability

While the daily routines are assumed to be relatively consistent, minor variations such as starting the day earlier or later are accounted for by matching a specific number of peaks, denoted as k. We adjust the entire dataset over d days to ensure that these peaks align across all days, as depicted in Fig. 3. The detailed procedures for aligning these signals, including the computation of necessary horizontal and vertical translations, are encapsulated in both Algorithm 2 and Algorithm 3 in the Appendix. This systematic alignment aids the RNN model in stabilizing the dataset for more effective learning and prediction.

3.5   Dynamic Clustering and Real-Time Adaptability

The method extends further by dynamically clustering signals based on their daily patterns and continuously adapting the centroids used for alignment and adjustment. This not only helps in handling immediate anomalies but also adapts to gradual shifts in consumer behavior, ensuring that the model remains effective over long periods.

By implementing these techniques, our proposed method significantly enhances the ability of the models to predict and disaggregate energy consumption accurately, leveraging both the historical consistency and the inherent variability of household energy usage patterns. This dual approach ensures that our models are not only effective in recognizing standard patterns but are also resilient in adapting to new or evolving behaviors, thereby providing a robust framework for energy management systems.

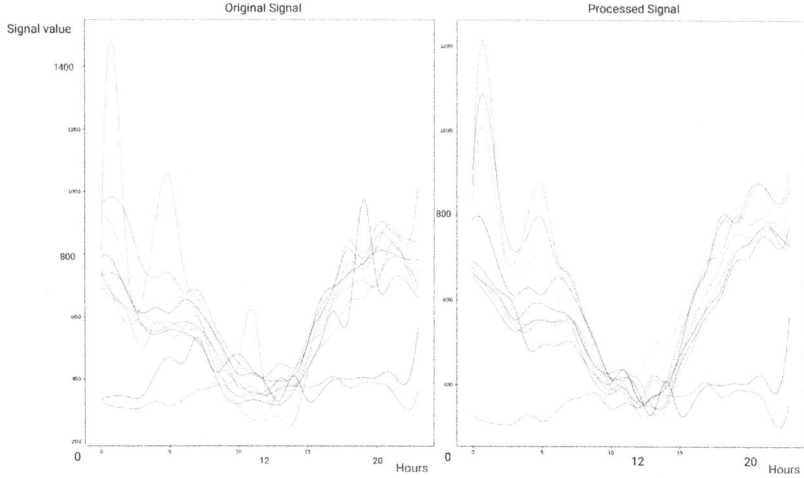

Fig. 3. Resulting signals after processing

3.6 Hyperparameter Configuration

To ensure optimal training and synthesis of data for our NILM models, we carefully selected the following hyperparameters that were used in our experiments:

- **Number of Clusters (c) and Peaks (k)**: We set the number of clusters to 3 and the number of peaks to 4 for aligning signals in our data generation process. This setting is intended to balance computational efficiency while capturing significant patterns.
- **Weight (w) for Vertical Translation**: A weight of 0.1 was chosen for vertical scaling during signal alignment to preserve the underlying distribution while incorporating necessary variability.
- **Desired Size**: To augment our dataset effectively, we aimed to expand it to approximately twice its original size, targeting a size of 3000 samples.
- **Batch Size**: A batch size of 32 was employed for training the RNN, offering a balance between memory usage and model update granularity.
- **Learning Rate**: We used an initial learning rate of 0.001 with a decay mechanism to gradually reduce the rate, enhancing the model's ability to fine-tune and converge more reliably.

- **Loss Function**: Dice loss was utilized as the loss function to effectively handle the imbalance in the dataset concerning different appliance states. The Dice loss function is defined as follows:

$$\text{Dice Loss} = 1 - \frac{2 \times \text{Intersection}(y_{\text{true}}, y_{\text{pred}}) + \text{smooth}}{\text{Union}(y_{\text{true}}, y_{\text{pred}}) + \text{smooth}}$$

  where *smooth* is a small constant added to avoid division by zero.
- **Class Weights**: Given the imbalance in the dataset with respect to different appliance states, class weights were dynamically calculated to ensure balanced training exposure across all classes.

These hyperparameters were crucial in shaping the performance of our models.

4   Experiments and Results

4.1   IRISE Dataset

IRISE dataset, a part of the larger REMODECE dataset, focuses on energy consumption data from French households [8,25]. This dataset was chosen due to its comprehensive nature, containing detailed energy usage information for various household appliances. The dataset includes data collected over a year with a sampling rate of 10 min for 100 households, providing a robust basis for validating our proposed method. The IRISE dataset is particularly suitable for NILM tasks as it includes diverse appliance types, enabling the evaluation of both anomaly detection and energy disaggregation performance.

Further detailing the dataset's use, the IRISE dataset serves a dual purpose in our research: appliance identification and future usage prediction. The identification component aims to discern individual appliance loads from aggregated power consumption data non-intrusively, focusing primarily on high energy-consuming appliances.

4.2   Grenoble INP Dataset

The second dataset comprises more recent data collected before and after the COVID-19 pandemic, from a research lab at Grenoble INP, France, in Grenoble, France. It provides a rich context for applying and testing regression models across a variety of household appliances and scenarios. This dataset was used for evaluating regression models, demonstrating that individual regression algorithms yield favorable outcomes for various appliances based on different train-test methods. [24,26] This dataset was used to validate the energy disaggregation capability of our method by predicting the signal as a whole as well as the On/Off states. The training was conducted using data from 2017 and 2018, and the testing was performed on data from 2019. This setup ensured that the models were evaluated on unseen data, providing a robust assessment of their generalization capabilities.

This detailed description of the Grenoble INP dataset specifically caters to evaluating the energy disaggregation capabilities by simulating realistic household scenarios. It ensures a robust validation of the model's performance, highlighting its potential and effectiveness in real-world applications.

4.3 Methodology

This section outlines our experimental approach to validate the efficacy of dynamic data augmentation in Non-Intrusive Load Monitoring (NILM). The methodology is divided into distinct phases, each targeting a specific aspect of NILM, from data augmentation to performance evaluation.

Data Augmentation Strategies. Our experiments utilized a range of data augmentation techniques to enhance the robustness of NILM models against the variability of household appliance usage:

- **Original**: Utilizes the baseline model without any data augmentation.
- **Threshold**: Implements traditional data augmentation by adding a random value, ranging from 0 to the mean of the aggregate signal. [27,29]
- **Adaptive**: Employs advanced data augmentation using adaptive thresholds based on the statistical distribution of the signal. [28]
- **Behavior-Driven**: Applies our proposed dynamic data augmentation method, designed to adapt to changes in consumption patterns dynamically.

Experiment Setup. The methodology was implemented through two main experimental setups:

Appliance State Prediction. This phase focused on predicting the on/off states of various household appliances such as dishwashers, electrical cookers, fridges, lights, and white goods. The models were evaluated based on several metrics:

- Accuracy
- F1 Score
- Mean Absolute Error (MAE)
- Root Mean Square Error (RMSE)

Energy Disaggregation Analysis. This part of the experiment aimed to disaggregate the total energy consumption into individual appliance contributions using Lasso Regressor and Recurrent Neural Network (RNN) models. The performance was primarily evaluated using the Mean Absolute Error (MAE) metric.

Each of these phases was carefully designed to assess the capability of the augmented data in improving the predictive accuracy of NILM systems under varied conditions.

Table 1. Performance comparison of different methods on the IRISE dataset

Appliance	Method	Accuracy	F1 Score	MAE	RMSE
Dish Washer	Baseline	80.55%	89.16%	0.447	0.164
	Threshold	71.11%	83.00%	0.4399	0.289
	Adaptive	81.72%	89.88%	0.481	0.1828
	Behavior-Driven	**96.60%**	**98.26%**	**0.043**	**0.034**
Electrical Cooker	Baseline	90.16%	94.80%	0.42	0.098
	Threshold	81.15%	89.53%	0.424	0.1885
	Adaptive	95.18%	97.52%	0.419	0.048
	Behavior-Driven	**95.24%**	**97.55%**	**0.047**	**0.047**
Fridge	Baseline	78.81%	88.07%	0.4465	0.2119
	Threshold	95.68%	97.78%	0.4049	0.0432
	Adaptive	86.18%	92.53%	0.4508	0.1382
	Behavior-Driven	**95.81%**	**97.85%**	**0.0426**	**0.0419**
Light	Baseline	72.12%	83.69%	0.4727	0.2789
	Threshold	76.71%	86.73%	0.4111	0.2329
	Adaptive	93.35%	96.54%	0.4482	0.0665
	Behavior-Driven	**96.35%**	**98.16%**	**0.042**	**0.0366**

4.4   Results and Discussion

Analysis of NILM Performance on IRISE Dataset. The results from
our initial experiments on the IRISE dataset, as summarized in Table 1, clearly
showcase the enhancements brought about by our proposed method. Compared
to traditional data augmentation techniques, our approach demonstrated superior prediction accuracy and robustness across various appliances for both the
appliance energy prediction and appliance state prediction as well. This improvement underscores the effectiveness of our dynamic data augmentation technique
in handling diverse and non-static consumption patterns typical in residential
settings.

Energy Disaggregation Experiments. In our second set of experiments,
we specifically addressed energy disaggregation tasks, predicting the energy
consumption profiles for individual appliances. Table 2 presents these results,
emphasizing the systematic improvements our method offers over traditional
approaches. Particularly, the ability to fine-tune the signal processing based on
historical data significantly enhances the prediction accuracy.

Detailed Appliance-Specific Analysis. For instance, regarding the dishwasher, our
method not only improved accuracy to 96.60% and F1 score to 98.26%, but it also

reduced the Mean Absolute Error (MAE) to 0.043 and the Root Mean Square Error (RMSE) to 0.034. These figures represent a substantial enhancement over the original method, which achieved an accuracy of 80.55% and an F1 score of 89.16%.

Such marked improvements were also noted in the performance metrics for other appliances. The electrical cooker and fridge, for example, showed not only higher accuracy and F1 scores but also demonstrated the method's consistency and reliability across different types of appliances. Our method particularly excelled in the lighting category, where it consistently outperformed the threshold and adaptive augmentation techniques, reinforcing the adaptability of our approach to various electrical loads and usage patterns.

Table 2. MAE Scores for Various Appliances

Appliance	Method	Lasso Regressor	RNN
Ventilation	Baseline	30.365	18.88
	Threshold	30.43	19.62
	Behavior-Driven	**29.53**	**18.42**
Other Electricity	Baseline	97.35	118.41
	Threshold	97.34	112.82
	Behavior-Driven	**94.76**	**111.82**
Sockets Plug	Baseline	10.17	9.16
	Threshold	10.165	10.18
	Behavior-Driven	**9.00**	**7.08**
Lighting	baseline	14.814	6.771
	threshold	14.809	5.43
	Behavior-Driven	**14.27**	**6.12**
Heating	Baseline	102.64	55.62
	Threshold	102.55	64.5
	Behavior-Driven	**94.94**	**48.83**
Cooling	Baseline	51.15	**23.12**
	Threshold	**51.14**	23.45
	Behavior-Driven	51.85	25.13

Methodological Contributions. These improvements can be attributed to the nuanced capability of our dynamic data augmentation method, which adapts to changes in appliance usage patterns more effectively than static models. By utilizing statistical distributions from historical consumption data and emphasizing recent behavioral trends, our method generates synthetic datasets that closely mimic real-world variations. This leads to better model training and, consequently, enhanced performance in practical scenarios.

The core advantage of our method lies in its sophisticated handling of synthetic data generation. Training on the generated data results in NILM models that are both robust and sensitive to subtle changes in energy consumption, crucial for accurate disaggregation and appliance state classification.

5   Limitations

While our proposed method shows promising results in enhancing NILM performance through behavior-driven data augmentation, there are several limitations that need to be addressed:

5.1   Sudden Changes in Consumer Behavior

Our approach assumes that consumer behavior follows a relatively consistent pattern over time. However, in environments such as hotels, hospitals, or other facilities with rapid and unpredictable changes in consumer behavior, the method may perform less effectively. These sudden changes can disrupt the prediction intervals and clustering, leading to inaccurate disaggregation and anomaly detection.

5.2   Scalability with Large Datasets

Despite advancements, NILM faces several challenges. Dynamic appliance usage patterns can vary significantly over time, necessitating continuous model updates to maintain accuracy. Current data augmentation techniques may not fully capture the complexity of real-world energy consumption, especially in environments with rapid changes such as hotels or commercial buildings.

The dynamic generation of synthetic data and clustering based on consumer behavior can become computationally intensive, especially with larger datasets. As the size of the data increases, the time required to process and generate synthetic data also increases, potentially making the method less scalable for very large datasets especially with limited resources. Advances in transfer learning and domain adaptation, as explored by Pan et al. [17], could also be beneficial for applying NILM models across different households with minimal retraining.

5.3   Fixed Parameters for Clustering

The current implementation uses fixed parameters for clustering and matching peaks. While these parameters work well under certain conditions, they may not be optimal for all scenarios. Adaptive parameter tuning could potentially enhance the performance but is not currently implemented.

5.4   Handling of Extreme Outliers

Although our method improves the robustness of NILM models, it may still struggle with extreme outliers that significantly deviate from normal usage patterns. These outliers can skew the distribution and affect the accuracy of the generated synthetic data and prediction intervals.

Addressing these limitations is crucial for improving the robustness and applicability of our method in various real-world scenarios. Future work could focus on developing adaptive algorithms that can handle rapid changes in behavior, optimizing computational efficiency, and enhancing the method's ability to deal with diverse and large-scale datasets.

6 Conclusion

In this paper, we proposed a novel dynamic data augmentation technique for Non-Intrusive Load Monitoring (NILM) that dynamically adjusts to changes in appliance usage patterns. Our method was validated through two sets of experiments using the IRISE dataset and another comprehensive dataset. The first set focused on predicting on/off states of various appliances, showing significant improvements in accuracy and robustness, while the second set concentrated on energy disaggregation, demonstrating superior performance with lower MAE scores for multiple appliances. These improvements highlight the potential of our method in generating realistic and varied training data, enhancing the model's ability to capture the inherent variability in household energy consumption.

Future work will focus on further optimizing our dynamic data augmentation technique and exploring its application to other NILM datasets and real-world scenarios. Additionally, we aim to investigate the integration of our method with other advanced machine learning models to further improve NILM performance. The results presented in this paper pave the way for more accurate and reliable energy monitoring solutions, demonstrating the significant impact of dynamic data augmentation in advancing NILM technologies.

Acknowledgments. This research was supported by The Natural Sciences and Engineering Research Council of Canada (NSERC). The authors express their gratitude for the financial support that made this study possible.

A   Appendix

Algorithm 1. Translate Signal

Require: *signal*: Array of signal values, *shift*: Integer (horizontal shift), *vertical_shift*: Integer, n: Maximum allowed signal value
Ensure: Adjusted signal array

1: Perform a circular shift of *signal* by *shift* positions.
2: Add *vertical_shift* to *signal*.
3: Clip the signal values to ensure they lie within the range $[0, n - 1]$.
4: **return** Adjusted *signal*

Algorithm 2. Selective Vertical Translation

Require: *signal*: Array of signal values, *centroid*: Array of centroid values, *weight*: Weight for adjustment, *split_point*: Index to split signal
Ensure: Vertically adjusted signal array

1: Split *signal* and *centroid* at *split_point* into *left_side*, *right_side*, *left_centroid*, and *right_centroid*.
2: Compute vertical shifts for both parts: *left_vertical_shift* and *right_vertical_shift*.
3: Apply computed shifts to the corresponding parts.
4: Clip adjusted signal values to ensure non-negativity.
5: Concatenate adjusted *left_side* and *right_side*.
6: **return** Concatenated adjusted signal

Algorithm 3. Align Signals by Peaks

Require: *signals*: Matrix of signal arrays, c: Number of clusters, k: Number of peaks, w: Weight for vertical translation, *split_point*: Optional split point
Ensure: Aligned signal matrix

1: Identify the peak positions for each signal in *signals*.
2: Flatten *signals* and perform clustering to assign signals to c groups.
3: **for** each cluster **do**
4:     Calculate reference peak positions using the centroid of the group.
5:     **for** each signal in the cluster **do**
6:         Calculate necessary horizontal and vertical translations.
7:         Apply horizontal translation.
8:         Apply selective vertical translation.
9:     **end for**
10:    Store adjusted signals.
11: **end for**
12: **return** Matrix of aligned signals

Algorithm 4. Generate Synthetic Data

Require: X_{train}, Y_{train}: training datasets
Require: c: number of clusters, k: number of peaks, w: weight for vertical translation
Require: desired_size: target size for the augmented dataset
Ensure: $X_{\text{augmented}}$, $Y_{\text{augmented}}$: augmented datasets

1: $d, n \leftarrow \text{shape}(X_{\text{train}})$
2: $X_{\text{augmented}} \leftarrow X_{\text{train}}$
3: $Y_{\text{augmented}} \leftarrow Y_{\text{train}}$
4: **while** $\text{len}(X_{\text{augmented}}) <$ desired_size **do**
5:     *aligned_signals* $\leftarrow$ align_signals_by_peaks($X_{\text{train}}, c, k, w$)
6:     *additional_size* $\leftarrow \min(\text{desired\_size} - \text{len}(X_{\text{augmented}}), \text{len}(aligned\_signals))$
7:     $X_{\text{augmented}} \leftarrow \text{np.vstack}((X_{\text{augmented}}, aligned\_signals[: additional\_size]))$
8:     $Y_{\text{augmented}} \leftarrow \text{np.vstack}((Y_{\text{augmented}}, Y_{\text{train}}[: additional\_size]))$
9: **end while**
10: **return** $X_{\text{augmented}}$, $Y_{\text{augmented}}$

References

1. Hart, G.W.: Nonintrusive appliance load monitoring. Proc. IEEE **80**(12), 1870–1891 (1992)
2. Zeifman, M., Roth, K.: Nonintrusive appliance load monitoring: review and outlook. IEEE Trans. Consum. Electron. **57**(1), 76–84 (2011)
3. Nour, M., Le Bunetel, J.C., Ravier, P., Raingeaud, Y.: Data augmentation strategies for high-frequency NILM datasets. IEEE Trans. Instrum. Meas. **72**, 6501809 (2023)
4. Linh, N.V., Arboleya, P.: Deep learning application to non-intrusive load monitoring. IEEE Proc. (2019)
5. Kelly, J., Knottenbelt, W.J.: Neural NILM: deep neural networks applied to energy disaggregation. CoRR, abs/1507.06594 (2015)
6. Buddhahai, B., Wongseree, W., Rakkwamsuk, P.: An energy prediction approach for a nonintrusive load monitoring in home appliances. IEEE Trans. Consumer Electron. (2019)
7. Parson, O., Ghosh, S., Weal, M., Rogers, A.: Non-intrusive load monitoring using prior models of general appliance types. In: Proceedings of the Twenty-Sixth AAAI Conference on Artificial Intelligence, pp. 356–362 (2012)
8. Basu, K., Debusschere, V., Bacha, S.: A classification framework for power events in non-intrusive load monitoring. IEEE Trans. Smart Grid **6**(4), 1949–1958 (2014)
9. Chang, H.-H., Chen, K.-C., Chen, Y.-C.: A new method for load identification of non-intrusive energy monitoring systems. IEEE Trans. Smart Grid **3**(1), 99–104 (2012)
10. Lin, Y.-S., Hsieh, C.-T., Lin, H.-T.: Development of a non-intrusive load monitor for demand-side management. IEEE Trans. Smart Grid **3**(4), 1880–1887 (2012)
11. Rashid, R., Sekercioglu, Y.A., Fitzpatrick, P., Khan, J.Y.: Evaluation of non-intrusive load monitoring algorithms for real-time applications. IEEE Access **7**, 3030–3045 (2019)
12. Dong, M., Meira, P.C., Xu, W.: Non-intrusive load monitoring with iterative and dynamic time warping. IEEE Trans. Smart Grid **4**(1), 411–420 (2013)
13. Massidda, L., Marrocu, L., Pilo, F.: Convolutional and recurrent neural networks for energy disaggregation of smart meters. IEEE Trans. Smart Grid **12**(2), 1630–1640 (2021)
14. Zhou, B., Xie, H., Yang, J., Wu, G.: Non-intrusive load monitoring using attention-based deep neural networks. IEEE Access **7**, 28204–28215 (2019)
15. Linh, N.V., Arboleya, A.: Deep learning for energy disaggregation: combining data filtering and power state clustering techniques. IEEE Access **7**, 145384–145395 (2019)
16. Pereira, L., Nunes, N.: Performance evaluation in non-intrusive load monitoring: datasets, metrics, and tools–a review. Renew. Sustain. Energy Rev. **82**, 1633–1645 (2018)
17. Pan, S.J., Yang, Q.: A survey on transfer learning. IEEE Trans. Knowl. Data Eng. **22**(10), 1345–1359 (2010)
18. Zaremba, W., Sutskever, I., Vinyals, O.: Recurrent neural network regularization. arXiv preprint arXiv:1409.2329 (2014)
19. Linh, L., Arboleya, M.: A comprehensive review on NILM algorithms for energy disaggregation. IEEE Access **7**, 168–179 (2019)
20. Zhang, C., Zhong, M., Wang, Z., Goddard, N., Sutton, C.: Sequence-to-point learning with neural networks for non-intrusive load monitoring. arXiv preprint arXiv:1612.09106 (2017)

21. Kolter, Z., Jaakkola, T.: Energy disaggregation using sequence-to-sequence learning of long short-term memory recurrent neural networks. In: Proceedings of the 2nd ACM International Conference on Embedded Systems for Energy-Efficient Built Environments, pp. 55–64 (2015)
22. Nguyen, K., Arboleya, A.: Deep learning for non-intrusive load monitoring: a review and outlook. IEEE Trans. Smart Grid **10**(4), 3820–3829 (2019)
23. Ciancetta, F., Bucci, G., Fiorucci, E., Mari, S., Fioravanti, A.: A new convolutional neural network-based system for NILM applications. IEEE Trans. Instrum. Meas. **69**(12), 1–12 (2020)
24. Akbar, M.K., Amayri, M., Bouguila, N.: A novel non-intrusive load monitoring technique using semi-supervised deep learning framework for smart grid. Build. Simul. **17**, 441–457 (2024)
25. Basu, K., Debusschere, V., Bacha, S.: Residential appliance identification and future usage prediction from smart meter. In: IECON 2013 - 39th Annual Conference of the IEEE Industrial Electronics Society, pp. 4994–4999 (2013). https://doi.org/10.1109/IECON.2013.6699944.
26. Akbar, M.K., Amayri, M., Bouguila, N., Delinchant, B., Wurtz, F.: Evaluation of regression models and Bayes-ensemble regressor technique for non-intrusive load monitoring. Sustain. Energy, Grids Networks 38, 101294 (2024)
27. Shorten, C., Khoshgoftaar, T.M.: A survey on image data augmentation for deep learning. J. Big Data **6**(1), 1–48 (2019). https://doi.org/10.1186/s40537-019-0197-0
28. Park, D.S., et al.: SpecAugment: a simple data augmentation method for automatic speech recognition. In: Interspeech 2019 (2019). arXiv:1904.08779
29. Delfosse, A., Hebrail, G., Zerroug, A.: Deep learning applied to NILM: is data augmentation worth for energy disaggregation? In: ECAI 2020, IOS Press, pp. 2972–2977 (2020)

Data-Driven Approach to Assess and Identify Gaps in Healthcare Set up in South Asia

T. Fatima[1], R. Elahi[1], S. W. Zahra[1], H. M. Abubakar[1], T. Zafar[3],
Z. Tahseen[2], M. T. Quddoos[4], and U. Nazir[1(✉)]

[1] Center for AI Research (CAIR), School of Computer and Information Technology,
Beaconhouse National University, Lahore, Pakistan
`{f2021-711,f2021-557,f2021-136,f2021-641,usman.nazir}@bnu.edu.pk`
[2] 4XPillars, Munich, Germany
`z.tahseen@4xpillars.com`
[3] Allama Iqbal Medical College, Lahore, Pakistan
`tehreemzafar687@gmail.com`
[4] Center for Urban Informatics, Technology and Policy (CITY) at LUMS,
Lahore, Pakistan
`muhammad.quddoos@lums.edu.pk`

Abstract. Primary healthcare is a crucial strategy for achieving universal health coverage. South Asian countries are working to improve their primary healthcare system through their country specific policies designed in line with WHO health system framework using the six thematic pillars: Health Financing, Health Service delivery, Human Resource for Health, Health Information Systems, Governance, Essential Medicines and Technology, and an addition area of Cross-Sectoral Linkages [11]. Measuring the current accessibility of healthcare facilities and workforce availability is essential for improving healthcare standards and achieving universal health coverage in developing countries. Data-driven surveillance approaches are required that can provide rapid, reliable, and geographically scalable solutions to understand a) which communities and areas are most at risk of inequitable access and when, b) what barriers to health access exist, and c) how they can be overcome in ways tailored to the specific challenges faced by individual communities. We propose to harness current breakthroughs in Earth-observation (EO) technology, which provide the ability to generate accurate, up-to-date, publicly accessible, and reliable data, which is necessary for equitable access planning and resource allocation to ensure that vaccines, and other interventions reach everyone, particularly those in greatest need, during normal and crisis times. This requires collaboration among countries to identify evidence based solutions to shape health policy and interventions, and drive innovations and research in the region.

Keywords: Health Inequities · Global Accessibility Health ·
Nighttime lights · Google Maps

T. Fatima, R. Elahi, Z. Tahseen, M. T. Quddoos, and U. Nazir—These authors contributed equally to this work.

1   Introduction

Even while the Sustainable Development Goal 3 (SDG 3) is being achieved, 381 million people, or 4.9% of the world's population, still live in severe poverty and do not have access to even the most basic healthcare [23,29]. Mortality rates are highest in sub-Saharan Africa and Southern Asia, essentially due to insufficient access to primary healthcare services [29]. It is incumbent to address discrepancies in order to close this gap and guarantee equitable access to healthcare. To achieve the vision of "Health for All", we must address the systemic barriers that perpetuate health inequities [4,20]. Inequity has many dimensions [24] as shown in Fig. 1; barriers include, but are not limited to: a) lack of access to infrastructure, e.g., health facilities [25]; b) access to communication infrastructure in remote locations, e.g., poor road networks and communication links [28]; and c)

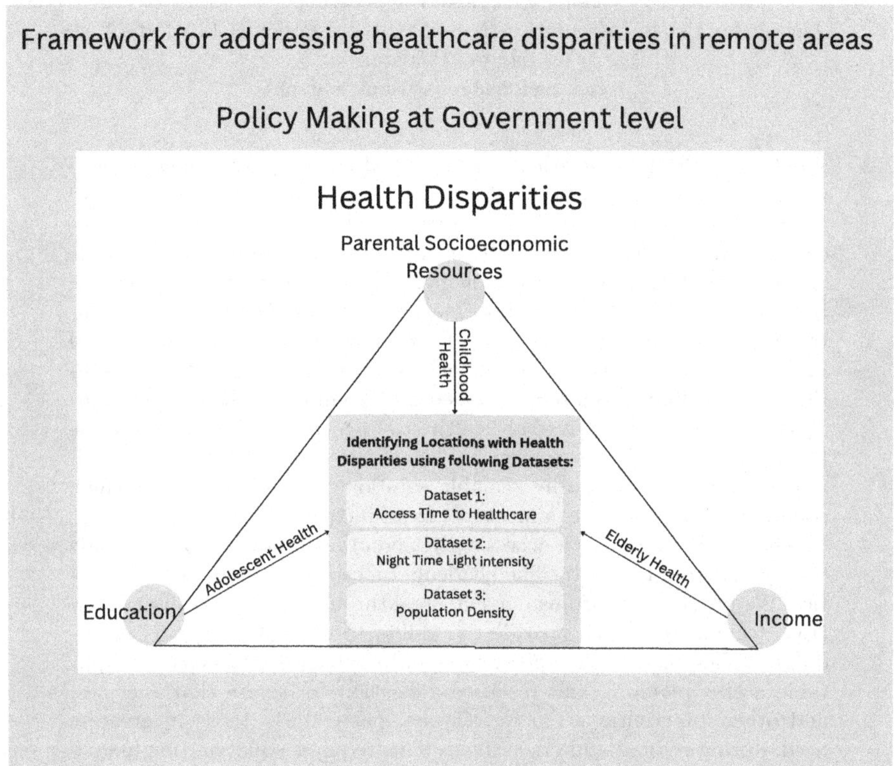

Fig. 1. Health Disparities due to: a) limited access to essential infrastructure, such as health facilities, despite interventions like safe medicines and vaccines; b) challenges in accessing communication infrastructure in remote areas, including poor road networks and communication links; and c) the influence of parental socioeconomic resources, education, income, affordability, and intersectional determinants, such as gender, age, and lifestyle, on access to healthcare.

affordability and intersectional determinants of access, such as gender, age, and lifestyle factors [21].

Insufficient and ineffective healthcare methods exacerbate the lack of appropriate access to healthcare [9], and many low- and middle-income nations lack real-world health data infrastructure due to overworked healthcare systems. In heavily populated areas, particularly urban slums and remote rural areas. Manual monitoring and data gathering in remote locations can be unacceptably time-consuming, costly, prone to human error and malpractice and unsustainable given limited resources [13].

Data-driven surveillance approaches are required that can provide rapid, reliable, and geographically scalable solutions to understand: a) which communities and areas are most at risk of inequitable access and when [10]; b) what barriers to health access exist; and c) how these barriers can be overcome in ways tailored to the specific challenges faced by individual communities [16]. It can help to reduce health data poverty and disparities. Better knowledge, achieved through better data and robust decision-support systems, can help address everyone's health requirements in an equitable manner [6].

The goal of the Centers for Disease Control and Prevention's (CDC) Data Modernization Initiative (DMI) is to increase the effectiveness and efficiency of health data collecting, sharing, and analysis. On the other hand, there isn't much research on integrating social determinants of health (SDOH) into state and federal DMI initiatives. The terms SDOH refer to an individual's living, learning, and working environments, as well as the ways in which these environments impact their health. Examples of SDOH include safe housing, access to wholesome food, employment possibilities, and transportation.

We propose to harness current breakthroughs in Earth-observation (EO) technology, which provides the ability to generate accurate, up-to-date, publicly accessible, and reliable data, which is required for equitable access planning and resource allocation to ensure that safe medicines, vaccines, and other interventions reach everyone, particularly those in greatest need, during normal times [7,27]. This data can also be used in emergency scenarios such as pandemics and natural catastrophes, which disproportionately affect underserved groups [17]. Therefore, this data creation can help identify requirements and track progress towards increasing equal access to healthcare worldwide. This is particularly important in resource-constrained areas, both in normal and emergency situations [2]. Figure 1 illustrates how government-level policies can address health inequalities, a key focus of SDG 3 [23,29]. It emphasizes the role of parents' socioeconomic resources in influencing children's health, which in turn influences the health of the young and old, as well as education and income levels. Our approach can identify health gaps in peripheral areas and improve surveillance, especially in resource-limited areas. By harnessing the power of EO and artificial intelligence, we can create a fairer and more sustainable healthcare system that leaves no one behind.

Our major contributions to mapping health equities in South Asia, as compared to AccessMod 5 (see Appendix A), include the following:

- Remote areas and potential improvement neighborhoods with missing nighttime light intensity: We use nighttime light intensity data to find unmet needs in the most remote areas. With this method we can also identify geographical areas with poorly served health services in terms of infrastructure development
- Localizing health disparities in underserved remote areas: We focus on pinpointing specific health disparities within the identified under-served areas. This localization helps to understand the specific health needs and challenges faced by these communities, allowing for targeted interventions.

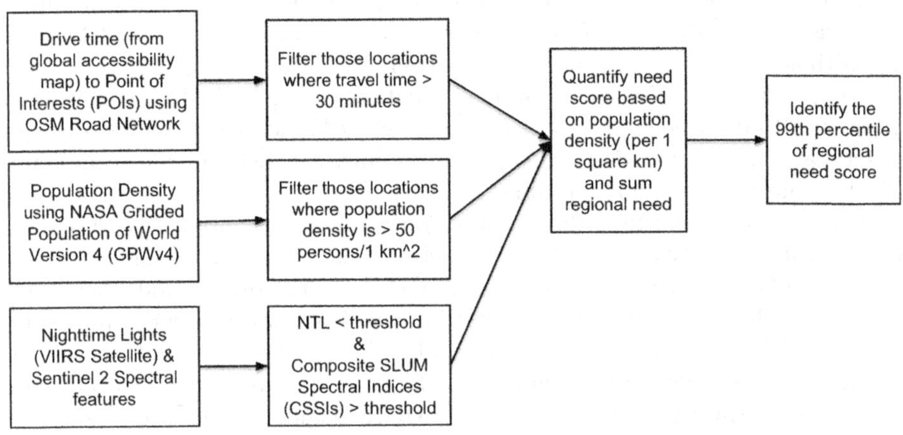

Fig. 2. Proposed approach for mapping health inequities.

2   Mapping Primary Healthcare Infrastructure Gaps

Strategists, academics, and public health practitioners strive to improve the overall health of the population by addressing health disparities based on geographic origin, race/ethnicity, socioeconomic status (SES), and other social factors. Health inequalities refer to systematic variations in health status among various socioeconomic classes in a society, which are socially produced, theoretically avoidable, and often considered undesirable in a civilized society [33]. The concept of health inequality makes no moral judgments about whether observed disparities are equitable or right. A health inequity, also known as a health disparity, is a subtype of health inequality that refers to an unjust discrepancy in health status. Allowing health inequalities to persist is unjust, according to one common definition [32]. In this sense, health inequities are systematic differences in health that could be avoided by reasonable means [16]. Social group differences in health, such as those based on race or religion, are considered health inequities because they reflect an unfair distribution of health risks and resources. The key distinction between inequality and inequity is that the former is simply a dimensional description employed whenever quantities are unequal, while the latter requires passing a moral judgment that the inequality is wrong. For example, the term"health inequality" can reflect racial/ethnic differences in US infant mortality rates, which are roughly three times greater for non-Hispanic blacks than whites [22].

Age-related health disparities, such as those between the ages of 20 and 60, are seen as health inequalities and are perceived as unavoidable rather than unjust. By contrast, disparities in health are investigated among populations worldwide according to factors such as income, caste, gender, education, race/ethnicity, and employment. To investigate these disparities, researchers compare socioeconomic variables and examine variations in health outcomes at the group level. For example, comparing the mean body mass index (BMI) of the wealthy to that of the impoverished can help explain socioeconomic differences. In order to reduce disparities brought about by the unequal distribution of social determinants of health, policy development and resource allocation must acknowledge and address social group health inequalities [1]. The World Health Organization proposes that health indicators be reported by groups, or equity stratifiers, in order to monitor health disparities. By concentrating on social groupings, we may comprehend present health inequalities within a historical and cultural framework and subsequently offer an explanation of the mechanisms underlying health inequality [19]. For instance, knowledge of American racism and slavery past can help explain current racial/ethnic health disparities. Similarly, studying the political and theological history of India's caste system allows us to better appreciate how it influences people's social position, employment, education levels, and health outcomes. Viewing health inequalities through the lens of social groups can help direct actions, enable surveillance of key equity concerns, and increase our knowledge of health by making connections that were not initially clear. Health inequalities along racial, ethnic, and socioeconomic lines exist in both low- and high-income nations, and they may be expanding, emphasizing the necessity of investigating group-level health discrepancies [3]. Understanding socially patterned health inequalities requires defining meaningful social groups. Each community has its own distinct method of classifying and separating people into social divisions.

The second related topic is whether absolute or relative position matters for health. This is especially relevant when it comes to poverty, which can be defined in two ways: absolute (comparing a given income to a static benchmark) and relative (comparing a given income to the general distribution of incomes in a community). Absolute poverty definitions are based on a set monetary barrier known as a poverty line, though this barrier is normally determined by the year, nation, and household size. Those with incomes less than the criteria are deemed destitute. In contrast, relative poverty is defined by comparing a particular income to the income distribution in a population. For example, one could argue that a person is comparatively impoverished if their income is less than thirty percent of the national per capital income [1]. This suggests that the concept of poverty is not static and may change over time. Wealth, income, and education are a few factors that contribute to an individual's socioeconomic position (SEP). Education can be defined as the highest level attained or the total number of years spent in school. The main factor affecting the material resources component of SEP is income, which is often calculated as household gross income divided by the number of participants. Wealth is defined as income plus all acquired material resources.

Occupation-based indicators are also often utilized. For example, the Registrar General's Social Classes classify jobs according to prestige into six hierarchical groupings, which are frequently classified as manual vs. non-manual. The Erikson and Goldthorpe Class Schema classifies jobs according to employment relations features, while the UK National Statistics Socio-Economic Classification follows similar concepts. Other categories, which concentrate on social interaction patterns within occupational groupings, include the Cambridge Social Interaction and Stratification Scale and Wright's Social Class Scheme. The following are used to illustrate societal patterns: housing issues, overpopulation, unemployment, and composite or proxy indicators. Geographic environment, not simply social group, influences health. Health inequalities can be impacted by people's physical and social settings. Geographic health inequities may be monitored by comparing health outcomes across various areas, taking into account both place (particular geographic locations) and space (broader geographical settings) [12,15]. Researchers must select how to present observed differences while tracking health disparities across time. Inequalities between groups can be expressed as absolute or relative differences. Absolute differences represent the actual disparity in health outcomes between groups, whereas relative differences assess the proportional gap in relation to the baseline health state [14,18].

Our proposed technique for mapping health disparities uses a variety of data sources and approaches to quantify and identify areas with severe health needs. As shown in the flowchart (Fig. 1), we employ travel time from global accessibility maps to places of interest (POIs), population density data from NASA's Gridded Population of the World Version 4 (GPWv4), and nighttime light data from VIIRS satellites and Sentinel 2 spectral characteristics. This method selects places based on travel time (>30 min), population density (>50 per km^2), and thresholds for nighttime lights and composite SLUM spectral indices. By measuring regional need ratings based on population density and aggregating them, we can discover the 99th percentile of areas with severe health needs.

This strategy is based on data that support established patterns of population and geographic health disparities. This enables the development of targeted interventions and policies to address health inequalities. Our approach provides a comprehensive framework for monitoring and addressing health inequities both at scale and at scale, combining multiple data sources with a regional and geographic focus.

The rest of the paper is organized as follows: Sect. 3 describes how to implement the proposed approach, including the integration of open source datasets such as VIIRS NTL data, GPWv4 and Accessibility to Healthcare 2019. It also presents the analytical findings and visualizations that resulted from these datasets. Section 5 provides an overview of the results of our analysis and highlights the urgent need to address global health disparities and the value of our data strategy to guide targeted action. Chapter 6 describes actions to address health disparities, including a multifaceted strategy that combines community-based solutions with modern data analytics and earth observation technology. Section 4 concludes the paper with a summary of the main conclusions, results and evaluation. Finally, Sect. 7 discusses how policies can be implemented to reduce health inequalities. It supports evidence-based interventions that support equitable health outcomes and access worldwide.

Algorithm 1: Mapping Health Inequities in South Asia

Input   : datasets ← [Accessibility to Healthcare 2019], [The Gridded Population of World Version 4 (GPWv4), Revision 11], [Annual global VIIRS V2.2 nighttime lights dataset]
Output : Mapping Health Inequities in South Asia
1 **Function** Main(*datasets*):
2   time ← Filter locations with AccessibilityTime > 30 min
3   pop ← Filter locations PopulationDensity >= 50 $\frac{persons}{1km^2}$
4   th ← 20 ;   //Set your desired threshold value here
5   povertyAreas ← Filter locations with nighttime lights < th
6   needPerPixel ← Quantify need based on population density using time, pop & povertyAreas
7   needScore ← Sum up the need around 25 km radius around each 1 km pixel
8   highNeedScore ← Identify the 99th percentile of regional need score
9   regionalNeedScore ← calculate mean regional need score of each region in ascending order
10   **return** *Mapping Health Inequities in South Asia*;

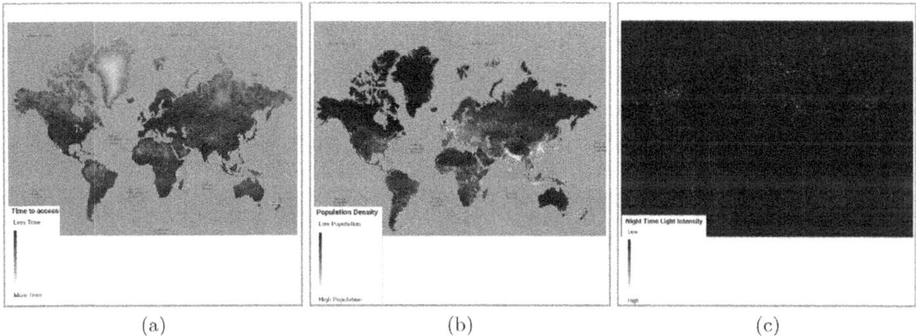

Fig. 3. Datasets: (a): Travel time to the nearest hospital or clinic, (b): Population density using NASA Gridded Population of World version 4 (GPWv4), (c): VIIRS Nighttime Day/Night Annual Band Composites V2.2. Darker areas indicate shorter travel times to the nearest facility, lower population density, and lower nighttime light intensity compared to brighter areas.

3   Proposed Approach

3.1   Open Source Datasets

The study utilizes three following primary open-source datasets (see Fig. 3) to identify global regions where new healthcare facilities are most needed.

1. Accessibility to Healthcare 2019
2. The Gridded Population of World Version 4 (GPWv4), Revision 11
3. VIIRS Nighttime Day/Night Annual Band Composites V2.2

Accessibility to Healthcare 2019. The global accessibility map for 2019 is the result of collaboration between the University of Twente in the Netherlands, Google, Telethon Kids Institute in Perth, Australia, and MAP at the University of Oxford [31]. It shows the land-based travel time (in minutes) from any place to the nearest medical facility. It also includes the duration of "walking-only" or using non-motorized modes of transportation exclusively. Using massive data collection efforts from Google Maps, OpenStreetMap, and university researchers, the most complete database of healthcare institution locations has been produced to date.

This project expands upon earlier research by [30]. Weiss et al. used datasets for national borders, railways, rivers, lakes, seas, topographic conditions (slope and elevation), landcover types, and roads (including the first-ever global use of Open Street Map and Google roads datasets). Each of these datasets was given a travel speed, or speeds, in terms of the amount of time needed to cross each kind of pixel. After that, the datasets were integrated to create a "friction surface", which is a map in which each pixel is given a nominal overall speed of motion according to the types that exist there. An upgraded friction surface was made for the current project to take advantage of newer developments in the OSM roads data.

GPWv4.11. The Gridded Population of World Version 4 (GPWv4), Revision 11 represents the global distribution of human population per (approximately) 1km grid cells. A proportionate distribution of the population from census and administrative units is used to distribute the population among the cells. In terms of relative spatial distribution, these population density grids' estimations of the number of people per 1 km grid cell are in line with national censuses and population registries [5].

VIIRS Nighttime Day/Night Annual Band Composites V2.2. Annual global VIIRS nighttime lights dataset is a time series produced from monthly cloud-free average radiance grids for 2022 [8]. After removing pixels from the sun, moon, and clouds in the first filtering stage, preliminary composites with lights, flames, aurora, and backdrop were produced. To create rough annual composites, monthly increments of the rough annual composites are created and then blended.

3.2 Finding Health Disparities

The proposed methodology for finding health disparities as shown in Fig. 2 and algorithm 1 involve following steps:

1. Importing Datasets: The analysis environment receives the datasets for accessibility, population density, and nighttime light intensity.
2. Filtering Areas with Limited Accessibility: Locations where traveling to the nearest medical facility takes more than thirty minutes are marked. These regions are typical of those with poor access to medical care.
3. Combining accessibility with population density and nighttime light intensity: areas with high population density (50 people per square kilometer or more), poor socioeconomic resources (night light intensity below the threshold) and poor access. to health services are identified for further study (travel time more than 30 \ n min). This policy ensures that the focus is always on densely populated areas with limited access to health services.

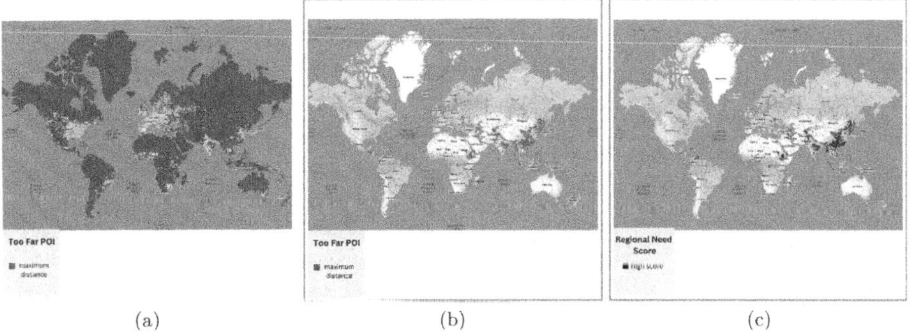

Fig. 4. Evaluation Results: (a): Location with too far Points of interest (health facilities is greater than 30 min); (b): Filter those locations where travel time is greater than 30 min & Population density is greater than 50 persons per 1 km^2; (c): Regional Need score based on population density per 1 km^2 - 99th percentile.

4   Results and Evaluation

4.1   Analysis

The analysis is done in multiple steps in order to determine which regions have the greatest need for healthcare facilities and to quantify that need: :

1. Quantifying Need Per Pixel. Based on acccooibility and population density, a need score is assigned to each region (or pixel). In highly populated locations with limited access to healthcare, the need score is greater (See Fig. 4 (a) & (b)).
2. Aggregating Need Over a Region: A regional need score is produced by adding up the requirements within a 25-kilometer radius. Rather than focusing only on individual pixels, this aggregation aids in identifying larger areas of need (See Fig. 4 (c)).

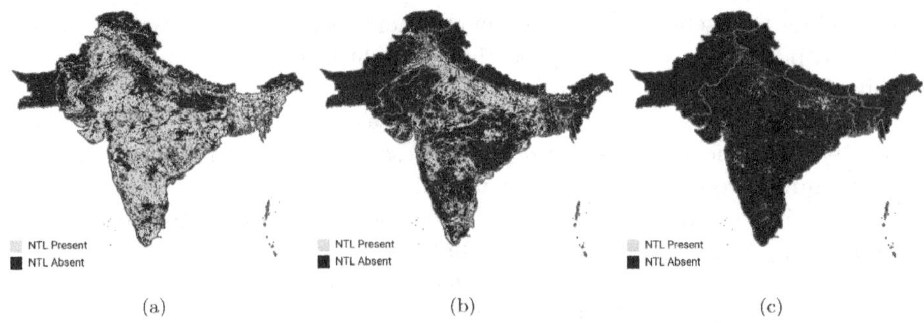

Fig. 5. Categorization of areas based on Nighttime Light (NTL) intensity: (a) Areas with $0 \leq NTL \leq 10$; (b) Areas with $10 \leq NTL \leq 20$; (c) Areas with $20 \leq NTL \leq 30$.

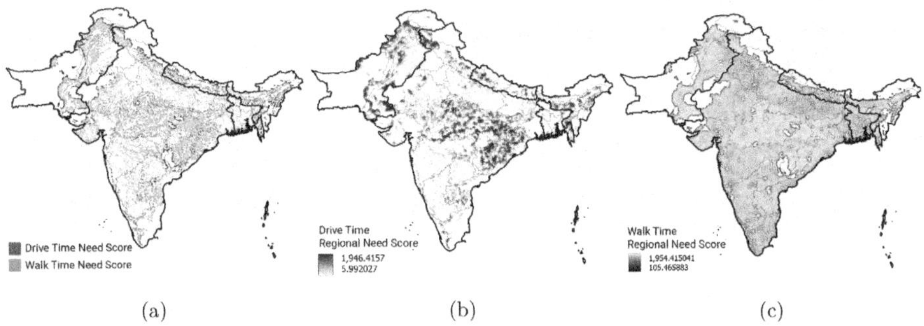

Fig. 6. (a) Regional need score to nearest hospital/clinic based on drive time and walk time; (b) Drive time regional need score; (c) Walk time regional need score.

3. Identifying High-Need Regions: By calculating the 99th percentile of the regional need ratings, the areas with the largest requirements are found. The top 1% of regions with the greatest demand for healthcare facilities are highlighted by applying this criterion.
4. Selecting Target Pixels: Target pixels are those that have a need score higher than the 99th percentile. These are the places with the greatest needs.
5. Identifying South Asian Regions: Pixels with a need score greater than the 99th percentile are considered targets. These are the areas that require the most assistance.

The final step involves visualizing the results to communicate the findings effectively:

4.2 Setting Map Center and Displaying Layers

A particular region serves as the focal point of the map, and the results are presented through a variety of layers. Among these layers are the following:

– Travel time to healthcare facilities (Refer to Fig. 3: (a))

(a)   (b)   (c)

Fig. 7. Under-served remote areas with (a): Drive time to nearest hospital/clinic is greater than 30 min and nighttime light intensity is between 0 and 10; (b): Drive time to nearest hospital/clinic is greater than 30 min and nighttime light intensity is between 10 and 20; (c): Drive time to nearest hospital/clinic is greater than 30 min and nighttime light intensity is between 20 and 30.

- Population density (Refer to Fig. 3: (b))
- Nighttime lights intensity for whole world is shown in Fig. 3: (c)) and we are focusing on South Asia in this study (see Fig. 5).
- Areas with poor healthcare access (Refer to Fig. 4: (a))
- Densely populated areas with poor healthcare access (Refer to Fig. 4: (b))
- Regional need scores (Refer to Fig. 4: (c))
- South Asian regions with the highest need (Refer to Fig. 6: (a, b, c))

The regions of the world and South Asia that will benefit most from the construction of new health facilities are clearly identified and illustrated in Fig. 4 and Fig. 6, providing important information for stakeholders and policy makers.

5 Summary of Findings

This study used three open-source datasets (Accessibility to Healthcare 2019, the Gridded Population of the World Version 4 (GPWv4), Revision 11, and VIIRS Nighttime Day/Night Annual Band Composites V2.2-to map) to assess health disparities on a worldwide scale specifically in South Asia (see Fig. 7). Our research identifies areas where healthcare facilities are most urgently required by combining data on travel time to medical facilities, population density, and nighttime light intensity. The findings underline the need of addressing health inequities, particularly in densely populated areas with limited access to healthcare. We identify the 1% of regions with the most health services by combining the needs scores calculation with the larger regions. This method ensures that treatments are directed where they are most needed, leading to a more efficient allocation of resources. The production of accurate, up-to-date and publicly available information has been facilitated by advanced technologies such as data science and Earth observation. These technologies not only help us better understand health disparities, but also provide a scalable, long-term solution for continuous development and monitoring. Our data-driven strategy can help reduce health disparities by enabling targeted treatment, especially in resource-constrained areas.

6 Strategies for Reducing Health Disparities

Health disparities between people and regions are the result of a complex interaction of social, economic and environmental variables. Identifying these gaps requires a multi-modal approach that includes data-driven insights and community-based solutions. We will review the following strategies based on our proposed strategy to identify where health disparities exist using advanced data analytics and ground observation technology:

6.1 Targeted Resource Allocation

Make use of the high-need areas that have been selected as focus centers for resource distribution. These areas have been identified by data analysis of socioeconomic factors, population density, and healthcare facility accessibility. Governments and healthcare groups can successfully lessen inequities in access to healthcare by allocating resources, such as medical staff, infrastructure, and health education initiatives, to these regions.

6.2 Addressing Socioeconomic Disparities with NTL Data

Statistics on nighttime lights (NTLs) can be used to determine how differently affluent people receive healthcare. Examining NTL data reveals disparities in the availability and standard of healthcare facilities according to socioeconomic level, especially in areas such as South Asia where health systems may prioritize the well-off over the underprivileged. Authorities can determine whether locations are devoid of access to high-quality healthcare for poor populations by superimposing NTL data with socioeconomic and health facility maps. This understanding can direct focused initiatives and legislative modifications to enhance impoverished populations' access to and quality of healthcare [26].

6.3 Leveraging Technology and Telemedicine

By using telemedicine and technology, we may provide healthcare services to underserved and rural places. By providing consultations, diagnosis, and treatment choices without requiring in-person travel, telemedicine helps close the gap that exists between patients and healthcare practitioners. This strategy can greatly improve access to healthcare and is especially helpful in areas with a weak healthcare system.

6.4 Encouraging Cross-Sector Collaboration

To address the socioeconomic determinants of health, we can encourage cooperation across the housing, transportation, education, and healthcare sectors. Since larger social and economic injustices frequently serve as the foundation for health disparities, a concerted effort including several sectors can have a more comprehensive and long-lasting effect on community health.

6.5 Regulation and Law Enforcement Actions

Encourage the adoption of evidence-based policies that advance fair access to and use of healthcare services. Regulations that address health inequities should be developed and put into effect by policymakers using data-driven insights. This entails making investments in healthcare infrastructure in places with a high need, helping low-income people pay for healthcare, and making sure that everyone has access to basic medical services.

7 Policy Framework, Shared Challenges and Joint Actions Proposal

7.1 Policy Framework

Primary healthcare is essential to efficient healthcare systems, and developing effective augmentation methods requires a thorough grasp of national health systems. A thorough understanding of the current health infrastructure is necessary to start making significant improvements, and integrating primary healthcare into health system reorganization is essential for effective reforms. One of the most important goals is Universal Health Coverage (UHC), which guarantees fair access to necessary medical treatment without financial hardship. Despite its importance, politicians frequently fail to recognize the specific contributions that primary healthcare makes to population health, which results in inadequate investment in facilities and placing a low priority on community-based specialty training. Recognized as a cost-effective and low-risk strategy, strengthening primary healthcare aims to achieve universal population coverage by 2030, emphasizing the need for regional collaboration and advocacy. Such initiatives promote prudent investment in community health services and professional training within primary healthcare settings, thereby enhancing healthcare accessibility and supporting UHC goals. The need to strengthen primary healthcare is underlined by global cooperation, which is particularly important in regions such as South Asia, which includes Bangladesh, India, Nepal, Pakistan and Sri Lanka - a quarter of the world's population. Given the shared natural resources and increased vulnerability to the effects of climate change, harmonization of health policies and promotion of cooperation between South Asian countries are crucial. Through collaboration, evidence based solutions can be found, effective health policies can be designed, and initiatives tailored to regional needs can be developed. Collaboration enables innovation in health financing, fighting non-communicable diseases and overcoming shared challenges South Asian countries participate in the World Health Organization's (WHO) offices for the South-East Asia Region (SEARO) and the Eastern Mediterranean Region (EMRO), advocating for regional Universal Health Coverage (UHC) strategies. Through the use of resources and knowledge from around the globe, WHO initiatives-which emphasize the significance of primary healthcare in achieving long-term health outcomes-support regional collaboration. This cooperative endeavor aims to enhance climate resilience, promote the welfare of South Asia's sizable populace, and enhance health results.

7.2 Shared Challenges

The following list of common issues that South Asian nations face calls for immediate attention and coordinated response from the region's decision-makers:

- Population Growth: Over the past few decades, South Asia's population has grown exponentially. India, Pakistan, and Bangladesh are among the world's most populous nations. According to UN estimates, the population of the region will reach astonishing numbers by 2050 if the current growth rate is allowed to continue unchecked. Numerous factors, such as high birth rates, limited access to family planning services, cultural norms, and religious beliefs, are blamed for this growth. It is necessary to take proactive measures to lower the overall fertility rate and raise the prevalence of contraception in order to combat population growth. Primary healthcare systems must be strengthened in light of demographic changes. This development is hampered, nevertheless, by low funding, inadequate planning, and a lack of knowledge about the role that primary healthcare plays in population health.
- Inequitable Access and Urban-Rural Disparities: In South Asian countries, there are notable geographic differences in the coverage of provinces, districts, and urban-rural areas. Research indicates that populations with lower incomes are probably less healthy, less nourished, less immunized, and less likely to use family planning. Achieving fair access to healthcare requires addressing these disparities.
- Lack of Private Sector Regulation: In South Asia, the private sector is frequently not sufficiently regulated, which causes disparities and inefficiencies in the provision of healthcare. Improving healthcare access and quality requires enforcing compliance and fortifying regulatory frameworks.
- Healthcare Workforce: The shortage of healthcare workers is a major issue for many South Asian nations, especially in rural and underdeveloped areas. Despite efforts to increase the proportion of physicians, dentists, nurses, and paramedics to the general population, disparities and shortages continue to affect the region's overall healthcare system.
- Climate Crisis: One of the areas most impacted by climate change is South Asia. Droughts, heat waves, unpredictable rainfall, glacier melting, heat waves, and other extreme weather events have caused massive economic losses

and presented significant difficulties for both the populace and governments. Morbidity and mortality are higher in children, the elderly, women, and the homeless, especially those who are impoverished and suffering from diseases. To address the climate crisis and lessen its effects on vulnerable populations, international cooperation and comprehensive policies are needed.

– Weak Healthcare Ecosystem: Developing a strong healthcare ecosystem is a major challenge for South Asia, which includes nations like India, Pakistan, Bangladesh, Nepal, Sri Lanka, Bhutan, and the Maldives. Even though the region's health indicators have improved, there are still several systemic problems that prevent healthcare services from being delivered effectively in the area. Inadequate funding, undeveloped supply chains for medical equipment, a lack of pharmaceutical company involvement in the management of non-communicable diseases (NCDs), and a dearth of healthcare facility developers are major factors in the weak healthcare ecosystem.

  • Healthcare Financing: Throughout South Asia, there is a severe lack of investment in the health sector. The amount of money allotted by the government to healthcare is frequently insufficient, usually much less than the 5 percent of GDP that is advised to be used in order to significantly advance Universal Health Coverage (UHC). For instance, the amount spent on healthcare in nations like Bangladesh and Pakistan is only a small portion of GDP-roughly 2-3 percent. Due to lack of funding, there are not as many resources available for developing infrastructure, training the healthcare workforce, or buying necessary medical supplies. Many health initiatives lack sufficient funding due to the insufficiency of donor funding and the private sector.

  • Medical Equipment Suppliers and Manufacturers: South Asia's medical equipment supply chain is undeveloped and dispersed. Due to a lack of manufacturing capacity locally, a significant amount of medical equipment and devices are imported. This reliance raises expenses and causes delays in the supply of essential equipment. In addition, variations in regulatory requirements for medical devices lead to discrepancies in terms of both safety and quality. Improving the availability of medical equipment requires stricter regulatory frameworks and stronger local manufacturing capabilities.

  • Pharmaceutical Companies and NCD Medicines:South Asia is experiencing a surge in non-communicable diseases (NCDs), including diabetes, hypertension, and cardiovascular disorders. But the pharmaceutical industry hasn't been able to meet this increasing demand. Local pharmaceutical firms frequently place a greater emphasis on generic drugs than on cutting-edge NCD therapies. There is a shortage of reasonably priced NCD drugs, especially in rural and low-income areas. Improving regulatory supervision and pharmaceutical companies' ability to create and market NCD medications are essential for better NCD management.

  • Healthcare Facilities Developers: Hospitals, clinics, and diagnostic centers are among the healthcare facilities that have not developed equally throughout South Asia. While access to these services may be somewhat better in urban areas, rural and isolated areas are still woefully under-privileged. Funding shortages, land acquisition concerns, and bureaucratic roadblocks frequently impede investments in healthcare infrastructure. Furthermore, the upkeep and modernization of current facilities are commonly disregarded, which results in declining service quality. Promoting public-private partnerships and optimizing regulatory procedures can expedite the establishment of healthcare facilities and enhance the availability of superior healthcare services.

7.3   Joint Actions for Strengthening Healthcare Ecosystems in South Asia

The healthcare ecosystems in South Asian nations need to be strengthened, and this will require a concerted and cooperative effort. Together, a data-driven strategy, policymakers' alignment, resource sharing, regional coordination, quick response capabilities, and a heavy emphasis on data science will enhance the region's resilience, healthcare delivery, and results. Through cooperation and shared commitment, South Asia can achieve a more robust and equitable healthcare system for all of its citizens. By implementing these coordinated actions, South Asian countries can significantly strengthen their healthcare ecosystems.

- Data-Driven Approach
    * Establish Regional Health Data Repositories: Construct a single, easily accessible repository for health data that can be used by all of South Asia's nations to gather, store, and process health data. Data on the prevalence of diseases, healthcare resources, patient outcomes, and other pertinent metrics ought to be included in this repository.
    * Implement Standardized Health Information Systems: To guarantee data consistency and interoperability, adopt standardized health information systems throughout the region. This will make it easier for health data from different nations to be shared and compared.
    * Utilize Advanced Analytics: Employ advanced data analytics and machine learning techniques to identify trends, predict outbreaks, and optimize resource allocation. This will enable a proactive rather than reactive approach to healthcare management.
- Align Policymakers and Local Bodies
    * Regular Inter-Governmental Health Summits: Arrange for policymakers from South Asian nations to convene on a regular basis to tackle shared healthcare issues, exchange optimal methodologies, and harmonize on regional health policies.
    * Collaborative Policy Frameworks: Develop and implement collaborative policy frameworks that address cross-border health issues such as infectious disease control, healthcare financing, and workforce development.
    * Engage Local Authorities: Make sure local organizations actively participate in the creation and execution of policies, taking into account their particular perspectives and practical experience.

- Sharing Resources and Transferability
    * Resource Sharing Agreements: Create official contracts for the sharing of essential healthcare resources like drugs, medical supplies, and knowledge. This will facilitate access to necessary services and assist in addressing shortages.
    * Cross-Border Healthcare Training Programs: Implement cross-border training programs for healthcare professionals to build capacity and ensure the transferability of skills and knowledge across the region.
    * Telemedicine and Remote Consultation Services: Increase the availability of telemedicine services to facilitate the sharing of resources and offer online consultations, particularly in underprivileged areas.
- Coordination Among Regions
    * Regional Health Coordination Bodies: Set up regional health coordination bodies to facilitate cooperation and coordination among South Asian countries. These bodies can oversee joint initiatives, monitor progress, and address emerging health challenges.
    * Joint Research and Development Initiatives: Encourage collaborative R&D projects centered around non-communicable diseases, infectious diseases, and technological advancements in healthcare.
    * Emergency Response Coordination: To effectively manage health crises such as pandemics, natural disasters, and other emergencies, develop a coordinated emergency response framework.
- Rapid Action Teams
    * Establish Regional Rapid Response Teams: Form multidisciplinary rapid response teams that can be deployed quickly to any South Asian country facing a health crisis. These teams should include medical professionals, epidemiologists, logistics experts, and communication specialists.
    * Regular Drills and Simulations:Hold frequent drills and simulations to make sure rapid response teams are capable of cross-border operations and are ready for a variety of scenarios.
    * Real-Time Communication Channels: Set up real-time communication channels for rapid action teams to coordinate efforts, share information, and mobilize resources swiftly.

- Data Scientists Team
    * Regional Data Science Consortium: Create a consortium of data scientists from South Asian countries to collaborate on health data analysis, predictive modeling, and decision support systems.
    * Training and Capacity Building: Fund educational initiatives to create a strong pool of data scientists with backgrounds in epidemiology and health informatics.
    * Data-Driven Decision Support Tools: Create and implement tools for decision-making that use data analytics to guide the allocation of resources, clinical procedures, and policy decisions

A   AccessMod

AccessMod, a developer's software, is limited to the Linux operating system, which restricts its accessibility for users on other platforms. It requires multiple datasets in various formats, such as *.shp*, *.shx*, *.dbf*, *.tif*, and *.csv*, all of which must have the same projection. This reliance on uniform projections increases complexity and makes the software prone to handling errors, especially when dealing with projection issues on the Base DEM (Digital Elevation Model). Moreover, AccessMod faces significant challenges in computing 'Access to facilities' due to these projection problems, making it difficult to accurately localize health disparities, particularly in underserved and remote areas.

In contrast, the approach outlined in our paper offers a more flexible and accessible solution. It utilizes open-source, publicly available datasets, including VIIRS Nighttime Lights, NASA's Gridded Population of the World (GPWv4), and global accessibility maps, which are easy to integrate and do not require complex file format handling or projection synchronization. This method simplifies the data processing pipeline, reducing errors and improving efficiency. Additionally, our approach is specifically designed to overcome the localization challenges that AccessMod struggles with. By using Earth observation data, population density, and travel time to healthcare facilities, we can effectively pinpoint health disparities in underserved areas, enabling more accurate and targeted interventions. This makes our method more scalable and adaptable for real-world applications in addressing health inequities.

References

1. Arcaya, M.C., Arcaya, A.L., Subramanian, S.V.: Inequalities in health: definitions, concepts, and theories. Glob. Health Action **8**(1), 27106 (2015)
2. Bank, W.: World development report 2019: the changing nature of work. The World Bank (2018)
3. Braveman, P., Tarimo, E.: Social inequalities in health within countries: not only an issue for affluent nations. Soc. Sci. Med. **54**(11), 1621–1635 (2002)
4. Carlsen, L., Bruggemann, R.: The 17 united nations' sustainable development goals: a status by 2020. Int. J. Sustain. Dev. World Ecol. **29**(3), 219–229 (2022)
5. Center for International Earth Science Information Network - CIESIN - Columbia University: gridded population of the world (GPW), v4
6. On Assuring the Health of the Public in the 21st Century, C.: The Future of the Public's Health in the 21st Century. National Academy Press (2003)
7. Citaristi, I.: United nations office for outer space affairs-unoosa. In: The Europa Directory of International Organizations 2022, pp. 247–248. Routledge (2022)
8. Elvidge, C., Zhizhin, M., Ghosh, T., Hsu, F., Taneja, J.: Annual time series of global VIIRS nighttime lights derived from monthly averages: 2012 to 2019. Remote Sens. **13**, 922 (2021)
9. Hazell, L., Shakir, S.A.: Under-reporting of adverse drug reactions. Drug Saf. **29**(5), 385–396 (2006)
10. Ibrahim, H., Liu, X., Zariffa, N., Morris, A.D., Denniston, A.K.: Health data poverty: an assailable barrier to equitable digital health care. The Lancet Digital Health **3**(4), e260–e265 (2021)

11. Indicators, A.: Monitoring the building blocks of health systems. WHO Document Production Services, Geneva, Switzerland (2010)
12. Jones, K., Moon, G.: Medical geography: taking space seriously. Prog. Hum. Geogr. **17**(4), 515–524 (1993)
13. Kiberu, V.M., Matovu, J.K., Makumbi, F., Kyozira, C., Mukooyo, E., Wanyenze, R.K.: Strengthening district-based health reporting through the district health management information software system: the Ugandan experience. BMC Med. Inform. Decis. Mak. **14**, 1–9 (2014)
14. King, N.B., Harper, S., Young, M.E.: Use of relative and absolute effect measures in reporting health inequalities: structured review. BMJ **345** (2012)
15. Macintyre, S., Ellaway, A.: Ecological approaches: rediscovering the role of the physical and social environment. Soc. Epidemiol. **9**(5), 332–348 (2000)
16. Marmot, M., Allen, J., Bell, R., Bloomer, E., Goldblatt, P.: Who European review of social determinants of health and the health divide. The Lancet **380**(9846), 1011–1029 (2012)
17. Nazir, U., Quddoos, M.T., Uppal, M., Khalid, S.: Predicting malaria outbreaks using earth observation measurements and spatiotemporal deep learning modelling: a South Asian case study from 2000 to 2017. The Lancet Planet. Health **8**, S17 (2024)
18. Oliver, A., Healey, A., Le Grand, J.: Addressing health inequalities. The Lancet **360**(9332), 565–567 (2002)
19. Organization, W.H.: Handbook on health inequality monitoring: with a special focus on low-and middle-income countries. World Health Organization (2013)
20. Organization, W.H., et al.: MDGs, millennium development goals to SDGs, sustainable development goals. World Health Organ. **204** (2015)
21. Organization, W.H., et al.: Addressing the social determinants of health: the urban dimension and the role of local government (2021)
22. People, H.: Conclusion and future directions: CDC health disparities and inequalities report-united states **62**(3), 184 (2013)
23. Sachs, J., Kroll, C., Lafortune, G., Fuller, G., Woelm, F.: Sustainable development report 2022. Cambridge University Press (2022)
24. Statistics, W.: Monitring health for the SDGs sustainable development goals. Geneva: World Health Organization (2017)
25. Tichenor, M., et al.: Interrogating the world bank's role in global health knowledge production, governance, and finance. Glob. Health **17**, 1–12 (2021)
26. Toharudin, T., et al.: Investigating adolescent vulnerability in Indonesia: a socio-remote sensing big data analytics study using night light data. IEEE Access (2024)
27. Topol, E.J.: High-performance medicine: the convergence of human and artificial intelligence. Nat. Med. **25**(1), 44–56 (2019)
28. Union, I.T.: Measuring digital development: facts and figures 2020. International Telecommunication Union (2020)
29. United Nations Department of Economic and Social Affairs: The Sustainable Development Goals Report 2023: Special Edition. UN (2023)
30. Weiss, D.J., et al.: A global map of travel time to cities to assess inequalities in accessibility in 2015. Nature **553**(7688), 333–336 (2018)
31. Weiss, D., et al.: Global maps of travel time to healthcare facilities. Nat. Med. **26**(12), 1835–1838 (2020)
32. Whitehead, M.: The concepts and principles of equity and health. Int. J. Health Serv. **22**(3), 429–445 (1992)
33. Whitehead, M.: A typology of actions to tackle social inequalities in health. J. Epidemiol. Commun. Health **61**(6), 473–478 (2007)

A Robust Tensor Decomposition Model for Traffic Data Imputation with Capped Frobenius Norm in Smart City

Linfang Yu[1,2], Hao Wang[1,2(✉)], Yuxin He[4], Chi-Sing Leung[3], and Yang Wen[1,2]

[1] Guangdong Key Laboratory of Intelligent Information Processing, Shenzhen University, Shenzhen, China
haowang@szu.edu.cn
[2] Guangdong Multimedia Information Service Engineering Technology Research Center, Shenzhen University, Shenzhen, China
[3] Department of Electronic Engineering, City University of Hong Kong, Hong Kong, China
[4] College of Urban Transportation and Logistics, Shenzhen Technology University, Shenzhen, China

Abstract. Data missing is a frequent issue in smart city systems, resulting in poor accuracy and reliability in related applications. Traditional models for traffic data imputation are often under the assumption of outlier-free data, limiting their effectiveness in real-world scenarios with outliers. In this work, we devise a robust tensor completion method for traffic data imputation (STTC-CF) based on tensor ring decomposition and Capped Frobenius norm to enhance robustness against missing data and outliers. Subsequently, the half-quadratic (HQ) optimization technique is utilized to transform the original problem into a tractable form. The solution to this reformulated problem is attained through alternating optimization combined with the alternating direction multiplier method (AO-ADMM). Extensive testing on three real-world traffic datasets demonstrates that our proposed method surpasses several state-of-the-art algorithms in traffic data imputation accuracy across various simulated scenarios.

Keywords: Traffic data imputation · Tensor completion · Robust method · Outliers

1 Introduction

Urban transportation data is crucial for effective transportation services and smart city applications like traffic management, infrastructure planning, and

This paper is supported by the National Natural Science Foundation of China (Grant No. 62206178 and 72301180) and Stable Support Plan for Higher Education Institutions in Shenzhen (Project No. 20231121221536001).

traffic flow forecasting. However, sensor malfunctions and network connectivity issues often cause missing data, compromising system reliability. Using incomplete data for prediction and analysis can lead to inaccurate outcomes, making accurate data recovery essential for smart cities.

Traffic data exhibits complex spatiotemporal variations and interconnections. To capture this information, this paper structures traffic data into a three-dimensional tensor of $space \times time \times day$, rather than the traditional spatiotemporal matrix. The low-rank assumption, effective due to daily periodicity and adjacent similarity in traffic data, is utilized in many studies to facilitate data completion.

Current research in this field can broadly be divided into two categories. The first focuses on rank minimization, seeking a low-rank approximation for incomplete tensors. Despite the common use of the tensor nuclear norm (NN), which can lead to excessive shrinkage, alternatives like truncated nuclear norm (TNN) [3] and truncated tensor Schatten p-norm (TSpN) [7] have been introduced to address this issue. However, rank minimization methods usually exhibit sensitivity to model parameters, necessitating intricate parameter adjustments. The second category is based on tensor decomposition, factorizing the incomplete tensor through a series of factors. Various models, such as CANDECOMP/PARAFAC (CP) decomposition with Bayesian inference, have been explored [2]. Recently, tensor ring (TR) decomposition has gained attention for tensor completion, showing a promising performance by factorizing an Nth-order tensor into N 3rd-order tensors. In traffic data imputation, TR decomposition has been applied with enhanced representation capabilities, leading to excellent data recovery performance [12].

Apart from the global structure, it is crucial to take local spatiotemporal priors into account when modeling traffic data. For instance, spatiotemporal consistency is integrated into the low-rank part through the manifold embedding approach, encoding the relation using a k-NN graph with a Gaussian kernel [9]. In contrast, the spatial adjacency matrix is constructed by the geographical road network [10].

However, many of the aforementioned approaches assume a Gaussian distribution for the objective function, but real-world traffic data inevitably contains anomalies. These anomalies can be non-structural, such as sensor errors, or structural, like rare accidents or temporary road controls. They are sparse and often manifest as outliers in the data. Recently, researchers have started addressing these anomalies in traffic data modeling. The study in [9] introduces a sparse tensor for anomaly detection and takes the spatially sparse and temporally smooth structure of urban traffic data into account by modeling the anomalies with ℓ_1-norm, thus enhancing the data recovery ability of the model. [4] builds a robust model for traffic data recovery by modeling unstructured anomalies with ℓ_1-norm constraint. [5] propose a method for robust data recovery and traffic extreme event detection, employing $\ell_{2,1}$-norm regularization to model structured anomalies. However, the performance of these methods typically relies on the assumption of abnormal distribution and the influence of the sparsity control parameter.

To overcome the problem mentioned above, we propose a novel spatial-temporal tensor completion model against anomalies and introduce a kernel method to adaptively adjust parameters related to anomalies. Note that the anomaly we consider in this work is the outlier. The main contributions of this work are summarized as follows:

1. We propose a robust spatial-temporal tensor completion with a Capped Frobenius norm termed STTC-CF for traffic data imputation considering outliers. STTC-CF uses the Capped Frobenius norm as an approximation error measure based on the tensor ring decomposition model. Hence, it is capable of resisting outliers.
2. We transform the non-convex optimization problem into a Frobenius norm term with a convex regularization term via half-quadratic (HQ) technique. By applying Alternating Optimization with the Alternating Direction Method of Multipliers (AO-ADMM), the multi-variable optimization is separated into a series of solvable convex subproblems.
3. Experiments on real-world traffic data show that our method achieves robustness in traffic data imputation tasks under various scenarios.

The rest of this paper is structured as follows. Section 2 introduces notations and provides background information. In Sect. 3, we formulate the proposed model and derive its solution. Section 4 presents the experimental evaluation based on real-world datasets. The findings and conclusions are summarized in Sect. 5.

2   Notations and Preliminaries

2.1   Notations

Bold calligraphic letters, bold upper cases, bold lower cases, and italics denote tensors, matrices, vectors, and scalars, respectively. Consider a Nth order tensor $\mathcal{X} \in \mathbb{R}^{I_1 \times I_2 \times \cdots \times I_N}$, its Frobenius norm is defined as $\|\mathcal{X}\|_F = \sqrt{\sum_{i_1,i_2,\ldots,i_n} x_{i_1 i_2 \ldots i_n}^2}$, and $\mathbf{X}_{(n)} \in \mathbb{R}^{I_n \times I_1 \cdots I_{n-1} I_{n+1} \cdots I_N}$ and $\mathbf{X}_{[n]} \in \mathbb{R}^{I_n \times I_{n+1} \cdots I_N I_1 \cdots I_{n-1}}$ represent two types mode-n unfolding of $\mathcal{X}$. Given a matrix $\mathbf{X} \in \mathbb{R}^{I_1 \times I_2}$, $\mathbf{X}^\top$ is the transpose of $\mathbf{X}$, and $\mathrm{tr}(\mathbf{X}) = \sum_{i=1}^{\min\{I_1, I_2\}} x_{ii}$ is the trace of $\mathbf{X}$.

2.2   Tensor Completion Model via Tensor Ring Decomposition

Definition 1 (Multilinear Product). *Given two tensors $\mathcal{X}_1 \in \mathbb{R}^{I_1 \times I_2 \times J}$ and $\mathcal{X}_2 \in \mathbb{R}^{J \times I_3 \times I_4}$, the multilinear product of $\mathcal{X}_1$ and $\mathcal{X}_2$ is denoted as $\mathcal{X}_{1,2} \in \mathbb{R}^{I_1 \times I_2 I_3 \times I_4}$, and the lateral slice matrix of $\mathcal{X}_{1,2}$ is defined as*

$$\mathbf{X}_{1,2}((i_3 - 1)I_2 + i_2) = \mathbf{X}_1(i_2)\mathbf{X}_2(i_3), \tag{1}$$

where $\mathbf{X}_{1,2} \in \mathbb{R}^{I_1 \times I_4}$, $\mathbf{X}_1(i_2)$ and $\mathbf{X}_2(i_3)$ are i_2-th and i_3-th lateral slice matrix of $\mathcal{X}_1$ and $\mathcal{X}_2$, respectively.

Definition 2 (Tensor Ring Decomposition [14]). *A Nth order tensor $\mathcal{X} \in \mathbb{R}^{I_1 \times I_2 \times \cdots \times I_N}$ can be factorized over a sequence of N 3rd-order core tensors $\mathcal{G} = \{\mathcal{G}_1, \mathcal{G}_2, \ldots, \mathcal{G}_N\}$, where $\mathcal{G}_n \in \mathbb{R}^{R_n \times I_n \times R_{n+1}}, n = 1, 2, \ldots, N$, which can be expressed in an element-wise form given by as follows:*

$$\mathcal{X}(i_1, i_2, \ldots, i_N) = \sum_{r_1, r_2, \ldots, r_N = 1}^{R_1, R_2, \ldots, R_N} \prod_{n=1}^{N} \mathcal{G}_n(r_n, i_n, r_{n+1}) \tag{2}$$

where $R = [R_1, R_2, \ldots, R_N] \ll \min[I_1, I_2, \cdots, I_N]$ is the TR rank. For simplicity, (2) is represented as $\mathcal{X} = \mathrm{TR}(\mathcal{G}_1, \mathcal{G}_2, \cdots, \mathcal{G}_N)$. Besides, the matrix form of (2) can be rewritten as

$$\mathbf{X}_{[n]} = \mathbf{G}_{n(2)} \mathbf{G}_{\neq n[2]}, \tag{3}$$

where $\mathbf{G}_{\neq n[2]}$ represents the mode-2 unfolding matrix of $\mathcal{G}_{\neq n}$, and $\mathcal{G}_{\neq n}$ is the subchain resulted from the multilinear product of all core tensors except $\mathcal{G}_n$.

2.3   Traffic Data Imputation Model Through Tensor Completion

Consider an incomplete traffic dataset from I_1 sensors over I_2 time points and I_3 days, represented as a third-order partially observed tensor $\mathcal{M}_\Omega \in \mathbb{R}^{I_1 \times I_2 \times I_3}$, where $\Omega \in \mathbb{R}^{I_1 \times I_2 \times I_3}$ denotes the observed entries, with $\Omega_{ijk} = 1$ for observed and $\Omega_{ijk} = 0$ for missing data. Due to the global low-rank property of traffic data across space and time, we use TR decomposition with low-rank core tensors for tensor completion. Thus, the problem is formulated as [12]

$$\min_{\mathcal{G}_1, \mathcal{G}_2, \mathcal{G}_3, \mathcal{X}} \|\mathcal{X} - \mathrm{TR}(\mathcal{G}_1, \mathcal{G}_2, \mathcal{G}_3)\|_{\mathrm{F}}^2$$
$$\text{s.t. } \mathcal{X}_\Omega = \mathcal{M}_\Omega, \tag{4}$$

where $\mathcal{X} \in \mathbb{R}^{I_1 \times I_2 \times I_3}$ is the recovered tensor, and the constraint $\mathcal{X}_\Omega = \mathcal{M}_\Omega$ implies that $\mathcal{X}$ contains partial observed entries in $\mathcal{M}_\Omega$.

3   Methodology

In this section, we propose a new objective function by combining TR decomposition with Capper Frobenius norm measure and spatiotemporal regularizations. We use the HQ technique and AO-ADMM as solvers to solve the optimization problem.

3.1   Capper Frobenius Norm for Outliers
Tensor Capper Frobenius Norm

Definition 3. *Given a tensor $\mathcal{R} \in \mathbb{R}^{I_1 \times I_2 \times I_3}$, the Capper Frobenius norm of $\mathcal{R}$ is defined as:*

$$\|\mathcal{R}\|_\phi = \sqrt{\sum_{i=1}^{I_1} \sum_{j=1}^{I_2} \sum_{k=1}^{I_3} \min\{r_{ijk}^2, e^2\}}, \tag{5}$$

where e is a capped threshold.

222   L. Yu et al.

Consider $\mathcal{R} = \mathcal{X} - \mathrm{TR}(\mathcal{G}_1, \mathcal{G}_2, \mathcal{G}_3)$ representing the approximation error, where the parameter e determines the boundary between normal errors and outliers. Elements with values below e are normal errors, while those exceeding e are outliers. Without outliers, both the Capped Frobenius norm and the Frobenius norm perform well. However, with outliers, the Frobenius norm becomes sensitive to increased approximation errors and does not handle traffic data imputation effectively. In contrast, the Capped Frobenius norm limits the influence of outliers by controlling e and performs well for data imputation. We use an adaptive strategy to adjust e, described as follows:

$$e = \arg\min_{r_{ijk}} \psi, \tag{6}$$

where ψ represents the outlier group which is distinguished from $\mathcal{R}$. To obtain outlier group ψ, we adopt Laplacian kernel to increase the magnitude of the difference between normal error and outliers, resulting in:

$$\hat{\mathcal{R}} = \exp\left(-\frac{\mathrm{abs}(\mathcal{R})}{\sigma}\right) \tag{7}$$

where $\mathrm{abs}(\mathcal{R})$ represents absolute operation for each element of $\mathcal{R}$, and σ is the kernel width and determined by Silverman'rule [8]. After that, we get the outlier group ψ by $\psi = \{r_{ijk} | \hat{r}_{ijk} < \epsilon\}$, where ϵ is a pre-setting parameter. Although ϵ must be set to determine e, our model is not highly sensitive to ϵ and e because the kernel amplifies the difference between outliers and normal errors. Taking traffic speed as an example, after applying the Laplacian kernel to the approximation error, the magnitude of normal errors exceeds 10^{-30}, while the magnitude of outliers is typically below 10^{-50}. Consequently, e can be selected from the range between 10^{-50} and 10^{-30}, and within this interval, the choice of ϵ does not substantially affect the experimental outcomes.

3.2 Spatio-Temporal Feature Learning

According to the formulation (3), the matrix form of TR decomposition for the traffic tensor is represented as follows:

$$\mathbf{X}_{[n]} = \mathbf{G}_{n(2)} \mathbf{G}_{\neq n[2]}, \quad n = 1, 2, 3. \tag{8}$$

For the mode-1 unfolding matrix $\mathbf{X}_{[1]} \in \mathbb{R}^{I_1 \times I_2 I_3}$, its rows preserve the spatial attribute of the traffic tensor $\mathcal{X}$. Correspondingly, the mode-2 unfolding matrix $\mathbf{G}_{1(2)}$ of core tensor $\mathcal{G}_1$ can be considered as the latent features of the spatial attribute, and the remaining core tensors $\mathcal{G}_2, \mathcal{G}_3$ are considered the weight of the latent features. Generalizing this concept to other modes, the core tensors $\mathcal{G}_2$ and $\mathcal{G}_3$ represent the primary feature of time mode and day mode, respectively. To improve the accuracy of data imputation and the representational ability of core tensors, we introduce spatial and temporal knowledge into core tensors. Similar to the operation in [9], the prior information in the three modes can be retained in the core tensors $\mathcal{G}_n, n = 1, 2, 3$ through the graph Laplacian term defined as:

$$\mathrm{tr}(\mathbf{G}_{n(2)}^{\top} \mathbf{L}_n \mathbf{G}_{n(2)}), \tag{9}$$

where $\mathbf{L}_n \in \mathbb{R}^{I_n \times I_n}$ is the Laplacian matrix calculated based on the rows of $\mathbf{X}_{[n]}$.

3.3   Model Formulation

Using the Capped Frobenius norm to measure approximation error and incorporating spatial-temporal priors, we formulate the STTC-CF problem as

$$\min_{\mathcal{G}_1,\mathcal{G}_2,\mathcal{G}_3,\mathcal{X}} \|\mathcal{X} - \mathrm{TR}(\mathcal{G}_1,\mathcal{G}_2,\mathcal{G}_3)\|_\phi^2 + \sum_{n=1}^{3} \alpha_n \mathrm{tr}(\mathbf{G}_{n(2)}^\top \mathbf{L}_n \mathbf{G}_{n(2)}) \tag{10}$$

$$\text{s.t. } \mathcal{G}_1,\mathcal{G}_2,\mathcal{G}_3 \geq 0, \mathcal{X}_\Omega = \mathcal{M}_\Omega,$$

where α_n is the regularization parameter, and the non-negative constraint on core tensors conforms to the non-negativity of traffic data. The STTC-CF model presented in (10) not only reduces the influence of general Gaussian noise but also effectively resists the negative impact of outliers. In addition, STTC-CF takes advantage of the local spatiotemporal information within traffic data. However, problem (10) is difficult to directly solve due to the non-convexity and non-smoothness of the Capper Frobenius norm. In the following subsection, we will discuss the solution method for this problem.

3.4   Solutions

Optimization via Half-Quadratic Theory. To solve problem (10), we use Half-Quadratic (HQ) optimization theory. With the introduction of an auxiliary variable $\mathcal{A} \in \mathbb{R}^{I_1 \times I_2 \times I_3}$, the formulation (10) can be rephrased as.

$$\min_{\mathcal{G}_1,\mathcal{G}_2,\mathcal{G}_3,\mathcal{X},\mathcal{A}} \|\mathcal{X} - \mathrm{TR}(\mathcal{G}_1,\mathcal{G}_2,\mathcal{G}_3) - \mathcal{A}\|_F^2 + \varphi_e(\mathcal{A}) + \sum_{n=1}^{3} \alpha_n \mathrm{tr}(\mathbf{G}_{n(2)}^\top \mathbf{L}_n \mathbf{G}_{n(2)}) \tag{11}$$

$$\text{s.t. } \mathcal{G}_1,\mathcal{G}_2,\mathcal{G}_3 \geq 0, \mathcal{X}_\Omega = \mathcal{M}_\Omega,$$

where $\mathcal{A}$ can be considered as the outlier component in the traffic tensor, and $\varphi_e(\mathcal{A})$ has the formulation as

$$\varphi_e(\mathcal{A}) = \sum_{i=1}^{I_1}\sum_{j=1}^{I_2}\sum_{k=1}^{I_3}[e^2 - (e - \min\{|a|, e\})^2]. \tag{12}$$

The derivation of (12) is provided in Appendix A.1.

Optimization via AO-ADMM. Based on the AO framework, let $\mathcal{Y}^k = \mathcal{X}^k - \mathcal{A}^k$, problem (11) can be solved by alternately optimizing the following subproblems:

$$\mathcal{G}_1^{k+1} = \arg\min_{\mathcal{G}_1 \geq 0}\|\mathcal{Y}^k - \mathrm{TR}(\mathcal{G}_1,\mathcal{G}_2^k,\mathcal{G}_3^k)\|_F^2 + \alpha_1 \mathrm{tr}(\mathbf{G}_{1(2)}^\top \mathbf{L}_1 \mathbf{G}_{1(2)}) + \beta\|\mathcal{G}_1 - \mathcal{G}_1^k\|_F^2 \tag{13a}$$

$$\mathcal{G}_2^{k+1} = \arg\min_{\mathcal{G}_2 \geq 0}\|\mathcal{Y}^k - \mathrm{TR}(\mathcal{G}_1^{k+1},\mathcal{G}_2,\mathcal{G}_3^k)\|_F^2 + \alpha_2 \mathrm{tr}(\mathbf{G}_{2(2)}^\top \mathbf{L}_2 \mathbf{G}_{2(2)}) + \beta\|\mathcal{G}_2 - \mathcal{G}_2^k\|_F^2 \tag{13b}$$

$$\mathcal{G}_3^{k+1} = \arg\min_{\mathcal{G}_3 \geq 0}\|\mathcal{Y}^k - \mathrm{TR}(\mathcal{G}_1^{k+1},\mathcal{G}_2^{k+1},\mathcal{G}_3)\|_F^2 + \alpha_3 \mathrm{tr}(\mathbf{G}_{3(2)}^\top \mathbf{L}_3 \mathbf{G}_{3(2)}) + \beta\|\mathcal{G}_3 - \mathcal{G}_3^k\|_F^2$$

$$\tag{13c}$$

$$\mathcal{A}^{k+1} = \arg\min_{\mathcal{A}}\|\mathcal{X}^k - \mathrm{TR}(\mathcal{G}_1^{k+1},\mathcal{G}_2^{k+1},\mathcal{G}_3^{k+1}) - \mathcal{A}\|_F^2 + \varphi_e(\mathcal{A}), \tag{13d}$$

224   L. Yu et al.

$$\mathcal{X}^{k+1} = \underset{\mathcal{X}_\Omega = \mathcal{M}_\Omega}{\arg\min} \|\mathcal{X} - \mathrm{TR}(\mathcal{G}_1^{k+1}, \mathcal{G}_2^{k+1}, \mathcal{G}_3^{k+1}) - \mathcal{A}^{k+1}\|_F^2, \tag{13e}$$

where $\mathcal{Y}^k = \mathcal{X}^k - \mathcal{A}^k$ and $\beta\|\mathcal{G}_n - \mathcal{G}_n^k\|_F^2$ imposed on the block update (13a-13c) is to make sure that variables converge to a stationary point.

Firstly, we address the core tensors update subproblems. Equations (13a), (13b) and (13c) have the similar structure, thus, they can be generalized as follows:

$$\mathcal{G}_n^{k+1} = \underset{\mathcal{G}_n \geq 0}{\arg\min} \|\mathcal{Y}^k - \mathrm{TR}(\mathcal{G}_{n-1}^{k+1}, \mathcal{G}_n, \mathcal{G}_{n+1}^k)\|_F^2 + \alpha_n \mathrm{tr}(\mathbf{G}_{n(2)}^\top \mathbf{L}_n \mathbf{G}_{n(2)}) + \beta\|\mathcal{G}_n - \mathcal{G}_n^k\|_F^2. \tag{14}$$

Referring to the matrix form (8), we rewrite (14) as follows:

$$\mathcal{G}_n^{k+1} = \underset{\mathbf{G}_{n(2)} \geq 0}{\arg\min} \|\mathbf{Y}_{[n]}^k - \mathbf{G}_{n(2)}\mathbf{G}_{\neq n[2]}\|_F^2 + \alpha_n \mathrm{tr}(\mathbf{G}_{n(2)}^\top \mathbf{L}_n \mathbf{G}_{n(2)}) + \beta\|\mathbf{G}_{n(2)} - \mathbf{G}_{n(2)}^k\|_F^2 \tag{15}$$

Subsequently, to separate the approximation term, regularization term and non-negative constraint, we introduce auxiliary variables $\{\mathbf{Q}\} := \{\mathbf{Q}_1, \mathbf{Q}_2, \mathbf{Q}_3\} \in \mathbb{R}^{I_n \times I_n}$ and $\{\mathbf{Z}\} := \{\mathbf{Z}_1, \mathbf{Z}_2, \mathbf{Z}_3\} \in \mathbb{R}^{I_n \times I_n}$. After that, problem (15) is recast as

$$\underset{\mathbf{G}_{n(2)}, \mathbf{Q}_n, \mathbf{Z}_n}{\min} \|\mathbf{Y}_{[n]}^k - \mathbf{G}_{n(2)}\mathbf{G}_{\neq n[2]}\|_F^2 + \alpha_n \mathrm{tr}(\mathbf{Q}_n^\top \mathbf{L}_n \mathbf{Q}_n) + \mathcal{I}(\mathbf{Z}_n) + \beta\|\mathbf{G}_{n(2)} - \mathbf{G}_{n(2)}^k\|_F^2$$
$$\text{s.t. } \mathbf{Q}_n = \mathbf{G}_{n(2)}, \mathbf{Z}_n = \mathbf{G}_{n(2)}, \tag{16}$$

where $\mathcal{I}(\mathbf{Z}_n)$ is an indicator function, which indicates 0 for elements z_{ij} that are non-negative, and 1 for elements z_{ij} that are negative. The augmented Lagrangian function of problem (16) is given by

$$\mathcal{L}(\mathbf{G}_{n(2)}, \mathbf{Q}_n, \mathbf{Z}_n, \mathbf{U}_n, \mathbf{V}_n) = \|\mathbf{Y}_{[n]}^k - \mathbf{G}_{n(2)}\mathbf{G}_{\neq n[2]}\|_F^2 + \alpha_n \mathrm{tr}(\mathbf{Q}_n^\top \mathbf{L}_n \mathbf{Q}_n) + \mathcal{I}(\mathbf{Z}_n)$$
$$+ \beta\|\mathbf{G}_{n(2)} - \mathbf{G}_{n(2)}^k\|_F^2 + \langle \mathbf{U}_n, \mathbf{G}_{n(2)} - \mathbf{Q}_n \rangle + \frac{\eta}{2}\|\mathbf{G}_{n(2)} - \mathbf{Q}_n\|_F^2 \tag{17}$$
$$+ \langle \mathbf{V}_n, \mathbf{G}_{n(2)} - \mathbf{Z}_n \rangle + \frac{\mu}{2}\|\mathbf{G}_{n(2)} - \mathbf{Z}_n\|_F^2,$$

where $\langle \mathbf{A}, \mathbf{B} \rangle = \mathrm{tr}(\mathbf{A}^\top \mathbf{B})$, $\mathbf{U}_n, \mathbf{V}_n \in \mathbb{R}^{I_n \times I_n}$ are the Lagrange multipliers and $\eta, \mu > 0$ are the trade-off parameters for the augmented terms. According to the ADMM framework, the subproblem (16) yields the following iterative procedures:

$$\mathbf{G}_{n(2)}^{l+1} := \underset{\mathbf{G}_{n(2)}}{\arg\min} \mathcal{L}\left(\mathbf{G}_{n(2)}, \mathbf{Q}_n^l, \mathbf{Z}_n^l, \mathbf{U}_n^l, \mathbf{V}_n^l\right), \tag{18a}$$

$$\mathbf{Q}_n^{l+1} := \underset{\mathbf{Q}_n}{\arg\min} \mathcal{L}\left(\mathbf{G}_{n(2)}^{l+1}, \mathbf{Q}_n, \mathbf{Z}_n^l, \mathbf{U}_n^l, \mathbf{V}_n^l\right), \tag{18b}$$

$$\mathbf{Z}_n^{l+1} := \underset{\mathbf{Z}_n \geq 0}{\arg\min} \mathcal{L}\left(\mathbf{G}_{n(2)}^{l+1}, \mathbf{Q}_n^{l+1}, \mathbf{Z}_n, \mathbf{U}_n^l, \mathbf{V}_n^l\right), \tag{18c}$$

$$\mathbf{U}_n^{l+1} = \mathbf{U}_n^l + \eta(\mathbf{G}_{n(2)}^{l+1} - \mathbf{Q}_n^{l+1}), \tag{18d}$$

$$\mathbf{V}_n^{l+1} = \mathbf{V}_n^l + \mu(\mathbf{G}_{n(2)}^{l+1} - \mathbf{Z}_n^{l+1}), \tag{18e}$$

where l denotes the l-th iteration of ADMM algorithm. Besides, (18a-18c) have the following solutions:

$$\mathbf{G}_{n(2)}^{l+1} = \left[\mathbf{Y}_{[n]}^k \mathbf{G}_{\neq n[2]} + \mu\mathbf{Z}_n^l - \mathbf{V}_n^l + \eta\mathbf{Q}_n^l - \mathbf{U}_n^l + \beta\mathbf{G}_{n(2)}^k\right]$$
$$\left[(\mathbf{G}_{\neq n[2]})^\top \mathbf{G}_{\neq n[2]} + (\mu + \eta + \beta)\mathbf{I}\right]^{-1}, \tag{19a}$$

$$\mathbf{Q}_n^{l+1} = (\eta\mathbf{I} + \alpha\mathbf{L}_n)^{-1}(\eta\mathbf{G}_{n(2)}^{l+1} + \mathbf{U}_n^l), \tag{19b}$$

$$\mathbf{Z}_n^{l+1} = \mathcal{P}_+\left(\mathbf{G}_{n(2)}^{n+1} + \frac{\mathbf{V}_n^l}{\mu}\right), \tag{19c}$$

where $\mathcal{P}_+(x) = \max\{x, 0\}$. When $\mathbf{G}_{n(2)}^{l+1}$ converges, the solutions of subproblems (13a), (13b) and (13c) are obtained from $\mathcal{G}_n^{k+1} = \text{folding}\left(\mathbf{G}_{n(2)}^{l+1}\right)$. The termination criterion is described as follows:

$$\epsilon_1 = \max\left\{\frac{\|\mathbf{G}_{n(2)}^{l+1} - \mathbf{G}_{n(2)}^l\|_F^2}{\|\mathbf{G}_{n(2)}^l\|_F^2}, \frac{\|\mathbf{G}_{n(2)}^{l+1} - \mathbf{Q}_n^{l+1}\|_F^2}{\|\mathbf{G}_{n(2)}^{l+1}\|_F^2}, \frac{\|\mathbf{G}_{n(2)}^{l+1} - \mathbf{Z}_n^{l+1}\|_F^2}{\|\mathbf{G}_{n(2)}^{l+1}\|_F^2}.\right\} \leq 10^{-3}. \tag{20}$$

The detailed ADMM algorithm for (15) is summarized in Algorithm 1.

Algorithm 1. ADMM for core tensor $\mathcal{G}_n^{k+1}$ update

---

Input: $\mathcal{X}^k$, $\mathcal{A}^k$, $\mathcal{G}_{n-1}^{k+1}$, $\mathcal{G}_n^k$, $\mathcal{G}_{n+1}^k$, $\mathbf{L}_n$, α_n.
1: Initialize $\mathbf{G}_{n(2)}^0$=unfolding$(\mathcal{G}_n^k)$, $\mathbf{Q}_n^0, \mathbf{Z}_n^0, \mathbf{U}_n^0, \mathbf{V}_n^0, \eta, \mu$.
2: Calculate $\mathcal{G}_{\neq n}$ via (1) with $\mathcal{G}_{n-1}^{k+1}$ and $\mathcal{G}_{n+1}^k$ and obtain $\mathbf{G}_{\neq n[2]}$ =unfolding$(\mathcal{G}_{\neq n})$
3: **repeat**
4:     Update $\mathbf{G}_{n(2)}^{l+1}$ via Eq. (19a).
5:     Update $\mathbf{Q}_n^{l+1}$ via Eq. (19b).
6:     Update $\mathbf{Z}_n^{l+1}$ via Eq. (19c).
7:     Update $\mathbf{U}_n^{l+1}$ via Eq. (18d).
8:     Update $\mathbf{V}_n^{l+1}$ via Eq. (18e).
9:     $\eta = \min\{1.1\eta, 10^8\}, \mu = \min\{1.1\mu, 10^8\}$,
10:     $l = l + 1$.
11: **until** The termination criterion (20) is satisfied.
Output: $\mathcal{G}_n^{k+1}$ =folding$\left(\mathbf{G}_{n(2)}^{l+1}\right)$.

Next, we focus on tackling subproblem (13d). Let $\mathcal{R}^{k+1} = \mathcal{X}^k - \text{TR}(\mathcal{G}_1^{k+1}, \mathcal{G}_2^{k+1}, \mathcal{G}_3^{k+1})$. Subproblem (13d) is rewritten as

$$\mathcal{A}^{k+1} = \arg\min_{\mathcal{A}}\|\mathcal{R}^{k+1} - \mathcal{A}\|_F^2 + \varphi_e(\mathcal{A}), \tag{21}$$

Its solution is given by

$$\mathcal{A}^{k+1} = \mathcal{T}_{e^{k+1}}(\mathcal{R}^{k+1}) = r_{iik}\text{sign}(|r_{iik}| - \min\left\{|r_{ijk}|, e^{k+1}\right\}) \tag{22}$$

which is derived in Appendix A.2.

Finally, for the update $\mathcal{X}$, we can easily get the solution

$$\mathcal{X}^{k+1} = \text{TR}\left(\mathcal{G}_1^{k+1}, \mathcal{G}_2^{k+1}, \mathcal{G}_3^{k+1}\right)_{\Omega_c} + \mathcal{M}_\Omega, \tag{23}$$

where Ω_c is the complement set of Ω.

Based on the aforementioned derivation, the proposed STTC-CF algorithm is summarized in Algorithm 2.

Algorithm 2. STTC-CF Algorithm

Input: $\mathcal{M}_\Omega, \mathbf{L}_1, \mathbf{L}_2, \mathbf{L}_3, R_1, R_2, R_3, \alpha_1, \alpha_2, \alpha_3.$
1: Initialize $\mathcal{G}_1^0, \mathcal{G}_2^0, \mathcal{G}_3^0, \mathcal{A}^0, \mathcal{X}^0.$
2: **repeat**
3:     **for** $n = 1, 2, 3$ **do**
4:         update $\mathcal{G}_n^{k+1}$ via Algorithm 1
5:     **end for**
6:     Update $\mathcal{A}^{k+1}$ via (22).
7:     Update $\mathcal{X}^{k+1}$ via (23).
8:     Calculate $\epsilon_2 = \frac{\|\mathcal{X}^{k+1} - \mathcal{X}^k\|_F^2}{\|\mathcal{X}^k\|_F^2}.$
9:     $k = k + 1.$
10: **until** $\epsilon_2 \leq 10^{-4}$ or $k > 500$
Output: $\mathcal{X}.$

3.5   Complexity Analysis

In this section, we analyze the computational complexity of STTC-CF. The primary computational cost in each iteration relies on updating the variables $\mathcal{G}_n, \mathbf{Q}_n, \mathbf{Z}_n, \mathbf{U}_n, \mathbf{V}_n, \mathcal{X}, \mathcal{A}$. Assuming that the number N of inner loop iterations for updating $\mathcal{G}_n$ using the ADMM framework is fixed, the complexity of updating $\mathcal{G}_1, \mathcal{G}_2,$ and $\mathcal{G}_3$ is $\mathcal{O}(I_1 I_2 I_3 r_1 r_2 N)$, $\mathcal{O}(I_1 I_2 I_3 r_2 r_3 N)$, and $\mathcal{O}(I_1 I_2 I_3 r_1 r_3 N)$, respectively. Updating $\mathbf{Q}_1, \mathbf{Q}_2,$ and $\mathbf{Q}_3$ requires $\mathcal{O}(I_1^3 N)$, $\mathcal{O}(I_2^3 N)$, and $\mathcal{O}(I_3^3 N)$, respectively. Similarly, updating $\mathbf{Z}_1, \mathbf{Z}_2,$ and $\mathbf{Z}_3$ needs $\mathcal{O}(I_1 r_1 r_2 N)$, $\mathcal{O}(I_2 r_2 r_3 N)$, and $\mathcal{O}(I_3 r_1 r_3 N)$, respectively. The updates for $\mathbf{U}_1, \mathbf{U}_2,$ and $\mathbf{U}_3$ require $\mathcal{O}(I_1 r_1 r_2 N)$, $\mathcal{O}(I_2 r_2 r_3 N)$, and $\mathcal{O}(I_3 r_1 r_3 N)$, respectively. Likewise, the updates for $\mathbf{V}_1, \mathbf{V}_2,$ and $\mathbf{V}_3$ need $\mathcal{O}(I_1 r_1 r_2 N)$, $\mathcal{O}(I_2 r_2 r_3 N)$, and $\mathcal{O}(I_3 r_1 r_3 N)$, respectively. The updates both $\mathcal{A}$ and $\mathcal{X}$ are $\mathcal{O}(I_1 I_2 I_3 r_1 r_2 r_3)$. Therefore, the total computational complexity of STTC-CF per iteration is $\mathcal{O}(I_1 I_2 I_3 m_1 m_2 m_3)$, where m_1, m_2, m_3 represent the three largest numbers in $\{r_1, r_2, r_3, N\}$.

Furthermore, we compare STTC-CF with the nonnegative matrix factorization-and-completion using the ADMM algorithm (NMF-ADMM) [13]. For a dataset of size I_n, the computational complexity of STTC-CF in each iteration is $\mathcal{O}(I_1 I_2 I_3 \, m_1 m_2 m_3)$, with ranks r_1, r_2, r_3 and inner loop N iterations. In comparison, the computational complexity of NMF-ADMM is $\mathcal{O}(I_1' I_2' r)$ with rank r, where $I_n = I_1 I_2 I_3 = I_1' I_2'$. The difference in computational complexity between STTC-CF and NMF-ADMM is influenced by the latent ranks and inner loop N iterations in STTC-CF. In fact, due to the low-rank nature of traffic data, the ranks r_1, r_2, r_3 are often small positive integers. In our experiments, Algorithm 1 generally converges within 20 iterations, meaning $N = 20$.

4   Experiments

4.1   Experiment Settings

Data Description. Three real-world traffic datasets are used for experiments in this work. The details of the dataset are described below.

- Guangzhou: The Guangzhou dataset provides average speeds recorded every 10 min over two months (August and September 2016) for 214 road links in Guangzhou, China. For matrix-based models, the data size is 214×8784, and for tensor-based models, it is $214 \times 144 \times 61$.
- PeMS-08: The PeMS-08 dataset contains traffic volume data from 170 loop detectors in San Bernardino over 62 d (July 1 to August 31, 2016) with a 5-minute interval. For matrix-based models, the data size is 170×17856, and for tensor-based models, it is $170 \times 288 \times 62$.
- Shenzhen: The Shenzhen dataset comprises average speed data recorded every 10 min for a week (September 1 to September 7, 2019) from 2545 road links in Shenzhen, China, based on GPS trajectory records. For matrix-based models, the data size is 2545×2016, and for tensor-based models, it is $2545 \times 288 \times 7$.

Missing Scenarios. In this study, we consider random missing (RM) and non-random missing (NM) scenarios with missing rates of 10%, 30%, 50%, 70%, and 90% to emulate the realistic data-missing situations. RM scenario replicates the situation of randomly occurring and instantaneous missing data caused by sensor error or measurement error. NM scenario emulates the situation where missing data occurs successively in certain road links over a continuous period.

Baseline Models. To assess the effectiveness of the proposed algorithm, we conduct the performance evaluation of traffic data imputation by contrasting our STTC-CF model with five classic or state-of-the-art models. These methods include two matrix-based models NMF via ADMM [6] and ST-NMF [11], and three tensor-based models BATF [1], BGCP [2] and LTRR-NLS [12].

Parameter Settings. The baseline models in our experiments were configured with parameter settings that adhered to the recommended values specified in the original papers. For our proposed model, we get $R_1 = R_2 = R_3$, $\alpha_1 = \alpha_2 = \alpha_3 = 0.1$. We choose the latent rank R from $\{5, 10, 15, 20, 25, 30, 35, 40\}$. We do not consider increasing the value of R as it would require significant computational resources, with marginal improvement in accuracy. All experiments were conducted on a Windows 11 machine with i7-10700 CPU at 2.90 GHz and 16 GB memory, and the models were implemented by Matlab R2022a.

Evaluation Metrics. The performance of traffic data recovery is evaluated using Mean Absolute error (MAE) and Root Mean Square Error (RMSE) defined by

$$MAE = \frac{1}{\|\Omega_c\|_1} \sum_{(i,j,k)|\Omega_c} |x_{ijk} - \hat{x}_{ijk}|, \tag{24a}$$

$$RMSE = \sqrt{\frac{1}{\|\Omega_c\|_1} \sum_{(i,j,k)|\Omega_c} (x_{ijk} - \hat{x}_{ijk})^2}, \tag{24b}$$

where x_{ijk} and $\hat{x}_{ijk}$ represent the ground truth value and the recovered value of entry (i, j, k), respectively.

4.2 Data Imputation Results and Analysis

Tables 1, 2 and 3 present the data imputation performance of our proposed STTC-CF and other baseline models on the three datasets. In the tables, the best results regarding the metrics are bold. Overall, we observed that the data representation significantly influences data imputation performance. Models based on tensor representations exhibit more efficient performance compared to those based on matrix representations. This may be attributed to the need for more factors in the model to capture global features. A third-order tensor, having an additional dimension to represent the global attribute of the data, better captures the algebraic structure of traffic data.

Another finding is that, in the RM scenario, our STTC-CF model performs best at low missing rates of 10% and 30%, followed by LTRR-NLS. At the 50% missing rate, STTC-CF model is second to LTRR-NLS. This is possibly due to the flexible tensor ring structure. At 70% and 90% missing rates, our STTC-CF model is second to BGCP. BGCP effectively describes the variation of data using flexible priors. Even at extreme missing rates, BGCP automatically adapts to the intrinsic information of the data through prior parameters. We embed spatial adjacency and temporal correlation into STTC-CF model, providing sufficient data information even at high missing rates, surpassing tensor models without prior information.

Additionally, we found that all models perform worst in the NM scenario compared to the RM scenario, indicating that NM has a greater disruptive effect on the data structure. Despite this, our STTC-CF model outperforms others in the NM scenario. Specifically, our model has the smallest values for the metrics. As the missing rate increases, the fluctuations in our model's metrics are smaller compared to other models, indicating lower sensitivity of STTC-CF to missing rates. This suggests that our model maintains stable performance even in scenarios where severe data is missing. This may be attributed to the limited ability of

Table 1. Imputation Performance Comparison on Guangzhou Dataset for RM and NM Cases with Different Missing Rates

	Methods	RM					NM				
		10%	30%	50%	70%	90%	10%	30%	50%	70%	90%
MAE	NMF	2.82	2.87	2.94	3.14	4.35	3.00	3.11	3.22	3.49	7.24
	ST-NMF	2.87	2.94	3.05	3.36	4.94	3.08	3.11	3.25	3.69	5.71
	BATF	2.67	2.67	2.68	2.70	2.77	2.98	3.02	3.10	3.70	4.63
	BGCP	2.55	2.55	2.56	**2.57**	**2.67**	2.94	2.98	3.10	3.44	4.46
	LTRR-NLS	2.27	2.30	**2.39**	2.61	11.70	3.44	3.40	3.58	5.06	15.52
	STTC-CF	**2.21**	**2.29**	2.45	2.65	3.31	**2.80**	**2.94**	**2.97**	**3.16**	**3.47**
RMSE	NMF	4.28	4.35	4.45	4.74	6.39	4.55	4.68	4.78	5.13	9.02
	ST-NMF	4.32	4.42	4.59	5.03	7.14	4.67	4.68	4.86	5.50	8.12
	BATF	4.06	4.07	4.08	4.10	4.20	4.52	4.59	4.71	6.79	7.82
	BGCP	3.88	3.88	3.89	3.92	**4.08**	4.47	4.56	4.71	5.52	7.19
	LTRR-NLS	3.29	3.35	**3.49**	**3.88**	35.62	5.15	5.09	5.46	6.89	17.74
	STTC-CF	**3.16**	**3.27**	3.58	3.96	4.81	**4.27**	**4.55**	**4.50**	**4.67**	**5.17**

Table 2. Imputation Performance Comparison on PeMS-08 Dataset for RM and NM Cases with Different Missing Rates

	Methods	RM					NM				
		10%	30%	50%	70%	90%	10%	30%	50%	70%	90%
MAE	NMF	19.84	20.69	22.10	25.18	45.64	21.10	27.01	31.67	50.39	137.00
	ST-NMF	19.97	20.86	22.52	26.35	40.16	23.89	23.69	26.26	31.53	-
	BATF	16.64	16.72	16.71	16.86	17.94	21.49	23.05	22.90	24.53	34.65
	BGCP	16.51	16.56	16.60	16.72	**17.47**	20.24	20.83	21.87	24.02	36.08
	LTRR-NLS	13.58	13.91	14.57	16.29	24.99	20.24	25.81	31.74	32.60	-
	STTC-CF	**13.08**	**13.65**	**14.46**	**15.54**	18.89	**17.83**	**18.86**	**20.22**	**23.16**	**31.86**
RMSE	NMF	35.28	36.68	39.19	43.88	71.34	40.58	50.78	54.92	80.79	183.12
	ST-NMF	34.24	35.48	38.12	43.97	60.70	38.72	40.69	44.43	50.49	-
	BATF	26.28	26.39	26.41	26.64	28.66	34.15	39.78	38.22	54.86	134.52
	BGCP	26.10	26.19	26.26	26.48	**27.96**	38.09	43.55	55.05	57.40	85.16
	LTRR-NLS	21.47	**22.06**	**23.21**	26.38	40.31	34.06	42.48	49.89	49.47	-
	STTC-CF	**21.02**	22.07	23.27	**24.98**	29.72	**30.80**	**33.19**	**34.70**	**39.63**	**51.51**

Table 3. Imputation Performance Comparison on Shenzhen Dataset for RM and NM Cases with Different Missing Rates

	Methods	RM					NM				
		10%	30%	50%	70%	90%	10%	30%	50%	70%	90%
MAE	NMF	6.49	6.52	6.55	6.65	7.13	6.79	6.94	7.19	7.24	10.34
	ST-NMF	6.61	6.60	6.61	6.65	6.87	6.72	6.82	6.86	7.04	8.05
	BATF	6.47	6.5	6.53	6.57	6.72	6.67	6.72	6.78	6.86	**7.23**
	BGCP	6.44	6.45	**6.47**	**6.51**	**6.66**	6.61	6.70	6.78	6.86	7.26
	LTRR-NLS	6.48	6.51	6.54	6.66	26.87	6.64	6.68	41.12	44.19	48.42
	STTC-CF	**6.40**	**6.41**	6.50	6.53	7.01	**6.59**	**6.60**	**6.69**	**6.78**	9.47
RMSE	NMF	8.88	8.92	8.96	9.08	9.71	9.30	9.48	9.77	9.84	12.93
	ST-NMF	9.03	9.02	9.03	9.08	9.35	9.17	9.33	9.38	9.62	10.92
	BATF	8.84	8.89	8.93	8.97	9.16	9.16	9.21	9.28	9.38	**9.86**
	BGCP	8.80	8.81	**8.84**	**8.89**	**9.08**	9.05	9.17	9.29	9.39	9.92
	LTRR-NLS	8.85	8.89	8.92	9.06	30.08	9.06	9.12	46.27	48.69	51.87
	STTC-CF	**8.78**	**8.80**	8.89	8.93	9.48	**8.99**	**9.05**	**9.18**	**9.22**	11.12

prior local spatiotemporal information to recover data in NM structural missing scenarios. In such cases, our model with the Capped Frobenius norm as the error measure reduces interference from outliers in the decomposition process, further enhancing data imputation capabilities.

4.3   Visualization of Data Recovery

Figure 1 provides an intuitive visualization of STTC-CF's imputation performance. We randomly select one week of PeMS-08 data with 50% NM values as shown in Fig. 1(b). We can see that the data are severely corrupted with the unobservable. We employ STTC-CF for imputation, and the result is shown in

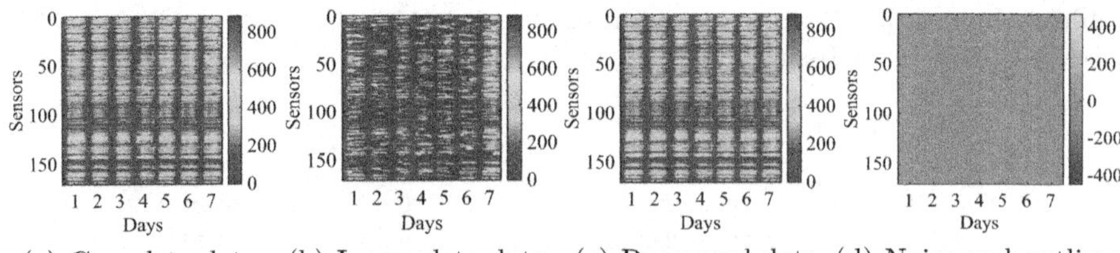

(a) Complete data (b) Incomplete data (c) Recovered data (d) Noise and outliers

Fig. 1. Visualization of STTC-CF for PeMS-08 with 50% missing rate in NM scenario.

1(c). The noise and outliers are presented in 1(d). STTC-CF exploits the observation and priors for tensor decomposition and uses the Capped Frobenius norm to extract outliers, successfully recovering the original data.

4.4 Data Imputation Against Outliers

To evaluate the capability of our model STTC-CF in handling outliers, we conduct traffic data imputation experiments resistant to outliers on the Guangzhou dataset with a 30% missing rate under the RM. Random outliers (RO) and non-random outliers (NO) are added separately to the data, with the magnitude of outliers ranging from 1.5 to 3 times that of the original data and the rates of 5%, 10%, 15%, 20%, 25%.

The comparison results of our model STTC-CF with other baseline models are presented in Table 4. It can be observed from Table 4 that the performance of our model STTC-CF significantly outperforms other comparative models. In experiments without outliers (Sect. 4.2), models such as BGCP and LTRR-NLS demonstrate superior performance compared to our STTC-CF model in

Table 4. Imputation Performance Comparison on Guangzhou Dataset for RO and NO Cases with Different Outlier Rates

	Methods	RO					NO				
		5%	10%	15%	20%	25%	5%	10%	15%	20%	25%
MAE	NMF	4.28	4.80	5.60	6.66	7.81	5.59	7.43	8.78	10.10	11.10
	ST-NMF	3.52	3.99	4.68	5.22	5.69	4.56	5.75	6.95	7.70	9.13
	BATF	3.49	5.25	7.45	9.94	12.63	4.14	7.18	9.41	11.68	14.30
	BGCP	3.42	5.18	7.41	9.90	12.58	5.07	7.16	7.53	11.45	14.14
	LTRR-NLS	3.11	3.88	4.80	5.77	6.84	6.13	6.83	11.11	15.27	16.58
	STTC-CF	**2.60**	**2.80**	**3.49**	**4.32**	**6.62**	**2.79**	**2.85**	**3.96**	**4.70**	**6.77**
RMSE	NMF	6.07	6.69	7.58	8.78	10.28	8.31	10.92	12.36	14.04	15.56
	ST-NMF	5.06	5.66	7.44	8.10	8.08	6.92	9.04	9.97	11.47	13.26
	BATF	4.91	6.54	8.64	11.07	13.74	6.34	10.58	13.34	15.90	18.84
	BGCP	4.76	6.40	8.55	10.93	13.64	7.65	10.48	10.80	15.66	18.43
	LTRR-NLS	4.53	5.33	6.34	7.36	8.51	15.36	14.03	16.78	21.88	23.54
	STTC-CF	**4.00**	**4.19**	**5.00**	**5.35**	**8.03**	**4.38**	**4.39**	**5.77**	**6.18**	**8.38**

some cases. However, their advantages diminish when outliers are present in the dataset, and their performance rapidly declines with increasing outlier rates. In contrast, the variations in the metrics of our model STTC-CF are relatively small. Comparing Table 1 and Table 4, there is a significant disparity in the metrics of other comparative models between scenarios where data is unaffected and scenarios where data is contaminated by outliers, whereas the metrics of our model show minor differences. These phenomena may be attributed to the novel error measure we designed, which enables our model STTC-CF to resist Gaussian noise in the absence of outliers and to withstand outliers when they appear. Therefore, our model STTC-CF exhibits greater robustness in traffic data imputation tasks.

5   Conclusion

In this study, we introduce a spatio-temporal tensor completion method using the Capped Frobenius norm, specifically designed for imputing traffic data in smart cities. The proposed model utilizes TR decomposition with the Capped Frobenius norm to approximate the original incomplete data, aiming to extract global features and enhance robustness. To solve the non-convex optimization problem, we employ HQ technique and AO-ADMM framework as solvers. Our experimental analysis, conducted on real-world traffic data, encompasses two scenarios of missing data and two cases of outliers, each with different corruption levels. The results indicate that STTC-CF is robust, especially in the more common NM scenarios. In the future, we plan to use STTC-CF for prediction tasks. Integrating deep learning models will further showcase the effectiveness of STTC-CF in handling outliers.

A   APPENDIX

A.1   Derivation of (11)

According to HQ optimization theory, for $\phi_e(r) = \min\left\{r^2, e^2\right\}$, there exists a convex dual function $\varphi_e(a)$ with $a \in \mathbb{R}$ so that

$$\phi_e(r) = \inf_a (r - a)^2 + \varphi_e(a), \tag{25}$$

where the dual function has the following form:

$$\varphi_e(a) = \sup_r \phi_e(r) - (r - a)^2 = \sup_r \min\left\{r^2, e^2\right\} - (r - a)^2 = e^2 - (e - \min\{|a|, e\})^2 \tag{26}$$

Thus, we have:

$$\min_r \phi_e(r) = \min_{r,a}(r - a)^2 + \varphi_e(a). \tag{27}$$

Given a tensor $\mathcal{R} \in \mathbb{R}^{I_1 \times I_2 \times I_3}$. Consider the following problem:

$$\min_{\mathcal{R}} \|\mathcal{R}\|_\phi^2 = \min_{\mathcal{R}} \sum_{i,j,k}^{I_1, I_2, I_3} \phi_e(r_{ijk}) = \sum_{i,j,k}^{I_1, I_2, I_3} \min_{r_{ijk}} \phi_e(r_{ijk}). \tag{28}$$

Following (27), (28) is equivalent to

$$\min_{\mathcal{R}} \|\mathcal{R}\|_\phi^2 = \sum_{i,j,k}^{I_1,I_2,I_3} \min_{r_{ijk}} \phi_e(r_{ijk}) = \sum_{i,j,k}^{I_1,I_2,I_3} \min_{r_{ijk}} [(r_{ijk} - a_{ijk})^2 + \varphi_e(a_{ijk})] \qquad (29)$$

which has the tensor form:

$$\min_{\mathcal{R},\mathcal{A}} \|\mathcal{R} - \mathcal{A}\|_F^2 + \varphi_e(\mathcal{A}). \qquad (30)$$

A.2   Derivation of the Solution (22)

The optimization problem (27) is convex with respect to each individual variable r and a. For the following optimization subproblem of problem (27)

$$\min_{a}(r - a)^2 + \varphi_e(a), \qquad (31)$$

we substitute Eq. (26) into Eq. (31), resulting in:

$$\min_{a}(r - a)^2 + \varphi_e(a) = \min_{a}(r - a)^2 + e^2 - (e - \min\{|a|, e\})^2 \qquad (32)$$

which has the following solution:

$$a = r\text{sign}(|r| - \min\{|r|, e\}). \qquad (33)$$

For simplicity, the optimal solution (33) is represented by $a = \mathcal{T}_e(r)$. Therefore, the problem (30) has the solution as

$$\mathcal{A} = \mathcal{T}_e(\mathcal{R}) = r_{ijk}\text{sign}(|r_{ijk}| - \min\{|r_{ijk}|, e\}), \qquad (34)$$

where $i = 1, 2, \ldots, I_1, j = 1, 2, \ldots, I_2$, and $k = 1, 2, \ldots, I_3$.

References

1. Chen, X., He, Z., Chen, Y., Lu, Y., Wang, J.: Missing traffic data imputation and pattern discovery with a Bayesian augmented tensor factorization model. Transp. Res. Part C: Emerging Technol. **104**, 66–77 (2019)
2. Chen, X., He, Z., Sun, L.: A Bayesian tensor decomposition approach for spatiotemporal traffic data imputation. Transp. Res. Part C: Emerging Technol. **98**, 73–84 (2019)
3. Chen, X., Yang, J., Sun, L.: A nonconvex low-rank tensor completion model for spatiotemporal traffic data imputation. Transp. Res. Part C: Emerging Technol. **117**, 102673 (2020)
4. Feng, X., Zhang, H., Wang, C., Zheng, H.: Traffic data recovery from corrupted and incomplete observations via spatial-temporal TRPCA. IEEE Trans. Intell. Transp. Syst. **23**(10), 17835–17848 (2022)
5. Hu, Y., Work, D.B.: Robust tensor recovery with fiber outliers for traffic events. ACM Trans. Knowl. Discovery Data (TKDD) **15**(1), 1–27 (2020)

6. Huang, K., Sidiropoulos, N.D., Liavas, A.P.: A flexible and efficient algorithmic framework for constrained matrix and tensor factorization. IEEE Trans. Signal Process. **64**(19), 5052–5065 (2016)
7. Nie, T., Qin, G., Sun, J.: Truncated tensor Schatten p-norm based approach for spatiotemporal traffic data imputation with complicated missing patterns. Transp. Res. Part C: Emerging Technol. **141**, 103737 (2022)
8. Silverman, B.W.: Density Estimation for Statistics and Data Analysis. Routledge (2018)
9. Sofuoglu, S.E., Aviyente, S.: Gloss: Tensor-based anomaly detection in spatiotemporal urban traffic data. Signal Process. **192**, 108370 (2022)
10. Sure, P., Srinivasan, C.P., Babu, C.N.: Spatio-temporal constraint-based low rank matrix completion approaches for road traffic networks. IEEE Trans. Intell. Transp. Syst. **23**(8), 13452–13462 (2021)
11. Wang, Y., Zhang, Y., Wang, L., Hu, Y., Yin, B.: Urban traffic pattern analysis and applications based on spatio-temporal non-negative matrix factorization. IEEE Trans. Intell. Transp. Syst. **23**(8), 12752–12765 (2021)
12. Wu, P.L., Ding, M., Zheng, Y.B.: Spatiotemporal traffic data imputation by synergizing low tensor ring rank and nonlocal subspace regularization. IET Intell. Transport Syst. (2023)
13. Xu, Y., Yin, W., Wen, Z., Zhang, Y.: An alternating direction algorithm for matrix completion with nonnegative factors. Front. Math. China **7**, 365–384 (2012)
14. Zhao, Q., Zhou, G., Xie, S., Zhang, L., Cichocki, A.: Tensor ring decomposition. arXiv preprint arXiv:1606.05535 (2016)

A Federated Domain Generalization Method by Enhancing Knowledge Distillation with Stylistic Feature Dispatcher

Guangshuo Wang[1], Yuesheng Zhu[1(✉)], Guibo Luo[1(✉)], Jie Ling[2], and Long Xie[3]

[1] Shenzhen Graduate School, Peking University, Beijing, China
wanggsh@stu.pku.edu.cn, {zhuys,luogb}@pku.edu.cn
[2] School of Computer Science and Engineering, Sun Yat-Sen University, Guangzhou, China
lingjie.8@mail2.sysu.edu.cn
[3] College of Control Science and Engineering, Zhejiang University, Hangzhou, China
xielong@zju.edu.cn

Abstract. Federated learning aims to utilize multiple client domains without sharing data to train a global model, yet generalizing well to unseen domains remains challenging in this paradigm. Recent works indicate that large vision-language models (LVLM) can be used to improve the model's generalization ability by knowledge distillation. However, due to the unstructured nature of the text features, client models have poor adaptability to distribution shifts. In this paper, we propose Federated Stylistic Feature Dispatcher (FeSFD) to conduct alignment across clients, which improves generalization capacity in privacy-preserving scenarios. Specifically, client domains' statistical information is shared, facilitating feature-level augmentation with different domain styles to create novel feature distribution. Also, we design a consistency loss that draws augmented predicted results closer to the original results for restraining semantic understanding of the model. Finally, the dual knowledge distillation further excavates the generalization capacity of LVLM. Our model outperforms state-of-the-art FedDG methods through comprehensive experiments on PACS, Office-Home, and Digits-DG benchmarks. What's more, we achieve 3.96%, 1.53%, and 4.71% improvement on the above three datasets compared to the baseline, which proves the superiority of our model.

Keywords: Federated Domain Generalization · LVLM · Knowledge Distillation · Data Augmentation

1 Introduction

The field of deep learning has witnessed significant advancements, wherein these methodologies often rely on independent and identically distributional (IID) assumptions. Testing on Out-of-Distribution (OOD) datasets typically results in

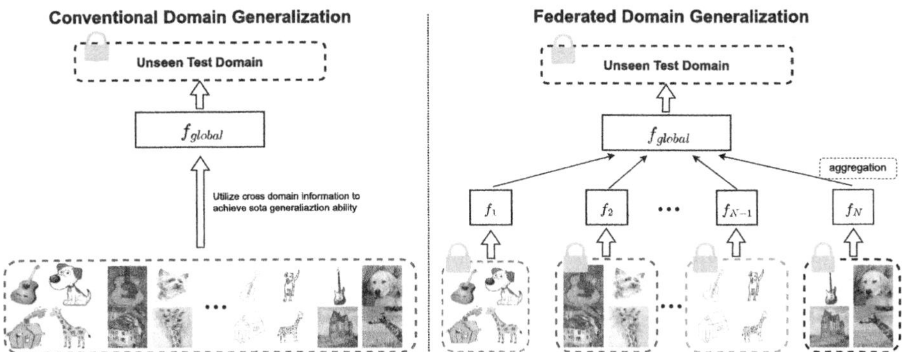

Fig. 1. Difference between traditional DG and federated DG.

noticeable performance degradation. For this, the Domain Generalization (DG) technique is proposed to generalize well to unseen data. Under the DG setting, the known and utilized domains for training are the source domains, the unknown domains are the target domains, and the categories for both domains remain the same. DG methods leverage data from source domains to improve generalization performance on the target domains. Nevertheless, the widespread adoption of distributed training has reduced the sharing of inter-domain data, posing challenges for data aggregation and centralization.

Federated Domain Generalization (FedDG) offers a decentralized and privacy-preserving training paradigm. Under FedDG setting [12] which is shown in Fig. 1, the training data possessed by each client serves as a source domain, characterized by non-shareable and non-IID data distributions across different client domains. After local training, each client model's parameters are sent to the server to conduct model aggregation, generating a new global model. Ultimately, the global model is employed for inference in unknown domains. Most traditional DG methods [6,9,10,25,27] necessitate simultaneous integration of information from multiple domains for domain-invariant learning. Therefore, these approaches become less applicable in the more realistic and challenging FedDG scenario.

The development of large vision-language models has progressed rapidly. In this article, we hope to use the characteristics of text to improve generalization ability. Texts express core information concisely and precisely, whereas images, constrained by complex backgrounds and texture information, are prone to confusion during information transmission. Therefore, text features are domain-invariant and helpful for improving generalization ability. Different client models can leverage unified text features for alignment, learning consistent and accurate representations without violating privacy-preserving protocols. Particularly, CLIP [16] proposes a straightforward pretraining method, building a bridge between natural language and visual content. In aims of harnessing LVLM to enhance generalization capabilities, we employ a knowledge distillation [8] approach, utilizing CLIP to assist in the training of a lightweight client model.

However, text inherently has unstructured problems, such as changes in expression that affect the model's understanding, and the meaning of a word depends on context. When facing different domain distributions and unable to interact between domains, the client model has poor adaptability in understanding and utilizing text semantics. To solve this, we propose the Federated Stylistic Feature Dispatcher (FeSFD). The dispatcher is stored on the global server and serves to share client domain styles with each other. During training, client models upload the statistical information of intermediate features to the style dispatcher. Subsequently, clients shall download other domains' information from the dispatcher to modify the low-frequency information of intermediate features, thereby generating new features that incorporate stylistic information from other domains. To better leverage the augmented information, we design a consistency loss function, constraining predictions of the manipulated features to draw near to the results of the original branch. This provides another effective guidance during the model's learning process of domain-invariant information. Naturally, we can simultaneously apply a dual knowledge distillation, comparing both the augmented and original predictions with CLIP teacher. Client encoders can consider a broader range of domain information through these operations without directly accessing plaintext images. Our method enables clients to better recognize and disregard information related to textures and backgrounds in images, focusing on higher-level semantic information. The main contributions can be summarized as follows:

- We propose the FeSFD, wherein client models share data distribution characteristics and conduct augmentation for feature alignment, promoting knowledge distillation from large vision-language models in the FedDG field.
- Building upon the aforementioned feature manipulation, we incorporate consistency loss and dual knowledge distillation to constrain the model's learning, aiding the model in focusing more on crucial semantic information.
- Comprehensive experiments on several datasets demonstrate that our results achieve state-of-the-art performance, validating the effectiveness of each module.

2   Related Work

2.1   Domain Generalization

Domain Generalization (DG) is a widely learned field. A majority of works follow the theory of [2] that models are generalizable to unseen target domains if they can extract invariant representations from different domains. Some DG methods utilize meta-learning to simulate the process of domain shift [1,6]. Another popular way is to enrich data diversity by image [26] or feature [10,27] augmentation. For instance, [10] obtains new domain attributes by calculating the statistical characteristics of multiple domains and enhancing its stylization through the encoder.

Many DG works take advantage of large vision-language models, especially CLIP [16], to gain better generalization ability. Some works make zero-shot predictions by CLIP, where it directly figures up the label closest to the input features. [5] endows style diversity and content consistency to the style word vectors of pseudo-words S_*. Other works focus on acquiring a lightweight model from CLIP using knowledge distillation. [9] achieves a more profound utilization of CLIP's text features by combining the absolute and relative loss of inter-domain information.

2.2   Federated Domain Generalization

Most conventional DG methods are no longer applicable under the schema of federated domain generalization (FedDG). [12] first proposed a solution for FedDG scenarios, transmitting data distribution information between clients to perform augmentation in continuous frequency space. [24] pointed out that not only should we improve the training ability of local models but also pay attention to the parameters aggregation process from local to global. [4] attempts to learn consistent distribution by building a shared style bank from which clients can generate novel domains. However, the limitations of [4] are centered around the issues of communication efficiency and additional computational complexity by introducing style transfer models. Our method improves these flaws by directly conducting augmentation at the feature level and only interacting with statistical information.

[13] first utilizes large vision-language models in FedDG field, which designs attention-based adapters for CLIP to retain clients' prior information for DG tasks. However, it is challenging to use CLIP's original large model directly in reality since decentralized devices that train client domains often have limited GPU memory and slow inference speed. In summary, we hope to use knowledge distillation to learn client models with reductive parameters from CLIP.

3   Method

In this chapter, we describe our method that conducts feature alignment across clients and learns domain-invariant information. Figure 2 shows the client model framework of our method. Firstly, we perform feature-level augmentation, which utilizes different domains' peculiarities through the style dispatcher. The client's style will also be updated at the same time. Then, we introduce consistency loss for preserving semantic information and dual knowledge distillation for better LVLM guidance. After each client finishes local training, their model parameters will be aggregated into the global model.

3.1   Preliminaries

Federated Domain Generalization (FedDG). Unlike the conventional scenario, FedDG can only train several client models on local datasets. Given source domains $\mathcal{D} = \{\mathcal{D}_1, ... \mathcal{D}_N\}$ where N denotes the number domains for training. Let (x, y) be a sample pair, x denotes a sample, and y is the corresponding one-hot label. Each client domain D_k contains M_k sample $(x_{k_i}, y_{k_i}) \in D_k, 0 < i < M_k$.

During local training, a baseline loss function $\mathcal{L}_{ce}$ measures the distance between the label and model prediction $f_k(x_k)$, where f_k denotes client model

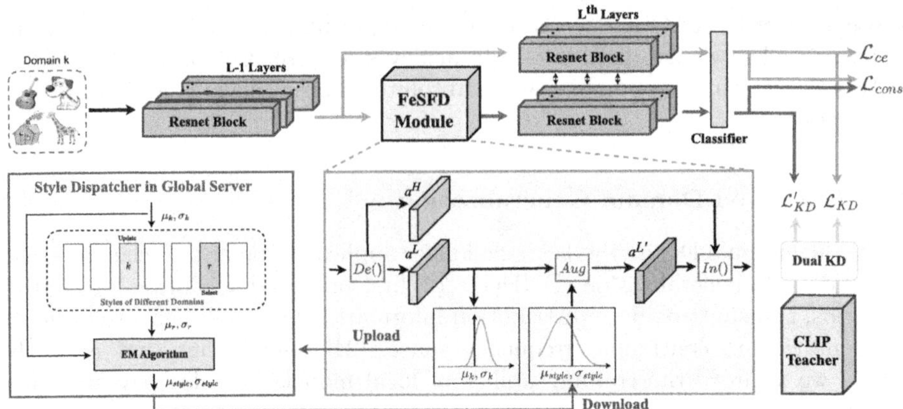

Fig. 2. Our client model architecture. FeSFD module generates a new branch of augmented features that exploit styles from other domain through style dispatcher in global server. The two branch go through resnet block and classifier that share weights. $\mathcal{L}_{cons}$ is the consistency loss for original and augmented predict results. Dual KD helps client model learn from CLIP teacher. $De()$ and $In()$ denote the decomposition and integration process for features at frequency level.

for corresponding k^{th} domain, containing an encoder to extract feature and a simple fully-connected layer as the classifier. Commonly, minimizing the cross-entropy loss $\mathcal{L}_{ce}$ is the training objective.

After a period of local iteration, the global model $f(\cdot)$ is updated by aggregating client models' parameters. In the FedDG scenario, the global model needs to handle computer vision tasks in unseen target domains.

Knowledge Distillation from LVLM. We exploit LVLM, specifically, CLIP in the local training process. By knowledge distillation, client models can gain better generalization ability from CLIP's rich knowledge. Following the recommended setting in CLIP, we use prompt template "A photo of {class}" as input and extract its text features from encoder $T(\cdot)$. Then, the image features produced by encoder $I(\cdot)$ will multiply text features to calculate cosine similarities $cos(\cdot)$. CLIP chooses the class whose text features is closest to the image features by $cos(\cdot)$ to make the zero-shot prediction. After that, the knowledge distillation loss can be described as:

$$\mathcal{L}_{KD} = KL(c(x), f_k(x), \tau) \tag{1}$$

where $c(x)$ denotes CLIP's prediction of x. $KL(\cdot, \cdot)$ is the KL divergence loss [8] to measure vectors' distance, helping the student model predict results closer to the teacher model. Following the default setting in [8], τ is set as 3, which denotes the temperature hyper-parameter to control the smoothness of CLIP's prediction.

3.2 Stylistic Feature Dispatcher and Augmentation

Previous works aggregate different domains' style information for augmentation, which is hampered in FDG's privacy-preserving scenario. To solve this problem, we introduce the federated stylistic feature dispatcher, enabling client models to conduct feature alignment from different domains without dataset leakage. Client models can upload their feature statistical information and obtain new styles from the global server.

This module is performed between the jth layer L_j and layer L_{j+1} of the backbone network, supposed $\boldsymbol{a}_j$ is the output from L_j. Inspired by [22], we obtain low frequency components $\boldsymbol{a}_j^L$ by executing average pooling and nearest neighbor upsampling to $\boldsymbol{a}_j$.

$$\boldsymbol{a}_j^L = Upsample(AvgPool(\boldsymbol{a}_j)) \tag{2}$$

And high frequency componet $\boldsymbol{a}_j^H$ comes from:

$$\boldsymbol{a}_j^H = \boldsymbol{a}_j - \boldsymbol{a}_j^L \tag{3}$$

In order to acquire style information of client domain, we compute the mean and variance of each channel batch-wisely:

$$\boldsymbol{\mu}_j^L = \frac{1}{V} \sum_{m=1}^{V} flat\left(\boldsymbol{a}_j^L\right)_m \tag{4}$$

$$\boldsymbol{\sigma}_j^L = \sqrt{\frac{1}{V} \sum_{m=1}^{V} \left(flat\left(\boldsymbol{a}_j^L\right)_m - \boldsymbol{\mu}_j^L\right)^2} \tag{5}$$

where $V = BH_jW_j$, and $flat(\cdot)$ performs flatten operation, converting the feature maps from a $B \times H_j \times W_j \times C$ shape to $(B \times H_j \times W_j) \times C$.

However, it is thriftless and inefficient for clients and server to communicate style messages due to large channel numbers. Thus, we compute the mean and variance of $\boldsymbol{\mu}_j^L$ and $\boldsymbol{\sigma}_j^L$ on the scale of channel to produce gaussian distribution, which offers lower communication cost.

$$\boldsymbol{S}_k = \left[\mu_{\mu_j^L},\ \sigma_{\mu_j^L},\ \mu_{\sigma_j^L},\ \sigma_{\sigma_j^L}\right] \tag{6}$$

$$\mu_k^{style} \sim \mathcal{N}\left(\mu_{\mu_j^L},\ \sigma_{\mu_j^L}\right) \tag{7}$$

$$\sigma_k^{style} \sim \mathcal{N}\left(\mu_{\sigma_j^L},\ \sigma_{\sigma_j^L}\right) \tag{8}$$

where $(\mu_{\mu_j^L},\ \sigma_{\mu_j^L})$ and $(\mu_{\sigma_j^L},\ \upsilon_{\sigma_j^L})$ refers to the mean and variance of $\boldsymbol{\mu}_j^L$ and $\boldsymbol{\sigma}_j^L$, respectively. $\boldsymbol{S}_k$ indicates the style information of D_k and $S = \{\boldsymbol{S}_k\}_{k=1}^K$ refers to style cloud.

Frequent and straightforward updates of style through the calculation of small batches may insult the stability of style information, and this method's discontinuity is also inefficient for data. We introduce the Exponential Moving

Average (EMA) algorithm to address this. EMA considers historical style messages, placing weight and significance on the recent data. EMA helps smooth the parameter renewal process by averaging the value over a period of time. As shown in Fig. 2, let r define the domain index that D_k chooses, the output style is formulated as:

$$S^{style} = \vartheta \cdot S_r + (1 - \vartheta) \cdot S_k \tag{9}$$

where S_r is mean and variance of μ_r and σ_r. μ^{style} and σ^{style} are sampled from S^{style}. ϑ is hype-parameters, which is set as 0.95.

After the upload of its style information, the client model downloads other domains' styles from server. The client model generates a new style by applying a transformation to the original low-frequency information, written as:

$$a_j^{L'} = \sigma^{style} \left(\frac{a_j^L - \mu_j^L}{\sigma_j^L} \right) + \mu^{style} \tag{10}$$

Lastly, the augmented features is a simple compound of composite low and high frequency components:

$$a_j' = a_j^H + a_j^{L'} \tag{11}$$

3.3 Semantic-Constraint Consistency Loss

We follow up on the augmentation network mentioned above and propose a semantic-constraint consistency loss. As discussed above, simple utilization of CLIP-based knowledge distillation has yet to achieve the desired efficacy of generalization ability successfully. In order to bridge the gap between different domains without insulting privacy-preserving protocol, we restrain the predicted results of augmented features and draw closer to the original branch. Analogous to Knowledge distillation, we measure the distance between two logits.

Another mentionable thing is that different from the classical distillation setting, we perform a sharpening operation on the original results. Instead of softening operation that pulls up the value of unmatched categories, it is more crucial for stylized prediction to focus on semantic information and enhance the learning process.

$$\mathcal{L}_{cons} = -\frac{1}{M_k} \sum_{i=1}^{M_k} f_k'(x_{k_i}) \log \left(f_k(x_{k_i}, \tau) \right) \tag{12}$$

where M_k is the number of samples (x_{k_i}, y_{k_i}) in D_k. Function $f_k(x, \tau)$ and $f_k'(x)$ denotes the predict results of original and augmented features. τ is set as 0.5 to perform square root of the predicted logits of f, denoting the sharpening approach.

Table 1. Results on PACS datasets. The best and second-best are bolded and underlined respectively.

Paradigm	Method	PACS				
		Art	Cartoon	Photo	Sketch	Avg
Regular DG	Jigen[3]	79.40	75.30	96.00	71.40	80.50
	DDAIG[25]	84.20	78.10	95.30	74.70	83.10
	L2A-OT[26]	83.30	78.20	96.20	73.60	82.80
	MixStyle[27]	84.10	78.80	96.10	75.90	83.70
	EISNet[18]	80.00	76.00	93.70	80.90	82.60
	RISE[9]	85.10	81.80	96.00	78.40	85.30
Federated DG	FedDG[12]	83.94	79.27	96.23	73.30	83.19
	CASC[23]	82.00	76.40	95.20	81.60	83.80
	FADH[21]	83.80	77.20	94.40	84.40	85.00
	COPA[19]	83.30	79.80	94.60	82.50	85.10
	CCST[4]	88.33	78.20	96.65	82.90	<u>86.52</u>
	Baseline[24]	81.28	76.73	93.97	82.57	83.64
Our Method	FeSFD	86.15	82.71	96.76	84.78	**87.60**
	Improvement	↑ 4.87	↑ 5.98	↑ 2.79	↑ 2.21	↑ 3.96

3.4 Dual Knowledge Distillation and Full Method

Following the FeSFD module, we add a knowledge distillation branch to the augmented predict results, written as $\mathcal{L}'_{KD}$. After the client model has seen different styles via the cloud, this branch is helpful for different features to align together and maintain semantic information. In the end, we train the network with the weighted sum of losses as follows:

$$\mathcal{L} = \mathcal{L}_{ce} + \lambda_1 \mathcal{L}_{KD} + \lambda_2 \mathcal{L}'_{KD} + \lambda_3 \mathcal{L}_{cons} \tag{13}$$

where λ_1, λ_2 and λ_3 are scalar hyper-parameters denoting the weight proportions of each loss.

4 Experiment

4.1 Evaluation on Three Benchmark Datasets

Datasets. We conduct experiments on three standard domain generalization datasets. PACS [11] contains 9,991 images with seven classes from four domains with different styles. Office-Home [17] contains 15,500 images with 65 classes from four domains. Digits-DG [26] contains four conventional digit datasets.

Implementation Details. We conduct leave-one-domain-out evaluations for all benchmarks, which select one domain at a time as the unseen domain while utilizing all other domains as the source clients for training. The split between the training and validation sets within each source domain follows the same configuration as in [7,25] for PACS, Office-Home, and Digits-DG. The entire

Table 2. Results on Office-Home datasets. The best and second-best are bolded and underlined respectively.

Paradigm	Method	Office-Home				
		Artist	Clipart	Product	Real-world	Avg
Regular DG	Jigen[3]	53.00	47.50	71.50	72.80	61.20
	DDAIG[25]	59.20	52.30	74.60	76.00	65.50
	L2A-OT[26]	60.60	50.10	74.80	77.00	65.60
	MixStyle[27]	58.70	53.40	74.20	75.90	65.50
	EISNet[18]	56.80	53.30	72.30	73.50	64.00
	RISE[9]	59.10	52.90	75.10	76.40	65.90
Federated DG	FedDG[12]	60.70	45.82	71.51	73.05	62.77
	FedAvg[15]	58.20	51.60	73.10	73.80	64.20
	FADH[21]	59.90	55.80	73.50	74.90	66.00
	COPA[19]	59.40	55.10	74.80	75.00	<u>66.10</u>
	CCST[4]	59.05	50.06	72.97	71.67	63.56
	Baseline[24]	58.57	54.39	73.39	74.73	65.27
Our Method	FeSFD	59.91	55.98	75.33	76.52	**66.80**
	Improvement	↑ 1.34	↑ 1.59	↑ 1.94	↑ 1.79	↑ 1.53

target domain is employed for testing. For fairness, we follow the protocol from [7] in local training. For PACS and Office-Home, we use Resnet-18 pre-trained on ImageNet. For Digits-DG, we use the same backbone as in [25], consisting of four convolutional layers. We follow a strong baseline of [24] to dynamically calibrate client domains in model weight aggregation. Hyperparameters j, $\lambda 1$, $\lambda 2$ and $\lambda 3$ are set as 2, 0.8, 0.2 and 6. Other hyperparameters and optimizers follow the default setting in [24].

Comparison Method. We select representative methods in both conventional and federated domain generalization fields. We choose six centralized DG methods: JiGen [3] enforces the encoder by solving a self-supervised jigsaw puzzle, DDAIG [25] generates new images through domain-adversarial learning, EISNet [18] learns from interdomain and intradomain supervision for images, FACT [20], Mix-style [27], and RISE [9] introduced in Sect. 2.1. We also choose six decentralized DG methods: FedAvg [15] is a milestone baseline, COPA [19] employs batch-wise mixed normalization and aggregation, FADH [21] trains the local models with domain hallucinations for robustness, CASC [23] utilizes layer weight to aggregate parameters, FedDG [12], GA [24] and CCST [4] introduced in Sect. 2.2.

4.2   Comparison with Existing Method

Table 1, Table 2 and Table 3 summarize a comparison of our model with current methods, detailing accuracies for each target domain and the overall average accuracy within each dataset. It is evident that FeSFD surpasses state-of-the-art methods across three datasets. Specifically, in PACS, our model shows a

Table 3. Results on Digits-DG dataset. The best and second-best are bolded and underlined respectively.

Paradigm	Method	Digits-DG				
		MNIST	SVHN	SYN	MNIST-M	Avg
Regular DG	Jigen[3]	96.50	63.70	74.00	61.40	73.90
	DDAIG[25]	96.60	68.60	81.00	64.10	77.60
	L2A-OT[26]	96.70	68.60	83.20	63.90	78.10
	EISNet[18]	96.40	56.00	60.50	87.90	75.20
	MixStyle[27]	96.50	64.70	81.20	63.50	76.50
Federated DG	FedDG[12]	96.30	61.20	66.70	90.00	78.60
	FedAvg[15]	96.93	62.19	90.04	57.75	76.71
	FADH[21]	97.70	73.30	90.60	65.30	<u>81.70</u>
	COPA[19]	97.00	71.61	90.66	66.52	81.49
	CCST[4]	96.16	66.58	88.79	62.76	78.57
	Baseline[24]	96.52	62.80	90.42	59.16	77.23
Our Method	FeSFD	96.92	73.60	91.31	67.85	**82.42**
	Improvement	↑ 0.40	↑ 10.8	↑ 0.89	↑ 8.69	↑ 5.19

1.08% improvement over the next best result, 0.7% and 0.93% in Office-Home and Digits-DG datasets, respectively.

Moreover, our model attains the best performance in most individual test domains. In PACS, performance across benchmarks is closely matched when photo is the target domain. The significant gains in other domains more robustly demonstrate our models' effectiveness. Simultaneously, while some models may achieve optimal results in one target domain, they do not maintain this performance uniformly and result in a lower average accuracy compared to our method, suggesting that they have yet to learn domain-invariant knowledge. For example, method of [12] exceeds our accuracy by 0.79% in the Art domain of Office-Home, yet it falls nearly 10% behind in the Clipart domain.

It is worth noticing that, despite RISE, a regular DG method, utilizes knowledge distillation similarly, our model leads by 2% in PACS, indicating a more efficient harnessing of the large vision-language model while adhering to the privacy protection paradigm.

4.3   Further Analysis

Ablation Study for Each Component. Table 4 demonstrates that each module within our model contributes to performance enhancement. We adopted FedAvg-GA as the strong baseline, as shown in the first row, where only L_{ce} is used. The ablation study are summarized as follows:

- Initially, the isolated addition of the $\mathcal{L}_{KD}$ and the stylization consistency loss modules yielded improvements of 2.57% and 1.14%, respectively.

Table 4. Ablation study for each component. Aug denotes our feature augmentation process.

$\mathcal{L}_{ce}$	Aug & $\mathcal{L}_{cons}$	$\mathcal{L}_{KD}$	$\mathcal{L}'_{KD}$	Acc
✓				83.64
✓	✓			84.78
✓		✓		86.21
✓	✓	✓		87.28
✓	✓		✓	87.12
✓	✓	✓	✓	**87.60**

Table 5. Ablation study for the block number FeSFD executes after.

Location	PACS				
	Art	Cartoon	Photo	Sketch	Avg
Block1	86.01	83.07	96.15	84.12	87.33
Block2	**86.15**	82.71	**96.76**	**84.78**	**87.60**
Block3	85.82	**83.12**	96.01	84.45	87.35
Block4	85.88	81.92	95.64	83.25	86.67

- In response to the foregoing, we note that the accuracy presented in the third row is 86.21%, which is slightly lower than the 86.52% reported by [4]. This indicates that simple knowledge distillation using CLIP alone is not effective enough.
- By integrating the $\mathcal{L}_{KD}$ with the stylization consistency loss module, the performance is further elevated to 87.28%, and the implementation of the dual knowledge distillation leads to an efficacy of 87.60%.

Ablation Study for Layer Choosed. Table 5 shows the generalization effect of different layers that the FeSFD module executes after. The results show that choosing the second block has the best effect, suboptimal for the third block. We believe that this is because the model has extracted sufficient semantic information after two blocks. At the same time, some low-level information is still retained after two blocks. The FeSFD module can better identify domain invariant features through style exchange after obtaining semantic information of category, thereby improving generalization ability.

Visualization Results of Extracted Features. Figure 3 shows the clustering visualization results of the model features. By t-SNE [14], we perform dimensionality reduction on the features extracted by the model encoder and the target domain is photo. In Figures (a) and (b), different colors represent categories. It

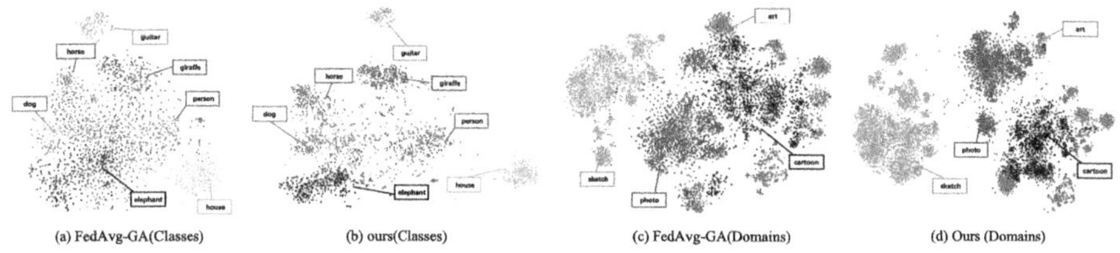

Fig. 3. The t-SNE visualization of feature representations extracted by the feature extractor of FedAvg-GA and our model.

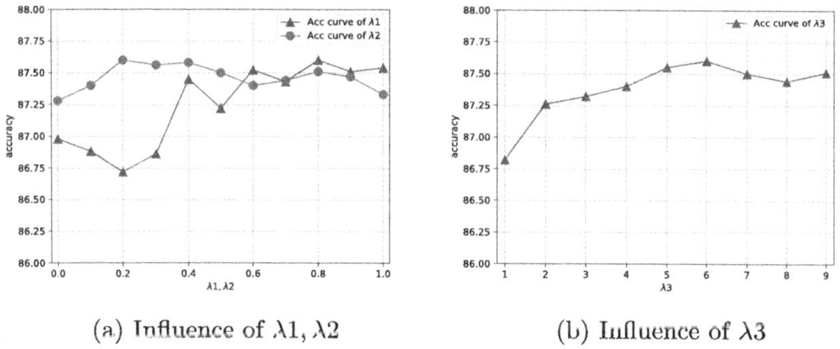

Fig. 4. Experiments on hyper-parameters.

is obvious that in the results our model obtained compared to baseline, the distance between classes is smaller and the interval between classes is more obvious. Specifically, Figure (a) mixed together person, dog, elephant, while Figure (b) can more significantly distinguish them. This indicates that our model feature extractor can perform tighter clustering of categories while well dividing different categories compared to [24]. In Figures (c) and (d), different colors represent domains. Also, our method can have a better clustering effect.

Parameter Sensitivity. Figure 4 demonstrates the performance accuracy of parameters under various settings. For the three parameters, we fixed two and altered only one for training and accuracy testing. For λ_1 and λ_2, as they increase, the model gains more information from the teacher, leading to a trend of improved accuracy. However, after reaching a peak, the effectiveness begins to decline, which is attributed to the neglect of the client's input, causing the model's learning to deviate. In the case of λ_3, the model, constrained by the feature enhancement branch, grows more robust with the increase of the parameter, achieving more robust learning effects. Additionally, our model demonstrates relatively stable performance even when the hyperparameters are varied.

5 Conclusion

In this work, we utilize the domain consistency of text features to improve the generalization ability of the model by distilling from LVLM knowledge. We propose a stylistic feature dispatching and augmenting method called FeSFD, solves the problem of poor adaptability of text features to non-iid data. Also, the consistency loss restricts semantic understanding and dual knowledge distillation better utilizes CLIP as a teacher. Experiments across three benchmarks shows that our method has strong generalization ability, and the ablation study certifies the effectiveness of each component.

Acknowledgments. This work is supported by Shenzhen Science and Technology Program (No. JCYJ20230807120800001), and 2023 Shenzhen sustainable supporting funds for colleges and universities (No. 20231121165240001). The authors sincerely appreciate the computing environment supported by the China Unicom Shenzhen Intelligent Computing Center.

References

1. Balaji, Y., Sankaranarayanan, S., Chellappa, R.: Metareg: towards domain generalization using meta-regularization. Adv. Neural Inform. Process. Syst. **31** (2018)
2. Ben-David, S., Blitzer, J., Crammer, K., Pereira, F.: Analysis of representations for domain adaptation. Adv. Neural Inform. Process. Syst. **19** (2006)
3. Carlucci, F.M., D'Innocente, A., Bucci, S., Caputo, B., Tommasi, T.: Domain generalization by solving jigsaw puzzles. In: Proceedings of the IEEE/CVF Conference on Computer Vision and Pattern Recognitionm pp. 2229–2238 (2019)
4. Chen, J., Jiang, M., Dou, Q., Chen, Q.: Federated domain generalization for image recognition via cross-client style transfer. In: Proceedings of the IEEE/CVF Winter Conference on Applications of Computer Vision, pp. 361–370 (2023)
5. Cho, J., Nam, G., Kim, S., Yang, H., Kwak, S.: Promptstyler: prompt-driven style generation for source-free domain generalization. In: Proceedings of the IEEE/CVF International Conference on Computer Vision, pp. 15702–15712 (2023)
6. Dou, Q., Coelho de Castro, D., Kamnitsas, K., Glocker, B.: Domain generalization via model-agnostic learning of semantic features. Adv. Neural Inform. Process. Syst. **32** (2019)
7. Gulrajani, I., Lopez-Paz, D.: In search of lost domain generalization. arXiv preprint arXiv:2007.01434 (2020)
8. Hinton, G., Vinyals, O., Dean, J.: Distilling the knowledge in a neural network. arXiv preprint arXiv:1503.02531 (2015)
9. Huang, Z., Zhou, A., Ling, Z., Cai, M., Wang, H., Lee, Y.J.: A sentence speaks a thousand images: domain generalization through distilling clip with language guidance. In: Proceedings of the IEEE/CVF International Conference on Computer Vision, pp. 11685–11695 (2023)
10. Jeon, S., Hong, K., Lee, P., Lee, J., Byun, H.: Feature stylization and domain-aware contrastive learning for domain generalization. In: Proceedings of the 29th ACM International Conference on Multimedia, pp. 22–31 (2021)
11. Li, D., Yang, Y., Song, Y.Z., Hospedales, T.M.: Deeper, broader and artier domain generalization. In: Proceedings of the IEEE International Conference on Computer Vision, pp. 5542–5550 (2017)

12. Liu, Q., Chen, C., Qin, J., Dou, Q., Heng, P.A.: Feddg: federated domain generalization on medical image segmentation via episodic learning in continuous frequency space. In: Proceedings of the IEEE/CVF Conference on Computer Vision and Pattern Recognition, pp. 1013–1023 (2021)
13. Lu, W., Xixu, H., Wang, J., Xie, X.: Fedclip: fast generalization and personalization for clip in federated learning. In: ICLR 2023 Workshop on Trustworthy and Reliable Large-Scale Machine Learning Models (2023)
14. Van der Maaten, L., Hinton, G.: Visualizing data using t-sne. J. Mach. Learn. Res. **9**(11) (2008)
15. McMahan, B., Moore, E., Ramage, D., Hampson, S., y Arcas, B.A.: Communication-efficient learning of deep networks from decentralized data. Artifi. Intell. Statist., 1273–1282 (2017)
16. Radford, A., et al.: Learning transferable visual models from natural language supervision. In: International Conference on Machine Learning, pp. 8748–8763. PMLR (2021)
17. Venkateswara, H., Eusebio, J., Chakraborty, S., Panchanathan, S.: Deep hashing network for unsupervised domain adaptation. In: Proceedings of the IEEE Conference on Computer Vision and Pattern Recognition, pp. 5018–5027 (2017)
18. Wang, S., Yu, L., Li, C., Fu, C.-W., Heng, P.-A.: Learning from extrinsic and intrinsic supervisions for domain generalization. In: Vedaldi, A., Bischof, H., Brox, T., Frahm, J.-M. (eds.) ECCV 2020. LNCS, vol. 12354, pp. 159–176. Springer, Cham (2020). https://doi.org/10.1007/978-3-030-58545-7_10
19. Wu, G., Gong, S.: Collaborative optimization and aggregation for decentralized domain generalization and adaptation. In: Proceedings of the IEEE/CVF International Conference on Computer Vision, pp. 6484–6493 (2021)
20. Xu, Q., Zhang, R., Zhang, Y., Wang, Y., Tian, Q.: A fourier-based framework for domain generalization. In: Proceedings of the IEEE/CVF Conference on Computer Vision and Pattern Recognition, pp. 14383–14392 (2021)
21. Xu, Q., Zhang, R., Zhang, Y., Wu, Y.Y., Wang, Y.: Federated adversarial domain hallucination for privacy-preserving domain generalization. IEEE Trans. Multimedia (2023)
22. Yoo, J., Uh, Y., Chun, S., et al.: Photorealistic style transfer via wavelet transforms. In: Proceedings of the IEEE/CVF International Conference on Computer Vision, pp. 9036–9045 (2019)
23. Yuan, J., Ma, X., Chen, D., Wu, F., Lin, L., Kuang, K.: Collaborative semantic aggregation and calibration for federated domain generalization. IEEE Trans. Knowl. Data Eng. (2023)
24. Zhang, R., Xu, Q., Yao, J., Zhang, Y., Tian, Q., Wang, Y.: Federated domain generalization with generalization adjustment. In: Proceedings of the IEEE/CVF Conference on Computer Vision and Pattern Recognition, pp. 3954–3963 (2023)
25. Zhou, K., Yang, Y., Hospedales, T., Xiang, T.: Deep domain-adversarial image generation for domain generalisation. In: Proceedings of the AAAI Conference on Artificial Intelligence, vol. 34, pp. 13025–13032 (2020)
26. Zhou, K., Yang, Y., Hospedales, T., Xiang, T.: Learning to generate novel domains for domain generalization. In: Vedaldi, A., Bischof, H., Brox, T., Frahm, J.-M. (eds.) ECCV 2020. LNCS, vol. 12361, pp. 561–578. Springer, Cham (2020). https://doi.org/10.1007/978-3-030-58517-4_33
27. Zhou, K., Yang, Y., Qiao, Y., Xiang, T.: Domain generalization with mixstyle. In: International Conference on Learning Representations (2020)

RetailEye: Supervised Contrastive Learning with Compliance Matching for Retail Shelf Monitoring

Mamoun Alghaslan[1], Khaled Almutairy[1], El-Sayed El-Alfy[1,2,3(✉)], and Abdul-Jabbar Siddiqui[2,3,4]

[1] Information and Computer Science Department, Dhahran, Saudi Arabia
[2] Computer Engineering Department, Dhahran, Saudi Arabia
[3] Interdisciplinary Research Center of Intelligent Secure Systems, Dhahran, Saudi Arabia
`alfy@kfupm.edu.sa`
[4] SDAIA-KFUPM Joint Research Center for Artificial Intelligence, King Fahd University of Petroleum and Minerals, Dhahran, Saudi Arabia

Abstract. By harnessing technological advancements in computer vision and artificial intelligence, retail entrepreneurs can not only meet their objectives but also position themselves for sustainable growth in a competitive marketplace. A critical area of focus is inventory management, particularly monitoring grocery products on shelves and identifying misplaced or out-of-stock items. However, automatically detecting and recognizing products in real-time retail environments presents significant challenges, including factors such as varied visual representations, unpredictable poses, partial or full occlusions, and variations of lighting reflections on glossy packaging, and a lack of unified resources. In this paper, we propose and evaluate a two-stage approach, termed RetailEye, which employs supervised contrastive learning with compliance matching and leverages the latest developments in deep learning. After evaluating different models for object detection and recognition, we designed our system based on YOLOv8s in the first stage and EfficientNetV2-S and ResNet18 in the second stage. The proposed model outperformed the one-stage approach with high detection and recognition accuracy. Additionally, we unveil a custom dataset specifically curated for this research, aimed at advancing the field of inventory management.

Keywords: Contrastive Learning · Video Analytics · Deep Learning · Product Detection and Recognition · Retail Management · Shelf Monitoring

1 Introduction

In retail stores, maintaining optimal inventory levels and ensuring accurate product placement on store shelves are critical aspects that directly impact customer satisfaction and overall business efficiency. The traditional methods of monitoring stock levels and repositioning misplaced items involve regular manual shelf inspection by retail workers, which is time-consuming, prone to human

M. Mahmud et al. (Eds.): ICONIP 2024, CCIS 2297, pp. 248–263, 2026.
https://doi.org/10.1007/978-981-96-7036-9_17

errors, labor-intensive, inflexible, and costly. Subsequently, they may lead to non-optimized operations and revenue losses.

The main objectives of entrepreneurs in the retail industry are to boost profit margins and attract a larger customer base [6]. Embracing recent technological advancements has enhanced the customer experience while also reducing operational costs. Advancements in computer vision and AI-driven distributed monitoring systems have significantly enhanced the feasibility of real-time inventory management by automating the detection and tracking of shelf conditions with greater accuracy and efficiency. These systems can effectively detect out-of-stock situations and assess product organization on racks in accordance with store's layout, commonly referred to as a planogram. Moreover, these advanced systems can provide valuable analytics that can help retailers reduce out-of-stock rates, enhance customer satisfaction, streamline operations and optimize business performance, and ultimately increase profitability.

There are two primary techniques for vision-based object detection and recognition, each with its own set of strengths and limitations. The traditional methods involve extracting manual features of the image based on domain knowledge, such as color histograms or texture patterns to classify grocery products. While this approach can be interpretable and computationally efficient, it does not generalize well due to the highly dynamic nature of product images. Manually extracted features struggle to adapt to the common challenges faced in real environments. On the other hand, deep learning models, specifically Convolutional Neural Networks (CNNs) and transformers, have recently demonstrated remarkable success in object detection and recognition. Deep learning can automatically learn representations and provide end-to-end solutions with different levels of abstraction, which is more efficient compared to manual feature engineering. Although deep learning methods are widely used nowadays, the performance of these methods is influenced by two factors. First, their effectiveness diminishes when working with small training datasets. Second, the previously learned knowledge from prior classes or tasks tends to be lost when the model is trained on newly introduced classes or tasks. Since supermarkets have numerous products that the model was trained on, retraining the model from scratch every time a new product appears in the market is impractical [5].

Even though deep learning methods outperformed traditional detection and recognition techniques, their performance is greatly influenced by the amount of training data available. Building a dataset for product recognition is an expensive task. The dataset should contain products captured from various perspectives and under different lighting conditions. It should also handle scenarios in which the products might be deformed or occluded by other items since they do not always maintain their intended form. Afterward, the captured images need to be manually annotated with all the necessary information to describe the products accurately, such as brand, flavor, type, size, and other relevant attributes. As a result, making a well-represented dataset becomes a challenge for supermarkets. Several studies have tried to mitigate these limitations. In [21,22], generative adversarial networks (GANs) have been successfully used to create synthetic context. In [4,11], a matching technique is described to classify products based on a reference database containing product images under a controlled environment with different viewpoints. If a new product is introduced into the model, it will be classified based on the most similar product on the database. While these solutions tackle the challenges of recognizing objects, detecting them given the complexity of the background and domain shift adds another layer of complexity to the problem. Therefore, an automatic shelf monitoring system is still very challenging to implement.

To address these challenges, we propose a novel end-to-end solution for the automatic shelf monitoring system following a two-stage approach that leverages advanced machine learning techniques for enhanced performance and accuracy. Additionally, it explores various methods and loss functions for model selection and tuning. In the first stage, our system employs a state-of-the-art object detection model specifically fine-tuned on a large-scale dataset of retail product images. This model can accurately localize products on the shelves by generating precise bounding boxes. In the second stage, these bounding boxes are processed by a robust recognition model, which utilizes a deep convolutional neural network (CNN) architecture. This CNN is fine-tuned using supervised contrastive learning to robustly classify products even in the presence of varying lighting conditions, occlusions, and packaging variations. By optimizing each stage independently, our solution ensures high detection and classification accuracy, making it a reliable tool for real-time shelf monitoring and inventory management.

The remainder of the paper is organized as follows. Section 2 briefly reviews related work. Section 3 describes the proposed methodology. Section 4 describes the conducted experiments and results and finally Sect. 5 concludes the paper.

2   Related Work

Traditional image processing techniques have a long history in textural, morphological, statistical, and multi-scale descriptors for compact representation of images due to their low costs and reasonable performance. These algorithms include, but are not limited to, Scale Invariant Feature Transform (SIFT), Speeded Up Robust Features (SURF), Binary Robust Invariant Scalable Keypoints (BRISK), Robust Independent Elementary Features (BRIEF), Features from Accelerated Segment Test (FAST), Oriented FAST and Rotated BRIEF (ORB), Gray Level Co-occurrence Matrix (GLCM), Hu's and Zernike's Moments, HARRIS-Stephens, Maximally Stable External Regions (MSER), MinEigen, Local Binary Pattern (LPB), Multilevel Binary Morphological Analysis, and Wavelets. Three examples of traditional methods are discussed in the following paragraphs.

Moorthy et al. [15] proposed a solution for detecting misplaced or missing on-shelf products in retail stores. Their approach involves storing a list of product cropped images to serve as reference, capturing target shelf images using a video or camera device, matching reference images with target images in grayscale, and identifying if the product is misplaced or missing. Image descriptors for both input and target images are extracted and matched using the SURF algorithm, then the resulting indices of the matched features are passed to the Estimate Geometric Transform function to determine the presence of the reference in the target, and lastly drawing a bounding box around it. This method has a limitation to front-facing products and cannot handle occlusion.

Out-of-stock (OOS) products can be detected by looking for empty spots on shelves. However, empty shelves do not always indicate an OOS product. The presence of labels is also an important factor in distinguishing whether an empty space is for an OOS product or simply an empty space. The proposed method by [18] consists of three parts: Stitching panoramic images using homography estimation and Fast Explicit Diffusion (FED), detecting labels using a cascade of weak classifiers measuring Local Binary Pattern (LBP), and detecting OOS by doing aisle segmentation, OOS segmentation, vertical separation of OOS candidates, filtering candidates by the CIE L*C*h* color space, and finally feature extraction and classification.

Another approach is based on RGBD (RGB images with Depth information) point cloud data captured using consumer-grade sensors has been implemented to estimate the availability of products and OOS, with no prior knowledge [14]. The proposed system was made to monitor perishable fresh products stored in countertops, baskets, or crates, although it can monitor any type of layout. Top-mounted and front-facing camera setups were experimented with a reference model that is first calibrated on empty shelves, regardless of the camera orientation or floor layout; meaning it could be flat, or inclined in any direction. Multiple images are taken of the reference shelf floor and averaged together to reduce the Maximum Likelihood Estimation (MLE) error. Once the products have been placed, another calibration is done and depth points sticking up of the shelf floor are used to estimate the On-Shelf-Availability (OSA). Segmentation is applied to distinguish points belonging to products from the original reference plane using a threshold that is automatically set from the sensor's Root Mean Square Error (RMSE).

Recent advances in object detection techniques have attained remarkable success. However, detecting densely packed objects, especially in scenes containing many identical objects closely packed together, such as products on shelves, remains a challenging task. To tackle this issue, an innovative deep learning-based method is proposed in [5], with two key components. The first component uses Soft-IoU score as a Jaccard index to estimate the similarity between the predicted bounding box and the ground truth box. It is computed by incorporating a fully-convolutional layer added as a third head on top of the Region Proposal Network (RPN). This Soft-IoU value helps to resolve the issue of distinguishing overlapping objects that are closely positioned and may have similar appearances. The second component is the EM-Merger unit, which transforms the predicted bounding boxes along with their corresponding Soft-IoU scores into a Mixture of Gaussian's representations. The Expectation-Maximization (EM) algorithm is then employed to cluster these representations into groups, effectively separating overlapping or adjacent detected bounding boxes that were initially identified as separate objects.

The most difficult part of any machine learning solution is collecting the dataset. Often, real-world datasets are unlabeled, which means that significant human effort is required for annotation. This is particularly the case in retail stores, where collecting and labeling the dataset is an expensive process. There are various techniques to tackle this problem with satisfactory performance, one of which is semi-supervised learning (SSL), where a large amount of unlabeled data is utilized in combination with a smaller set of labeled data to train a model. In [23], a proposed approach fine-tunes three pre-trained models (RetinaNet, YOLOv3, and YOLOv4) on the labeled set based on mAP (mean average precision), F1-score, and recall evaluation metrics. Subsequently, the unlabeled set is assigned pseudo-labels generated from the best-performing model's predictions. These pseudo-labels serve as ground truth annotations during the training process for the unlabeled data. The best model is then retrained using both the labeled data and the unlabeled data with the assigned pseudo-labels. The final trained model is employed to detect 'Product', 'Empty Shelf', and 'Almost Empty Shelf' areas on the shelves.

In [8], the authors focused on the quality of the data, starting from collection, cleaning, and annotation before modeling. They collected a dataset of 1000 images following well-defined guidelines and annotated the images. The main goal of this work is to propose an end-to-end, real-time, computationally efficient pipeline solution for empty-shelf detection. They compared two different versions of both EfficientDet and YOLOv5 models to balance accuracy and inference run-time trade-offs. Additionally, they conducted extensive latency and throughput analysis of the models, utilizing several quantization and inference run-time optimization techniques. The observations are as follows. First, the time spent on image decoding and preprocessing must be optimized using parallelization. Second, the choice of batch size affects the model's maximum throughput. Third, memory requirements for model deployment should be determined carefully as they significantly impact both latency and throughput. Lastly, runtime optimizations like OpenVINO can boost the model's performance.

3   Methodology

A high-level overview of the proposed system pipeline for retail store shelf monitoring is shown in Fig. 1. It consists of two stages. In the first stage, we evaluated several state-of-the-art detection methods to localize and generate bounding boxes around products on the shelves. We evaluated YOLOv8s, RetinaNet, Soft-IoU and EM-Merger, and SAPD on the SKU-110K dataset [5], which includes tightly-packed images of various scales, orientations, lighting conditions and noise levels from thousands of stores.

Based on the pilot testing, YOLOv8s [9] demonstrated superior performance achieving an mAP_{50-90} score of 0.558. While it was comparable to SAPD [24] in this metric, YOLOv8s showcases strong performance, making it an optimal choice for accurately localizing products on shelves. Additionally, YOLOv8s has a lightweight architecture (7.9M parameters) and fast inference speed (30FPS). These characteristics are important for real-world applications where computational resources and efficiency are essential considerations. Therefore, we select

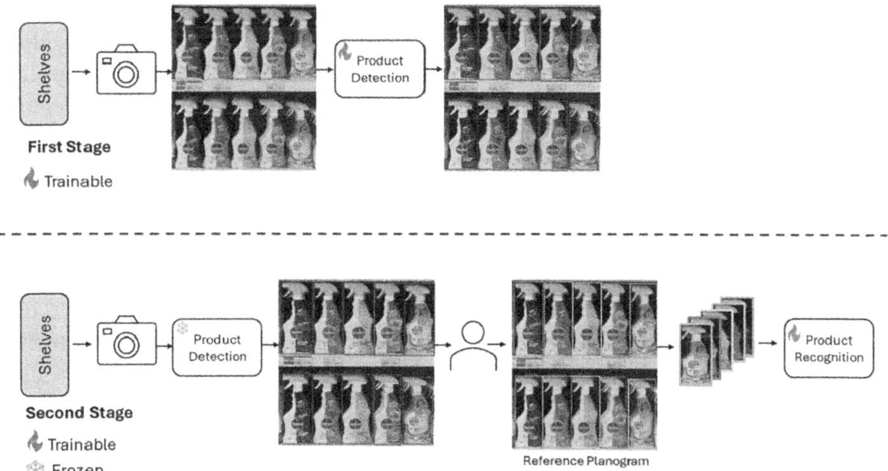

Fig. 1. High-level overview of the shelf monitoring system pipeline. The first stage involves training the detection model framework whereas the second stage is trained for product recognition utilizing detected products from the captured shelf images.

YOLOv8s as the primary model for product detection in our proposed shelf monitoring system.

Subsequently, these generated boxes will be input into the second stage, which is a recognition model for classification. The recognition model is a pretrained CNN model that is fine-tuned using supervised contrastive learning on products cropped from shelves instead of products taken under a controlled environment. This approach enables fine-grained classification, overcoming challenges of limited and under-represented datasets and domain shifts when new products are launched. This final output will be a comprehensive shelf image with all products detected and classified. This output will then be compared with a reference image of the shelf, adhering to the planogram layout of the rack. The rationale of the two-stage process allows for independent optimization of each component, ensuring the best performance of the detection and recognition models.

Recognition Stage Backbone: The backbone consists of a convolutional neural network serving as the feature extractor, responsible for capturing essential patterns and features from the input data. The choice of the backbone architecture can impact the performance and efficiency of the recognition model. More complex architectures may capture complex features but require more computational resources. On the other hand, simpler architectures may be more computationally efficient but could struggle to capture fine-grained features. We initially considered four options as potential backbone architectures: ResNet, EfficientNet, Vision Transformer (ViT), and MobileNet. ResNet is known for its strong generalization capabilities and its ability to capture features of varying complexities. EfficientNet strikes a balance between accuracy and computational efficiency, making it suitable for resource-constrained scenarios. ViT models excel in capturing global context but may require a large amount of training data. MobileNet is designed for resource-constrained devices and embedded applications, offering a good balance between accuracy and computational efficiency. The choice of backbone architecture should consider the specific requirements and available resources of the shelf monitoring system, such as computational resources, and available labeled data. Each mentioned backbone architecture has its advantages and trade-offs that should be carefully considered during the selection process.

Supervised Contrastive Learning: It aims to enhance the discriminative power of representations by leveraging labeled data. This approach contrasts representations of positive pairs (instances belonging to the same class) while simultaneously pushing representations of negative pairs (instances belonging to different classes) apart in an embedding space. By optimizing a contrastive loss function, which encourages similar representations for positive pairs and dissimilar representations for negative pairs, the model learns to capture meaningful and semantically rich features. Unlike traditional supervised learning methods that rely solely on cross-entropy loss, supervised contrastive learning enhances robustness against label corruption and dataset biases. The most commonly used loss functions in contrastive learning is InfoNCE, which is defined as follows:

$$L_{\text{InfoNCE}} = -\log\left(\frac{\exp\left(z_i.z_p/\tau\right)}{\sum_{n\in\mathcal{N}}\exp(z_i.z_n/\tau)}\right)$$

where $s_i = z_i.z_p/\tau$ represents the similarity score between the representation of the original sample i and its positive sample p, and $z_i.z_n/\tau$ represents the similarity score between the representation of the original sample and each of its negative sample $n \in N$. To extend InfoNCE to a supervised setting with multiple positives, SupCon (Supervised Contrastive) loss is introduced [10]:

$$L_{\text{SupCon}} = \frac{1}{N}\sum_{i=1}^{N}\frac{1}{|\mathcal{P}_i|}\sum_{p\in\mathcal{P}_i}\log\left(\frac{\exp(z_i.z_p/\tau)}{\sum_{a\in A_i}\exp(z_i.z_a/\tau)}\right)$$

where $\mathcal{P}_i$ and $\mathcal{A}_i$ are the positive examples and negative examples of x_i, respectively, $z_i = f(x_i)$ represents the embedding or feature vector of x_i, τ is a temperature parameter for normalization, and N is the number of samples in the batch. To enhance the control over positive-negative pair distances and mitigate biases, an enhanced version of InfoNCE, called ε-SupInfoNCE, is proposed in [1], as illustrated in Fig. 2. It is defined by:

$$L_{e\text{-SupInfoNCE}} = -\sum_{i}\log\left(\frac{\exp(s_i^+)}{\exp(s_i^+ - \varepsilon) + \sum_{j}\exp(s_j^-)}\right) + \lambda R_{\text{FairKL}}$$

where λ is a regularization parameter that controls the strength of the FairKL debiasing regularization term while ε is the margin between positive and negative samples. By preventing the use of biased features and improving control over

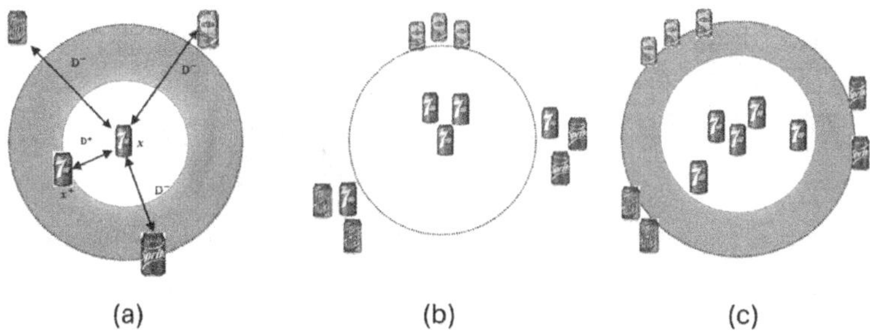

Fig. 2. Illustration of ε-SupInfoNCE that increases the distance between a positive sample and the nearest negative sample, creating a margin that better separates them. This margin prevents biased clusters from mixing positive and negative samples, as shown in scenario (c) compared to (b).

positive-negative pair distances, ε-SupInfoNCE enables the model to learn more robust and unbiased representations. This refinement can lead to better generalization and performance, especially in product recognition. Once the recognition model is fine-tuned using the supervised contrastive learning loss function, we do not use the classification layer anymore. Instead, the encoder layer of the model will be used to extract features of the input image. The extracted features will then be used later for planogram compliance optimization, which is explained next.

Planogram Compliance Control: Planograms are visual guides used in retail to arrange products on shelves effectively. They help promote products, improve customer navigation, and ultimately boost sales [2,13,19]. Traditionally, ensuring planogram compliance involves manual inspection, which is time-consuming and prone to errors. Various automated methods have been explored to address this, including inventory-based systems, RFID tags, and depth cameras. However, these approaches often fall short in terms of accuracy or cost-effectiveness. Our focus is on leveraging computer vision, a promising approach that has received less attention [12]. Our proposed solution involves installing fixed cameras in front of target shelving sections. Initially, a reference image is captured when products are correctly positioned. Each product in this image is annotated with a bounding box and labeled manually. These bounding boxes are then encoded into feature vectors and stored in a vector database for future comparison using the FAISS library [3]. When a test image is captured from the same camera, our detection model generates bounding boxes around products. To determine which products are correctly positioned, these boxes are compared with the reference boxes by calculating their Intersection over Union (IoU) values. The IoU value measures the overlap between the bounding boxes, indicating the degree of alignment between the detected and reference positions. Boxes with IoU values below a certain threshold (e.g., 15%) are considered misaligned and discarded. Subsequently, each correctly positioned product from the test image is cropped and encoded into a feature vector. These vectors are then compared with

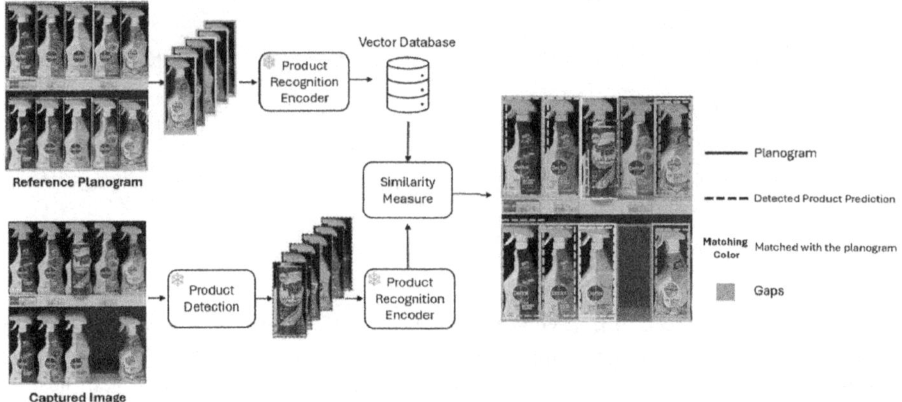

Fig. 3. Illustration of the compliance matching algorithm.

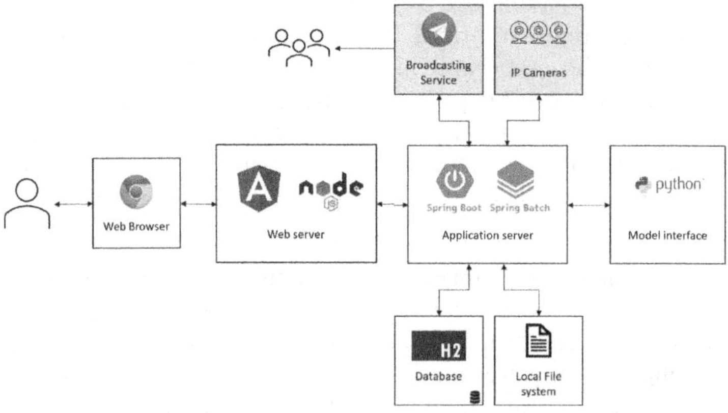

Fig. 4. High-level structure of the developed web-based application.

those in the reference database. If one of the top-5 matching vectors is found with the same desired label, the product is considered correctly placed; otherwise, it's deemed misplaced. Figure 3 illustrates the algorithm for compliance matching.

Out-of-Stock Evaluation: We evaluate out-of-stock situations by measuring the alignment between the detected boxes in the test image and the reference boxes using a predefined IoU threshold value. if the IoU value falls below 15%, indicating poor alignment, we classify the corresponding reference box of the product as empty (out-of-stock). This algorithm works well most of the time as long as the reference and the test images are captured from the same camera position.

Web-Based Application: After training and evaluating the model, we developed a full-stack web-based application as outlined in Fig. 4. Through a web browser, the user interacts with the web server, developed in Angular and

Node.js, which in turn connects to an application server built using Spring Boot. The architecture is designed for easy integration with Broadcasting services or IP cameras, enhancing the system's applicability in real-time monitoring and response scenarios. This flexibility is key in adapting to various operational environments and requirements. The front-end is developed using Angular, a platform and framework for building single-page client applications using HTML and TypeScript. Angular's modular nature allows developers to build large-scale, well-structured applications that are easy to maintain over time as well as providing intuitive navigation and interaction mechanisms for system users. This allows users to efficiently manage and interact with the system, from monitoring real-time data to configuring system settings and parameters.

Fig. 5. An example image of the curated SOMAM dataset

4   Experimental Work

4.1   Data Collection and Preparation

SOMAM Dataset: This dataset was collected by us from Danube, which is a local supermarket in Saudi Arabia, during October and November 2023, to address the lack of publicly available datasets that fit our vision-based shelf monitoring goals. SOMAM stands for Shelf Out-of-stock And Misplaced Monitoring. The images were captured using a mobile camera mounted on a tripod stand, positioned parallel in front of a shelf for a flat perspective. The dataset focuses on five shelves, containing a total of approximately 150 products. Each product was removed once as if OOS (Out-Of-Stock), then an image was captured, and then placed back in the original position, resulting in 150 images, each depicting a single OOS product. Next, 32 products were randomly selected, and each product was intentionally misplaced in five different places, each with a captured image, resulting in 160 images. A total of 310 images were collected for quick experimentation and evaluation. The dataset was annotated using Roboflow[1].

---

[1] https://roboflow.com/annotate.

During the annotation process, we considered products that share similar visual appearance but differ by size, flavor, or quantity to have unique labels. Thus, each unique product has a unique class label. Figure 5 shows an example image of the dataset with all products on the correct place.

SHAPE Dataset: While preparing the camera-ready version of this paper, we found another dataset published online in June 2024, called is SHAPE (SHelf mAnagement Product datasEt)[2]. It provides a rich resource for fine-grained SKU classification tasks, offering 50,000 images across 17,000 distinct SKUs from 62 product categories [16]. Each image is carefully cropped to the boundaries of the product package and labeled with its corresponding European Article Number (EAN). These images were collected from 2,000 supermarkets in Italy using various smartphone cameras, ensuring diversity in image capture conditions. The dataset utilizes the Global Product Classification (GPC) system, which organizes products hierarchically into categories based on shared properties. For the purposes of our work, we leveraged the SHAPE dataset to train and evaluate the product recognition models. While it is a robust dataset for training and evaluating recognition models, it does not include shelf images with products positioned and labeled on shelves, which is required for our shelf monitoring algorithm, with each unique product assigned a distinct label. This limitation is addressed in our collected dataset.

4.2   Implementation Details and Results

We developed our models using Python and PyTorch library. Whenever possible, we utilized existing models and supplemented them with our custom code as needed. All experiments were run on a single RTX-4090 GPU.

Baseline (One-Stage Model): Initially, we built a baseline model to establish a benchmark for comparison with more complex models. This baseline serves as a reference point to assess the performance of other proposed models. We created a version of SOMAM dataset with three classes only: out-of-stock (OOS), misplaced (MIS), and product, to focus on the target classes (MIS and OOS) and disregard all products that are correctly positioned on the shelves. In this way, the products can be localized and classified in one shot, which can also be compared to our proposed algorithm as described in the methodology section. We fine-tuned YOLOv5, YOLOv7, and YOLOv8s for 200 epochs. The model training performance is collected using a Tensorboard. Table 1 shows the comparison results of the three YOLO versions. The YOLOv8s model exhibits strong overall baseline performance on the test set. For misplaced products (MIS), the model achieves perfect precision but with a lower recall of 0.727, suggesting potential improvements in localization. For OOS, the model demonstrates a good performance with a precision of 0.962, perfect recall, and high mAP values of 0.995 and 0.78 for mAP_{50} and mAP, respectively. The 'Product' class is not our primary focus, as we are specifically interested in detecting OOS and MIS. However,

---

[2] https://figshare.com/articles/dataset/SHAPE_-_SHelf_mAnagement_Product_datasEt/24100704.

Table 1. Comparison of three YOLO versions on the test set as a baseline (MIS: MISplaced, OOS: Out-Of-Stock, Prc: Precision, Rec: Recall)

Method	Class	Prc	Rec	mAP_{50}	mAP_{50-90}
YOLOv5	Overall	0.362	0.408	0.550	0.416
	MIS	0.356	0.160	0.246	0.139
	OOS	0.089	0.067	0.409	0.255
	Product	0.640	0.999	0.994	0.853
YOLOv7	Overall	0.916	0.893	0.918	0.672
	MIS	1	0.680	0.759	0.420
	OOS	0.753	1	0.995	0.752
	Product	0.996	0.999	0.999	0.844
YOLOv8s	Overall	0.987	0.909	0.957	0.767
	MIS	1	0.727	0.882	0.648
	OOS	0.962	1	0.995	0.78
	Product	0.998	1	0.995	0.873

(a)

(b)

Fig. 6. Example illustrating the model's difficulty in detecting a misplaced product in (a) visually complex scene, (b) when surrounded with visually similar items.

including this class is necessary to enable the model to learn the correct location of each product. In conclusion, YOLOv8s shows reliable performance, especially in detecting Out-of-stock items. While it achieves a high precision for misplaced products, there is room for further improvement in recall scores.

Additionally, we identified instances where the baseline model failed to correctly detect the classes. Notably, our investigation revealed a specific challenge related to the detection of misplaced products, particularly those situated in front of items with similar shapes or colors. The model struggled in scenarios where the misplaced products shared visual similarities with neighboring items.

Table 2. Performance comparison of two-stage contrastive learning models with a baseline one-stage model in terms of precision, recall and F1 score

Method	Misplaced product			Out-of-stock		
	Prc	Rec	F1	Prc	Rec	F1
Baseline (YOLOv8s)	1.0	0.7270	0.8419	0.9620	1.0	0.9806
YOLOv8s + MobileNetV3-Small	0.9233	0.9331	0.9135	1.0	1.0	1.0
YOLOv8s + MobileNetV3-Large	0.9315	0.9234	0.9099	1.0	1.0	1.0
YOLOv8s + ResNet18	0.9395	0.9895	0.9549	1.0	1.0	1.0
YOLOv8s + EfficientNetV2-S	0.9687	0.9135	0.9308	1.0	1.0	1.0
YOLOv8s + EfficientNetV2-B0	0.8568	0.9226	0.8685	1.0	1.0	1.0

Table 3. Comparison of recognition models on SHAPE dataset using different models with different loss functions

Ref	Model	Loss	Accuracy		
			Top-1	Top-5	Top-10
[16]	MobileNetV3-Large	Triplet	0.93	0.96	0.97
	MobileNetV3-Small	Triplet	0.86	0.95	0.97
	EfficientNetV2-B0	Triplet	0.91	0.96	0.97
Ours	MobileNetV3-Large	ε-SupInfoNCE	0.9323	0.9877	0.9926
	MobileNetV3-Small	ε-SupInfoNCE	0.8979	0.9582	0.9680
	EfficientNetV2-B0	ε-SupInfoNCE	0.9410	0.9803	0.9926

A visual representation of these challenges is illustrated in Fig. 6. We believe that it's necessary to augment the dataset with a variety of samples, particularly images with different product arrangements. Although the current model demonstrates a good performance for this task, a larger and more diverse dataset would provide it with a richer set of cases to learn from.

Performance Evaluation (Two-Stage Model with Contrastive Learning): Our proposed shelf compliance algorithm follows a deterministic process and depends heavily on the performance of both detection and recognition models, where errors in the earlier stages can negatively impact the algorithm's overall accuracy. For the detection stage, we employed the YOLOv8s model to generate proposals and experimented with variants of recognition models, namely MobileNet [17], ResNet [7], and EfficientNet [20]. When comparing our two-stage compliance algorithm to the baseline method, it demonstrated superior recall while maintaining high precision as shown in Table 2. Notably, it achieved perfect precision and recall in identifying out-of-stock products, whereas the baseline method produced slightly lower precision at 96.2% for the same task, which could result from different reasons including increased model complexity or overlap of extracted features for misplaced products in the two-stage method. These findings indicate that our two-stage method, which utilizes contrastive learning to compare feature vectors, is more effective than relying solely on detection models for identifying compliance issues. By incorporating reference images and comparing feature vectors, our method improves recall without sacrificing precision.

We also used the SHAPE dataset to compare the performance of the recognition models using the ε-SupInfoNCE supervised contrastive learning loss. Specifically, we compared three models: two versions of the MobileNetV3 network and one EfficientNetV2 network. These models were evaluated against the results reported by the authors of the SHAPE dataset [16]. Our models were trained using the AdamW optimizer with a learning rate of 0.0075 and a cosine decay schedule over 30 epochs, with a batch size of 64. As shown in Table 3, the use of the ε-SupInfoNCE loss function led to superior performance across all models compared to the Triplet loss. This improvement is likely due to ε-SupInfoNCE's more effective handling of sample distances, allowing for better differentiation between classes.

5 Conclusion and Future Work

In conclusion, the retail industry is witnessing a transformative shift driven by technological advancements aimed at enhancing the customer experience and operational efficiency. Real-time inventory management facilitated by computer vision technology offers substantial benefits by automating tasks such as shelf monitoring and product recognition. However, the complexity of real-world environments poses challenges such as varied visual representations and lighting conditions, necessitating robust solutions to ensure accurate performance. Our proposed solution for automatic shelf monitoring employs a two-stage deep learning approach to enhance performance and accuracy. The first stage utilizes a state-of-the-art object detection model, fine-tuned on a comprehensive dataset of retail product images, to accurately localize products on store shelves by generating precise bounding boxes. In the second stage, these bounding boxes are processed by a robust recognition model using a deep convolutional neural network (CNN) architecture fine-tuned with supervised contrastive learning. This two-stage process allows for the independent optimization of detection and recognition, ensuring high accuracy in real-time shelf monitoring and inventory management, thereby reducing out-of-stock rates and enhancing customer satisfaction. Moreover, a custom dataset, SOMAM, has been curated for further related work advancing the field. Future research can enhance the robustness of automatic shelf monitoring by focusing on several key areas, e.g. improving image quality through advanced preprocessing techniques to mitigate blurring and perspective distortion, and using incremental learning or one-shot learning to update the model without requiring complete retraining.

Acknowledgment The authors would like to acknowledge the support during this work provided by King Fahd University of Petroleum and Minerals, and the Interdisciplinary Research Center for Intelligent Secure Systems (IRC-ISS), Saudi Arabia, under Grant No. INSS2204.

References

1. Barbano, C.A., Dufumier, B., Tartaglione, E., Grangetto, M., Gori, P.: Unbiased supervised contrastive learning. arXiv preprint arXiv:2211.05568 (2022)
2. Battiato, S., Gallo, G., Schettini, R., Stanco, F. (eds.): ICIAP 2017. LNCS, vol. 10484. Springer, Cham (2017). https://doi.org/10.1007/978-3-319-68560-1
3. Douze, M., et al.: The faiss library (2024)
4. Georgiadis, K., et al.: Products-6k: a large-scale groceries product recognition dataset. In: 14th Pervasive Technologies Related to Assistive Environments Conference, pp. 1–7 (2021)
5. Goldman, E., Herzig, R., Eisenschtat, A., Goldberger, J., Hassner, T.: Precise detection in densely packed scenes. In: Proceedings of IEEE/CVF Conference on Computer Vision and Pattern Recognition, pp. 5227–5236 (2019)
6. Guimarães, V., Nascimento, J., Viana, P., Carvalho, P.: A review of recent advances and challenges in grocery label detection and recognition. Appl. Sci. **13**(5), 2871 (2023)
7. He, K., Zhang, X., Ren, S., Sun, J.: Deep residual learning for image recognition. In: Proceedings of IEEE conference on Computer Vision and Pattern Recognition, pp. 770–778 (2016)
8. Jha, D., Mahjoubfar, A., Joshi, A.: Designing an efficient end-to-end machine learning pipeline for real-time empty-shelf detection. arXiv preprint arXiv:2205.13060 (2022)
9. Jocher, G., Chaurasia, A., Qiu, J.: Yolo by ultralytics (2023). https://github.com/ultralytics/ultralytics, Accessed 30 February 2023
10. Khosla, P., et al.: Supervised contrastive learning. Adv. Neural. Inf. Process. Syst. **33**, 18661–18673 (2020)
11. Klasson, M., Zhang, C., Kjellström, H.: Using variational multi-view learning for classification of grocery items. Patterns **1**(8) (2020)
12. Laitala, J., Ruotsalainen, L.: Computer vision based planogram compliance evaluation. Appl. Sci. **13**(18), 10145 (2023)
13. Liu, W., et al.: SSD: single shot multibox detector. In: Leibe, B., Matas, J., Sebe, N., Welling, M. (eds.) ECCV 2016. LNCS, vol. 9905, pp. 21–37. Springer, Cham (2016). https://doi.org/10.1007/978-3-319-46448-0_2
14. Milella, A., Petitti, A., Marani, R., Cicirelli, G., D'orazio, T.: Towards intelligent retail: automated on-shelf availability estimation using a depth camera. IEEE Access **8**, 19353–19363 (2020)
15. Moorthy, R., Behera, S., Verma, S., Bhargave, S., Ramanathan, P.: Applying image processing for detecting on-shelf availability and product positioning in retail stores. In: Proceedings of 3rd International Symposium on Women in Computing and Informatics, pp. 451–457 (2015)
16. Pietrini, R., Paolanti, M., Mancini, A., Frontoni, E., Zingaretti, P.: Shelf management: a deep learning-based system for shelf visual monitoring. Expert Syst. Appl. **255**, 124635 (2024)
17. Qian, S., Ning, C., Hu, Y.: Mobilenetv3 for image classification. In: IEEE 2nd International Conference on Big Data, Artificial Intelligence and Internet of Things Engineering (ICBAIE), pp. 490–497 (2021)
18. Rosado, L., Gonçalves, J., Costa, J., Ribeiro, D., Soares, F.: Supervised learning for out-of-stock detection in panoramas of retail shelves. In: 2016 IEEE International Conference on Imaging Systems and Techniques (IST), pp. 406–411. IEEE (2016)

19. Saran, A., Hassan, E., Maurya, A.K.: Robust visual analysis for planogram compliance problem. In: 14th IAPR International Conference on Machine Vision Applications (MVA), pp. 576–579. IEEE (2015)
20. Tan, M., Le, Q.: Efficientnetv2: smaller models and faster training. In: International Conference on Machine Learning, pp. 10096–10106. PMLR (2021)
21. Tonioni, A., Serra, E., Di Stefano, L.: A deep learning pipeline for product recognition on store shelves. In: IEEE International Conference on Image Processing, Applications and Systems (IPAS), pp. 25–31 (2018)
22. Wei, X.S., Cui, Q., Yang, L., Wang, P., Liu, L.: Rpc: a large-scale retail product checkout dataset. arXiv preprint arXiv:1901.07249 (2019)
23. Yilmazer, R., Birant, D.: Shelf auditing based on image classification using semi-supervised deep learning to increase on-shelf availability in grocery stores. Sensors **21**(2), 327 (2021)
24. Zhu, C., Chen, F., Shen, Z., Savvides, M.: Soft anchor-point object detection. In: Vedaldi, A., Bischof, H., Brox, T., Frahm, J.-M. (eds.) ECCV 2020. LNCS, vol. 12354, pp. 91–107. Springer, Cham (2020). https://doi.org/10.1007/978-3-030-58545-7_6

Solving Expensive Dynamic Multi-objective Problem via Cross-Problem Knowledge Transfer

Ziqi Cheng[1], Xinyu Xue[1], Huajin Tang[2,3], and Liang Feng[1(✉)]

[1] College of Computer Science, Chongqing University, Chongqing, China
{202214021023t,20194292}@stu.cqu.edu.cn, liangf@cqu.edu.cn
[2] The State Key Lab of Brain-Machine Intelligence, Zhejiang University,
Hangzhou, China
htang@zju.edu.cn
[3] College of Computer Science and Technology, Zhejiang University,
Hangzhou, China

Abstract. Expensive Dynamic Multi-Objective Optimization Problems (EXDMOPs) pose significant challenges due to their dynamic and costly nature, as both the objective and constraint functions evolve over time with limited fitness evaluations. Traditional methods treat EXDMOPs as multiple independent and static expensive multi-objective problems, which often ignore the experiences obtained in solving the EXDMOPs or reuse experiences within a single EXDMOP, resulting in inefficient optimization performance. Taking this cue, in this paper, we introduce a novel approach leveraging cross-problem knowledge to enhance the ability to solve EXDMOPs. Unlike existing knowledge transfer methods within a single dynamic problem, our proposed method leverages archived solutions from well-optimized tasks to construct an effective initial population through cross-problem task selection and knowledge transfer. Specifically, a multivariate Gaussian distribution and the Wasserstein distance are adopted to choose the most appropriate task for knowledge transfer. Cross-problem knowledge transfer devises a Support Vector Machine (SVM) classifier to transfer solutions, enabling the construction of high-quality initial populations when dynamic occurs. To assess the performance of the proposed method, extensive empirical studies were conducted on commonly used EXDMOP benchmarks. The results confirm the method's ability to enhance both the efficiency and effectiveness in solving EXDMOPs.

Keywords: Cross-problem knowledge transfer · Expensive dynamic multi-objective optimization · Evolutionary optimization

1　Introduction

In practical scenarios, most optimization tasks involve multi-objective problems (MOPs), where various conflicting goals must be addressed simultaneously, such as vehicle routing in logistics [1], phylogenetic tree inference in biology [2], water resource-planning in engineering [3], and cloud services load balancing in

© The Author(s), under exclusive license to Springer Nature Singapore Pte Ltd. 2026
M. Mahmud et al. (Eds.): ICONIP 2024, CCIS 2297, pp. 264–278, 2026.
https://doi.org/10.1007/978-981-96-7036-9_18

finance [4]. These issues escalate to dynamic multi-objective problems (DMOPs) when aspects like objective or constraint functions change over time or fluctuate with varying environmental conditions. When these DMOPs also require significant resources for evaluating potential solutions—making them costly—they are termed expensive dynamic multi-objective optimization problems (EXDMOPs). A typical example is found in global mineral processing operations, which are dynamic by nature, there are frequent changes not only in objective and constraint functions but also in equipment capacities and operational conditions. The process is further complicated and made costly by the complex physical and chemical reactions involved, necessitating considerable time and financial investment [5].

To solve EXDMOP, while it can be approached as multiple independent static expensive MOPs, the constrained number of evaluations often leads to suboptimal performance. To improve the optimization performance when solving EXDMOPs, it is beneficial to consider either explicit or implicit correlations across different stages of optimization in EXDMOPs [6–8]. Utilizing knowledge from previous stages can enhance current optimization efforts [9]. Dynamic Multi-Objective Evolutionary Algorithms (DMOEAs) have been effective in this context, demonstrating significant results by leveraging historical optimization insights. However, applying DMOEAs directly to EXDMOPs poses challenges due to limited fitness evaluations, which hinder the algorithms' ability to effectively track the current Pareto optimal set (POS) and Pareto optimal front (POF). For instance, Feng et al. [10] developed an evolutionary search approach based on autoencoding to navigate the evolving directions of the POS effectively during the evolutionary process. Yet, the autoencoder struggles with learning a robust mapping matrix due to sparse function evaluations, leading to a less effective initial population. Similarly, Zhou et al. [11] enhanced prediction accuracy through a kernelized autoencoder that considers both decision and objective spaces. However, the approach requires extensive evaluations to validate the predicted objective function, which is not feasible for EXDMOPs due to their expensive nature. Jiang et al. [12] introduced a classifier-based transfer method to create initial populations, significantly enhancing their quality. Despite its potential, this method also demands many fitness evaluations, making it less practical for EXDMOP scenarios. These examples highlight the ongoing challenge of balancing computational cost and optimization effectiveness in the dynamic and costly landscape of EXDMOPs.

Keeping the above in mind, drawing inspiration from recent advancements in utilizing large-scale models for complex optimization tasks [13] and leveraging classifier to enhance the quality of the search population [14], this paper introduces a cross-problem knowledge transfer approach aimed at creating an effective initial population for solving EXDMOPs. Specifically, we propose retaining high-quality solutions from previously optimized EXDMOPs. Using these archived solutions, we develop a task selection strategy to pinpoint the most appropriate previously solved task (i.e., EXDMOPs) and employ its dominated and non-dominated solutions to learn a general classifier. This classifier is designed to generate or predict high-quality populations for new dynamic challenges in EXDMOPs, thereby potentially increasing the efficacy and efficiency of existing DMOEAs. Unlike earlier methods that rely on continuous fitness evaluations to update the initial population, our approach minimizes the need for extensive evaluations, making it a more efficient alternative in expensive and dynamic optimization scenarios. To assess the proposed method, we carry out extensive empirical studies using commonly used EXDMOP benchmarks, comparing our approach against both existing state-of-the-art expensive dynamic and static multi-objective optimization evolutionary algorithms (i.e., EXDMOEA and EXMOEA). The obtained results indicate that our cross-problem transfer method significantly enhances both the efficiency and effectiveness in solving EXDMOPs.

The remainder of the paper is structured as follows. Section 2 introduces the definition of EXDMOP and reviews the related work. The detailed design of the proposed cross-problem knowledge transfer method is presented in Sect. 3. Furthermore, Sect. 4 provides empirical analysis comparing the proposed method against dynamic and static EXMOEA on DMOP benchmarks. Section 5 draws concluding remarks and outlines future research directions.

2   Preliminary

In this section, we begin by defining EXDMOP and discuss key concepts associated with it. We then provide a concise overview of related work, offering context and highlighting the contributions to the field.

2.1   Expensive Dynamic Multi-Objective Optimization Problem

The formulation of EXDMOP is generally presented as follows [15]:

$$\begin{cases} \min F(\mathbf{x}, t) = \langle f_1(\mathbf{x}, t), f_2(\mathbf{x}, t), \ldots, f_m(\mathbf{x}, t) \rangle, \\ s.t. \ \mathbf{x} \in \Omega \end{cases} \tag{1}$$

where $\mathbf{x}$ represents a vector with n dimensions within the decision domain Ω, and f_i corresponds to the i-th objective function evaluated at time step t. $F(\mathbf{x}, t)$ consists of m distinct f_i values. In EXDMOPs, fitness evaluation incurs such significant computational costs that only a few fitness evaluations are accessible. In DMOPs, the set of non-dominated solutions at time step t is referred to as the dynamic Pareto optimal set (DPOS$_t$), and the $F(\mathbf{x}, t)$ associated with DPOS$_t$ are termed the dynamic Pareto optimal front (DPOF$_t$), both of which are defined as follows:

Definition 1. *Non-dominated Solutions in Dynamic Environment.* *$\mathbf{x_1} \prec_t \mathbf{x_2}$ (where $\mathbf{x_1}, \mathbf{x_2} \in \Omega$) indicates that $\mathbf{x_2}$ dominates $\mathbf{x_1}$ at time step t. In this case, $\mathbf{x_2}$ is called a non-dominated solution, where:*

$$\begin{cases} \forall i \in \{1, \ldots, m\} \ f_i(\mathbf{x_1}, t) \geq f_i(\mathbf{x_2}, t) \\ \exists j \in \{1, \ldots, m\} \ f_j(\mathbf{x_1}, t) > f_j(\mathbf{x_2}, t) \end{cases} \tag{2}$$

Definition 2. *Dynamic Pareto Optimal Set (DPOS).* *The POS at time step t, represented by $DPOS_t^*$, is defined as the set of all non-dominated solutions:*

$$DPOS_t^* = \{\mathbf{x}_i^* \mid \nexists \mathbf{x}_j \in \Omega, \mathbf{x}_i^* \prec_t \mathbf{x}_j, \mathbf{x}_i^* \in \Omega\}. \tag{3}$$

Definition 3. *Dynamic Pareto Optimal Front (DPOF).* *The POF at time step t, represented by $DPOF_t^*$, consists of the objective vectors corresponding to $DPOS_t^*$:*

$$DPOF_t^* = \{F(\mathbf{x}_i, t) \mid \mathbf{x}_i \in DPOS_t^*\}. \tag{4}$$

2.2 Related Work

In recent years, many studies have been proposed to enhance optimization performance by leveraging the transfer of problem-solving experiences. While these studies have primarily focused on tackling dynamic DMOP, relatively few have specifically addressed EXDMOP. In particular, Fan et al. [16] introduced TL-MOEAD/D-EGO, a transfer learning-based approach employing transfer component analysis (TCA) [17] to construct the surrogate model for the current optimization process, leveraging knowledge extracted from earlier time steps. Liu et al. [18] proposed DKRVEA, which combines transfer learning and prediction. This approach not only transfers similar knowledge to enrich the training dataset for surrogate model construction but also integrates the population prediction strategy (PPS) [19] to generate competitive solutions, thereby enhancing the quality of solutions in the sampling pool. Moreover, Zhang et al. [20] proposed TrSA-DMOEA, another transfer learning-based approach. They applied a transfer method focused on knee points and employed the geodesic flow kernel (GFK) [21] to predict new knee points, which are then used to construct new surrogate models.

Although these approaches have been successful, it is important to highlight that they often transfer knowledge within a single dynamic problem to assist the optimization in the current time step. However, in EXDMOPs, the limited number of fitness evaluations may result in poor-quality and very few non-dominated solutions in the previous time step. Consequently, transferring these solutions could be ineffective for the current optimization process. To address this issue, in this paper, we propose a cross-problem knowledge transfer method for solving EXDMOP, which obtains high-quality solutions by transferring knowledge from appropriate and well-optimized EXDMOPs.

3   Proposed Method

In this section, the detailed design of the proposed cross-problem knowledge transfer method is presented. In particular, Sect. 3.1 illustrates how our proposed method integrates with the existing evolutionary solver to address EXD-MOPs and provides an overview of the method's workflow. Subsequently, the introductions of the key components in our proposed method are provided in Sect. 3.2.

3.1   Outline of the Proposed Method

Figure 1 illustrates the complete solving process for EXDMOPs, also shown in Algorithm 1. Initially, at time step $t = 1$, the initial population is generated using Latin hypercube sampling (LHS), and the solver proceeds to obtain the Pareto solutions $\mathbf{x_i}$ for the current time step, as seen in lines 1–2 of Algorithm 1. When dynamic changes occur ($t = t + 1$), particularly when $t \geq 2$, the proposed cross-problem knowledge transfer method, which includes *cross-problem task selection* and *cross-problem transfer*, kicks in to generate the initial population. The initial population is then considered as the population of the evolutionary search, thereby enhancing solver performance during optimization, as shown in lines 5–7 in Algorithm 1. This iterative process continues until the termination condition is satisfied, ultimately yielding Pareto solutions across all time steps. A detailed exposition of the proposed method is presented in Sect. 3.2.

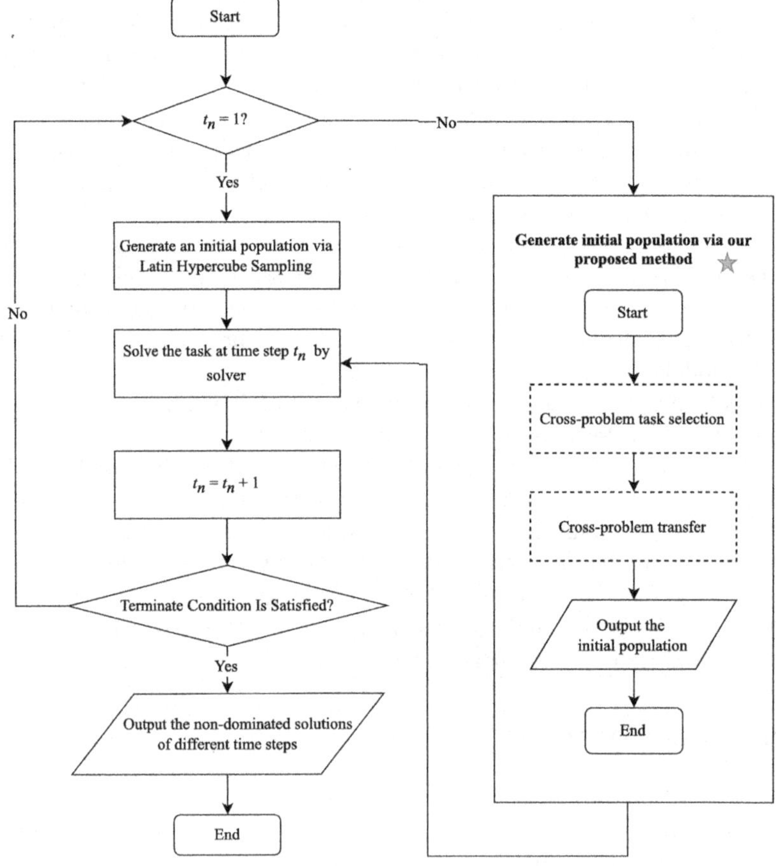

Fig. 1. Outline of the proposed method for solving EXDMOPs.

Algorithm 1. Pseudocode of cross-problem knowledge transfer method

---

Input: $\mathcal{T}_t$: the expensive dynamic optimization problem.
Output: The non-dominated solutions of different time steps.

1: Generate the initial population POP_{init} of N individuals by Latin Hypercube Sampling; initialize $t_n = 1$.
2: Perform evolutionary optimization with POP_{init} to solve the task and get the non-dominated solutions at time step t_n;
3: **while** *Terminate Condition Is Not Satisfied* **do**
4:     $t_n = t_n + 1$;
5:     Select the appropriate task $\mathcal{T}_s$ via the proposed cross-problem task selection;
6:     Set the source task as $\mathcal{T}_s$, generate initial population POP_{CR} via the proposed cross-problem transfer;
7:     Perform evolutionary optimization with POP_{CR} to solve the task and get the non-dominated solutions at time step t_n;
8: **end while**
9: **return** the non-dominated solutions of different time steps;

3.2 Details in the Proposed Method

Figure 2 illustrates the flowchart of the proposed method, which comprises two main components: Cross-Problem Task Selection and Cross-Problem Transfer. A detailed description of these components is provided below.

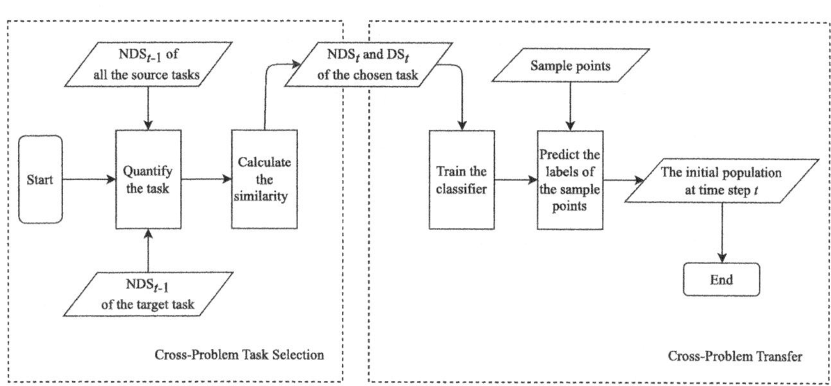

Fig. 2. The flowchart of cross-problem knowledge transfer method

Cross-Problem Task Selection: In this paper, we refer to the task being solved as the target task and the previously well-optimized tasks as the source tasks. Before performing knowledge transfer, the selection of similar source tasks is crucial. Multi-objective optimization solutions are obtained from the Pareto optimal set, in which each solution contains multiple fitness values, making it difficult to assess similarity using simple methods such as the method based on fitness rank correlation [22]. Therefore, we assess the similarity of each task based on the solution space. In particular, as shown in Fig. 2, suppose that the current time step is t, NDS_{t-1} represents the non-dominated solutions in time step $t-1$. Subsequently, a multivariate Gaussian distribution [23] is employed to capture the characteristics of the NDS_{t-1} of the target task and all the source tasks, which is given by:

$$\mathcal{N}(\mu, \Sigma) \tag{5}$$

$$\mu_i = \frac{1}{k} \sum_{j=1}^{k} \mathbf{x_i} \tag{6}$$

$$\Sigma = \frac{1}{k-1}(\mathbf{x_i} - \frac{1}{k} \sum_{j=1}^{k} \mathbf{x_j})(\mathbf{x_i} - \frac{1}{k} \sum_{j=1}^{k} \mathbf{x_j})^T \tag{7}$$

where $\mathbf{x_i}$ is the i-th individual in NDS_{t-1}, k denotes the number of NDS_{t-1} and Σ represents the covariance matrix. Once the target task and all the source tasks have been quantified, the Wasserstein distance [24] is adopted to compute the distance between the target task and source tasks:

$$\mathcal{W}[\mathcal{N}_t, \mathcal{N}_s] = ||\mu_t - \mu_s||^2 + Tr(\Sigma_t) + Tr(\Sigma_s) - 2Tr((\Sigma_t^{1/2} \Sigma_s \Sigma_t^{1/2})^{1/2}) \tag{8}$$

where $\mathcal{N}_t$, $\mathcal{N}_s$, μ_t, μ_s, Σ_t, and Σ_s represent target task distribution, source task distribution, the mean of target task distribution, the mean of source task distribution, the covariance matrix of target task distribution, and the covariance matrix of source task distribution, respectively. In general, the closer the distance between distributions, the more similar the tasks. We describe the similarity between the target task and source tasks as follows:

$$sim(t, s) = \frac{1}{\mathcal{W}[\mathcal{N}_t, \mathcal{N}_s]} \tag{9}$$

Based on Eq. 9, the source task with the largest $sim(t, s)$ is selected for knowledge transfer at time step t.

Cross-Problem Transfer: Once the appropriate task is selected, the knowledge transfer stage commences, as illustrated in Fig. 2. In this stage, we extract the non-dominated and dominated solutions at time step t from the source task to construct the training set $\mathcal{TS}$:

$$\mathcal{TS} = \{(\mathbf{x}, 1) | \mathbf{x} \in NDS_t\} \cup \{(\mathbf{x}, 0) | \mathbf{x} \in DS_t\} \tag{10}$$

where DS_t denotes the dominated solutions, labeled as 0, and NDS_t denotes the non-dominated solutions, labeled as 1, respectively. This converts the knowledge of the source task into a binary classification problem, where the relationship between non-dominated and dominated solutions defines a classification boundary. After training with $\mathcal{TS}$, the classifier could effectively identify the optimal classification hyperplane, thus encoding the knowledge of the source task. In this paper, the Support Vector Machine (SVM) [25] is employed as a classifier due to its ability to perform well with a small number of samples. This capability is especially valuable when dealing with EXDMOPs, where only a few solutions are accessible [26]. If the problem at hand entails higher complexity, alternative classifiers with enhanced accuracy, such as neural networks [27], can be considered. Nevertheless, it is imperative to acknowledge that this approach requires a longer training and prediction time, warranting careful consideration and resource allocation.

Once the classifier is constructed, a significant number of sample points are randomly generated to cover the entire solution space, ensuring a comprehensive evaluation of their Pareto potential. In this study, for simplicity, we generate $100*N$ sample points, where N represents the population size. These sample points are then input into the SVM classifier to predict their labels. From those predicted as 1 by the classifier, N sample points are randomly selected as the initial population for the target task at time step t. Last but not least, if the number of sample points predicted as 1 is less than N, additional random sampling is conducted. The initial population is then considered as the population of the evolutionary search at time step t_n.

Distinct from existing methods that rely solely on historical solutions within the same dynamic problem, our cross-problem approach leverages knowledge from tasks that exhibit similarities and have been well-optimized. By strategically selecting and transferring knowledge from these source tasks, our method enriches the initial population with high-quality solutions. This proactive transfer approach mitigates the reliance on potentially inadequate and poor-quality historical data alone, thereby could enhance the robustness and effectiveness of the optimization process in EXDMOP scenarios.

4 Empirical Study

4.1 Experimental Setup

Test Problem: As far as we know, no specific test problems currently exist for EXDMOPs. According to [20], we adopt the DF test suite [28] with a limitation on the number of evaluations per time step to simulate these scenarios. In this paper, we use all the bi-objective problems in the DF test suite (i.e., DF1-DF9) as the test problems. When one problem is solved, the other problems are considered as well-optimized and are placed in the task pool. In particular, we employed RMMEDA [29] to optimize these problems in the task pool over 1000 iterations per time step with a population of 100 individuals. Subsequently, the dominated and non-dominated solutions of these problems are stored for further experimentation.

Each of the DF1-DF9 problems exhibits unique dynamic characteristics. In DF2, only the POS changes over time, while the POF remains constant; in contrast, both POS and POF vary over time in the other eight problems. During the optimization process, t is used to control the dynamics of these test functions:

$$t = \frac{1}{n_t} \left\lfloor \frac{\tau}{\tau_t} \right\rfloor \tag{11}$$

where τ, τ_t, and n_t denote the total number of evaluations, the frequency of change, and the severity of change, respectively. In this paper, according to [20], τ_t is set as $2*N$, where N represents the population size, which is set as 100. τ is set as $40*N$, ensuring 20 changes occur in each run. Each test problem is independently run ten times at three different severity levels of change: n_t=5, n_t=10, and n_t=20. A lower n_t value corresponds to greater change severity in EXDMOPs.

Performance Metric: The modified inverse generational distance(MIGD) [30], which takes into account both solution convergence and diversity, is used to assess algorithm performance. A lower MIGD value suggests that the solutions are nearer to the true POF and exhibit a good spread within the objective space. The MIGD is given by:

$$MIGD(Q^{t*}, Q^t) = \frac{1}{|T|} \sum_{t \in T} IGD(Q^{t*}, Q^t) \tag{12}$$

Here, T represents the number of time steps. Q^{t*} denotes the reference points on the true POF, and Q^t refers to the POF generated by the algorithm at time step t. Specifically, the IGD is defined as follows:

$$IGD(Q^{t*}, Q^t) = \frac{\sum_{p^* \in Q^{t*}} min_{p \in Q^t} ||p^* - p||}{|Q^{t*}|} \tag{13}$$

Here, $||p^* - p||$ represents the Euclidean distance between the point $p^* \in Q^{t*}$ and the point $p \in Q^t$, while $|Q^{t*}|$ denotes the number of points in Q^{t*}.

Compared Algorithms: We compare the proposed cross-problem knowledge transfer method with the state-of-the-art EXDMOEA, i.e., DKRVEA [18]. To ensure fairness, the data augmentation method in DKRVEA, which transfers historical knowledge from the previous time step within the same problem, is replaced with our proposed method (labeled as Ours-DK). Additionally, we conduct a comparison with KRVEA [31], which has demonstrated excellent performance in static expensive multi-objective optimization. In this comparison, KRVEA treats EXDMOPs as multiple independent static expensive multi-objective optimization problems without any knowledge transfer. The population generated randomly at each time step in KRVEA is replaced with a population constructed by our proposed method (labeled as Ours-KR) to investigate whether our method enhances the solver's capacity to handle dynamic problems.

For the experiments, we maintain consistency in the configuration of other evolutionary operators and parameters, following the settings used for DKRVEA in [18] and for KRVEA in [31]. The experimental setup includes the following key parameters:

- The dimension of the decision vector: $d = 10$.
- Population size: $N = 100$.
- Independent number of runs: $runs = 10$.
- Maximum fitness evaluations: $Max_{FES} = 20 * N$.
- Three severity levels of change: $n_t = 5, n_t = 10, n_t = 20$.

4.2   Results and Discussion

This section presents a comparative analysis of the proposed cross-problem knowledge transfer method against DKRVEA and KRVEA, focusing on optimization quality and computational efficiency.

Table 1 presents the statistical analysis of the averaged MIGD and its standard deviation for the compared algorithms across three severity levels of change (i.e., $n_t = 5, 10$, and 20) over 10 independent runs on DF1-DF9. Superior performances are highlighted in bold within the table to facilitate easy identification of the best-performing methods. A Wilcoxon rank-sum test at a 95% confidence level is conducted to ensure the statistical significance of the results. The symbols "+", "−", and "≈" represent the proposed cross-problem knowledge transfer method is statistically better, worse, or comparable to the other algorithms, respectively. Furthermore, the summary of the significance tests for the obtained MIGD values by each algorithm on DF1-DF9, considering all three severity levels (totaling 27 instances), is provided in the bottom row of each comparison group. +, −, and ≈ represent the number of instances where the proposed method outperforms, underperforms, or is comparable to the other algorithms, respectively.

Overall, as observed in Table 1, each solver equipped with the proposed cross-problem transfer method (i.e., Ours-DK and Ours-KR) consistently outperformed its counterpart (i.e., DKRVEA and KRVEA) in terms of the averaged MIGD value on the majority of the DMOPs under all three severity levels of change. These results confirm that the proposed cross-problem transfer method effectively enhances the performance of the solvers. In particular, on DF1-DF8, Ours-DK obtained superior averaged MIGD values compared to DKRVEA under almost all three severity levels of change. A similar result can be observed in the comparison between Ours-KR and KRVEA on DF1, DF2, and DF4-DF8. Furthermore, in a total of 27 DF instances, Ours-DK and Ours-KR demonstrated competitive averaged MIGD values compared to DKRVEA and KRVEA on 23 and 22 instances, respectively. Moreover, the outcomes of the Wilcoxon rank-sum test (as shown in the bottom row labeled "+/≈/−" in Table 1) demonstrate that the solvers equipped with the proposed cross-problem knowledge transfer method (i.e., Ours-DK and Ours-KR) achieved significantly superior MIGD values on 20 and 21 instances against DKRVEA and KRVEA, respectively. These results show the proposed cross-problem knowledge transfer method is capable of identifying the appropriate problems and extracting their knowledge to help

Table 1. MIGD values (mean and standard deviations) for DF1-DF9 with $n_t = 5, 10, 20$ obtained by the compared algorithm (highlighted in bold are superior performances; "+", "−", and "≈" indicate statistically significant superiority, inferiority, or equivalence of the proposed method compared to other algorithms.).

Problems	n_t	Ours-DK	DKRVEA	Ours-KR	KRVEA
DF1	5	**8.900e−03(1.562e−03)**+	1.650e−02(3.940e−03)	**8.900e−03(5.586e−04)**+	1.750e−02(1.109e−03)
	10	**7.900e−03(1.011e−03)**+	1.730e−02(1.215e−03)	**6.900e−03(5.716e−04)**+	1.900e−02(8.422e−04)
	20	**7.400e−03(4.986e−04)**+	2.070e−02(4.235e−03)	**7.300e−03(7.565e−04)**+	1.910e−02(6.911e−04)
DF2	5	**1.190e−02(8.702e−04)**+	1.540e−02(3.116e−03)	**1.200e−02(1.164e−03)**+	1.740e−02(1.586e−03)
	10	**1.280e−02(5.914e−04)**+	1.840e−02(5.129e−03)	**1.490e−02(8.233e−04)**+	1.710e−02(6.229e−04)
	20	**1.230e−02(5.574e−04)**+	2.070e−02(3.458e−03)	**1.250e−02(6.044e−04)**+	1.710e−02(4.515e−04)
DF3	5	**3.420e−02(3.332e−02)**≈	3.820e−02(3.402e−02)	2.270e−02(2.894e−02)≈	**1.680e−02(7.871e−03)**
	10	**9.600e−03(6.737e−03)**+	5.090e−02(5.502e−02)	1.060e−02(4.358e−03)−	**6.400e−03(7.072e−04)**
	20	**8.400e−03(5.430e−03)**≈	2.860e−02(3.102e−02)	**8.200e−03(3.042e−03)**−	8.900e−03(8.055e−03)
DF4	5	**4.691e−01(2.144e−01)**+	1.103e+00(5.236e−01)	**2.104e−01(2.467e−01)**+	1.183e+00(2.709e−01)
	10	**5.054e−01(1.158e−01)**+	2.438e+00(5.743e−01)	**2.430e−01(9.163e−02)**+	2.340e+00(5.669e−01)
	20	**6.066e−01(2.139e−01)**+	1.887e+00(5.715e−01)	**3.276e−01(1.882e−01)**+	2.656e+00(6.158e−01)
DF5	5	**6.660e−02(1.313e−02)**+	1.113e−01(3.033e−02)	**5.970e−02(1.016e−02)**+	2.263e−01(2.627e−02)
	10	**6.020e−02(2.507e−02)**+	1.364e−01(4.533e−02)	**2.800e−02(7.240e−03)**+	2.624e−01(2.987e−02)
	20	**4.470e−02(1.452e−02)**+	1.153e−01(3.702e−02)	**8.800e−03(3.197e−03)**+	2.510e−01(3.931e−02)
DF6	5	1.870e+01(3.856e+00)−	**1.274e+01(2.059e+00)**	1.963e+01(2.027e+00)+	2.180e+01(1.361e+00)
	10	**2.466e+00(4.319e−01)**+	1.346e+01(2.051e+00)	**1.998e+00(5.220e−01)**+	2.124e+01(1.460e+00)
	20	**1.251e+00(3.250e−01)**+	1.476e+01(1.590e+00)	**6.651e−01(1.352e−01)**+	2.145e+01(1.336e+00)
DF7	5	**3.510e−02(5.630e−03)**+	4.350e−02(6.156e−03)	**3.600e−02(3.699e−03)**+	5.660e−02(8.710e−03)
	10	**3.000e−02(1.786e−03)**+	4.620e−02(1.741e−02)	**3.440e−02(5.548e−03)**+	5.640e−02(7.573e−03)
	20	**4.120e−02(5.812e−03)**≈	4.390e−02(1.308e−02)	**4.220e−02(1.072e−02)**+	5.530e−02(8.452e−03)
DF8	5	**5.190e−02(1.267e−02)**+	2.614e−01(5.260e−02)	**4.550e−02(8.753e−03)**+	4.184e−01(4.598e−02)
	10	**7.850e−02(1.813e−02)**+	3.522e−01(6.637e−02)	**5.910e−02(1.212e−02)**+	4.089e−01(7.853e−02)
	20	**6.260e−02(2.000e−02)**+	3.685e−01(1.224e−01)	**4.340e−02(1.384e−02)**+	3.914e−01(5.494e−02)
DF9	5	1.075e+00(9.524e−02)−	**6.871e−01(1.119e−01)**	2.273e+00(3.078e−01)−	**5.680e−01(4.444e−02)**
	10	1.099e+00(1.675e−01)−	**7.310e−01(1.163e−01)**	1.003e+00(1.838e−01)−	**5.733e−01(6.040e−02)**
	20	1.101e+00(2.320e−01)−	**6.110e−01(9.432e−02)**	1.105e+00(2.226e−01)−	**5.710e−01(4.128e−02)**
+/≈/−		20/3/4		21/1/5	

the solver improve its ability to solve a majority of instances. However, it is also observed that Ours-DK loses to DKRVEA on DF9, similar to Ours-KR losing to KRVEA. This is because DF9 is the sole problem with multiple disconnected Pareto front (PF) segments, which means that there are no appropriate tasks in the task pool for effective knowledge transfer. The proposed method selected the task with the smallest discrepancy. The transfer of knowledge from this selected task resulted in negative transfer, thereby reducing the solver's performance in solving EXDMOPs. Similarly, on DF3, Ours-KR also loses to KRVEA. However, we found that the distribution of the selected well-optimized task exhibited the shortest distance from the distribution of the task being solved compared to other tasks within the pool. Consequently, we believe that the negative transfer is due to the lack of useful tasks for this task in the task pool.

In addition to comparing solution quality, we also present convergence curves plotting averaged MIGD values (over 10 runs) achieved by Ours-KR and KRVEA. Specifically, Fig. 3 shows the convergence curves obtained by Ours-KR and KRVEA on DF1-DF9 under $n_t = 10$. The y-axis in the figure represents the averaged MIGD value, while the x-axis denotes the index of dynamic changes.

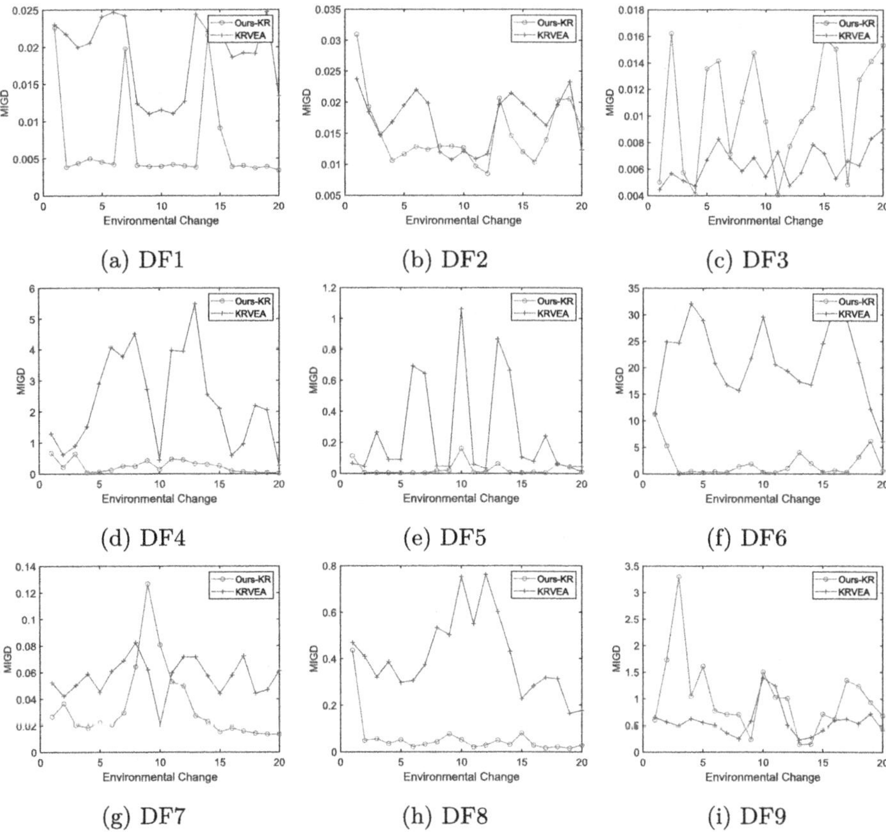

Fig. 3. Comparison of the convergence curves of averaged MIGD (across 10 runs) between Ours-KR and KRVEA on DF1-DF9 under the severity of change: $n_t = 10$ (x-axis: Dynamic change index, y-axis: MIGD).

As observed in Fig. 3, for DF1, DF2, and DF4-DF8, when changes occur, the proposed method demonstrates rapid adaptation to the new POS and POF through cross-problem knowledge transfer, thereby achieving lower MIGD values and reduced fluctuations. In particular, on EXDMOPs, such as DF4, DF5, DF6, etc., at time step $t_n = 1$, Ours-KR and KRVEA obtained comparable MIGD values. However, subsequent to performing the proposed cross-domain knowledge transfer (i.e., $t_n \geq 2$), Ours-KR method yielded lower MIGD values and minor fluctuations, which further validates the efficacy of the proposed cross-problem knowledge transfer method. In the case of DF3 and DF9, the lack of appropriate source tasks in the task pool leads to diminished performance following knowledge transfer, as evidenced by increased fluctuations and higher MIGD values in the curves.

5   Conclusion

This paper has introduced an innovative method for cross-problem knowledge transfer to tackle EXDMOPs. This method enables us to keep track of changing POS and POF with limited fitness evaluations when dynamic changes occur. Specifically, the proposed method consists of two main components: cross-problem task selection and cross-problem knowledge transfer. For cross-problem task selection, the solution distributions of both the task being solved and the well-optimized tasks are characterized using a multivariate Gaussian distribution. The Wasserstein distance is utilized to quantify the discrepancy between these distributions, with the aim of identifying the most similar task that can provide valuable knowledge. With regard to the cross-problem knowledge transfer, a binary classifier is trained to leverage knowledge from the selected problem and apply it to the one being solved, thereby enhancing algorithm performance. To assess the effectiveness of the proposed approach, we have conducted comprehensive empirical studies on a total of 27 problem instances, including nine commonly used bi-objective EXDMOPs with three different severity levels of change. The cross-problem knowledge transfer method has been compared with a state-of-the-art EXDMOEA and an algorithm known for its performance in static expensive multi-objective optimization. In the comparisons, the proposed method achieved superior averaged MIGD values compared to the algorithms presented in [18,31] on 23 and 22 out of totally 27 instances, respectively, thereby substantiating the effectiveness of cross-problem knowledge transfer method.

However, the proposed method obtained poor results on the test problem, where appropriate tasks for knowledge transfer were not available. Therefore, in future work, our focus will delve deeper in refining the knowledge transfer strategy to control its occurrence and mitigate negative transfers. In addition, we intend to implement the proposed method in practical contexts, specifically in problems such as dynamic portfolio optimization.

Acknowledgments. This work is mainly supported by National Key R&D Program of China (2022YFC3801700).

References

1. Wang, S., Mei, Y., Zhang, M.: A multi-objective genetic programming algorithm with α dominance and archive for uncertain capacitated arc routing problem. IEEE Trans. Evol. Comput. **27**(6), 1633–1647 (2022)
2. Handl, J., Kell, D.B., Knowles, J.: Multiobjective optimization in bioinformatics and computational biology. IEEE/ACM Trans. Comput. Biol. Bioinf. **4**(2), 279–292 (2007)
3. Marler, R.T., Arora, J.S.: Survey of multi-objective optimization methods for engineering. Struct. Multidiscip. Optim. **26**, 369–395 (2004)
4. Yang, P., Zhang, L., Liu, H., Li, G.: Reducing idleness in financial cloud services via multi-objective evolutionary reinforcement learning based load balancer. Sci. China Inf. Sci. **67**(2), 120102 (2024)

5. Ding, J., Chai, T., Wang, H., Chen, X.: Knowledge-based global operation of mineral processing under uncertainty. IEEE Trans. Industr. Inf. **8**(4), 849–859 (2012)
6. Gupta, A., Ong, Y.S., Feng, L.: Multifactorial evolution: toward evolutionary multitasking. IEEE Trans. Evol. Comput. **20**(3), 343–357 (2015)
7. Feng, L., et al.: Evolutionary multitasking via explicit autoencoding. IEEE Trans. Cybernet. **49**(9), 3457–3470 (2018)
8. Tan, K.C., Feng, L., Jiang, M.: Evolutionary transfer optimization-a new frontier in evolutionary computation research. IEEE Comput. Intell. Mag. **16**(1), 22–33 (2021)
9. Yazdani, D., Cheng, R., Yazdani, D., Branke, J., Jin, Y., Yao, X.: A survey of evolutionary continuous dynamic optimization over two decades–part a. IEEE Trans. Evol. Comput. **25**(4), 609–629 (2021)
10. Feng, L., Zhou, W., Liu, W., Ong, Y.S., Tan, K.C.: Solving dynamic multiobjective problem via autoencoding evolutionary search. IEEE Trans. Cybernet. **52**(5), 2649–2662 (2020)
11. Zhou, W., Feng, L., Tan, K.C., Jiang, M., Liu, Y.: Evolutionary search with multiview prediction for dynamic multiobjective optimization. IEEE Trans. Evol. Comput. **26**(5), 911–925 (2021)
12. Jiang, M., Wang, Z., Guo, S., Gao, X., Tan, K.C.: Individual-based transfer learning for dynamic multiobjective optimization. IEEE Trans. Cybernet. **51**(10), 4968–4981 (2020)
13. Wu, X., Wu, S.h., Wu, J., Feng, L., Tan, K.C.: Evolutionary computation in the era of large language model: Survey and roadmap. arXiv preprint arXiv:2401.10034 (2024)
14. Hao, H., Zhang, X., Zhou, A.: Enhancing saeas with unevaluated solutions: a case study of relation model for expensive optimization. Sci. China Inf. Sci. **67**(2), 1–18 (2024)
15. Liang, J., et al.: An evolutionary multiobjective method based on dominance and decomposition for feature selection in classification. Sci. China Inf. Sci. **67**(2), 120101 (2024)
16. Fan, X., Li, K., Tan, K.C.: Surrogate assisted evolutionary algorithm based on transfer learning for dynamic expensive multi-objective optimisation problems. In: 2020 IEEE Congress on Evolutionary Computation (CEC), pp. 1–8. IEEE (2020)
17. Pan, S.J., Tsang, I.W., Kwok, J.T., Yang, Q.: Domain adaptation via transfer component analysis. IEEE Trans. Neural Netw. **22**(2), 199–210 (2010)
18. Liu, Z., Wang, H.: A data augmentation based kriging-assisted reference vector guided evolutionary algorithm for expensive dynamic multi-objective optimization. Swarm Evol. Comput. **75**, 101173 (2022)
19. Zhou, A., Jin, Y., Zhang, Q.: A population prediction strategy for evolutionary dynamic multiobjective optimization. IEEE Trans. Cybernet. **44**(1), 40–53 (2013)
20. Zhang, X., Yu, G., Jin, Y., Qian, F.: An adaptive gaussian process based manifold transfer learning to expensive dynamic multi-objective optimization. Neurocomputing **538**, 126212 (2023)
21. Gong, B., Shi, Y., Sha, F., Grauman, K.: Geodesic flow kernel for unsupervised domain adaptation. In: 2012 IEEE Conference on Computer Vision and Pattern Recognition, pp. 2066–2073. IEEE (2012)
22. Zhou, L., Feng, L., Zhong, J., Zhu, Z., Da, B., Wu, Z.: A study of similarity measure between tasks for multifactorial evolutionary algorithm. In: Proceedings of the Genetic and Evolutionary Computation Conference Companion, pp. 229–230 (2018)

23. Huang, S., Zhong, J., Yu, W.J.: Surrogate-assisted evolutionary framework with adaptive knowledge transfer for multi-task optimization. IEEE Trans. Emerg. Top. Comput. **9**(4), 1930–1944 (2019)
24. Zhang, J., Zhou, W., Chen, X., Yao, W., Cao, L.: Multisource selective transfer framework in multiobjective optimization problems. IEEE Trans. Evol. Comput. **24**(3), 424–438 (2019)
25. Burges, C.J.: A tutorial on support vector machines for pattern recognition. Data Min. Knowl. Disc. **2**(2), 121–167 (1998)
26. Wei, F.F., et al.: A classifier-assisted level-based learning swarm optimizer for expensive optimization. IEEE Trans. Evol. Comput. **25**(2), 219–233 (2020)
27. Pan, L., He, C., Tian, Y., Wang, H., Zhang, X., Jin, Y.: A classification-based surrogate-assisted evolutionary algorithm for expensive many-objective optimization. IEEE Trans. Evol. Comput. **23**(1), 74–88 (2018)
28. Jiang, S., Yang, S., Yao, X., Tan, K.C., Kaiser, M., Krasnogor, N.: Benchmark functions for the cec'2018 competition on dynamic multiobjective optimization. Newcastle University, Tech. rep. (2018)
29. Zhang, Q., Zhou, A., Jin, Y.: Rm-meda: a regularity model-based multiobjective estimation of distribution algorithm. IEEE Trans. Evol. Comput. **12**(1), 41–63 (2008)
30. Muruganantham, A., Tan, K.C., Vadakkepat, P.: Evolutionary dynamic multiobjective optimization via kalman filter prediction. IEEE Trans. Cybernet. **46**(12), 2862–2873 (2015)
31. Chugh, T., Jin, Y., Miettinen, K., Hakanen, J., Sindhya, K.: A surrogate-assisted reference vector guided evolutionary algorithm for computationally expensive many-objective optimization. IEEE Trans. Evol. Comput. **22**(1), 129–142 (2016)

XImgCom: Fine-Tuned Text-Guided X-Ray Image Synthesis for Airport Logistics Based on Hypercomplex Attention

Zhao Li[1], Donghui Lian[1], Xuan Peng[2], Wenning Huang[2], Xianghui Zeng[2], Dingzhou Zhu[1], and Guoheng Huang[1(✉)]

[1] School of Computer Science and Technology, Guangdong University of Technology, Guangzhou 510006, China
`2112205232@mail2.gdut.edu.cn`, `kevinwong@gdut.edu.cn`
[2] Logistics Co., Ltd. of Guangdong Airport Authority, Guangdong, China
`{pengxuan,huangwenning,cengxianghui}@gdairport.com`

Abstract. With the rapid development of technology, its applications in various industries have become increasingly widespread, particularly in the field of security detection. X-ray security screening, as a critical non-contact detection technology, plays an essential role in security inspections at public places such as airports, train stations, and borders. However, the application of advanced technology in X-ray security screening faces a significant challenge: the lack of real and large-scale X-ray security images for model training. Traditional methods of acquiring X-ray security images are time-consuming and labor-intensive, constrained by safety and ethical requirements. To address this issue, we propose a novel method utilizing pre-trained controllable diffusion models to synthesize realistic X-ray images for training purposes. Our approach incorporates a Hypercomplex Spatial Channel Attention (HSCA) module with hypercomplex attention and quaternion computations within the encoder of a Variational Autoencoder (VAE). This innovative attention module enhances the synthesis effect by improving foreground detail preservation and attribute accuracy. Extensive experiments demonstrate that our method effectively generates high-quality X-ray images that closely align with real-world scenarios, significantly enhancing the performance of X-ray security screening models.

Keywords: X-ray Screening Detection · Diffusion Models · Hypercomplex Attention · Quaternion Computations · Prompt · X-ray Image Synthesis

1 Introduction

With the rapid development of technology, its applications in various industries have become increasingly widespread, especially in the field of security detection and semantic segmentation [1–3]. The introduction of advanced technology has brought revolutionary changes to traditional security screening methods. Among

M. Mahmud et al. (Eds.): ICONIP 2024, CCIS 2297, pp. 279–292, 2026.
https://doi.org/10.1007/978-981-96-7036-9_19

them, X-ray security screening, as an important non-contact detection technology, plays an indispensable role in the security inspections of public places such as airports, train stations, and borders. However, the application of these technologies in X-ray security screening faces a crucial challenge: the lack of real and large-scale X-ray security images for model training [4–7].

The traditional method of acquiring X-ray security images often relies on actual detection processes at the scene, which is both time-consuming and labor-intensive, and limited by the safety and ethical requirements of actual operations. Therefore, efficiently obtaining a sufficient number of X-ray security images to support the training and optimization of models has become an urgent issue in this field.

In recent years, diffusion models [8–10], a novel contender arising as an alternative to the prevalent generative adversarial networks [11–14], have made significant progress in the field of image generation [15–18]. By simulating the transformation of noise into intricate image details, diffusion models empower the generation of rich and varied visuals, all while minimizing the reliance on vast amounts of authentic training data.

Despite their notable achievements in image synthesis and generation, diffusion models frequently encounter challenges in exercising precise control over foreground attributes and effectively preserving the distinct identity of foreground elements. Specifically, generative composition methods built on large pre-trained diffusion models struggle with issues like unnatural boundaries, inharmonious illumination, and unsuitable poses, which reduce the effectiveness of the synthesized images for practical applications [19–22].

Inspired by these advancements, we propose a novel method specifically designed for the synthesis of X-ray images. Our approach leverages the composition task from the pre-trained controllable diffusion model, which combines multiple partial images into a single composite image. This technique allows us to create detailed and realistic X-ray images that better represent the types of complex, overlapping objects typically encountered during security checks.

In addition to these innovations, our primary model integrates a Hypercomplex [23] Spatial Channel Attention (HSCA) module with hypercomplex attention and quaternion computations to enhance the synthesis effect. This attention module is embedded within the encoder of a Variational Autoencoder (VAE), providing improved attention mechanisms that lead to better preservation of foreground details and attributes. To summarize, the main contributions of this paper are as follows:

1. Utilizing pre-trained controllable diffusion models to synthesize realistic X-ray security screening images for the purpose of training X-ray security screening models.
2. Improved Foreground Identity Preservation: By enhancing the composition framework with hypercomplex attention and quaternion computations, we ensure better preservation of foreground details and attributes, making the synthesized images more accurate for detection purposes.

3. Controlled Image Composition: Introducing a controllable image composition technique that allows selective adjustment of foreground illumination and pose to ensure the synthesized images align closely with real-world scenarios.

By embedding this advanced attention module within the VAE encoder, our method not only achieves superior image synthesis but also addresses key challenges in generating realistic and diverse X-ray images for security screening applications.

2   Related Work

2.1   Image Composition

The aim of image synthesis is to integrate foreground elements from a given image with background images, thereby creating the synthetic images. However, the integrity of synthetic images may be troubled with inconsistencies between the foreground and background, such as unnatural boundaries, inharmonious illumination, and unsuitable poses. These obstacles have spurred the creation of diverse tasks, each tailored to tackle a specific issue.

Image blending focuses on improving unnatural boundaries between the foreground and background, allowing foreground and background seamlessly blending together. Adjusting the illumination of the foreground to align with the background achieves image harmonization. Despite their success in appearance adjustment, these methods cannot address geometric inconsistencies between the foreground and background. Techniques have been developed to estimate warping parameters for geometric correction to cope with inconsistent camera viewpoints. However, these methods commonly forecast limited transformations, including affine and perspective, which fall short when confronted with intricate scenarios like synthesizing foreground objects from novel viewpoints or extending their applicability to non-rigid objects.

More recently, generative image composition has aimed to address all concerns through a single comprehensive model, producing synthetic images using the end-to-end method. Representative works include PbE [24] and ObjectStitch [25]. Yet, the synthesized foreground objects exhibit poor fidelity and there is a lack of control over their specific attributes, resulting in suboptimal outcomes.

Currently, most of the research on X-ray contraband image synthesis uses convolution method [26], directly uses traditional image processing technology or threat image projection (TIP) method [27], which makes the synthesized images very unrealistic. So we now propose a method based on the diffusion model to solve this problem

2.2   Hypercomplex Attention and Quaternion Computations

Hypercomplex attention and quaternion computations have recently emerged as powerful techniques for improving the performance of neural networks, especially in the context of image synthesis [28–35]. Hypercomplex numbers, including quaternions, extend complex numbers to higher dimensions and have been shown to enhance the representational capacity of neural networks.

Quaternions [36], introduced in the 19th century by Hamilton, provide a way to encode rotations and other transformations in three dimensions. They have been extensively used in computer graphics and robotics for their efficiency in representing and computing 3D rotations. In the context of neural networks, quaternion computations enable the efficient processing of multi-dimensional data, preserving more information and capturing complex interactions between features.

Recent works have integrated quaternion algebra into neural networks, demonstrating improvements in tasks such as image classification, object detection, and image synthesis. By leveraging the properties of quaternions, these models can better handle the geometric and spatial relationships within the data, leading to more accurate and realistic results. Gaudet et al. [37] proposed a deep quaternion network, but only used quaternion multiplication to proxy the real multiplication of convolutions, and did not take advantage of the nature of the quaternion complex numbers. Zhu et al. [38] proposed quaternion convolutional neural networks for color images that can be naturally represented as quaternion matrices, and designed basic modules such as convolutional and fully connected layers in the quaternion domain, which can be used to build full quaternion convolution neural network, which is proved to outperform the real-valued CNN in classification and denoising tasks. Nguyen et al. [23] proposed a quaternion graph neural network based on the previous work to learn the quaternion space representation of the inner graph.

In our work, we integrate hypercomplex attention and quaternion computations within a diffusion model to enhance the synthesis of X-ray images. This approach allows us to preserve the detailed attributes of the foreground objects and ensure harmonious integration with the background, addressing the challenges of traditional image composition methods.

2.3 Prompt Learning

This topic stems from the realm of Natural Language Processing (NLP), where the impetus lies in leveraging pre-trained language models, exemplified by BERT [39] and GPT [40], as repositories of knowledge from which pertinent information is extracted to benefit downstream tasks. Specifically, with a pre-trained language model at hand, the challenge often assumes the form of a "fill-in-the-blank" cloze exercise, requiring the model to anticipate the masked token in a sentence, such as deciding between "positive" or "negative" for the masked portion in "No reason to watch. It was [MASK]" in the context of sentiment analysis. The crux of the matter revolves around crafting the underscored segment, known as the prompt (or template), in a manner that resonates with the model's comprehension patterns.

Rather than relying on manual prompt design, the pursuit of prompt learning research endeavors to automate this process by harnessing a manageable amount of labeled data. Jiang et al. [41] have employed text mining and paraphrasing techniques to create the diverse set of candidate prompts, from which the optimal prompts, boasting the highest training accuracy, are meticulously selected. Shin et al. [42] propose AutoPrompt, a gradient-based approach that selects from a vocabulary the best tokens that cause the greatest changes in gradients based on the label likelihood.

Our research aligns most closely with the realm of continuous prompt learning methodologies [43–45], where the core concept revolves around transforming prompts into a collection of continuous vectors, enabling end-to-end optimization tailored to a specific objective function. For a broader and more comprehensive overview of this field, we recommend referring to Liu et al. [46]. Prompt learning, a burgeoning research avenue in computer vision, has emerged as a recent focus of exploration. [47–51]. Our research is inspired by these learnable prompts, where we introduce a lightweight network into our model. This network extracts a token from the features of a hypercomplex encoder, which are then added to the prompt words.

3   Methods

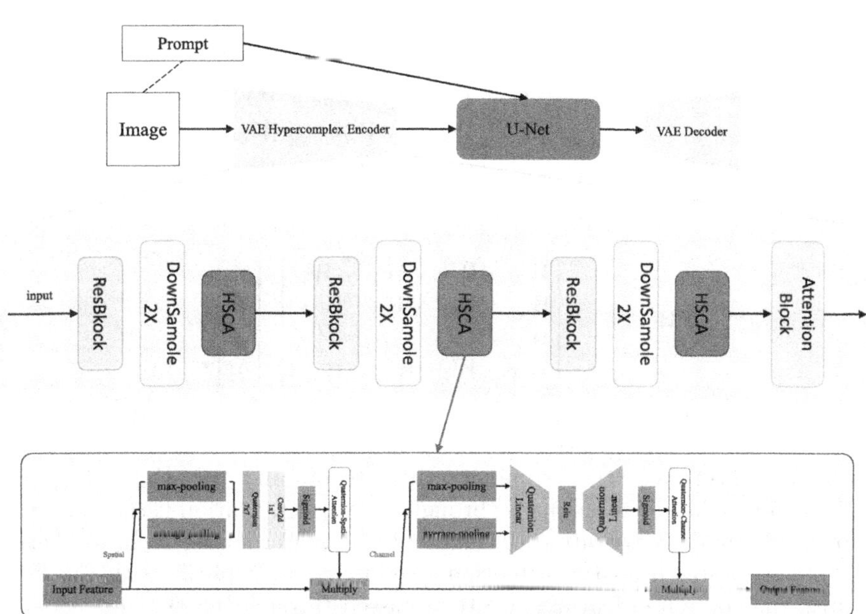

Fig. 1. Flowchart of XimgCom

3.1   Hypercomplex Spatial-Channel Attention Module

By observing real X-ray images of hazardous materials in airport logistics, we can see that they exhibit significant spatial information. When generating useful X-ray images, it is crucial to consider the consistency of spatial structure, color, and angle between different types of hazardous materials and the background. Although the VAE encodes image data into a low-dimensional latent space representation, capturing key features of the image, we found that this representation is insufficient for effectively extracting the necessary spatial structural information. Therefore, we designed an attention module that integrates hyper-complex convolution, allowing the VAE to better process and preserve the features of X-ray images during the decoding stage.

Quaternion Convolutional Neural Networks (QCNNs) is integrated into our Hypercomplex Spatial-Channel Attention Module(HSCA), which uses quaternion convolution instead of regular real-valued convolution for data representation and processing. As shown in Fig, convolution in the quaternion domain involves the convolution of quaternion-weighted matrices with quaternion vectors. Given a quaternion weighting matrix (i.e., quaternion convolution kernel) of $W = W_R + W_X i + W_Y j + W_Z k$, the quaternion vector $q = q_r + q_a i + q_b j + q_c k$. is used to represent the features input to the convolution layer. The convolution between two quaternions is accomplished by quadratic multiplication (Hamilton product):

$$
\begin{aligned}
W \otimes q = &(W_R q_r - W_X q_a - W_Y q_b - W_Z q_c) \\
&+ (W_X q_r + W_R q_a - W_Z q_b + W_Y q_c)i \\
&+ (W_Y q_r + W_Z q_a + W_R q_b - W_X q_c)j \\
&+ (W_Z q_r - W_Y q_a + W_X q_b + W_R q_c)k
\end{aligned}
\tag{1}
$$

and this convolutional computation can also be expressed in the form of a matrix as:

$$
W \otimes q = \begin{bmatrix} W_R & -W_X & -W_Y & -W_Z \\ W_X & W_R & -W_Z & W_Y \\ W_Y & W_Z & W_R & -W_X \\ W_Z & -W_Y & W_X & W_R \end{bmatrix} \begin{bmatrix} q_r \\ q_a \\ q_b \\ q_c \end{bmatrix}
\tag{2}
$$

As can be seen from Fig. 2, when a feature is input into the HSCA, it first passes through the spatial attention part, averages and maximizes all the channels of the feature and then combines them to obtain a two-channel feature, then expands it into four channels through a 1×1 convolution kernel, and then performs convolution operation through a 3×3 quaternion convolution kernel, and finally obtains a spatial attention weight and multiply it with the input features; the computed features are then passed through the channel attention part, and two feature vectors are obtained by maximizing and averaging the maximum and average values of each channel of the features respectively, which are then fed into a shared feed-forward network (two quaternion fully-connected layers are used in this paper), and then the two vectors are summed up and a channel attention weight is obtained by the sigmoid activation function, and the final weight is compared with the input channel attention weight. Finally, this

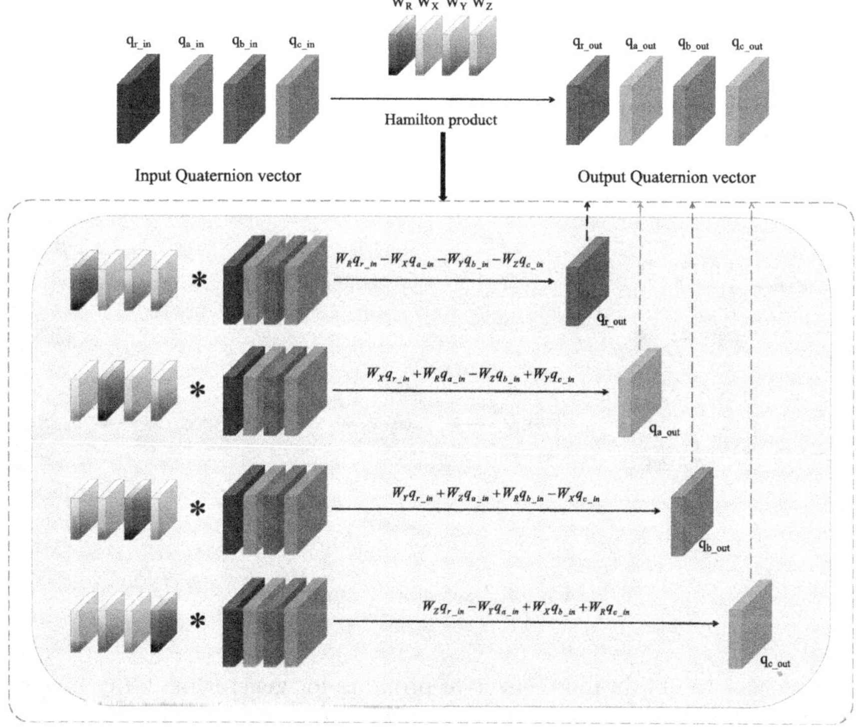

Fig. 2. Hamiltonian product of quaternions.

weight is multiplied with the features before the input channel attention to get our output features.

By inputting the spatial features into the quaternion convolution in the spatial attention part, we can better capture the spatial features and get a more appropriate spatial weight. Similarly, using the quaternion-based shared feedforward network in the channel attention part can better capture and combine the features, and more correctly weight each channel, so as to improve the integration of the channel information. We continuously insert HSCA modules into the encoder of the VAE. When compressing input image data into low-dimensional latent space representations, these modules better preserve the spatial, structural, and color features of the images. This ensures that the latent vectors remain consistent with the main features and structures of the images, thereby improving the quality and consistency of the generated images.

3.2 Learnable Prompt

The summary of our prompt method can see in Fig. 4. CoOp represents a highly data-efficient methodology that empowers the training of context vectors solely utilizing a minimal set of labeled images from a downstream dataset, signifi cantly reducing the need for extensive data. Additionally, CoCoOp illustrates that extracting tokens from image features and incorporating them into learnable prompt embeddings can notably enhance the model's capability to adapt to and excel in downstream tasks. This strategy harnesses the combined power of image-derived features and adaptable prompts to facilitate superior performance and generalization.

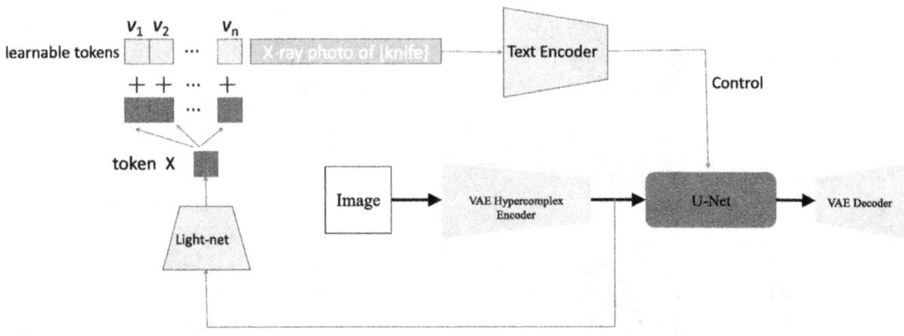

Fig. 3. Learnable Prompt

So in order to obtain more effective prompts for generating X-ray images of X-Ray Screening Detection, the prompt is comprised of both a fixed section and a learnable component. Additionally, we have constructed a lightweight network, LightNet, which is dedicated to extracting the latent features of images in order to obtain the token x. This network is remarkably straightforward, comprising only three linear layers. It represents a streamlined and parameter-efficient design, ensuring efficiency and performance.

$L_\theta(\cdot)$ expresses the Light-Net with parameters θ, every learnable prompt token is now got by $v_n(x) = v_n + X$ where $X = L_\theta(x)$ and $n \in \{1, 2, \ldots, n\}$. The token for the every class comes from as follows input,

$$P_i(x) = \{v_1(x), v_2(x), \ldots, v_n(x), \text{X-ray}, \text{photo}, \text{of}, c_i\}$$

During training, we update the learnable prompt $\{v_m\}_{m=1}^{M}$ with the Light-Net's parameters θ. In our work, this Light-Net is built with a three-layer structure (Linear-ReLU-Linear-ReLU-Linear). The input to the Light-Net is simply the output features produced by the hypercomplex encoder.

4   Experiments

4.1   Dataset

Nowadays, most of the existing X-ray contraband image datasets are used for security contraband detection, such as GDXray [52], SIXray [53], PIXray [54], etc. Our task was to synthesize X-ray security images of prohibited items. None of these datasets was suitable for us, and the datasets suitable for our task were few and hard to find, so we constructed our own dataset for synthesizing X-ray security images, COMXray. This dataset was collected over a period of six months at the airport, where checked cargo goes through X-ray screening. COMXray contains over 3000 X-ray images of contraband and a total of 6 categories of contraband. Since it is difficult to obtain X-ray images of multiple types of contraband in the real world, our current dataset only contains the six most common types of contraband for research. In subsequent work, we will continue to expand the dataset.

4.2   Comparison Experiments

In this chapter, we compare alternative methods for generating hazmat X-ray images to validate the effectiveness of our model. We have chosen two methods to compare with our model: a direct combination of foreground (hazmat) and background (baggage) (called AddIm) and a method of image composition using a diffusion model [55]. The main purpose of our research is to synthesize object detection training data. Therefore, we think that using the metrics of object detection to measure the effect of synthetic images in the comparative experiment is more in line with our research purpose.

As depicted in Table 1, we trained the YOLOv6 [56] model for object detection on real airport logistics hazardous material X-ray images, achieving an average precision of 88% for detecting various hazardous material categories. We then

Fig. 4. Visualization of different methods. In each row, we show the airport logistics hazardous materials X-ray screening charts generated by combining different types of hazardous materials with the background under different methods.

Table 1. Illustration of the comparison study of individual models

Methods	Precision ↑	Map ↑	Recall ↑
Baseline	0.936	0.906	0.942
AddIm	0.685	0.679	0.693
ControlCom [55]	0.724	0.718	0.729
XimgCom(Ours)	**0.822**	**0.789**	**0.801**

Table 2. Metrics in different experimental setups.

HSCA	Prompt	FID↓	LPIPS↓	AHIQ↑
✓	✓	**83.242**	**0.145**	**0.458**
✓		85.637	0.157	0.443
	✓	88.460	0.149	0.349

tested all generation methods using the trained YOLOv6 model. As shown in the table, in terms of map, recall, and precision, our method most closely approximates the real indicators across various hazardous material categories. This demonstrates that our method effectively integrates hazardous materials with the background in generated images, reducing issues with abrupt differences in color, orientation, and spatial hierarchy.

We complete our idea using PyTorch and train our model on the NVIDIA RTX4090 GPU with random seed set as 30. The training set comes from our private dataset COMXray that contains over 3000 images with six categories. We use the Adam optimizer, the learning rate is $1e^{-6}$ and the batchsize is 4. We trained our model for 100 epochs. Many parts of our model use pre-trained weights and freeze them. The only parts that need to be trained are the vae encoder and learnable prompt.

4.3 Ablation Study

In this section, we conduct an ablation study of XImgCom to investigate the impact of various model configurations on hazmat generation. Our aim is to demonstrate the robustness and superior repair performance of our model. In order to minimise experimental errors and improve the credibility of the model, we ensured consistency of the experimental environment, data and conditions, with the only variable being the frame configuration.

We conducted ablation study on all the constituent modules of the model. We selected three commonly used metrics, FID [57], LPIPS [58], and AHIQ [59], to evaluate the effectiveness of image generation, as shown in Table 2, our model is optimal in all three metrics. It is demonstrated that the learnable cue word method and hypercomplex can be effectively combined to better integrate the channel information and learn semantic information to generate near-realistic hazardous materials X-ray maps.

5   Conclusion

In this paper, we presented a novel approach for synthesizing realistic X-ray images of hazardous materials for use in airport logistics. Our method leverages a Hypercomplex Spatial-Channel Attention Module (HSCA) integrated into the encoder of a Variational Autoencoder (VAE) and a learnable prompt module. By incorporating quaternion convolutions within the HSCA, we effectively capture and preserve the spatial and structural features of the X-ray images, ensuring harmonious integration with the background. This results in high-quality composite images that maintain consistency in spatial structure, color, and angle between the foreground objects and the background.

Furthermore, the learnable prompt module enhances the generation process by dynamically adjusting prompts based on extracted image features, thus improving the adaptability and performance of the model without requiring extensive textual input or finetuning during inference. Through comprehensive experiments, we demonstrated that our method significantly outperforms traditional image composition techniques, generative models, and image harmonization approaches in terms of precision, recall, and mean average precision (mAP).

The superior performance of our model in generating realistic X-ray images highlights its potential for enhancing security screening processes in airport logistics by providing high-fidelity training data for object detection models. Of course, our research still has certain limitations. For example, our model is relatively large, the synthesis process is relatively slow and the impact of model hyperparameters on model performance is not explored. So, future work will explore further improvements in model efficiency and the application of this approach to other domains requiring high-quality synthetic images.

Acknowledgement. This study is based on the cooperative project with Logistics Co., Ltd. of Guangdong Airport Authority. We express our gratitude to Logistics Co., Ltd. of Guangdong Airport Authority for providing data and computing resource support, which enables our research to proceed smoothly. Once again, our thanks go to it.

References

1. Liu, X., et al.: Weakly supervised semantic segmentation via saliency perception with uncertainty-guided noise suppression. Vis. Comput., 1–16 (2024)
2. Li, H., et al.: Psanet: prototype-guided salient attention for few-shot segmentation. Vis. Comput., 1–15 (20240
3. Tang, H., et al.: Rm-unet: Unet-like mamba with rotational ssm module for medical image segmentation. Signal Image Video Process. (2024)

4. Zhang, X., et al.: Enhancing explainability in multimodal large language models. arXiv, From redundancy to relevance (2024)
5. Wei J., Zhang, X.: Dopra: decoding over-accumulation penalization and re-allocation in specific weighting layer. arXiv (2024)
6. Zhang, X., Zishan, X., Tang, H., Chaochen, G., Zhu, S., Guan, X.: When parameter-efficient fine-tuning with multimodal meets shadow removal, Shadclips (2024)
7. Dong, Y., Chen, X., Shen, Y., Kwok-Po Ng, M., Qian, T., Wang, S.: Multi-modal mood reader: pre-trained model empowers cross-subject emotion recognition. arXiv preprint, arXiv: 2405.19373 (2024)
8. Croitoru, F.-A., Hondru, V., Ionescu, R.T., Shah, M.: Diffusion models in vision: A survey. IEEE Trans. Pattern Analy. Mach. Intell. **45**(9), 10850–10869 (2023)
9. Chen, X., Lei, B., Pun, C.-M., Wang., S.: Brain diffuser: an end-to-end brain image to brain network pipeline. In: PRCV, pp. 16–26 (2023)
10. Zhou, T., Chen, X., Shen, Y., Nieuwoudt, M., Pun, c.-M., Wang, S.: Generative ai enables eeg data augmentation for alzheimer's disease detection via diffusion model. In: ISPCE-ASIA, pp. 1–6 (2023)
11. Goodfellow, I.: Generative adversarial networks. Commun. ACM **63**(11), 139–144 (2020)
12. Zuo, Q., et al.: Brain functional network generation using distribution-regularized adversarial graph autoencoder with transformer for dementia diagnosis. Comput. Model. Eng. Sci. **137**(3), 2129 (2023)
13. Zuo, Q., Li, R., Shi, B., Hong, J., Zhu, Y., Chen, X., Yixian, W., Guo, J.: U-shaped convolutional transformer gan with multi-resolution consistency loss for restoring brain functional time-series and dementia diagnosis. Front. Comput. Neurosci. **18**, 1387004 (2024)
14. Zuo, Q., et al.: Diffgan-f2s: Symmetric and efficient denoising diffusion gans for structural connectivity prediction from brain fmri. arXiv preprint, arXiv: 2309.16205 (2023)
15. Zhang, X., Chen, F., Wang, C., Tao, M., Jiang, G.-P.: Sienet: siamese expansion network for image extrapolation. IEEE Signal Process. Lett. **27**, 1590–1594 (2020)
16. Zhang, X.F., Gu, C.C., Zhu, S.Y.: Memory augment is all you need for image restoration. arXiv (2023)
17. Gong, C., et al.: Generative ai for brain image computing and brain network computing: a review. Front. Neurosci. **17**, 1203104 (2023)
18. Luo, S., Xu, R., Chen, X., Li, Z., Pun, C.-M., Wang, S.: Docdeshadower: frequency-aware transformer for document shadow removal. arXiv preprint, arXiv: 2307.15318 (2023)
19. Xu, Z., Zhang, X., Chen, W., Liu, J., Xu, T., Wang, Z.: Muraldiff: diffusion for ancient murals restoration on large-scale pre-training. IEEE Trans. Emerging Topics Comput. Intell. (2024)
20. Ding, B., Zhang, X., Yu, Z., Zhao, C., Yao, J., Hui, Z.: Ll-diff: low-light image enhancement utilizing langevin sampling diffusion. Inter. J. Pattern Recogn. Artifi. Intell. (2024)
21. Huang, G., Chen, X., Shen, Y., Wang, S.: Mr image super-resolution using wavelet diffusion for predicting alzheimer's disease. In: BI, pp. 146–157 (2023)
22. Jiang, H., Chen, X., Jin, C., Wang, S.: Structural brain network generation via brain denoising diffusion probabilistic model. In: International Conference on AI in Healthcare, pp. 264–277 (2024)
23. Nguyen, T.D., Phung, D., et al. Quaternion graph neural networks. In ACML, pp. 236–251 (2021)

24. Yang, B., et al.: Paint by example: Exemplar-based image editing with diffusion models. In: CVPR, pp. 18381–18391 (2023)
25. Song, Y., et akl.: Objectstitch: Object compositing with diffusion model. In: CVPR, pp. 18310–18319 (2023)
26. Duan, l., Wu, M., Mao, L., Yin, J., Xiong, J., Li, X.: Rwsc-fusion: region-wise style-controlled fusion network for the prohibited x-ray security image synthesis. In: Proceedings of the IEEE/CVF Conference on Computer Vision and Pattern Recognition, pp. 22398–2240 (2023)
27. Porta, R.R., Sterchi, Y., Schwaninger, A.: How realistic is threat image projection for x-ray baggage screening? Sensors **22**(6), 2220 (2022)
28. Chen, X., Cun, X., Pun, C.-M., Wang, S.: Shadocnet: learning spatial-aware tokens in transformer for document shadow removal. In: ICASSP, pp. 1–5 (2023)
29. Li, Z., Chen, X., Pun, C.-M., Cun, X.: High-resolution document shadow removal via a large-scale real-world dataset and a frequency-aware shadow erasing net. In: ICCV, pp. 12449–12458 (2023)
30. Luo, S., Chen, X., Chen, W., Li, Z., Wang, S., Pun, C.-M.: Devignet: high-resolution vignetting removal via a dual aggregated fusion transformer with adaptive channel expansion. In: AAAI (2024)
31. Li, Z., Chen, X., Wang, S., Pun, C.-M.: A large-scale film style dataset for learning multi-frequency driven film enhancement. In: IJCAI, pp. 1160–1168 (2023)
32. Guo, X., Chen, X., Luo, S., Wang, S., Pun, C.-M.: Dual-hybrid attention network for specular highlight removal. In: ACM MM (2024)
33. Li, Z., Chen, X., Guo, S., Wang, S., Pun, C.-M.: Wavenhancer: unifying wavelet and transformer for image enhancement. J. Comput. Sci. Technol. **39**(2), 336–345 (2024)
34. Chen, X., Pun, C.-M., Wang, S.: Cross-modal prompting for multi-task medical image translation. arXiv, Medprompt (2023)
35. Jiang, Y., Chen, X., Pun, C.-M., Wang, S., Feng, W.: Mfdnet: multi-frequency deflare network for efficient nighttime flare removal. Vis. Comput., 1–14 (2024)
36. Hamilton, W.R.: Elements of quaternions. Longmans, Green, & Company, London (1866)
37. Gaudet, C.J., Maida, A.S.: Deep quaternion networks. In: IJCNN, pp. 1–8 (2018)
38. Zhu, X., Xu, Y., Xu, H., Chen, C.: Quaternion convolutional neural networks. In: ECCV, pp. 631–647 (2018)
39. Devlin, J., Chang, M.-W., Lee, K.: and Kristina Toutanova. Pre-training of deep bidirectional transformers for language understanding. arXiv, Bert (2018)
40. Achiam, J., et al. Gpt-4 technical report. arXiv (2023)
41. Jiang, Z., Xu, F.F., Araki, J., Neubig, G.: How can we know what language models know? Trans. Associat. Comput. Linguist. **8**, 423–438 (2020)
42. Shin, T., Razeghi, Y., Logan IV, R.L., Wallace, E., Singh, S.: Eliciting knowledge from language models with automatically generated prompts. arXiv, Autoprompt (2020)
43. Lester, B., Al-Rfou, R., Constant, N.: The power of scale for parameter-efficient prompt tuning. arXiv (2021)
44. Li, X.L., Liang, P.: Prefix-tuning. Optimizing continuous prompts for generation. arXiv (2021)
45. Zhong, Z., Friedman, D., Chen, D.: Factual probing is [mask]: Learning vs. learning to recall. arXiv (2021)
46. Liu, P., Yuan, W., Jinlan, F., Jiang, Z., Hayashi, H., Neubig, G.: Pre-train, prompt, and predict: a systematic survey of prompting methods in natural language processing. ACM Comput. Surv. **55**(9), 1–35 (2023)

47. Ju, C., Han, T., Zheng, K., Zhang, Y., Xie, W.: Prompting visual-language models for efficient video understanding. In: ECCV, pp. 105–124 (2022)
48. Rao, Y., et al.: Denseclip: language-guided dense prediction with context-aware prompting. In: CVPR, pp. 18082–18091 (2022)
49. Yao, Y., Zhang, A., Zhang, Z., Liu, Z., Chua, T.-S., Sun, M.: Cpt: colorful prompt tuning for pre-trained vision-language models. AI Open **5**, 30–38 (2024)
50. Zhang, R., et al.: Pointclip: Point cloud understanding by clip. In: CVPR, pp. 8552–8562 (2022)
51. Zhou, K., Yang, J., Loy, C.C., Liu, Z.: Learning to prompt for vision-language models. Inter. J. Comput. Vis. **130**(9), 2337–2348 (2022)
52. Mery, D., et al.: Gdxray: the database of x-ray images for nondestructive testing. J. Nondestr. Eval. **34**(4), 42 (2015)
53. Miao, C., et al.: Sixray: A large-scale security inspection x-ray benchmark for prohibited item discovery in overlapping images. In: CVPR, pp. 2119–2128 (2019)
54. Ma, B., Jia, T., Min, S., Jia, X., Chen, D., Zhang, Y.: Automated segmentation of prohibited items in x-ray baggage images using dense de-overlap attention snake. IEEE Trans. Multimedia **25**, 4374–4386 (2022)
55. Zhang, B., et al.: Controlcom: Controllable image composition using diffusion model. ArXiv (2023)
56. Li, C., et al.: Yolov6: A single-stage object detection framework for industrial applications. ArXiv (2022)
57. Heusel, M., Ramsauer, H., Unterthiner, T., Nessler, B., Hochreiter, S.: Gans trained by a two time-scale update rule converge to a local nash equilibrium. In: NeurIPS (2017)
58. Zhang, R., Isola, P., Efros, A.A., Shechtman, E., Wang, O.:The unreasonable effectiveness of deep features as a perceptual metric. In: CVPR, pp. 586–595 (2018)
59. Lao, S., et al.: Attentions help cnns see better: attention-based hybrid image quality assessment network. In: CVPR Workshops, pp. 1140–1149 (2022)

Multiclass Semantic Segmentation of Satellite Imagery Using Convolutional Neural Networks

Mohammad Omar Faruk[1,2]([⊠]) [iD], Mohammad Anwar Hosen[1] [iD],
Michael Johnstone[1] [iD], Md. Rasel Hossain[2] [iD], and Faisal M Rahman[2] [iD]

[1] Institute for Intelligent Systems Research and Innovation, Deakin University,
75 Pigdons Road, Waurn Ponds, VIC 3216, Australia
{s224158311,anwar.hosen,michael.johnstone}@deakin.edu.au
[2] Department of Statistics, Noakhali Science and Technology University,
Sonapur Road, Noakhali 3814, Bangladesh
rasel.stat@nstu.edu.bd.bd, faisal1514@student.nstu.edu.bd

Abstract. Semantic Segmentation is a fundamental discipline of computer vision that aims to divide an image into distinct segments and provide semantic labels for every pixel. Semantic segmentation models consist of deep convolutional neural networks that continuously show striking improvements in the techniques of image analysis. This study applied different deep learning models to high-resolution satellite imagery of the Dubai metropolitan area. 72 images were utilized for this study, which were collected from the Pleiades-1A satellite. After augmentation, the total images for analysis were 504. The images considered in this study were carefully annotated and pre-processed to improve the training and reliability of the models. Five distinct deep learning models, VGG16, DeepLabV3+, Inception-ResNet-V2, Multi-UNet, and MobileNet-V2, were applied and compared to find the best-performed model among them. Inception-ResNet-V2 outperformed other models with 92% accuracy and a dice coefficient of 0.87, proving its superiority in extracting intricate urban features. The predicted mask image produced by the Inception-Resnet-V2 showed better results compared to other models' predicted masks. This study emphasised the importance of identifying the most suitable structure and training regimen that significantly extracts the information from an image. Future studies should focus on improving the model structure to increase the model accuracy and reduce the execution time substantially.

Keywords: Computer Vision · Semantic segmentation · Satellite Imagery · Convolutional Neural Networks · Deep Learning

1 Introduction

Semantic segmentation is an essential task in computer vision (CV), predicts pixel-level classifications for an image into multiple segments and assigns a semantic label to each segment [17]. It facilitates the analysis and comprehension

© The Author(s), under exclusive license to Springer Nature Singapore Pte Ltd. 2026
M. Mahmud et al. (Eds.): ICONIP 2024, CCIS 2297, pp. 293–307, 2026.
https://doi.org/10.1007/978-981-96-7036-9_20

of intricate images with irregular shapes, textures, or overlapping boundaries. In addition to image classification and object detection, applications of image segmentation in computer vision provide a multitude of novel functionalities, including autonomous driving [17], robotics, farming, medical imaging [14,22], and urban planning and development [13,30]. The primary objective of image segmentation is to organize the classification of pixels in images based on their semantic information, thereby encapsulating a pixel-wise classification task and facilitating object analysis and interpretation [3,7,21]. For image segmentation tasks, a variety of deep learning (DL) models have been employed. Several computer vision models and methods are used to find important characteristics in images, including U-Net, SegNet, DeepLabv3+, and MASK R-CNN [19].

Deep learning-based image segmentation has become firmly established in various domains in recent years, where it is widely used to separate homogeneous areas and conduct image segmentation [12]. A South Asian study called "Brick-Kiln-Belt" used high-resolution satellite images to find brick kilns. The ResNet-152 model and a YOLOv3-based object detector were used to classify the images [20]. High-tech remote sensing imagery continues to grapple with the challenge of selecting and integrating suitable features based on their spectral and spatial characteristics in the South Asian region. A. Josephine Atchaya et al. proposed the pre-trained AlexNet technique in 2023 to classify satellite images [1]. CNN-based semantic segmentation methods such as ResNet50, VGG16 and DeepLabV3+ generally have two strategies for obtaining global information. Multiscale pooling or dilated convolution for deep features initially enhances the receptive field. Secondly, the model implements the encoder-decoder architecture, and the context is linked through a skip connection [28]. Fully Convolutional Networks (FCNs), Inception-v3, U-Nets, MobileNetV2, Multi-UNet, VGG16, DeepLabV3+, Mask R-CNN, and Transformers represent all notable deep learning models that are used for satellite image segmentation [6,27].

The application of convolutional neural network (CNN) algorithm has resulted in substantial progress in image segmentation in recent years. The current investigation aims to determine the best model for the image segmentation task as its implementation expands with technological advancements, emphasizing its vital significance in the modern era. These models are trained on extensive, annotated datasets to verify the importance of image segments and identify semantic classifications [9]. One of the significant advancements in deep learning for image segmentation is the availability of extensively annotated datasets, such as ADE20K, ImageNet, Cityscapes, and COCO. The annotation of millions of photos in these datasets has enabled CNNs to acquire diverse visual representations for various item categories [23]. Satellite imagery, obtained through imaging satellites, has significantly changed our understanding of the Earth. The study of science, wildlife tracking, farming, disaster prevention, urban development, and national security constitute numerous applications of this technology [29]. This investigation aimed to assess various CV techniques for satellite imagery and identify the most effective predictive model among the implemented algorithms.

2   Materials and Methods

This investigation focuses on identifying the best feature detection deep learning models. The models are applied to identify and distinguish specific parts or patterns within a picture, and they are critical for various applications, including urban planning, environmental monitoring, and infrastructure building. Different data preprocessing steps for training the CV models are described below and presented in Fig. 1.

2.1   Dataset Acquisition

This research used secondary data from high-resolution satellite imagery. These images were procured from Pleiades-1A, a reputable source renowned for its detailed imagery captured on August 05, 2022 [18]. These images boasted a spatial resolution of 0.5 m, providing intricate insights into urban landscapes. The dataset acquisition phase ensured access to rich and comprehensive imagery crucial for subsequent analysis and model training. The dataset used in this paper contains 72 images. Fifty-six images are augmented to 504 (Augmented 448 images + original 56 images) images, which are used to train and validate the model.

2.2   Annotation Using Computer Vision Annotation Tool (CVAT)

The acquired satellite images underwent meticulous annotation using the CVAT. Annotation tasks involved precisely delineating and labelling various urban features and land cover classes. Annotators meticulously marked vegetation, built-up areas, informal settlements, impervious surfaces like roads and highways, barren land, water bodies, and even unlabelled regions, ensuring comprehensive coverage and accurate annotation across the dataset [9].

2.3   Image Augmentation

To enhance dataset diversity and model robustness, a systematic application of augmentation techniques was employed on annotated satellite images. These strategies, ranging from random rotation and flipping to brightness and contrast adjustments, are aimed at introducing variability while preserving semantic information. Leveraging the augmentations library, which offers a comprehensive suite of image transformation operations, crucial techniques like random cropping, rotation, and contrast enhancement were applied. This augmentation process not only expands the dataset but also aids in improving the model's generalisation and performance on unseen data and real-world scenarios, thus enhancing its robustness in computer vision tasks.

2.4   Image Patching

Following augmentation, each annotated image was partitioned into smaller, manageable tiles of dimensions 512×512 pixels, known as image patching. This strategic partitioning facilitated more efficient training and enabled models to capture localised spatial dependencies within the imagery. Consequently, every original image was dissected into multiple tiles, effectively expanding the dataset size and paving the way for more robust model training. This process was crucial for accommodating the computational requirements of deep learning models and optimising their performance on the dataset.

2.5   Computer Vision Techniques

Five different deep convolutional neural network computer vision models were used after the data had been pre-processed to find the best model for image segmentation. Below is a brief description of the models:

Multi-UNet: Multi-UNet is a more advanced version of the U-Net architecture, combining multiple pathways for hierarchical feature extraction to make it sound at finding features. contracting path is used in its architecture with convolutional and pooling layers for feature extraction and an expanding path with transposed convolutional layers for segmentation refinement. Maintaining spatial information across the network is ensured by including skip connections between the respective encoder and decoder layers. Additionally, dropout layers are strategically positioned to prevent overfitting during training. The final layer of the model employs SoftMax activation to generate pixel-wise class probabilities, enhancing its adaptability to various multi-class segmentation tasks [15].

Inception-ResNet-V2: A cheaper inception block was used rather than the original inception in the residual version of the inception network. A 1×1 convolution without activation filter-expansion layer following each inception block was utilized to scale up the filter bank's dimensionality prior to the accumulation to match the intensity of the input, which needs to compensate in reducing dimensionality encouraged by the inception block required to be compensated. Over the several versions of residual inception, Inception-ResNet-V2 only matched the Inveption-V2 network's raw cost. A simple technical difference of the Inception-ResNet from conventional models is that batch normalization was used on top of the traditional layers instead of summation [25].

DeepLabV3+: DeepLabV3+ is an encoder-decoder structural model with DeepLabv3 as the network's encoder. It optimizes the effect of edge information extraction of the target. Subsequently, the decoder is employed to retrieve the feature information and generate the expected results, enhancing the segmentation outcomes and preserving the edge details of the target. Xception serves as the foundational model for DeepLabv3+. The decoder module and ASPP module utilize Deep separable convolution, resulting in improved segmentation outcomes when combined with encoder-decoder networks [5].

VGG16: VGG16 is a deep convolutional neural network that utilizes convolutional layers to extract image characteristics. The components of this model are comprised of an increasing quantity of 3×3 filters. To preserve the spatial resolution, the convolutional layer's inputs are padded, and the stride is set to 1. Max pooling layers are utilized to merge the blocks. A max pooling operation

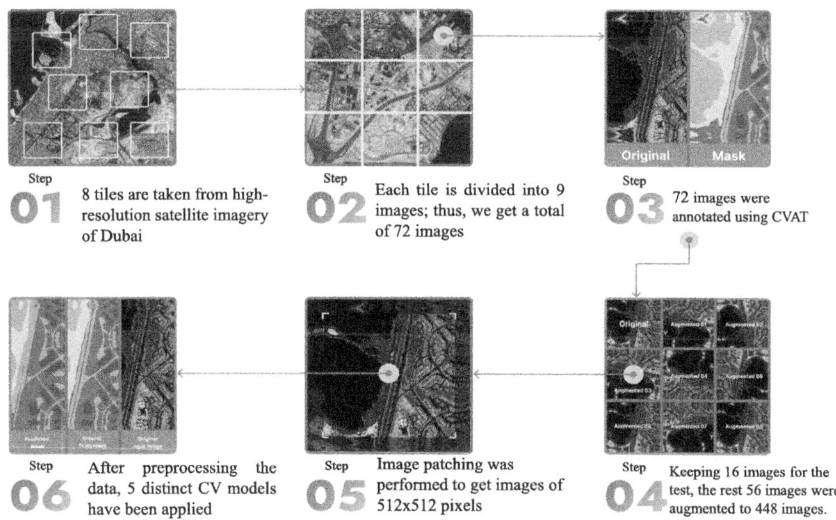

Fig. 1. Data preprocessing and analysis flow diagram.

with a stride of 2 is applied to more than 22 windows. Three fully connected layers follow the five convolutional layer blocks. A soft-max layer, the final layer, generates probabilities for every class [24].

MobileNetV2: The capacity of the MobileNetV2 architecture to discriminate between the input and output domains of the layer transformation and building blocks is an interesting feature. The layer transformation is a non-linear function that changes the input into the desired output. The MobileNetV2 architecture consists of an initial set of fully convolutional layers with a filter size 32, followed by 19 bottleneck residual layers. A non-linear Rectified Linear Unit 6 (ReLU6) is employed due to its resilient and accurate low-precision calculation capabilities. The network used a conventional 3×3 kernel and incorporated dropout and batch normalization techniques during the training process. A constant growth rate was used for all layers in the network, except the initial layer [16].

3 Results and Discussion

This study applied deep learning networks to detect urban features utilising the following image segmentation techniques: Multi-UNet, Inception-ResNet-V2, DeepLabV3+, VGG16, and MobileNetV2. The aim was to evaluate the performance of these models on the satellite in terms of the ability to detect urban features with high accuracy from satellite imagery.

3.1 Dataset

The dataset obtained from "Humans in the Loop (2022)" [22] serves as the cornerstone for this project. This dataset contains a diversified array of high-resolution photos covering various sites in Dubai. It contains 72 images grouped into 8 larger tiles. Each image captures comprehensive details of the metropolitan scene, including buildings, roads, plants, and other important characteristics. The use of data augmentation techniques helps to mitigate the lack of diversity by generating a wider range of training, which improves generalization. While a larger dataset would certainly provide more robustness, this study was a proof-of-concept, and the results demonstrate that the methodology is effective even with limited data. However, class imbalances in the dataset can significantly affect model performance, causing the model to favor dominant classes and leading to reduced recall for underrepresented classes. This bias can hinder the model's ability to generalise effectively and may skew evaluation metrics like accuracy, loss, and dice coefficient. The custom combination of the Focal Loss and Dice Loss using a weighted sum was used to address the class imbalance in the datasets [2].

3.2   Model Summary

The weights and biases that must be learned during training are known as the parameters. The elements that are to be adjusted throughout the training process to improve the model accuracy and minimise the loss are known as the parameters of the CNN models. The MobileNetV2 model employed in this study comprises 11,754,134 parameters, with 11,725,942 parameters being trainable (Table 1). In comparison, Table 1 shows that the non-trainable parameters amount to 28,192. This parameter count indicates the model's moderate size and complexity, making it suitable for feature detection tasks on the Dubai Imagery Dataset. The Inception-ResNet-V2 model has a significantly larger architecture than other models, with 36,793,414 parameters, consisting of 36,752,390 trainable parameters and 41,024 non-trainable parameters. The VGG16 model, on the other hand, consists of 17,831,494 parameters, with 17,796,710 trainable parameters and 34,784 non-trainable parameters. These statistics offer valuable insights into the scale and computational requirements of the models utilised in this study for feature detection tasks. Refer to Table 1 for a detailed summary of each model's architecture and parameter breakdown.

Table 1. Summary of Semantic Segmentation Model Architectures and Parameters.

Model	Model Summary		
	Total Parameters	Trainable Parameters	Non-trainable Parameters
Multi-UNet	1941190	1941190	0
Inception-ResNet-V2	36793414	36752390	41024
DeepLabV3+	17831494	17796710	34784
VGG16	25862662	25858822	3840
MobileNetV2	11754134	11725942	28192

3.3   Model Performance

Table 2 summarises the performance metrics obtained for each model applied to the Dubai dataset after training for 45 epochs. The number of fixed epochs in each model provides consistency in training and makes the results comparable across experiments. Future work could incorporate more sophisticated methods like adaptive learning schedules or hyperparameter tuning to refine performance further. However, the current setup provides a solid foundation and shows promising results with the available resources. Notably, all models achieved high accuracy, with Inception-ResNet-V2 demonstrating superior performance regarding dice coefficient, accuracy, and loss (Table 2).

Table 2. Performance Metrics of Semantic Segmentation Models.

Model	Dice Coefficient	Accuracy	Loss
Multi-UNet	0.85	0.89	0.87
Inception-ResNet-V2	**0.87**	**0.92**	**0.23**
DeepLabV3+	0.76	0.83	0.41
VGG16	0.76	0.84	0.41
MobileNetV2	0.75	0.85	0.43

3.4 Qualitative Analysis

Figures 2, 3, 4, 5 and 6 illustrate images from the Dubai Satellite Imagery Dataset along with the corresponding feature detection results obtained by each model. The images demonstrate the effectiveness of the Inception-ResNet-V2 model in detecting features such as buildings, roads, and vegetation. Compared to other models, the feature boundaries identified by Inception-ResNet-V2 appeared more defined and precise, contributing to its higher dice coefficient (0.87) and accuracy of 92%. This visual evidence supports the quantitative findings and demonstrates how well the model was able to capture important subtleties and characteristics in the image dataset.

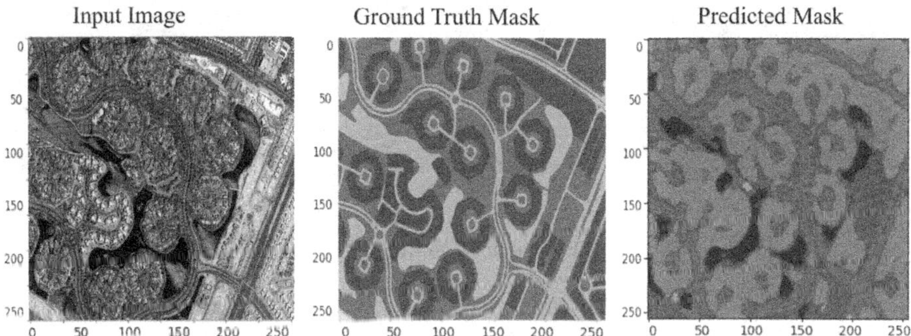

Fig. 2. Comparison of ground truth mask image and predicted mask image by Multi-UNet.

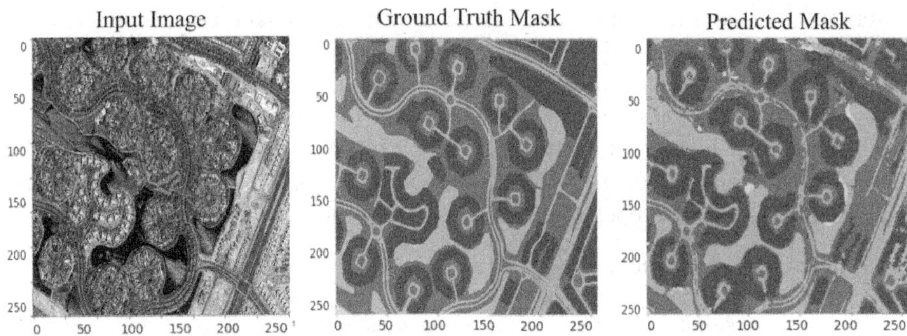

Fig. 3. Comparison of ground truth mask and predicted mask image by Inception-ResNet-V2.

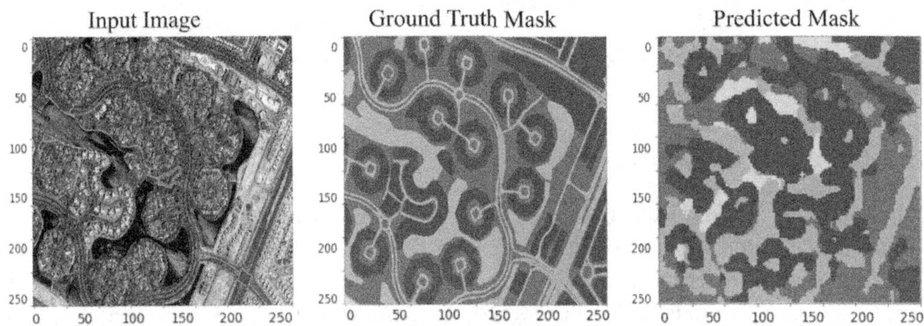

Fig. 4. Comparison of ground truth mask and predicted mask image by DeepLabV3+.

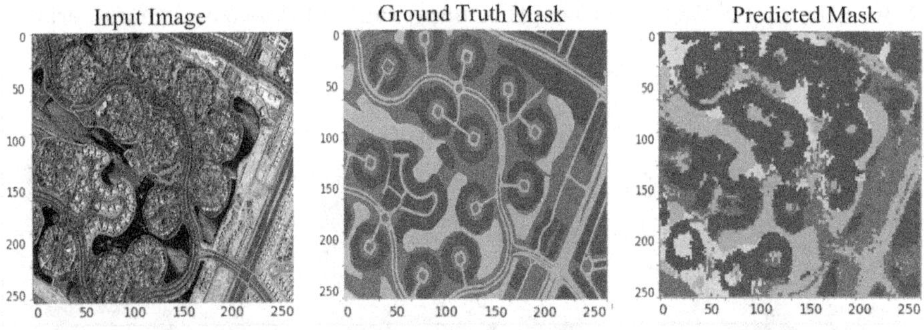

Fig. 5. Comparison of the ground truth mask and predicted mask image by VGG16.

3.5 Training Progress

Figure 7 illustrates the multi-UNet model's training progress over 100 epochs. The plot showcases the training loss values across epochs, providing insights into the model's convergence behaviour. A downward trend in training loss indicates that the model is learning to minimise errors and improve its per-

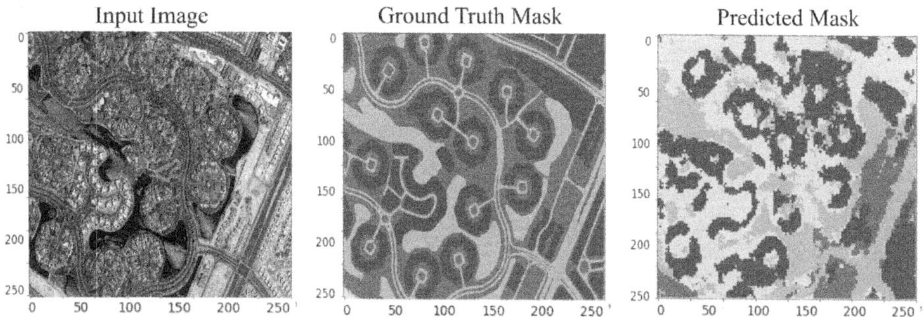

Fig. 6. Comparison of ground truth mask and predicted mask image by MobileNetV2.

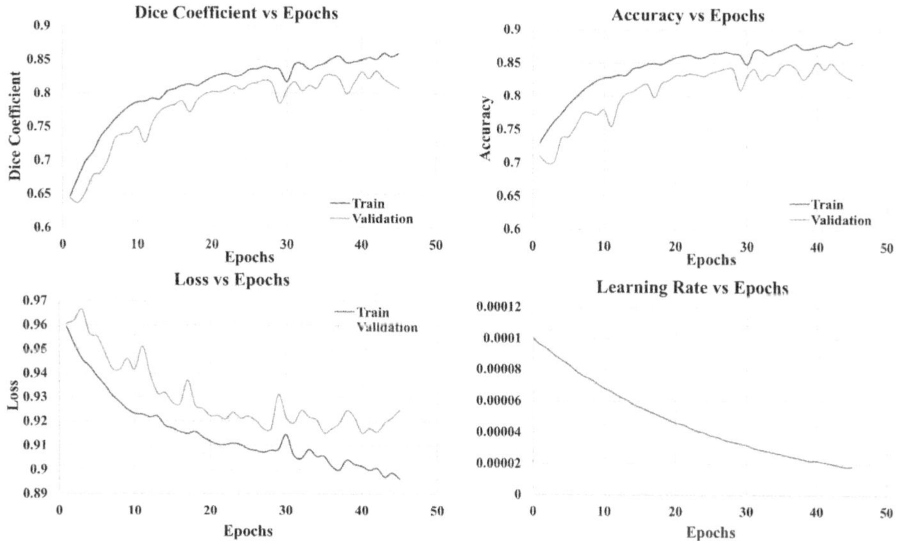

Fig. 7. Dice Coefficient, Accuracy, Loss, and Learning Rate vs Epoch of of Multi-UNet.

formance over successive epochs. The training progress of the Inception-ResNet-V2, DeepLabV3+, VGG16, and MobileNetV2 models over 45 epochs is shown in Figs 8, 9, 10 and 11. The plots depict the dice coefficient, accuracy, loss values, and learning rate across epochs, respectively, showcasing the model's convergence behaviour and providing insights into the model's performance trends throughout the training process. Using the Satellite Imagery Dataset, this study thoroughly assessed the capabilities of different deep learning models for picture segmentation. Through rigorous experimentation and analysis, distinct differences in the effectiveness of various architectures in accurately delineating features within aerial imagery are observed. Inception-ResNet-V2 emerged as the standout performer, achieving an impressive accuracy of 92% and a dice coefficient of 0.87. Figure 8 shows that the plot of Inception-Resnet-V2 steadily decreases

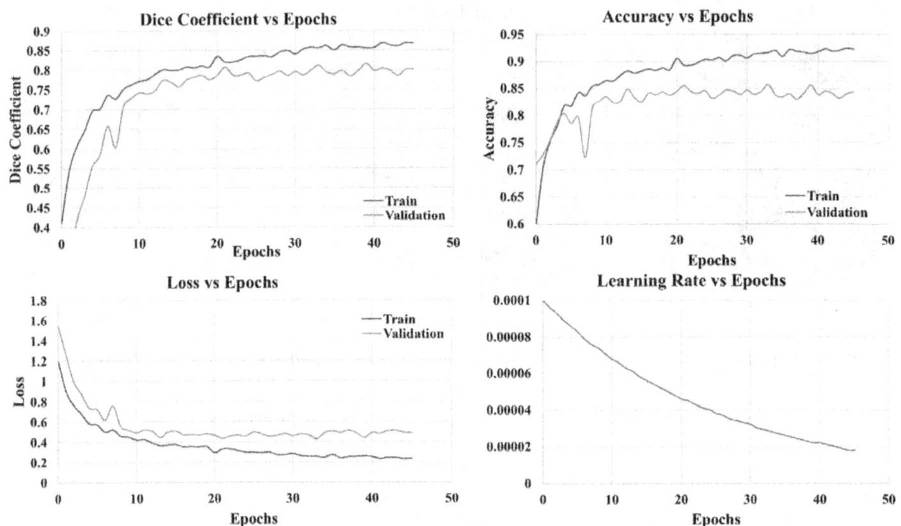

Fig. 8. Dice Coefficient, Accuracy, Loss, and Learning Rate vs Epoch of Inception-ResNet-V2.

its training loss and improves the dice coefficient and accuracy, reinforcing its effectiveness and its ability to capture intricate details and refine feature boundaries, which signifies its efficacy in handling complex feature detection tasks. The model's architecture, combined with inception modules and residual connections, helps mitigate the vanishing gradient problems and contributes to its superior performance by efficiently capturing intricate details and refining feature boundaries [29], like buildings and roads. In comparison, the Multi-UNet model also demonstrated competitive performance with an accuracy of 89% and a dice coefficient of 0.85, showcasing its potential for feature detection despite its smaller parameter size but struggled with finer details due to its simple architecture. Furthermore, the VGG16 model exhibited notable performance with a dice coefficient of 0.76, albeit with a larger parameter size, but lacked advanced mechanisms for detailed feature extraction, given its older, uniform convolutional architecture. DeepLabV3+, with its atrous convolutions, captured broad spatial contexts but lagged in boundary precision, resulting in 83% accuracy and a dice coefficient of 0.76. The MobileNetV2 model can balance efficiency and accuracy, making it appropriate for real-time applications or contexts with restricted computing resource.

While all models showed promise, Inception-ResNet-V2's superior accuracy and refined feature detection capabilities underline its suitability for applications requiring precise feature delineation in aerial imagery. These results provide valuable insights abouy the comparative performance of DL models for feature detection tasks and highlight the importance of selecting appropriate architectures. The generalizability of the models, particularly Inception-Resnet-V2, is a key

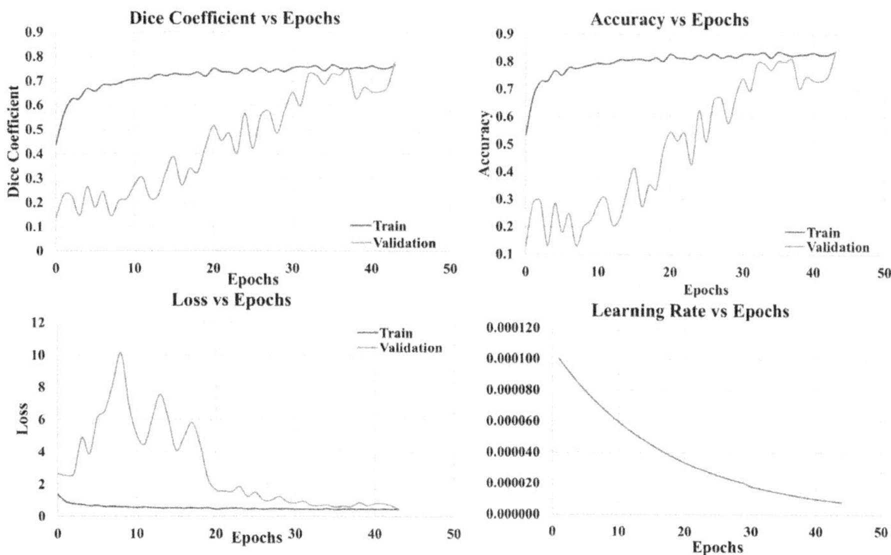

Fig. 9. Dice Coefficient, Accuracy, Loss, and Learning Rate vs Epoch of DeepLabV3+.

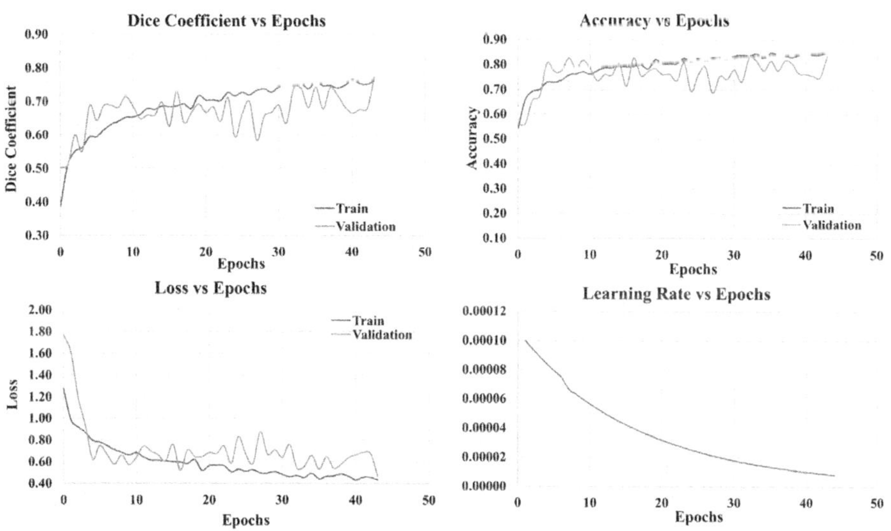

Fig. 10. Dice Coefficient, Accuracy, Loss, and Learning Rate vs Epoch of VGG16.

consideration given the reliance on a single dataset and fixed hyperparameters in this study. While further testing on diverse datasets is necessary, there is strong evidence that Inception-Resnet-V2 is likely to generate well across domains. Its architecture, combining inception modules and residual connections, has shown

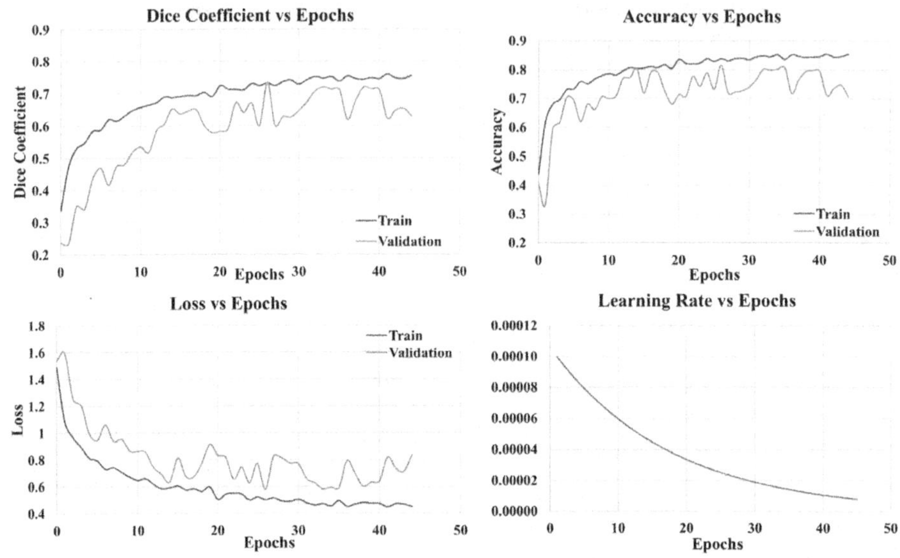

Fig. 11. Dice Coefficient, Accuracy, Loss, and Learning Rate vs Epoch of MobileNetV2.

robust performance in various fields like medical imaging and remote sensing by capturing multi-scale features and addressing the vanishing gradient problems [10,25]. This design has proven effective in handling complex tasks beyond the current dataset [4]. However, future work should explore additional datasets and hyperparameter optimization to improve the model's adaptability further. These findings contribute to advancing remote sensing applications and lay the groundwork for further research aimed at enhancing feature detection accuracy and robustness.

Inception-ResNet-V2 is a deep hybrid architecture that merges the inception module's feature-capturing capabilities with the efficiency of residual connections from ResNet [26]. By incorporating multiple branches within its modules, the model can extract spatial features at various scales using factorized convolutions, which help reduce computational costs [8]. The key innovation lies in the use of residual connections, which add the input of each block to its output, allowing for more effective gradient flow during backpropagation and mitigating the vanishing gradient problem often encountered in deep networks [11]. This enables Inception-ResNet-V2 to achieve greater depth while maintaining training stability. Despite its complexity, the model remains computationally efficient, balancing high accuracy with practical resource demands. In large-scale image segmentation tasks, it outperforms earlier models by leveraging its deeper architecture and stable learning, making it both powerful and efficient for high-performance tasks.

4   Conclusion

This study evaluated the performance of segmentation algorithms such as Multi-UNet, Inception-ResNet-V2, VGG16, MobileNetV2, and DeepLabV3+. The results demonstrate that the performance of the models differs substantially across metrics. The results showed that Inception-ResNet-V2 outperformed other CV algorithms with the highest accuracy, dice coefficient, and minimum loss. Those who demand exhaustive feature detection irrespective of class imbalance could benefit from the high recall of Inception-ResNet-V2.

This research is not free of constraints. Although the datasets utilised are exhaustive, they might not entirely encompass the diversity of global landscapes. Further research should strive to incorporate a more diverse range of satellite images, encompassing those of varying resolutions and captured under different atmospheric conditions, to assess the models' generalisability. Furthermore, the dynamic realm of deep learning presents novel frameworks and instructional approaches that have the potential to augment the efficacy of models designed for semantic segmentation. Subsequent investigations ought to delve into the incorporation of these developments, including transformer models and innovative convolutional neural network (CNN) architectures, into the process of satellite imagery segmentation. An additional area of interest is the investigation of ensemble methods, which take advantage of the respective assets of numerous models to attain enhanced precision and resilience beyond what a solitary model could offer. Furthermore, examining the utilisation of unsupervised and semi-supervised learning techniques may serve as a solution to the issue of inadequately annotated data in the analysis of satellite images.

5   Limitations and Future Directions

This study relies on a single dataset and fixed epochs, limiting generalisability and potentially under-optimising model performance. Future work should explore diverse datasets and hyperparameter optimisation to enhance robustness and performance. Experiment with hyperparameters like learning rates and batch sizes to optimise model performance and reduce overfitting. This study demonstrates Inception-ResNet-V2's effectiveness in feature detection, paving the way for advancements in remote sensing applications. Further research should refine models and explore broader datasets considering ensemble methods and unsupervised learning for comprehensive analysis.

Acknowledgments. We thank the Kaggle and Humans in the Loop website for its open-access data.

Disclosure of Interests. The authors declared no conflicts of interest

References

1. Atchaya, A.J., Anitha, J., Priya, A.G., Poornima, J.J., Hemanth, J.: Multilevel classification of satellite images using pretrained alexnet architecture. In: International Conference on Applied Machine Learning and Data Analytics, pp. 202–209. Springer (2022). https://doi.org/10.1007/978-3-031-34222-6_17
2. Azad, R., et al.: Loss functions in the era of semantic segmentation: A survey and outlook. arXiv preprint arXiv:2312.05391 (2023)
3. Chen, L., Dai, H., Zheng, Y.: Rafnet: reparameterizable across-resolution fusion network for real-time image semantic segmentation. IEEE Trans. Circ. Syst. Video Technol. (2023)
4. Chen, L.C.: Rethinking atrous convolution for semantic image segmentation. arXiv preprint arXiv:1706.05587 (2017)
5. Chen, L.-C., Zhu, Y., Papandreou, G., Schroff, F., Adam, H.: Encoder-decoder with atrous separable convolution for semantic image segmentation. In: Ferrari, V., Hebert, M., Sminchisescu, C., Weiss, Y. (eds.) ECCV 2018. LNCS, vol. 11211, pp. 833–851. Springer, Cham (2018). https://doi.org/10.1007/978-3-030-01234-2_49
6. Chen, Y., et al.: Hybrid attention fusion embedded in transformer for remote sensing image semantic segmentation. IEEE J. Selected Topics Appli. Earth Observat. Remote Sensing (2024)
7. Cheng, G., Lang, C., Wu, M., Xie, X., Yao, X., Han, J.: Feature enhancement network for object detection in optical remote sensing images. J. Remote Sensing (2021)
8. Chollet, F.: Xception: deep learning with depthwise separable convolutions. In: Proceedings of the IEEE Conference on Computer Vision and pattern Recognition, pp. 1251–1258 (2017)
9. Emek Soylu, B., Guzel, M.S., Bostanci, G.E., Ekinci, F., Asuroglu, T., Acici, K.: Deep-learning-based approaches for semantic segmentation of natural scene images: A review. Electronics **12**(12), 2730 (2023)
10. He, K., Zhang, X., Ren, S., Sun, J.: Deep residual learning for image recognition. In: Proceedings of the IEEE Conference on Computer Vision and Pattern Recognition, pp. 770–778 (2016)
11. He, K., Zhang, X., Ren, S., Sun, J.: Identity mappings in deep residual networks. In: Leibe, B., Matas, J., Sebe, N., Welling, M. (eds.) ECCV 2016. LNCS, vol. 9908, pp. 630–645. Springer, Cham (2016). https://doi.org/10.1007/978-3-319-46493-0_38
12. Hesamian, M.H., Jia, W., He, X., Kennedy, P.: Deep learning techniques for medical image segmentation: achievements and challenges. J. Digit. Imaging **32**, 582–596 (2019)
13. Hosen, M.A., et al.: Nn-based prediction interval for nonlinear processes controller. Int. J. Control Autom. Syst. **19**(9), 3239–3252 (2021)
14. Hosen, M.A., Khosravi, A., Nahavandi, S., Creighton, D.: Prediction interval-based neural network modelling of polystyrene polymerization reactor-a new perspective of data-based modelling. Chem. Eng. Res. Des. **92**(11), 2041–2051 (2014)
15. Ibtehaz, N., Rahman, M.S.: Multiresunet: rethinking the u-net architecture for multimodal biomedical image segmentation. Neural Netw. **121**, 74–87 (2020)
16. Jing, J., Wang, Z., Rätsch, M., Zhang, H.: Mobile-unet: an efficient convolutional neural network for fabric defect detection. Text. Res. J. **92**(1–2), 30–42 (2022)
17. Ku, T., Yang, Q., Zhang, H.: Multilevel feature fusion dilated convolutional network for semantic segmentation. Int. J. Adv. Rob. Syst. **18**(2), 17298814211007664 (2021)

18. In the Loop H: Semantic segmentation dataset (2022). https://humansintheloop. org/resources/datasets/semantic-segmentation-dataset-2/, Accessed 20 September 2024
19. Manakitsa, N., Maraslidis, G.S., Moysis, L., Fragulis, G.F.: A review of machine learning and deep learning for object detection, semantic segmentation, and human action recognition in machine and robotic vision. Technologies **12**(2), 15 (2024)
20. Nazir, U., Mian, U.K., Sohail, M.U., Taj, M., Uppal, M.: Kiln-net: a gated neural network for detection of brick kilns in south Asia. IEEE J. Selected Topics Appli. Earth Observat. Remote Sensing **13**, 3251–3262 (2020)
21. Qiu, T., Xiao, Y., Yang, Q., Jiang, X., Zhang, T.: Semantic segmentation based on vision transformer via interactive attention. In: 2023 IEEE International Conference on Systems, Man, and Cybernetics (SMC), pp. 413–418. IEEE (2023)
22. Rezaei, A., Asadi, F.: Systematic review of image segmentation using complex networks. arXiv preprint arXiv:2401.02758 (2024)
23. Sharada, K., Alghamdi, W., Karthika, K., Alawadi, A.H., Nozima, G., Vijayan, V.: Deep learning techniques for image recognition and object detection. In: E3S Web of Conferences, vol. 399, p. 04032. EDP Sciences (2023)
24. Simonyan, K., Zisserman, A.: Very deep convolutional networks for large-scale image recognition. arXiv preprint arXiv:1409.1556 (2014)
25. Szegedy, C., Ioffe, S., Vanhoucke, V., Alemi, A.: Inception-v4, inception-resnet and the impact of residual connections on learning. In: Proceedings of the AAAI Conference on Artificial Intelligence, vol. 31 (2017)
26. Szegedy, C., Vanhoucke, V., Ioffe, S., Shlens, J., Wojna, Z.: Rethinking the inception architecture for computer vision. In: Proceedings of the IEEE Conference on Computer Vision and Pattern Recognition, pp. 2818–2826 (2016)
27. Wang, Y., Wang, N., Xia, C.: Image semantic segmentation algorithm of deeplabv3+ based on attention mechanism and strip pooling. In: 2023 IEEE 3rd International Conference on Electronic Technology, Communication and Information (ICETCI), pp. 293–297. IEEE (2023)
28. Wang, Z., Xia, M., Weng, L., Hu, K., Lin, H.: Dual encoder-decoder network for land cover segmentation of remote sensing image. IEEE J. Selected Topics Appli. Earth Observat. Remote Sensing (2023)
29. Young, O.R., Onoda, M.: Satellite earth observations in environmental problem-solving. Satellite Earth observations and their impact on society and policy, pp. 3–27 (2017)
30. Zhao, J., et al.: Autonomous driving system: a comprehensive survey. Expert Syst. Appli., 122836 (2023)

PPDA: A Privacy Preserving Framework for Distributed Graph Learning

Nikita Malik[1(✉)], Nimesh Agrawal[2], and Sandeep Kumar[1,2,3]

[1] Bharti School of Telecommunication Technology and Management, Delhi, India
`{bsz218185,ksandeep}@iitd.ac.in`
[2] Department of Electrical Engineering, Delhi, India
`eey217515@iitd.ac.in`
[3] Yardi School of Artificial Intelligence, Indian Institute of Technology, Delhi, India

Abstract. Graph-based learning finds wide application in social networks, robotics, communication, and medicine. However, datasets often contain sensitive information, hindering their use in graph learning applications due to privacy concerns. Existing privacy-preserving methods pre-process the data to extract user-side features, and only these features are used for subsequent learning. Unfortunately, these methods are vulnerable to adversarial attacks to infer private attributes. We present a novel privacy-respecting framework for distributed graph learning and graph-based machine learning. In order to perform graph learning and other downstream tasks on the server side, this framework aims to learn features as well as distances without requiring actual features while preserving the original structural properties of the raw data. The proposed framework is quite generic and highly adaptable. We demonstrate the utility of the Euclidean space, but it can be applied with any existing method of distance approximation and graph learning for the relevant spaces. Through extensive experimentation on both synthetic and real datasets, we demonstrate the efficacy of the framework in terms of comparing the results obtained without data sharing to those obtained with data sharing as a benchmark. This is, to our knowledge, the first privacy-preserving distributed graph learning framework.

Keywords: Privacy · Distance approximation · anonymized embeddings · Graph learning · Graph-based clustering

1 Introduction

Graphs are essential for understanding complex structures within seemingly unstructured high-dimensional data. They're widely used in various real-world scenarios like social networks, patient-similarity networks, and scientific research. However, as data becomes more distributed across different locations, ensuring privacy becomes a challenge. This is crucial in areas like social networks and patient data, where privacy breaches can have serious consequences. In some cases, it's even illegal to share sensitive data, like genomic information, between entities [4]. Patient data, collected from various sources such as hospitals and mobile devices, is multifaceted and intricate. Protecting privacy while analyzing

© The Author(s), under exclusive license to Springer Nature Singapore Pte Ltd. 2026
M. Mahmud et al. (Eds.): ICONIP 2024, CCIS 2297, pp. 308–331, 2026.
https://doi.org/10.1007/978-981-96-7036-9_21

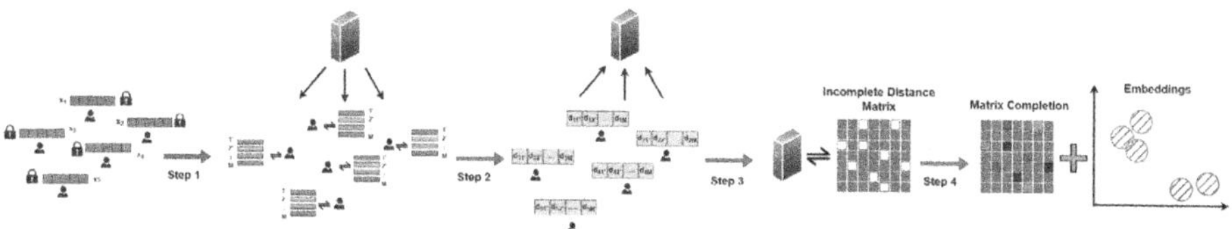

Fig. 1. The overall framework of PPDA. Each user in the distributed network has their attributes privately stored with them. In Step (**1**), a centralized server populates all the users with a predefined set of feature vectors called '**Anchors**'. Anchor features have the same distribution as the underlying user data. In Step (**2**), all the users compute the pairwise distance with all the anchors and push it to the server. To this end, at the server, we have an Incomplete Distance Matrix with all the user-anchor distances known and user-user distances unknown as seen in Step (**3**). Finally, in Step (**4**), we use anchored/non-anchored MDS to obtain the completed distance matrix along with embeddings, which are used further for graph learning and downstream tasks.

such data is paramount for extracting meaningful insights [13]. There's a growing need for distributed machine learning, driven by increased privacy awareness. In a distributed setting, client feature information is highly sensitive and must remain confidential without being shared among clients or with the server. Yet, acquiring similarity information between clients is essential, especially without exposing individual features. In this paper, we introduce PPDA, a novel privacy-preserving distributed graph learning framework aimed at maintaining user privacy while facilitating graph-based learning and analysis of distributed data without centralized data sharing. This privacy-preserving similarity inference approach would be critical for many applications, including finance and medical, patient similarity networks and genetic studies. Our privacy model excels in learning graph structures or embeddings without sharing sensitive data, making it indispensable for scenarios necessitating a graph framework. However, it is unsuitable for situations where similarity between entities or client information is itself a sensitive attribute.

Learning a graph represented by an adjacency matrix (A) requires pairwise node distances or the sample covariance matrix S [17]. However, computing inter-user distances while preserving privacy is challenging. To address this, as illustrated in Fig. 1, we initialize users/clients with known attributes, referred to as 'Anchors'. Each client computes their distances to the anchors and shares them with the server after local communication. Since user-to-user distances remain unknown, the server receives an incomplete distance matrix. We employ matrix completion techniques to estimate inter-user distances and learn a feature representation of users, maximizing retention of the underlying neighborhood structure in the raw data. So in summary, our work follows the following key objectives: (1) Secure distance learning: Achieve pairwise distance learning without disclosing features, (2) Graph learning: Leverage this secure distance information to construct graphs, and (3) Graph embedding: Generate graph embeddings for advanced machine learning tasks. These goals are rigorously validated in the experiment section. While we demonstrate using the Euclidean space, PPDA is adaptable to various embedding spaces such as Riemannian, spherical, or hyperbolic space [38]. Extensive experimentation on synthetic and real datasets demonstrates the scalability and compatibility of our method with existing pipelines for matrix completion and graph learning frameworks.

Notations. The lowercase bold letters refer to vectors, while uppercase letters represent matrices. The element located at row i and column j of a matrix X is written as X_{ij}. The transpose of matrix X is denoted as X^T, and its pseudo-inverse as $X^\dagger$. The vector of all ones, with dimensions suited to the context, is represented by $\mathbf{1}$. The Frobenius norm of matrix X is defined as $\|X\|_F^2 = \sum_{ij} X_{ij}^2$, and the Euclidean norm of vector $\boldsymbol{x}$ is denoted by $\|\boldsymbol{x}\|_2$.

2 Related Work

Current privacy-protection methods often fall short in balancing privacy and utility. While techniques like feature conversion aim to safeguard data, recent advancements show that adversaries can still reverse engineer features back to raw data, compromising privacy [10,24]. Even FL, widely adopted for distributed learning, has its limitations. Research indicates that model gradients can occasionally be leveraged to reconstruct client data, highlighting potential privacy risks [40]. Privacy-Preserving Data Mining (PPDM) techniques aim to extract knowledge from large datasets while preserving sensitive information [1]. Traditional methods like k-anonymity [35], t-closeness [22], and l-diversity [23] are tailored for centralized data storage and may not suit our needs. Differential Privacy (DP) [33] and Homomorphic Encryption (HE) [32] offer strong privacy guarantees but have drawbacks. DP may sacrifice utility, while HE can be computationally intensive, limiting their practicality in some scenarios. Moreover, these methods are vulnerable to adversarial attacks such as poisoning attacks [30] and other attacks [3,15,29]. In [30], it is well established that the DP-Federated Learning frameworks are not inherently robust and are vulnerable to a carefully designed attack method. Furthermore, they reveal that it is challenging for existing robust FL methods to defend against attacks on DPFL. This can be attributed to the fact that the local gradients of DP-FL are perturbed by random noise, and the selected central gradients inevitably incorporate a higher proportion of poisoned gradients compared to conventional FL. Moreover, in [3,15,29], it is well examined that standalone differential privacy and homomorphic encryption techniques can't guarantee data privacy against attacks. It is important to consider the measure of privacy protection, especially in cases when facing some adversarial attack. For the proposed framework, the only parameter that is sent from the client to the server is the client-to-anchor distances. In the worst case, the attacker can estimate the pairwise similarities among the nodes but it cannot estimate the client attributes as guaranteed by Theorem 1 of the paper. On the other hand, assuming that we use HE for encrypting the features of the client before sending it to the server. It becomes infeasible to perform operations on encrypted data for pairwise distance computation involved in graph learning.

Various methods for privacy-preserving distance computation have been proposed [31]. Efficient secure computation approaches for tasks like the Similar Patient Query problem exist [2], yet they overlook distributed data. A recent study [27] introduces a method for distributed graph learning considering communication costs but neglects data privacy. Given these challenges, there's a clear need for a privacy-respecting distributed graph learning framework to tackle the demand for secure and efficient data analysis in distributed environments.

3   Background

In this section, we give a brief background of graphs and graph learning including probabilistic graphical models. Then we describe the problem statement and the parameters of our method.

3.1   Graph

Let $\mathcal{G} = (V, E, A, X)$ represent a graph with features, where V is the set of N vertices, E is the set of edges, A is the adjacency matrix, and X is the feature matrix. For an undirected graph without self-loops, the entries of A are defined as $A_{ij} > 0$ if $(i, j) \in E$, and $A_{ij} = 0$ if $(i, j) \notin E$. The feature matrix X has dimensions $N \times d$, where each row vector $\boldsymbol{x}_i \in \mathbb{R}^d$ represents the feature vector associated with a node in $\mathcal{G}$. The Laplacian matrix Θ and the Adjacency matrix A are commonly used to represent graphs, with Q being a diagonal matrix where $Q_{ii} = \sum_j A_{ij}$ representing the node degrees of $\mathcal{G}$.

3.2   Graph Learning

A basic premise about data residing on the graph is that it changes smoothly between connected nodes. To quantify this assumption, if two signals $\boldsymbol{x}_i$ and $\boldsymbol{x}_j$ are from the smooth set and are located on well-connected nodes (i.e. A_{ij} is large), then it is expected that their distance $\|\boldsymbol{x}_i - \boldsymbol{x}_j\|$ will be small, which leads to a small value for $\mathrm{tr}(X^T \Theta X)$, where Θ is the laplacian matrix defined as $Q - A$ [19]. The matrix $\Theta \in \mathbb{R}^{N \times N}$ is a valid graph Laplacian matrix if it belongs to the following set:

$$\mathcal{S}_\Theta = \left\{ \Theta \in \mathbb{R}^{N \times N} \mid \Theta_{ij} = \Theta_{ji} \leq 0 \text{ for } i \neq j; \Theta_{ii} = -\sum_{j \neq i} \Theta_{ij} \right\} \tag{1}$$

The graph Laplacian matrix is a widely used tool in graph analysis and has many important applications in various fields. Some examples of its use include embedding, manifold learning, spectral sparsification, clustering, and semi-supervised learning [5,36,41]. Under this smoothness assumption, to learn a graph from data, [17] proposed an optimization formulation in terms of adjacency matrix (A) and pairwise distance matrix (K), where $K_{ij} = \|\boldsymbol{x}_i - \boldsymbol{x}_j\|^2 = d_{ij}$ represents the Euclidean distance between samples $\boldsymbol{x}_i$ and $\boldsymbol{x}_j$:

$$\min_{A \in \mathcal{S}_A} \sum_{i=1}^{N} A_{ij} d_{ij} + f(A) \tag{2}$$

$$\mathcal{S}_A = \left\{ A \in \mathbb{R}_+^{N \times N} : A = A^T, \mathrm{diag}(A) = 0 \right\} \tag{3}$$

where $f(A)$ is the regularization term for limiting A to satisfy certain properties, and $\mathcal{S}_A$ in eq. (3) refers to the set for valid Adjacency matrix.

3.3   Probabilistic Graphical Models

Conditional dependencies among variables are captured by Gaussian graphical modeling (GGM), simulating the Gaussian property of multivariate data [39]. For an unidrected graph, let $\boldsymbol{x} = (x_1, \ldots, x_N)^T$ represent an N-dimensional zero mean multi-variate random variable with $S \in \mathbb{R}^{N \times N}$ as the sample covariance matrix (SCM) derived from d observations. In GGM, an edge between two nodes signifies their conditional dependence given other nodes. The method learns a graph through the optimization problem [14,20]:

$$\max_{\Theta \in S_{++}^N} \log \det(\Theta) - \mathrm{tr}(XX^T\Theta) - \alpha h(\Theta) \tag{4}$$

Here, $X \in \mathbb{R}^{N \times d}$ is the feature matrix, $\Theta \in \mathbb{R}^{N \times N}$ is the graph matrix to estimate, N is the number of nodes, $h(\cdot)$ is a regularization term, $S_{++}^N \in \mathbb{R}^{N \times N}$ represent the set of positive definite matrices, and $\alpha > 0$ is the regularization hyperparameter. The SCM $S = XX^T$ can be expressed using the distance matrix K as:

$$S = \frac{-1}{2} \left(K - \mathbf{1}(\mathrm{diag}(XX^T))^T - \mathrm{diag}(XX^T)\mathbf{1}^T \right) \tag{5}$$

where $\mathrm{diag}(A)$ denotes a column vector of the diagonal entries of A and $\mathbf{1}$ is the column vector of all ones. When the observed data follows a zero mean multivariate Gaussian distribution, the optimization in (4) corresponds to the penalized maximum likelihood estimation (MLE) of the inverse covariance matrix, or precision matrix, which is also referred to as a Gaussian Markov Random Field (GMRF) [20]. Our objective is to approximate distances without revealing features to compute the SCM using Eq. (5), ensuring privacy. The estimated distances and corresponding S are then utilized for graph learning and related tasks.

4   Proposed Method

In this section, we formalize our problem setting, and then we introduce the proposed framework. Finally, we provide an illustration of the framework using Multi-dimensional Scaling (MDS) for embedding.

4.1   Proposed Privacy Preserving Distance Approximation (PPDA)

Consider a distributed setting of N isolated clients, where our approach assumes that the data point held by each client is IID (Independent and Identically Distributed) data. Let $X_{NA} = [x_1, x_2, \ldots, x_N]^T \in \mathbb{R}^{N \times d}$ be the feature representation of these clients referred to as *Non-anchors (NA)*, where $x_i \in \mathbb{R}^d$ is the feature vector of dimension d for i^{th} user. A centralized server now generates M reference nodes represented by $\{a_1, a_2, \cdots, a_M\}$ referred here as *Anchors (A)* and sends the features associated with these anchors to all the clients $i \in \{1, 2, \cdots, N\}$. For instance, this server could be owned by companies such as Google or Facebook, healthcare institutions, or a government agency. This is shown as Step **(1)** in Fig. 1. Let $X_A = [p_1, p_2, \cdots, p_M]^T \in \mathbb{R}^{M \times d}$ be the feature representation of these anchors. Now each client i have access to its own feature and anchor features i.e. $\{x_i, p_1, p_2, \cdots, p_M\}$. Each client then computes the exact distance with each anchor given by $(d_{ia_1}, d_{ia_2}, \cdots, d_{ia_M})$ $\forall i \in \{1, 2, \cdots, N\}$ and sends these distances back to server shown as Step **(2)** in Fig. 1. As a result, we have an incomplete distance matrix, with all client-to-client distances unknown. This matrix is then used to perform matrix completion to obtain an inter-client distance matrix for learning a graph and conducting graph-based downstream tasks.

4.2   Construction of Incomplete Distance Matrix for PPDA

As mentioned in Sect. 4 , using the pairwise distance between clients and anchors, we construct a distance matrix D given in Eq. (6), which can be partitioned into three block matrices namely: K being the unknown client-to-client distance, D_{12} being the client to anchor distance and D_{22} being the anchor to anchor distance.

$$D = \begin{bmatrix} K & D_{12} \\ D_{12}^T & D_{22} \end{bmatrix} \tag{6}$$

4.3   Anchor Generation

In the proposed approach, the distribution of anchor data plays a crucial role. The anchor serves as shared information accessible to all clients. Anchor data is either generated randomly or sourced from open datasets to ensure privacy protection. The proper scheduling of the anchors has a significant impact on the overall performance and accuracy of the framework. There are several factors to consider when developing the anchor scheduling strategy, including:

Number of anchors: The number of anchors used in the framework has a direct impact on the algorithmic performance. Too few anchors may not preserve the structural information while ensuring privacy, while too many anchors may lead to overfitting and may violate privacy.

Selection criteria: The criteria used to select anchors can also impact the performance of the system. Selecting anchors from the same probability distribution as of the underlying user data may be more effective than selecting them at random. For example, the data distribution of patient similarity networks or social networks will depend on factors including a number of patients/users or similarity of patients/connection between users.

The pseudo-code for PPDA is listed down in Algorithm 1. It is evident that the proposed framework is straightforward as it only requires a set of anchors to approximate the inter-client distance, thereby providing privacy protection at no extra cost.

Algorithm 1. PPDA: Privacy Preserved Distance Approximation

Input: M anchors $X_A \in \mathbb{R}^{M \times d}$, embedding dimension $(\widetilde{d})$

Output: Embedding $\left(\widetilde{X}\right)$ in dimension $\widetilde{d}$, Estimated user to user distance matrix $K \in \mathbb{R}^{N \times N}$

1: Populate all the users with anchors X_A from the server.
2: Compute anchor-anchor distance matrix D_{22}
3: **for** every user $i \in \{1, 2, \cdots, N\}$ **do**
4:     Compute user to M anchors distance to get matrix D_{12}
5: **end for**
6: Send the incomplete distance matrix formed as in (6) back to the server.
7: Perform distance matrix completion to obtain K along with $\widetilde{X} \in \mathbb{R}^{N \times \widetilde{d}}$

4.4   Demonstrating PPDA Using MDS

Utilizing the measurements of distances among pairs of objects, MDS (multidimensional scaling) finds a representation of each object in d - dimensional space such that the distances are preserved in the estimated configuration as closely as possible. To validate the goodness-of-fit measure, MDS optimizes the loss function (known as "Stress"(σ)) given by:

$$\sigma(X) = \min_{X} \sum_{i < j \leq N} w_{ij} \left(\delta_{ij} - d_{ij}(X)\right)^2 \tag{7}$$

$$w_{ij} = \begin{cases} 1, & \text{if } \delta_{ij} \text{ is known} \\ 0, & \text{if } \delta_{ij} \text{ is missing} \end{cases}, \quad W = \begin{bmatrix} 0_{N,N} & 11_{N,M} \\ 11_{N,M}^T & 11_{M,M} \end{bmatrix} \tag{8}$$

where X represents the computed configuration, $d_{ij}(X) = \|\boldsymbol{x}_i - \boldsymbol{x}_j\|$ is the Euclidean distance between nodes i and j, δ_{ij} is the measured distance computed privately and weights w_{ij} are defined in Eq. (8). Placing the weights of unknown inter-user distance to zero, the weight matrix W can be partitioned into block matrices as shown in Eq. (8), where $11_{N,M}$ is a matrix of ones with shape $N \times M$. De Leeuw [8] applied an iterative method called SMACOF (Scaling by Majorizing a Convex Function) to estimate the configuration X. As the objective is a non-convex function, SMACOF minimizes the stress using the simple quadratic function $\tau(X, Z)$ which bounds $\sigma(X)$ (the complicated function)

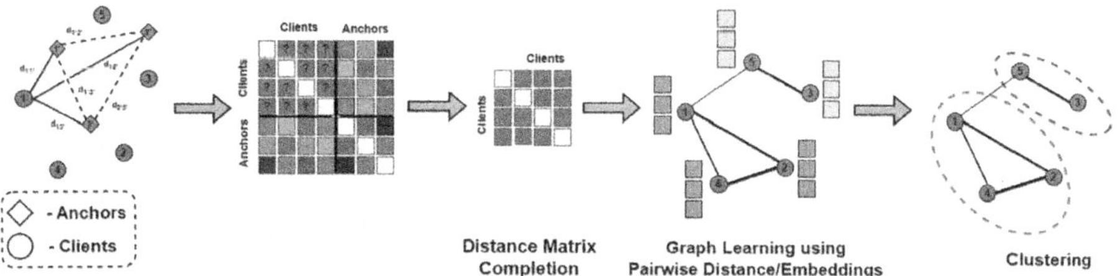

Fig. 2. The overall pipeline for Private Graph-based Clustering. For the client's and anchor's attributes lying in any feature space, we first compute all the pairwise distances such that only inter-client distance is unknown. Next, we form an incomplete distance matrix in block form as shown in the figure. Using any existing methods for distance matrix completion we estimate unknown distances as well as the client's feature representation. Based on this pairwise distance/embedding, a graph structure among the clients is learnt, which is further used for graph-based clustering or any other downstream task.

from above and meets the surface at the so-called supporting point Z as given in Eq. (9).

$$\sigma(X) \leq \tau(X, Z) = \sum_{i<j} w_{ij}\delta_{ij}^2 + \sum_{i<j} w_{ij}d_{ij}^2(X) - 2\sum_{i<j} w_{ij}\delta_{ij}{}^2 \frac{(\boldsymbol{x}_i - \boldsymbol{x}_j)^T (\boldsymbol{z}_i - \boldsymbol{z}_j)}{\|\boldsymbol{z}_i - \boldsymbol{z}_j\|} \tag{9}$$

Equation (9) can be written in matrix form as:

$$\tau(X, Z) = C + \text{tr}\left(X^T V X\right) - 2\,\text{tr}\left(X^T B(Z)Z\right) \tag{10}$$

The iterative solution which guarantees monotone convergence of stress [7] is given by Eq. (11), where $Z = X^{k-1}$:

$$X^{(k)} = \min_X \tau(X, Z) = V^\dagger B(X^{(k-1)})X^{(k-1)} \tag{11}$$

This algorithm offers flexibility to embed features in any dimension other than d, which enables the handling of high-dimensional data and also meets privacy constraints. As V is not of full rank, hence the Moore-Penrose pseudoinverse $V^\dagger$ is used. The elements of the matrix $B(X)$ and V are defined in Eq. (12).

$$b_{ij} = \begin{cases} -\frac{w_{ij}\delta_{ij}}{d_{ij}(X)} & d_{ij}(X) \neq 0, i \neq j \\ 0 & d_{ij}(X) = 0, i \neq j \\ -\sum_{j=1,j\neq i}^{N} b_{ij} & i = j \end{cases}, \quad v_{ij} = \begin{cases} -w_{ij} & i \neq j \\ -\sum_{j=1,j\neq i}^{N} v_{ij} & i = j \end{cases} \tag{12}$$

Next, as the features of anchors are being shared among all users, it is shown that only user-level embedding can be computed as a function of X_A [9], utilizing the incomplete distance matrix given in Eq. (6).

4.5   Using Anchored-MDS

For a network of N users with M anchors, with total $(N + M)$ unique nodes, we can partition X from Eq. (11) as: $X = \begin{bmatrix} X_{NA} \ X_A \end{bmatrix}^T$, with $X_{NA} = [\boldsymbol{x}_1, \cdots, \boldsymbol{x}_N]^T \in \mathbb{R}^{N \times d}$, $X_A = [\boldsymbol{x}_{N+1}, \cdots, \boldsymbol{x}_{N+M}]^T \in \mathbb{R}^{M \times d}$. Similarly, matrices V and $B(X)$ as described in (12) can also be partitioned into block matrices as shown in Eq. (13). It should be noted that V is the Laplacian of the weight matrix W defined in (8), whereas $B(X)$ is the Laplacian of weighted W. From this definition, V can be represented in the block form as given in (13).

$$V = \begin{bmatrix} V_{11} & V_{12} \\ V_{12}^T & V_{22} \end{bmatrix}, \quad B(X) = \begin{bmatrix} B_{11} & B_{12} \\ B_{12}^T & B_{22} \end{bmatrix}, \quad V = \begin{bmatrix} M * I & -11_{N,M} \\ -11_{N,M}^T & (N + M)I - 11_{M,M} \end{bmatrix} \tag{13}$$

where the size of matrices V_{11} and B_{11} are $N \times N$, V_{12} and B_{12} being $N \times M$, V_{22} and B_{22} of $M \times M$. Now, for the majorization function $\tau(X, Z)$ as in (10), it is established that using SMACOF, the embeddings of the non-anchors can be learnt as a function of X_A using the iterative solution given in Eq. (14):

$$X_{NA}^{(k)} = V_{11}^{\dagger} \left(B_{11} X_{NA}^{(k-1)} + (B_{12} - V_{12}) X_A \right) \tag{14}$$

The proof of this formulation has been elaborated in [9]. Using V_{12} and inverse of V_{11} as in (13), the simplified update rule for computing the embeddings is given by:

$$X_{NA}^{(k)} = \frac{\left(B_{11} X_{NA}^{(k-1)} + (B_{12} + 11_{N,M}) X_A \right)}{M} \tag{15}$$

However, in order to preserve the privacy of the user data, there is a restriction on the number of anchors (M) used for computing the embeddings using (15). We provide the following Theorem that characterizes the connection between M and the dimensionality (d) of the data.

Note: In our framework, neither the non-anchors nor the server can access non-anchors data. However, the server can potentially learn the exact feature vectors of non-anchor nodes if the number of anchor nodes (M) exceeds $d - 1$, due to the available anchor locations and distances. Our Theorem 1 extends the results from [18] on distributed sensor localization, which use pairwise distances and at least three anchors in a 3D coordinate system. By choosing fewer than $d - 1$ anchors, exact feature recovery is prevented, ensuring our privacy objectives.

Theorem 1. *To avoid learning exact feature embeddings and ensure privacy, the number of anchors generated must be less than the original dimensionality of the data, i.e., $M < d$.*

Proof. Consider a network in d-dimensional Euclidean space $\mathbb{R}^d$, comprising anchors $A = \{A_1, A_2, \ldots, A_M\}$ and non-anchor nodes $P = \{P_1, P_2, \ldots, P_N\}$, with feature vectors $\boldsymbol{x}_i \in \mathbb{R}^d$. The locations of the anchors are known, while the locations of the non-anchors need to be estimated. Previous work [18] shows that in $\mathbb{R}^d$, a minimum of $(d+1)$ anchors with known locations is required to locate N non-anchor nodes. The use of anchors for distributed sensor localization is a well-researched area [18]. The study also assumed that all non-anchor points must be within the convex hull of the anchors. However, to accurately recover features in $\mathbb{R}^d$, at least d anchors are necessary, even if non-anchors are placed in any location. Thus, having fewer than d anchors, i.e., $M < d$, guarantees that exact feature embeddings cannot be obtained, ensuring privacy. Refer to the Sect. 6.8 for the experimental validation of Theorem 1.

Next, the focus will be on exploring the use of these privacy-preserving embeddings to perform graph-based downstream tasks such as clustering.

Algorithm 2. PPDA using MDS/Anchored-MDS

Input: M anchors $X_A \in \mathbb{R}^{M \times d}$, Incomplete Distance Matrix D, tolerance (ϵ), maximum epochs (t), embedding dimension ($\widetilde{d}$)
Output: Embeddings (X/X_{NA}) in dimension d or $\widetilde{d}$, Estimated user to user distance matrix $K \in \mathbb{R}^{N \times N}$
1: Randomly initialize $X^{(0)}/X_{NA}^{(0)}$
2: **while** $\sigma\left(X^{(k-1)}\right) - \sigma\left(X^{(k)}\right) < \epsilon$ or $k = t$ **do**
3:     Compute matrix D using block matrices D_{12} and D_{22}
4:     Compute matrix B
5:     Update X, X_{NA} as in equation (11), (15) respectively
6:     Compute Stress (σ) using (7)
7:     Set $k \leftarrow k + 1$
8: **end while**

---

5   PPDA for Graph-Based Clustering

Graph-based clustering is a method of clustering data points into groups (or clusters) based on the relationships between the data points. This is typically done by constructing a graph as discussed in Sect. 3.2. Clustering algorithms can then be applied to the graph to identify clusters of closely related data points. Graph-based clustering is often used in data mining and machine learning applications, as it can provide a more powerful and flexible way of identifying patterns and structures in data compared to traditional clustering methods. In [26], Nie et al. proposed a new graph-based clustering method based on the rank constraint of Laplacian. The authors exploited an important property of the Laplacian matrix which tells that if $rank(\Theta) = N - k$, then the graph is an ideal representation for partitioning the data points into the k clusters without using K-means or other discretization techniques that are often required in traditional graph-based clustering methods such as spectral clustering [25]. Utilizing this characteristics, given the initial adjacency (A), they aim to learn a new adjacency matrix P whose corresponding laplacian $\Theta_P = Q_P - \frac{P+P^T}{2}$ is constrained to rank $(N-k)$. In addition, to prevent some rows of P from being all zeros, they impose an additional constraint on P to ensure that the sum of each row is equal to one. The solution to the optimization formulation as in (16) is defined as Constrained Laplacian Rank (CLR) for graph clustering.

$$\mathcal{L}_{\mathrm{CLR}} = \min_{\sum_j p_{ij}=1, p_{ij} \geq 0, \mathrm{rank}(\Theta_P)=N-k} \|P - A\|_F^2 \tag{16}$$

To build an initial graph, we solve the problem (17) similar to (2) which gives the estimate of A, utilizing privately estimated pairwise distance matrix K.

$$\min_{a_i^T \mathbf{1}=1, a_i \geq 0, a_{ii}=0} a_i^T k_i + \gamma \|a_i\|_2^2 \tag{17}$$

In order to perform clustering based on probabilistic graphical models, the authors in [20] proposed a comprehensive optimization framework for structured graph learning (SGL), incorporating eigenvalue constraints on the graph matrix Θ:

$$\begin{aligned} \underset{\Theta}{\mathrm{maximize}} \quad & \log \det(\Theta) - \mathrm{tr}(\Theta S) - \alpha h(\Theta) \\ \text{subject to} \quad & \Theta \in \mathcal{S}_\Theta, \lambda(\mathcal{T}(\Theta)) \in \mathcal{S}_\mathcal{T} \end{aligned} \tag{18}$$

where $\mathcal{S}_\Theta$ is the set of valid graph Laplacian matrix as in (1) and $\lambda(\mathcal{T}(\Theta))$ denotes the eigenvalues of $\mathcal{T}(\Theta)$ arranged in increasing order, where $\mathcal{T}(\cdot)$ is a transformation applied to the matrix Θ. As demonstrated in Sect. 4, we have the complete pairwise distance for each user estimated privately using PPDA and their corresponding embeddings as the algorithm's output, which can be used further to compute S.

Figure 2 shows the overall pipeline for privacy preserved distributed graph learning and graph-based clustering. As we just need adjacency matrix A for clustering, which can be computed using pairwise distance matrix K, this framework can be adapted for any graph-based clustering algorithm beyond CLR, expanding its range of applications.

5.1 PPDA for Privacy Preserving Data Collaboration

Our method, tailored for distributed graph learning in a privacy-preserving context, can be easily adapted to other scenarios. For instance, in [6], a distance matrix of the original data obtained from shared common data among organizations is utilized to learn neighbor information. This approach, aimed at data collaboration, is suitable for scenarios with limited data and significant data bias. Our proposed PPDA framework is compatible with such scenarios, allowing us to efficiently learn the inter-client structure and reconstructed embeddings of organizational data using PPDA (anchored or non-anchored MDS). These learned graphs can then be employed in various GNN-based downstream applications or other use cases.

6   Experiments

In this section, we perform extensive experiments on widely adopted real as well as synthetic datasets. The efficacy of our proposed framework is elucidated by showing that structural properties are retained even without data sharing. We also evaluate graph-based clustering on datasets that are sensitive to privacy. Firstly, we introduce the experimental settings in Sect. 6.1 and present the evaluation of the proposed framework in further subsections.

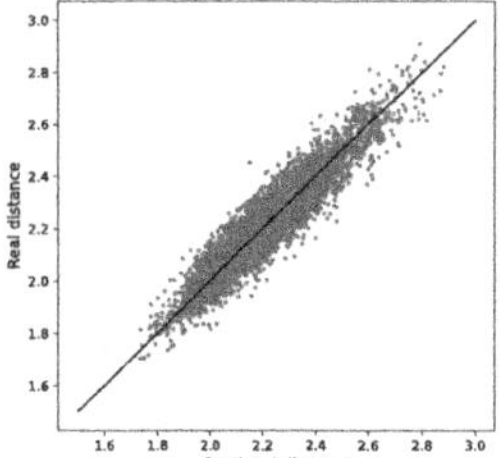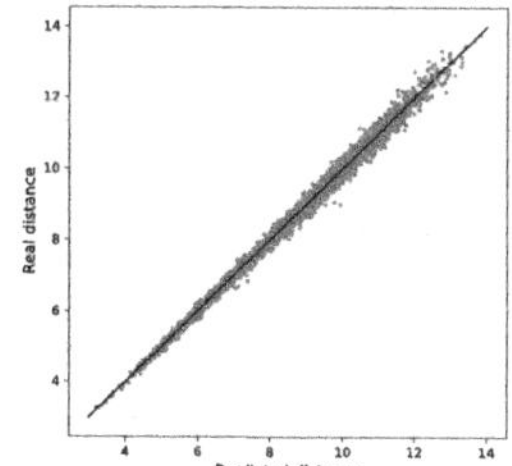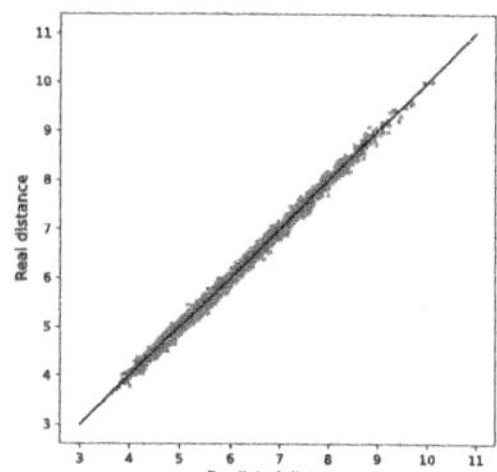

Fig. 3. True euclidean distance (vertical axis) vs. predicted euclidean distance (horizontal axis) for PPDA: ER with $N = 100$ and $M = 199$ (left), BA with $N = 100$ and $M = 399$ (middle) and RGG with $N = 100$ and $M = 699$ (right)

6.1   Experimental Setting

For conducting experiments involving synthetic data, we create various datasets based on diverse graph structures $\mathcal{G}$. We utilize three synthetic graphs with parameters N, M, m, p namely: (i) *Erdos Renyi (ER)* graph with $(N + M)$ nodes and $p = 0.1$ (ii) *Barabasi Albert (BA)* graph with $(N + M)$ nodes and $m = 2$ and (iii) *Random Geometric (RGG)* graph with $(N + M)$ nodes and $p = 0.1$, where N is the number of non-anchors and M is the number of anchors. Initially, we create an IGMRF model using the true precision matrix $\Theta_\mathcal{G}$ as its parameter, which complies with the Laplacian constraints in (1) and the above graph structures. Than, we draw total of d samples $\left\{\mathbf{x}_i \in \mathbb{R}^{N+M}\right\}_{i=1}^{d}$ from the IGMRF model as $\mathbf{x}_i \sim \mathcal{N}\left(0, \Theta_\mathcal{G}^{\dagger}\right)$. We sample M nodes from this same distribution to use as anchors. We use the metrics namely relative distance error (RE) for distance approximation and F-score (FS) to manifest neighborhood structure preservation. These performance measures are defined as:

$$\text{Relative Error} = \frac{\left\|\hat{D} - D_{\text{true}}\right\|_F}{\left\|D_{\text{true}}\right\|_F}, \quad \text{F-Score} = \frac{2\text{tp}}{2\text{tp} + \text{fp} + \text{fn}} \tag{19}$$

where $\hat{D}$ is the estimated distance matrix based on the PPDA algorithm and D_{true} is the original distance matrix. For each node, we consider their k nearest neighbors. True positive (tp) stands for the case when one of the other nodes is the actual neighbor and the algorithm detects it; false positive (fp) stands for the case when the other node was not a neighbor but the algorithm detects so; and false negative (fn) stands for the case when the algorithm failed to detect an actual neighbor. The F-score takes values in [0, 1] where 1 indicates perfect neighborhood structure recovery [12].

6.2   Experiments for PPDA

Synthetic Datasets: For learning a graph structure, we need the pairwise distance matrix or sample covariance matrix as discussed in Sects. 3.2 and 3.3. However, since we are not sharing the features, we estimate these parameters through PPDA and then learn the graph structure. It's worth noting that we are not recovering the exact features but we ensure to preserve the network structure. For this experiment, we use graph structures and samples generated as described in Sect. 6.1. We examine two cases for learning the embeddings: (i) in the original dimension and (ii) in a lower dimension. Table 1 summarizes the performance of PPDA for various numbers of non-anchors and anchors for the case (i). We consider various dimensions of data when generating samples for each synthetic graph to demonstrate the effectiveness of the algorithm across a broad range of dimensionality. For ER we fix $d = 200$, $d = 400$ for BA, and $d = 700$ for RGG. As stated in Theorem 1, the number of anchors (M) chosen for each dataset is $(d-1)$. For case (i) to learn X_{NA}, we run the Algorithm 2 for tolerance $\epsilon = 0.001$ and maximum epochs $t = 5000$. Note that, for computing F-score, we are considering $k = 10$ nearest neighbors. It can be observed from Table 1 that for $M >> N$ the relative distance error is significantly low, and the F-score is high, indicating that structural information is well preserved. The plot in Fig. 3 illustrates that the actual and estimated distance pairs are situated closely around the line $y = x$ (shown in black), indicating that PPDA provides accurate distance approximation. For case (ii), we generate isotropic gaussian blobs to perform clustering. We created 100 samples per cluster for a total of 5 clusters with $d = 1000$, where d is the number of features for each sample. We divided this dataset by keeping 10% of the samples as anchors and the rest as non-anchors. The embeddings X are then learned in a reduced dimension with $\tilde{d} = 3$. Despite the reduced dimension, PPDA is able to approximate the distances with a relative distance error of 0.1805. Furthermore, utilizing the SGL algorithm described in Sect. 5, we construct a k-component graph with $k = 5$ using the sample covariance matrix for original features as well as the learned embeddings. From Fig. 4, through visual inspection, it can be observed that even without feature sharing we are able to cluster the nodes exactly similar to one with original features, which is notable.

Practical Use Case: For many real-world applications across various domains, such as financial institutes, electronic health record mining, medical data segmentation, and pharmaceutical discovery, sharing data within an institution is permitted. This allows us to have known block entries in the fully incomplete distance matrix K. Let us denote the set of known blocks as $\{S_1, S_2, \cdots, S_R\}$. Hence

Table 1. Results of distance approximation error (RE) and F-score for embeddings in the original dimension

Datasets	Nodes (N)	Anchors (M)	Error (RE)	F-score
ER	100	199	0.0292	0.7503
	500	199	0.0264	0.6192
	1000	199	0.0258	0.5773
BA	100	399	0.0157	0.9406
	500	399	0.0167	0.9147
	1000	399	0.0174	0.8894
RGG	100	699	0.0127	0.9117
	500	699	0.0140	0.8200
	1000	699	0.0148	0.7220

Table 2. Distance approximation with partial inter-user distances known

Datasets	Nodes (N)	RE/FS (r = 0.1)	RE/FS (r = 0.3)	RE/FS (r = 0.5)
ER	100	0.025/0.789	0.021/0.831	0.017/0.878
	500	0.022/0.690	0.017/0.784	0.012/0.865
BA	100	0.014/0.948	0.012/0.957	0.010/0.961
	500	0.015/0.925	0.013/0.934	0.011/0.952
RGG	100	0.012/0.920	0.010/0.932	0.009/0.949
	500	0.013/0.834	0.011/0.864	0.010/0.898

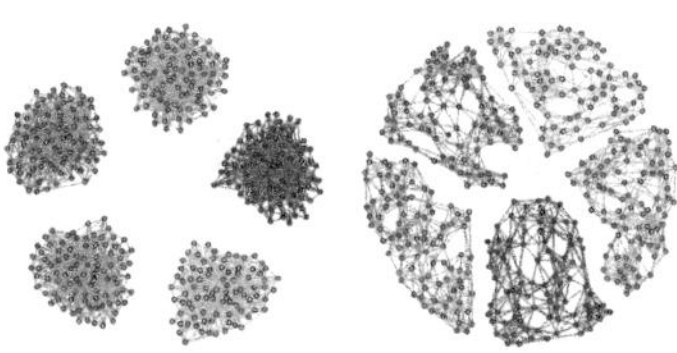

Fig. 4. The image shows the clustered graph estimated using SGL for synthetic dataset with 5 clusters with $d = 3$ for (a) original features (Left) (b) without knowing features (Right)

for the matrix K, K_{ij} is known $\forall \{x_i, x_j\}_{i \neq j} \in S_r$, where $r \in \{1, 2, \cdots, R\}$. To illustrate this use case, we assume that a certain percentage of total entries in the distance matrix for N nodes is available. For the same set of non-anchors and anchors used in case (i) of Sect. 6.2, we examine various ratios r between the known and overall entries of the distance matrix. From Table 2, it is clearly evident that as the number of known entries increases, there is a decrease in relative distance error (RE) and an increase in the corresponding F-score.

6.3   PPDA for Private Graph-Based Clustering

We conduct experiments on the widely adopted benchmark datasets to showcase the performance of private distributed graph learning and graph based clustering. For unsupervised data clustering, we explore the application of multiscale community detection on similarity graphs generated from data. Specifically, we focus on how to construct graphs that accurately reflect the dataset's structure in order to be employed inside a multiscale graph based clustering framework. We have evaluated our framework on 8 benchmark datasets from the UCI repos-

322   N. Malik et al.

Table 3. Dataset Statistics for 8 benchmark datasets from UCI repository

Data	Nodes (N)	Dim (d)	Classes (c)
Iris	150	4	3
Glass	214	9	6
Wine	178	13	3
Control Chart (CC)	600	60	6
Parkinsons	195	22	2
Vertebral	310	6	3
Breast tissue (BT)	106	9	6
Seeds	210	7	3

Table 4. Performance evaluation of Graph-based clustering with and without data sharing

Data	FS	NMI		ARI	
		Non-Private	Private	Non-Private	Private
Iris	0.8659	0.892	0.785	0.913	0.718
Glass	0.9337	0.354	0.341	0.199	0.173
Wine	0.9934	0.377	0.377	0.381	0.381
CC	0.8271	0.811	0.811	0.620	0.620
Parkinsons	0.9880	0.016	0.016	0.055	0.055
Vertebral	0.7612	0.468	0.459	0.338	0.331
BT	0.9938	0.341	0.341	0.162	0.162
Seeds	0.9745	0.701	0.619	0.739	0.549

itory (Table 3 summarizes dataset statistics) [11]. All datasets have ground truth labels, which we use to validate the results of the clustering algorithm.

Experimental Details: For every dataset, we assume that each sample belongs to one user in a distributed setting. As it is more efficient to sample anchors from a similar distribution as of original data, we split the given datasets keeping 10% of samples as anchors and rest as non-anchors. With only $(d-1)$ anchors, we are able to preserve neighborhood structure ensuring privacy, as is evident from the F-score presented in Table 4. We then employ the Constrained Laplacian Rank (CLR) algorithm for graph-based clustering. To construct the initial graph from the data we estimate the Adjacency matrix by solving the optimization formulation given in equation (17) as discussed in Sect. 5. To validate the graph construction under clustering, we compute Normalised Mutual Information (NMI) [34] and Adjusted Rand Index (ARI) [16] as quality metrics. Results obtained with attribute sharing are considered benchmarks. To set the number of clusters, we use the number of classes in the ground truth (c) as input to the CLR algorithm.

Table 4 provides strong evidence that the validation metrics are comparable whether the original data was shared or the attributes were kept private. These results establish a strong foundation for privacy-preserving graph-based clustering. Figure 5 shows the graph constructed along with clusters using raw data and with our proposed framework. It can be seen that the clustering performance is almost the same for both cases, which is remarkable.

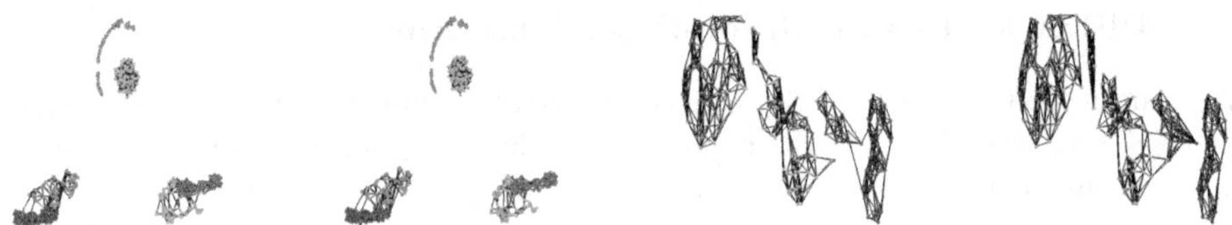

Fig. 5. Graph Clustering performed using CLR: Control chart (left): Non-private and Private; Seeds (right): Non-Private and Private.

6.4   Experiment on Sensitive Benchmark Data

Additionally, we also evaluated our framework on a privacy-critical medical dataset. We consider the RNA-Seq Cancer Genome Atlas Research Network dataset (PANCAN) [37] available at UCI repository [11]. This dataset is a random extraction of gene expressions of patients having different types of tumor namely: breast carcinoma (BRCA), kidney renal clear-cell carcinoma (KIRC), lung adenocarcinoma (LUAD), colon adenocarcinoma (COAD), and prostate adenocarcinoma (PRAD). The data set consists of 801 labeled samples with 20531 genetic features. This dataset's objective is to group the nodes into tumor-specific clusters based on their genetic characteristics. We take 10% of the total samples as anchors and use Algorithm 2 to learn the embeddings with $\tilde{d} = 500$. Figure 6 shows the clustered graph built using CLR with those cancer types being labeled with colors lightcoral, gray, blue, red, and chocolate respectively. The clustering performance is highly comparable with NMI being 0.9943 and 0.9841, ARI being 0.9958 and 0.9875 for shared and private data respectively.

6.5   Comparison with SOTA

The comparison of our proposed solution on the end goal can be made using the setting of Federated Learning.

Comparison with Federated Learning Framework: It is not an alternative to Federated Graph Machine Learning (FGML) but can be used where a graph structure is needed. Our approach sends the distance function to the server, preserving privacy without transmitting raw client features. This method can also be integrated into FGML settings, where the graph and embeddings are privately learned using PPDA

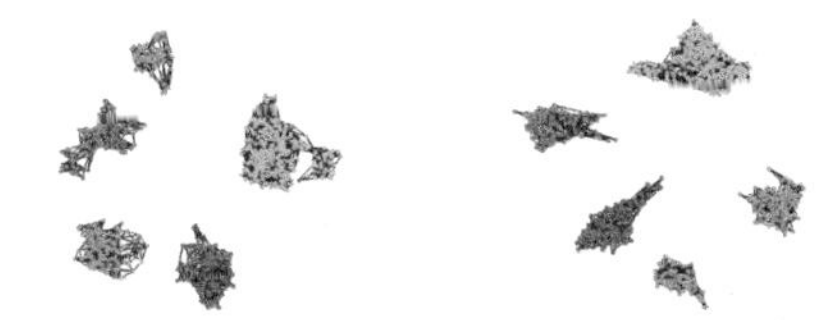

Fig. 6. The image shows the clustered graph estimated using CLR for the PANCAN dataset with the following metrics for non-private (left) /private (right) graphs: NMI = 0.9943/0.9841 and ARI = 0.9958/0.9875

and used for downstream ML tasks. Unlike traditional FGML, which relies on differential privacy, our method uses standard distance measures for graph learning, avoiding additional privacy mechanisms and enabling computation without revealing private data. Instead of sending raw data to the server, the traditional Federated Learning setting applies Local Differential Privacy (LDP) to the parameters. It involves two hyperparameters, i.e. δ which is the clipping threshold and λ is the Laplacian noise parameter. We apply LDP on the raw features and present the NMI metric used in the paper to evaluate the performance of graph-based clustering with different δ and λ values for two benchmark datasets. We observed that the use of a higher value of λ can result in a decrease in performance. This is because increasing the value of λ results in stronger Laplace noise being added to the raw features, which ultimately leads to less

324   N. Malik et al.

Table 5. NMI metrics for different λ values

Dataset	δ	Without LDP	NMI for different λ		
			$\lambda = 0.0025$	$\lambda = 0.005$	$\lambda = 0.01$
Iris	0.2	0.892	0.8106	0.6946	0.6183
	0.1	-	0.7169	0.6138	0.5060
Seeds	0.2	0.701	0.4367	0.3605	0.1828
	0.1	-	0.3383	0.0430	0.0112

Table 6. Privacy budget (ϵ) for different λ

δ	Privacy budget ϵ for different λ		
	$\lambda = 0.0025$	$\lambda = 0.005$	$\lambda = 0.01$
0.2	160	80	40
0.1	80	40	20

accurate performance. Additionally, by considering the upper bound on the privacy budget, which is expressed as $\epsilon = \frac{2\delta}{\lambda}$, we can see from Table 6 that as λ increases and δ decreases, the privacy budget ϵ becomes smaller, indicating better privacy protection but with a significant impact on utility. On the other hand, the performance of the proposed solution is lesser affected by noise as can be observed from Table 5 and Table 7.

6.6 Privacy Guarantees of the Proposed Method Against Adversarial Attacks

Case 1: Can the attacker estimate exact features?
For the proposed framework, the only parameter that is sent from the client to the server is the client-to-anchor distances. In the worst case, the attacker can estimate the pairwise similarities among the nodes but it cannot estimate the exact client attributes as guaranteed by Theorem 1.

Case 2: Can attackers collude to degrade the performance?
We present a threat model where we consider that some percentage of the clients are malicious. These clients send noisy client-to-anchor distances instead of sending actual parameters, thereby aiming to degrade the overall performance. For empirical evaluation, we consider Iris and Seeds Dataset. We randomly sampled some percentage of clients as attackers and added some constant noise to each client-to-anchor distance for these attackers. We evaluate the performance metrics like F-score (FS), Relative Distance Error (RE), and NMI for graph-based clustering introduced in the paper. The overall performance is presented in Table 7. It can be observed that as the number of attackers increases, there is a corresponding decrease in the accuracy of the distance estimation and performance of graph-based clustering.

6.7 Performance on Moderate and Large Graphs

To demonstrate the performance of the proposed solution on larger datasets, we evaluated our framework on some benchmark datasets from the UCI repository [1], namely Smartphone Dataset for Human Activity Recognition (HAR). Each instance belongs to one of the activity from (standing, sitting, laying, walking, walking upstairs, walking downstairs) serving as ground truth labels for the downstream task. There are 5744 instances with 561 attributes in moderate size data, while 10,299 instances with 561 attributes. Using the preprocessed

Table 7. Performance of different metrics under attack

Dataset	Metric	Percentage of Attackers				
		0%	5%	10%	20%	50%
Iris	FS	0.8659	0.8403	0.8104	0.7000	0.6167
	RE	0.0425	0.0735	0.0929	0.1528	0.1828
	NMI	0.7850	0.7339	0.7148	0.7087	0.5796
Seeds	FS	0.9745	0.9390	0.9038	0.7902	0.6589
	RE	0.0102	0.1251	0.1760	0.2471	0.3702
	NMI	0.6190	0.6140	0.3998	0.3571	0.3153

train-test split of each dataset, we consider 4261 and 7352 nodes as non-anchors respectively, while choosing the maximum possible number of anchors (i.e. 560) from the respective test datasets. Then we perform the node classification using GCN on the privately learned graph using the proposed solution as well as the graph using raw data. The performance evaluation for moderate and large datasets are shown in Table 8.

Table 8. Performance Comparison of GNN on HAR Dataset

Dataset	GNN Accuracy		RE	F-Score
	Non-Private	Private		
HAR (Moderate)	0.8566	0.8143	0.0634	0.8768
HAR (Large)	0.9504	0.9313	0.0703	0.8468

6.8   Experimental Validation of Theorem 1

It is not possible to recover the features exactly if we choose the number of anchors less than d. To demonstrate this, we created a curated dataset that includes certain private characteristics such as age, as well as categorical features like gender, occupation, etc. We produced data for 5 different clients (Table 9), each with 5 attributes, and we also created 4 reference samples as anchors. We then apply the proposed PPDA framework to learn the embeddings. Table 10 shows the embeddings obtained by proposed method. It is evident that the embedding generated by PPDA does not expose the sensitive details of the original data, thus preventing any breach of privacy.

6.9   Case Studies and Privacy Implications of Estimating Distance Through Anchors in Real-World Scenarios

Taking the subset of original samples as anchors is just for the empirical evaluation as it is crucial to choose anchors from the same distribution as of underlying

Table 9. Original Data

Clients	x_1	x_2	x_3	x_4	x_5
C1	1500	25	3	1	10
C2	2000	35	2	0	15
C3	1745	28	2	0	18
C4	1620	32	1	1	13
C5	1200	45	3	1	12

Table 10. Embeddings obtained by PPDA

Clients	x_1	x_2	x_3	x_4	x_5
C1	80.630	26.896	13.795	96.939	-39.321
C2	-189.837	-28.557	-61.648	-272.603	141.184
C3	-58.064	2.598	-26.164	-79.338	49.985
C4	-5.492	-8.256	6.935	12.915	-12.891
C5	210.016	14.741	75.140	328.450	-172.165

data. We believe that we only need to know the kind of data for the task to be performed underneath. For example, for the experiment demonstrated using Animal's Dataset in Fig. 7 we do not utilize the original samples. Instead, as all the features were having binary values, we randomly generated anchors having binary features. In fact, for *real-world applications like distributed healthcare settings, hospitals may want to collaborate to learn a Patient Similarity Network to build a patient diagnosis or prediction model while keeping patient data private. Each hospital can provide a subset of its patient data (not sensitive) as reference samples to the server. The server can use these reference samples as anchors and distribute them to the participating hospitals. Similarly, financial institutions may want to collaborate to detect fraudulent transactions based on graph learning. Each institution can provide a subset of its transaction data as reference samples to the server. The server can use these reference samples as anchors and distribute them to the participating institutions.*

6.10 Additional Experiment to Demonstrate PPDA

We consider the animal's dataset [21, 28] for privacy-preserving distributed graph learning. Each animal is represented as a node in a graph and the edges between nodes indicate similarities between animals based on answers to 102 questions such as *"is warm-blooded?"* and *"has lungs?"*. Using the proposed PPDA algorithm, we are able to preserve the neighborhood similarity with relative distance error of 0.1238 and F-score of 0.9402. Figure 7 shows the results of estimating the graph of the animals dataset using the SGL algorithm with original data and without sharing the data. The results are evaluated through visual inspection and it is expected that similar animals such as (ant, cockroach), (bee, butterfly), and (trout, salmon) are clustered together. It is clearly evident from the figure that PPDA is able to maintain the structure similarity even without sharing the data.

6.11 Performance on Node Classification

In this setting where feature sharing is prohibited, one can learn graph and embedding through PPDA as GNN requires graph and features as input. We then apply GNN for node classification on this privately learned graph through

Fig. 7. The image depicts the graph estimated for animals dataset using SGL for (left) original features and (right) without revealing features

our proposed method and benchmark this against the ideal setting where raw features are available. We adopted a few datasets used in the paper to showcase the performance of the node classification task. Considering the split of 80:20 for training and testing nodes, with 2-layer GCN architecture, the average accuracy of node classification over 10 runs is shown in the Table 11

Table 11. Comparison of Performance Metrics for Different Datasets

Dataset	Iris	Seeds	Control Chart
Non-Private	0.9703 ± 0.027	0.8684 ± 0.097	0.5361 ± 0.1184
Private	0.9667 ± 0.045	0.8578 ± 0.102	0.4989 ± 0.043

6.12 Further Analysis:

Noisy Parameters: To accomplish this, random noise uniformly distributed in the range $[0, c]$ is added to each client to anchor distance measurements that are to be sent to the server. Here c is set as $0.1, 0.3, 0.5$, and 0.7 respectively to examine different levels of noise. As shown in Fig. 8, increasing natural noise can lead to distortion in the original structure of the data in turn impacting the evaluation metrics. It can be observed that F-score (FS) decreases while Relative Distance Error (RE) increases with the increasing level of noise, which is to be expected.

Computational Efficiency: While increasing M tends to improve results, it also introduces a trade-off between privacy guarantees and approximation quality. For optimal privacy preservation, M should be less than $d-1$, with $M = d-1$ being the ideal setting. Table 12 summarizes the experimental results using the Random Geometric Graph (RGG) model for varying numbers of anchors in both low and high-dimensional settings. Thus, for low-dimensional data, $d-1$ anchors are optimal due to manageable computational costs. For high-dimensional data,

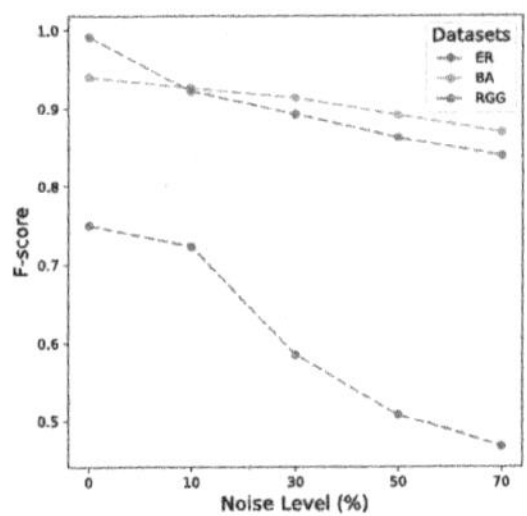

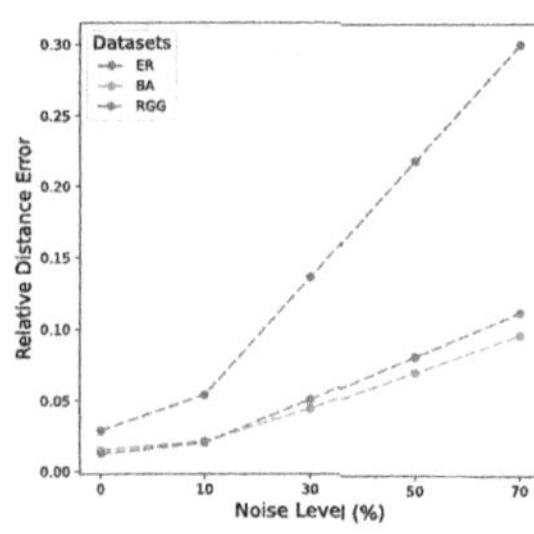

Fig. 8. Analysis of performance metrics for (left) FS and (right) Relative Distance Error (RDE) with noise-induced distance parameters for three synthetic data (ER, BA, and RGG) with different noise levels (0%, 10%, 30%, 50%, and 70%).

Table 12. Comparison of performance metrics across different parameter settings and time (sec)

Nodes (N)	Dim (d)	Anchors (M)	FS	RE	time (t)
500	100	99	0.79	0.02	16
		90	0.78	0.03	13
		50	0.64	0.04	5
		30	0.51	0.052	2
500	700	699	0.82	0.014	660
		350	0.78	0.019	312
		100	0.72	0.023	30
		50	0.65	0.038	10

using $M < d - 1$ anchors reduces computational costs, although using too few anchors significantly decreases performance, illustrating a trade-off between performance and computational cost.

Time Complexity: The time complexity of our proposed method, PPDA using anchored MDS for distance and graph embedding estimation, is $O((N + M)M(d^2)\log(M + N)d)$. The time complexity for calculating the incomplete distance matrix (Sect. 4.2) is $O(M^2d + MNd)$.

7   Conclusion

In this paper, we presented the novel privacy-preserving distance approximation framework, PPDA for distributed graph-based learning. By utilizing only the pairwise distance between clients and some set of anchors, PPDA is able to estimate the inter-client distances privately which is further used for graph learning. We also performed downstream task like graph-based clustering without accessing the actual features. Extensive experiments with both real and synthetic datasets demonstrate the efficacy of the proposed PPDA framework in preserving the structural properties of the original data. As an application to a privacy-sensitive dataset, the experiment on PANCAN produced promising results. It is, to our knowledge, the first work that can learn the graph structure for both analyzing smooth signals and building a probabilistic graphical model while respecting the privacy of the data.

References

1. Aggarwal, C.C., Yu, P.S.: A general survey of privacy-preserving data mining models and algorithms. In: Aggarwal, C.C., Yu, P.S. (eds.) Privacy-Preserving Data Mining, pp. 11–52. Springer US, Boston, MA (2008). https://doi.org/10.1007/978-0-387-70992-5_2

2. Asharov, G., Halevi, S., Lindell, Y., Rabin, T.: Privacy-preserving search of similar patients in genomic data. Cryptol. ePrint Arch. () (2017)

3. Aydin, F., Aysu, A.: Leaking secrets in homomorphic encryption with side-channel attacks. J. Cryptographic Eng., 1–11 (2024)

4. Aziz, M., Anjum, M.M., Mohammed, N., Jiang, X.: Generalized genomic data sharing for differentially private fedcrated learning. J. Biomed. Inform. **132**, 104113 (2022). https://doi.org/10.1016/j.jbi.2022.104113, https://www.sciencedirect.com/science/article/pii/S1532046422001290

5. Belkin, M., Niyogi, P., Sindhwani, V.: Manifold regularization: a geometric framework for learning from labeled and unlabeled examples. J. Mach. Learn. Res. **7**(11) (2006)

6. Chang, H., Ando, H.: Privacy-preserving data sharing by integrating perturbed distance matrices. SN Comput. Sci. **1**(3), 1–10 (2020)

7. De Leeuw, J.: Convergence of the majorization method for multidimensional scaling. J. Classif. **5**(2), 163–180 (1988)

8. De Leeuw, J.: Applications of convex analysis to multidimensional scaling (2005)

9. Di Franco, C., Bini, E., Marinoni, M., Buttazzo, G.C.: Multidimensional scaling localization with anchors. In: 2017 IEEE International Conference on Autonomous Robot Systems and Competitions (ICARSC), pp. 49–54. IEEE (2017)

10. Dosovltskiy, A., Brox, T.: Inverting visual representations with convolutional networks. In: Proceedings of the IEEE Conference on Computer Vision and Pattern Recognition, pp. 4829–4837 (2016)

11. Dua, D., Graff, C.: UCI machine learning repository (2017). http://archive.ics.uci.edu/ml

12. Egilmez, H.E., Pavez, E., Ortega, A.: Graph learning from data under Laplacian and structural constraints. IEEE J. Sel. Top. Signal Process. **11**(6), 825–841 (2017)

13. El Kassabi, H.T., Serhani, M.A., Navaz, A.N., Ouhbi, S.: Federated patient similarity network for data-driven diagnosis of COVID-19 patients. In: 2021 IEEE/ACS 18th International Conference on Computer Systems and Applications (AICCSA), pp. 1–6. IEEE (2021)

14. Friedman, J., Hastie, T., Tibshirani, R.: Sparse inverse covariance estimation with the graphical lasso. Biostatistics **9**(3), 432–441 (2008)

15. Guo, Q., Nabokov, D., Suvanto, E., Johansson, T.: Key recovery attacks on approximate homomorphic encryption with non-worst-case noise flooding countermeasures. In: Usenix Security (2024)

16. Hubert, L., Arabie, P.: Comparing partitions. J. Classif. **2**(1), 193–218 (1985)

17. Kalofolias, V.: How to learn a graph from smooth signals. In: Artificial Intelligence and Statistics, pp. 920–929. PMLR (2016)

18. Khan, U.A., Kar, S., Moura, J.M.: Distributed sensor localization in random environments using minimal number of anchor nodes. IEEE Trans. Signal Process. **57**(5), 2000–2016 (2009)

19. Kipf, T.N., Welling, M.: Semi-supervised classification with graph convolutional networks. arXiv preprint arXiv:1609.02907 (), (2016)

20. Kumar, S., Ying, J., de M. Cardoso, J.V., Palomar, D.P.: A unified framework for structured graph learning via spectral constraints. J. Mach. Learn. Res. **21**(22), 1–60 (2020). http://jmlr.org/papers/v21/19-276.html
21. Lake, B., Tenenbaum, J.: Discovering structure by learning sparse graphs (2010)
22. Li, N., Li, T., Venkatasubramanian, S.: t-Closeness: privacy beyond k-anonymity and l-diversity. In: 2007 IEEE 23rd International Conference on Data Engineering, pp. 106–115. IEEE (2006)
23. Machanavajjhala, A., Kifer, D., Gehrke, J., Venkitasubramaniam, M.: l-Diversity: privacy beyond k-anonymity. ACM Trans. Knowl. Disc. Data (TKDD) **1**(1), 3–es (2007)
24. Mahendran, A., Vedaldi, A.: Understanding deep image representations by inverting them. In: Proceedings of the IEEE Conference on Computer Vision and Pattern Recognition, pp. 5188–5196 (2015)
25. Ng, A., Jordan, M., Weiss, Y.: On spectral clustering: analysis and an algorithm. In: Advances in Neural Information Processing Systems, vol. 14 (2001)
26. Nie, F., Wang, X., Jordan, M., Huang, H.: The constrained Laplacian rank algorithm for graph-based clustering. In: Proceedings of the AAAI Conference on Artificial Intelligence, vol. 30 (2016)
27. Nobre, I.C.M., El Gheche, M., Frossard, P.: Distributed graph learning with smooth data priors. In: ICASSP 2022-2022 IEEE International Conference on Acoustics, Speech and Signal Processing (ICASSP), pp. 5852–5856. IEEE (2022)
28. Osherson, D.N., Stern, J., Wilkie, O., Stob, M., Smith, E.E.: Default probability. Cogn. Sci. **15**(2), 251–269 (1991)
29. Pullen, M.: Adversarial attacks on encrypted machine learning models. Int. J. Multidisc. Innov. Res. Methodol. **3**(2), 127–134 (2024). ISSN: 2960-2068
30. Qi, T., Wang, H., Huang, Y.: Towards the robustness of differentially private federated learning. In: Proceedings of the AAAI Conference on Artificial Intelligence, vol. 38, pp. 19911–19919 (2024)
31. Rane, S., Boufounos, P.T.: Privacy-preserving nearest neighbor methods: comparing signals without revealing them. IEEE Signal Process. Mag. **30**(2), 18–28 (2013)
32. Rivest, R.L., Adleman, L., Dertouzos, M.L., et al.: On data banks and privacy homomorphisms. Found. Secure Comput. **4**(11), 169–180 (1978)
33. Smith, A., Thakurta, A., Upadhyay, J.: Is interaction necessary for distributed private learning? In: 2017 IEEE Symposium on Security and Privacy (SP), pp. 58–77. IEEE (2017)
34. Strehl, A., Ghosh, J.: Cluster ensembles—a knowledge reuse framework for combining multiple partitions. J. Mach. Learn. Res. **3**(Dec), 583–617 (2002)
35. Sweeney, L.: k-anonymity: a model for protecting privacy. Internat. J. Uncertain. Fuzziness Knowl.-Based Syst. **10**(05), 557–570 (2002)
36. Von Luxburg, U.: A tutorial on spectral clustering. Stat. Comput. **17**(4), 395–416 (2007)
37. Weinstein, J.N., et al.: The cancer genome atlas pan-cancer analysis project. Nat. Genet. **45**(10), 1113–1120 (2013)
38. Wilson, R.C., Hancock, E.R., Pekalska, E., Duin, R.P.: Spherical and hyperbolic embeddings of data. IEEE Trans. Pattern Anal. Mach. Intell. **36**(11), 2255–2269 (2014)
39. Zhou, Y.: Structure learning of probabilistic graphical models: a comprehensive survey (2011). https://doi.org/10.48550/ARXIV.1111.6925, https://arxiv.org/abs/1111.6925

40. Zhu, L., Liu, Z., Han, S.: Deep leakage from gradients. In: Advances in Neural Information Processing Systems, vol. 32 (2019)
41. Zhu, X., Ghahramani, Z., Lafferty, J.D.: Semi-supervised learning using gaussian fields and harmonic functions. In: Proceedings of the 20th International Conference on Machine Learning (ICML-03), pp. 912–919 (2003)

MonoTCM: Semantic-Depth Fusion Transformer for Monocular 3D Object Detection with Token Clustering and Merging

Changyu Zeng[1,2,5], Zimu Wang[1,2], Jimin Xiao[1], Anh Nguyen[2], Kaizhu Huang[3], Wei Wang[1(✉)], and Yutao Yue[1,4,5(✉)]

[1] Xi'an Jiaotong-Liverpool University, Suzhou, China
{Changyu.Zeng17,Zimu.Wang19}@student.xjtlu.edu.cn,
{Jimin.Xiao,Wei.Wang03}@xjtlu.edu.cn
[2] University of Liverpool, Liverpool, UK
Anh.Nguyen@liverpool.ac.uk
[3] Duke Kunshan University, Suzhou, China
kaizhu.huang@dukekunshan.edu.cn
[4] The Hong Kong University of Science and Technology (Guangzhou), Guangzhou, China
yutaoyue@hkust-gz.edu.cn
[5] Institute of Deep Perception Technology, JITRI, Guangzhou, China

Abstract. Monocular 3D object detection presents significant challenges due to the inherent absence of depth and geometric information, rendering it more complex than 2D detection. This paper introduces MonoTCM, a Semantic-Depth Fusion Transformer that leverages a Token Clustering and Merging (TCM) module to enhance the efficiency and accuracy of monocular 3D object detection. The TCM module aggregates multi-scale grid-based tokens into clustering-based tokens, dynamically adjusting their shapes and sizes based on local density and distance metrics. This allows for finer granularity in critical areas while consolidating less informative regions. The aggregated tokens are subsequently decomposed into semantic and depth features, processed through dedicated transformer-based encoders, and integrated using a semantic-depth fusion decoder modeled after DETR. This approach enhances the model's ability to capture implicit global geometric information and provides a cost-effective solution for real-time intelligent driving applications. Experimental results demonstrate the superiority of MonoTCM in enhancing detection performance compared to other advanced methods, highlighting its potential to advance the field of monocular 3D object detection.

Keywords: Computer Vision · Monocluar 3D Object Detection · Depth Estimation

1 Introduction

With the robust development of autonomous driving and deep learning technologies, increasing attention of scientific research has been directed towards 3D

<table>
<tr><td>(a) Input Monocular Image</td><td>(b) Grid-based Tokens</td><td>(c) Clustering-based Tokens</td></tr>
</table>

Fig. 1. Demonstration of (a) input monocular image, (b) grid-based tokens, and (c) clustering-based tokens. The grid-based tokens exhibit a uniform square shape and are treated with equal importance. In contrast, the clustering-based tokens, generated by the Token Clustering and Merging (TCM) module, strategically assigns finely granulated tokens to critical areas such as vehicles contours, while coarser tokens are allocated to regions with minimal detail, such as the background.

detection in traffic scenes. Prior studies have yielded promising outcomes by utilizing multiple cameras [13,17] to provide multiview perspectives and LiDAR [4,25] for point cloud depth mapping. However, the synchronization of device communication and the real-time processing of large-scale datasets for intelli gent driving remain significant challenges. As a result, the research community is gravitating towards developing cost-effective and efficient 3D object detection methods to minimize the reliance on numerous devices, models, and extensive datasets. In this context, monocular 3D detection emerges as a particularly promising solution.

Nevertheless, the inherent difficulty in 3D detection, stemming from the scarcity of depth and geometric information, presents a more formidable challenge compared to its 2D counterparts. Existing techniques for monocular 3D detection are broadly classified into keypoint prediction [12,16], pseudo-LiDAR [4,25], and purely visual [7,19,25,31] approaches. However, the majority of these methods adhere to the 2D detection paradigm, which involves identifying the center of an object and subsequently aggregating the adjacent visual features. This strategy confines the model's attention to local details and fails to adequately encapsulate global context. Moreover, the tokens (image patches) generated based on traditional Convolution Neural Network (CNN) and Vision Trans former (ViT) are uniformly square shaped, as depicted in Fig. 1 (b). This suggests that each token is assigned equal significance, which contracts with human visual perception, where foreground objects should usually be prioritized over the background.

To address the aforementioned issues, we propose MonoTCM, which employs a Token Clustering and Merging (TCM) module to selectively aggregate grid-based tokens into clustering-based tokens for monocular 3D detection (see Fig. 1 (c)). Cluster based tokens are dynamically generated by merging regions with similar characteristics into a single token. Consequently, the tokens produced vary in shape and size according to local density and distance metrics. Typically, critical segments of the image are represented by more refined tokens, whereas redundant or less informative areas are consolidated into a single, coarser token.

Following the TCM module, the aggregated representations are then decoupled into semantic and depth features, which capture implicit geometric information from two distinct perspectives. These features are subsequently processed through their corresponding transformer-based encoders. A semantic-depth fusion decoder, modeled based on DEtection TRansformer (DETR) [3], is then utilized to sequentially integrate the semantic and depth information. A set of object queries is injected into the depth cross-attention layers to extract geometric hints and are fused with the semantic embedding to adeptly merge the two types of representations. This strategy enables MonoTCM to move beyond a focus on restricted local regions, instead encapsulating and integrating long-range, depth-guided visual features within the object queries for the prediction of both 2D and 3D properties. Taking single-view images as input, the proposed MonoTCM achieve remarkable performance on the test set of the KITTI dataset [8], improving upon the second-best MonoDETR by +0.33%, +0.49%, and +0.78% across the three difficulty levels of *BEV* detection.

The contributions of this paper are summarized as follows:

- We propose MonoTCM, a model that targets salient regions by consolidating insignificant tokens in the image, achieving high precision and reduced latency in monocular 3D object detection.
- We introduce a semantic-depth fusion decoder that integrates depth cues and semantic embedding into object queries in a sequential manner, capturing a comprehensive global scene-level representation derived from depth geometry and visual characteristics.
- MonoTCM achieves notable success on the monocular KITTI benchmark and outperforms other advanced methods.

2   Related Work

2.1   Monocular 3D Object Detection

Monocular 3D object detection is a task that involves identifying and localizing objects within a three-dimensional space using a single camera image. Initial approaches rely on hand-crafted features and geometric constraints but struggled with depth estimation. Recently, the advent of deep learning has enabled more sophisticated feature extraction and the development of end-to-end frameworks that are able t o predict both 2D and 3D properties simultaneously. Deep3DBox [21] is the pioneer in introducing geometric constraints for regression of stable 3D object properties. M3D-PRN [1] proposes a depth-aware convolution allowing the development of location-specific features and the understanding of 3D scene. Certain tasks, such as semantic segmentation and keypoint prediction, have been shown to aid in single-view 3D detection. Methods like SMOKE [16] and RTM3D [12] utilize a keypoint-based approach to estimate 3D object properties by predicting keypoints on the object and employing geometric constraints to infer 3D bounding boxes. Furthermore, the introduction of transformer-based architectures has further pushed the boundaries of monocular 3D detection by enabling more effective context aggregation and feature representation. MonoDTR [10] proposes the enhancement of depth-aware features with global integration of context features to improve 3D reasoning. MonoDETR [30] eliminates center-directed constraints and captures spatial cues at the scene level using a depth-guided transformer. MonoATT [32] leverages a novel ViT with heterogeneous tokens of varying shapes to facilitate dynamic monocular detection. Despite these advancements, monocular 3D object detection remains a challenging task due to the computational inefficiencies of traditional fine-grained grid-based approaches and their inability to perform global sensing. To tackle this issue, MonoTCM adopts the token clustering and merging technique to preserve the crucial regions and achieve semantic and depth feature aggregation via a cross-attention based fusion transformer, thereby enhancing the performance in monocular 3D detection.

2.2   Feature Aggregation Networks

Feature aggregation has been widely used in various domains within the field of computer vision. During its infancy phase, techniques such as Feature Pyramid Networks (FPNs) [14] and the Path Aggregation Network (PANet) [15] facilitated the aggregation of multi-scale and hierarchical features, resulting in marked enhancements in object detection. Subsequently, graph-based methodologies, such as PointNet++ [22] and DGCNN [28], have substantially elevated performance in point cloud segmentation through strategies like farthest point sampling and the aggregation of K-nearest neighbor features. The realm of self-supervised learning has also frequently adopted feature aggregation techniques. For instance, SimCLR [5] enhances consensus between augmented perspectives of input imagery, effectively gathering similar characteristics within the embedding space. Moreover, the practice of feature integration is widespread in multi-modal domains, with models such as CLIP [24] facilitating the clustering and merging of features across different modalities to support tasks like image captioning and cross-modal retrieval. Recently, aggregation strategies have been integrated into transformer architectures. The Tcformer [29], for example, incorporates progressive clustering with adaptable geometries during the feature extraction phase. MonATT [32] dynamically allocates more refined tokens to salient image regions to boost monocular 3D object detection. Our proposed MonoTCM synthesizes the aforementioned methodologies by computing the distance and significance scores between features at a multi-level patch token stage, preserving crucial tokens while aggregating those with lower importance score.

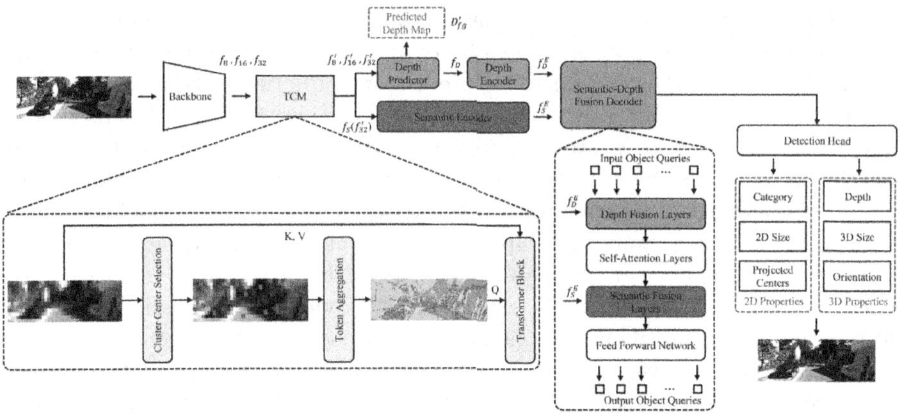

Fig. 2. Overall architecture for MonoTCM. The input monocular image is first processed by a backbone to yield multi-scale features, which are then fed into the Token Clustering and Merging (TCM) module for token aggregation. The resulting aggregated features are split into semantic and depth components, each passing through a dedicated encoders. The depth representation also undergoes a depth predictor block to estimate the depth of the foreground in the image. The refined representations from the encoders are sequentially integrated with object queries in the semantic-depth fusion decoder. The output queries, containing global semantic and depth information, are then passed to the detection head to predict the 2D and 3D properties of objects within the image.

3   Method

Figure 2 shows the overall architecture of MonoTCM. The input monocular image initially undergoes multi-scale feature extraction via a backbone network, followed by feature aggregation through the TCM module. Subsequently, semantic and depth encoders process these features, encoding them in terms of geometric structure and visual representation, respectively. The semantic-depth fusion decoder then sequentially integrates these encoded features, and the detection head outputs the 3D and 2D properties of objects. The details of each module will be described in the following subsections.

3.1   Multi-level Feature Extraction

Given a single view image $I \in \mathbb{R}^{H \times W \times 3}$ as input, where H and W refer to height and width, respectively, we adopt the classic ResNet-50 network [9] as the backbone for multi-level feature extraction. Subsequently, we collect the outputs of last three stages, f_8, f_{16}, and f_{32}, where the number donates the down-sampled ratio, as the input for the following TCM module.

3.2   Token Clustering and Merging

As depicted in Fig. 2, inspired by [29], the proposed Token Clustering and Merging (TCM) module comprises three key processes: **Cluster Center Selection**, **Token Aggregation**, and **Token Interaction**. Initially, the original grid-based tokens are grouped into clusters, with each cluster center being determined through the application of density and distance metrics. Subsequently, the aggregation will be conducted around the established cluster's center to merge similar regions into a single token. Finally, the aggregated tokens interact with the original grid-based ones through a cross-attention transformer.

Cluster Center Selection. In this phase, the local density and distance indicator metrics are adopted to determine the center location of each cluster. Initially, a modified version of density peak clustering based on k nearest neighbors (DPC-KNN) is utilized to calculate the token's local density ρ:

$$\rho_i = exp(-\frac{1}{k} \sum_{t_j \in KNN(t_i)} ||t_i - t_j||_2^2), \tag{1}$$

where $KNN(t_i)$ refers to the k-nearest neighbors associated with the token t_i.

Subsequently, the distance indicator σ is calculated as the smallest Euclidean distance between the arbitrarily chosen token and the token exhibiting the highest local density. In the case of the token with the highest local density, σ is defined as the maximum distance between this token and all other tokens.

$$\sigma_i = \begin{cases} min_{j:\rho_j>\rho_i}||x_i - x_j||, & \rho_j > \rho_i \\ max_j||x_i - x_j||, & otherwise \end{cases}. \tag{2}$$

Then, the product of $\rho \times \sigma$ is treated as the importance score (IS) that serves as a criterion for the selection of cluster centers. Tokens with higher scores are considered more probable candidates for cluster centers. In addition, tokens that are not designated as centers are then assigned to the nearest cluster center based on their feature distances.

Token Aggregation. To better preserve the feature in a cluster, we leverage the important score (IS) as the indicator for the token aggregation. Concretely, the tokens belonging to the same cluster are merged by averaging the internal tokens with the guidance of the importance scores:

$$y_i = \frac{\sum_{i \in C_i} e^{IS_j} x_j}{\sum_{j \in C_i} e^{IS_j}}, \tag{3}$$

where C_i denotes the set of the i-th cluster. x_j and IS_j represent the initial features of the individual tokens and their relative importance scores, respectively. The merged token y_i corresponds to the consolidated representation for the tokens in the i-th cluster.

Token Interaction. To facilitate the long range association between the aggregated tokens and original tokens, we employ the merged tokens as the queries Q and the original tokens as the key K and value V when processing through the transformer blocks. To enhance the contribution of crucial tokens, such as those corresponding to foreground regions, we incorporate the importance score IS as a bias term in the computation of the attention matrix:

$$Attention(Q, K, V) = softmax(\frac{QK^T}{\sqrt{d_k}} + IS)V, \tag{4}$$

where d_k represents the dimension of the query vectors in terms of the number of channels.

Through the incorporation of the importance score, the TCM module is able to focus on the more salient components of images. The overall process of TCM module could be expressed as follows:

$$f_8', f_{16}', f_{32}' = TCM(f_8, f_{16}, f_{32}), \tag{5}$$

the dimensions of multi-scale input and output (i.e., f_i and f_i'), is one-to-one correspondence.

3.3 Feature Decoupling and Depth Prediction

After passing through the TCM module, the resultant merged multi-level feature is divided into two distinct components: semantic and depth features. The semantic feature f_S (f_{32}') is the result of the last stage, preserving the deeper and more global visual appearance of the input image. And depth feature f_D is synthesized through the aggregation of three multi-level f_8', f_{16}', and f_{32}'. Concretely, f_8' and f_{32}' undergo resizing operations through convolutional layers and bilinear interpolation to match the dimensions of f_{16}', followed by their fusion via element-wise addition. Subsequently, two convolutional layers with a kernel size of 3 are employed to extract a fine-grained depth feature, which encapsulates hierarchical geometric information.

Then, a 1×1 convolutional layer is utilized to predict the foreground depth map $D_{fg} \in \mathbb{R}^{\frac{H}{16} \times \frac{W}{16} \times (B+1))}$ of the input image, where B means the number of bins within the specified depth range and the additional one represents the background. Departing from the traditional approach of regressing an exact metric depth for each monocular image, we draw inspiration from CaDNN [25] and formulate the depth prediction as a pixel-wise categorical depth distributions classification task: the depth range is segmented into several discrete bins, and the goal is to estimate the probability of each pixel belonging to one of these predefined depth bins. Moreover, we adopts discrete object-wise depth label generated from the image itself as the supervision rather than introducing the projected LiDAR depth map. This label-efficient approach enhances the robustness and efficiency of our depth prediction branch.

The objective of depth prediction is to produce sharper and more accurate features. To this end, we adopt the focal loss (FL) to encourage the predicted depth distribution to align closely with the correct depth bin:

$$L_D = \frac{1}{H' \times W'} \sum_{u=1}^{H'} \sum_{v=1}^{W'} FL(D_{fg}, D'_{fg}), \tag{6}$$

where $H' = \frac{H}{16}$ and $W' = \frac{W}{16}$ are the dimensions of predicted feature map. u and v are the pixel coordinate within this map. And D'_{fg} is the discrete object-wise depth label.

3.4   Semantic-Depth Fusion Transformer

Building upon MonoDETR [30], we employ the Semantic-Depth Fusion Transformer (SDFT) to fuse the semantic and depth features, which is made up of three modules: a semantic encoder, a depth encoder, and a fusion decoder.

Semantic and Depth Encoder. Initially, the semantic feature f_S and the depth feature f_D undergo distinct encoders, each tailored to their respective domains. The non-local geometric cues will be captured by the self-attention layers and feed-forward network. Through this process, long-range dependencies across diverse foreground areas are effectively learned. Consequently, the semantic feature f_S^E and the depth feature f_D^E are independently output, reflecting the unique encoding performed on the depth geometry and visual appearance.

Fusion Decoder. For the decoder, we utilize the DETR-base paradigm to fuse the depth feature f_D^E and semantic feature f_S^E sequentially. Specifically, a set of learnable queries $q \in \mathbb{R}^{N \times C}$, where N means the number of objects, is leveraged to infer the global context of the scene and the foreground spatial coordinates. Such queries first perform the Depth-Cross-Attention (DCA) operation with the depth feature f_D^E to generate depth-aware queries q' with some linear transformation:

$$q' = DCA(Q_q, K_D, V_D) = softmax(\frac{Q_q K_D^T}{\sqrt{C}})V_D, \tag{7}$$

where $Q_q = Linear(q), K_D, V_D = Linear(f_D^E)$. This mechanism enables each object query to adaptively extract spatial cues from depth-guided regions within the image, thereby enhancing the scene-level spatial comprehension. Subsequently, the depth-aware queries are processed through an inter-query Self-Attention (SA) layer to promote feature interaction between objects and a Semantic-Cross-Attention (SCA) layer to aggregate visual semantics from f_S^E. The final queries q_{out} are then output through a Feed Forward Network (FFN):

$$q'' = SA(Q_{q'}, K_{q'}, V_{q'}),$$
$$Q_{q'}, K_{q'}, V_{q'} = Linear(q'), \tag{8}$$

$$q''' = SCA(Q_{q''}, K_S, V_S),$$
$$Q_{q''} = Linear(q''), K_S, \quad V_S = Linear(f_S^E), \tag{9}$$

$$q_{out} = FFN(q'''). \tag{10}$$

In the two successive cross-attention computations, deep features and semantic features are employed sequentially as the key and value, respectively, with the queries, enabling a sophisticated fusion of these features. As a result, the final object queries produced by SDFT encapsulate both the geometric depth information and the holistic semantic features.

3.5   Detection Heads and Loss

The output object queries q_{out} are then fed into a sequence of the Multi-layer Proceptron (MLP)-based detection head to predict several attributes of the foreground objects. Following [3], we perform bipartite matching between each query and the ground truth by utilizing the Hungarian algorithm [11] to identify the global optimal matching. Concretely, we categorize the prediction sets into 2D and 3D properties and aggregate the losses of each group as L_{2D} and L_{3D}. Combined with the depth loss L_D, the total loss is formulated as:

$$Loss = \frac{1}{N_{gt}} \sum_{n=1}^{N_{gt}} (L_{2D} + L_{3D}) + L_D, \tag{11}$$

where N_{gt} is the number of ground truth objects in an image.

4   Experiments

4.1   Datasets and Metric

We evaluated the effectiveness of MonoTCM on the widely recognized KITTI dataset [8], encompassing 3,712 training, 3,769 validation, and 7,518 testing images. The detection performance is assessed across three difficulty levels: easy, moderate, and hard. The evaluation metrics include the average precision (AP) of bounding boxes in both 3D space (AP_{3D}) and bird-eye view (AP_{BEV}), which are computed at 40 recall positions.

4.2   Implementation Details

We constructed the model framework using PyTorch-based implementation. The backbone network employs a pre-trained ResNet-50 model [9] to extract features at various levels of image representation. Within the TCM module, we employed a single-stage transformer block for speedy token aggregation. For the subsequent semantic encoder and decoder, we utilized deformable attention due to its memory efficiency. In contrast, the depth encoder and decoder employed vanilla global attention to more effectively capture global geometric information. For the estimation of foreground depth, we constrained the range between 0 m to 60 m, divided into 80 bins. To optimize the MonoTCM model, we trained it for 195 epochs with a batch size of 12 and a learning rate of 2e-4. We used AdamW as the optimizer to facilitate the training process.

Table 1. Monocular performance of the car category on KITTI **test** sets. The top-performing method is highlighted in **bold** and the second-best results are marked by underline.

Method	Venue	Extra data	Test, AP_{3D}			Test, AP_{BEV}		
			Easy	Mod.	Hard	Easy	Mod.	Hard
PatchNet [20]	ECCV	Depth	15.68	11.12	10.17	22.97	16.86	14.97
D4LCN [7]	CVPR	Depth	16.65	11.72	9.51	22.51	16.02	12.55
Kinematic3D [2]	ECCV	Multi-frames	19.07	12.72	9.17	26.69	17.52	13.10
MonoRUn [4]	CVPR	LiDAR	19.65	12.30	10.58	27.94	17.34	15.24
CaDDN [25]	CVPR	LiDAR	19.17	13.41	11.46	27.94	18.91	17.19
AutoShape [18]	ICCV	CAD	22.47	14.17	11.36	30.06	20.08	15.59
SMOKE [18]	CVPR	None	14.03	9.76	7.84	20.83	14.49	12.75
MonoFlex [31]	CVPR	None	19.94	13.89	12.07	28.23	19.75	16.89
GUPNet [19]	ICCV	None	20.11	14.20	11.77	-	-	-
MonoDTR [10]	CVPR	None	21.99	15.39	12.73	28.59	20.38	17.14
MonoDETR [30]	ICCV	None	<u>23.65</u>	**15.92**	<u>12.99</u>	<u>32.08</u>	<u>21.44</u>	<u>17.85</u>
MonoTCM(Ours)	-	None	**23.68**	<u>15.47</u>	**13.30**	**32.41**	**21.93**	**18.63**
Improvement	-	v.s. second-best	+0.03	-	+0.31	+0.33	+0.49	+0.78

Table 2. Monocular performance of the car category on **validation** sets. The top-performing method is highlighted in **bold** and the second-best results are marked by underline.

Method	Venue	Extra data	Val, AP_{3D}			Val, AP_{BEV}		
			Easy	Mod.	Hard	Easy	Mod.	Hard
MonoDIS [26]	ICCV	None	11.06	7.60	6.37	18.45	12.58	10.66
MonoGRNet [23]	AAAI	None	11.90	7.56	5.76	19.72	12.81	10.15
MoVi-3D [27]	ECCV	None	14.28	11.13	9.68	22.36	17.87	15.73
M3D-RPN [1]	ICCV	None	14.53	11.07	8.65	20.85	15.62	11.88
SMOKE [18]	CVPR	None	14.76	12.85	11.50	19.99	15.61	15.28
MonoPair [6]	CVPR	None	16.28	12.30	10.42	24.12	18.17	15.76
Kinematic3D [2]	ECCV	Multi-frames	19.76	14.10	10.47	27.83	19.72	15.10
MonoRUn [4]	CVPR	LiDAR	20.02	14.65	12.61	-	-	-
D4LCN [7]	CVPR	Depth	22.32	16.20	12.30	31.53	22.58	17.87
GUPNet [19]	ICCV	None	22.76	16.46	13.72	31.07	22.94	19.75
CaDDN [25]	CVPR	LiDAR	23.57	16.31	13.84	-	-	-
MonoFlex [31]	CVPR	None	23.64	17.51	14.83	-	-	-
MonoDTR [10]	CVPR	None	24.52	18.57	15.51	33.33	25.35	21.68
MonoDETR [30]	ICCV	None	<u>28.84</u>	**20.61**	<u>16.38</u>	<u>37.86</u>	<u>26.95</u>	<u>22.80</u>
MonoTCM(Ours)	-	None	**28.92**	<u>20.59</u>	**17.31**	**38.22**	**26.97**	**23.03**
Improvement	-	v.s. second-best	+0.08	-	+0.93	+0.36	+0.02	+0.23

4.3   Quantitative and Qualitative Results

We first demonstrated the quantitative comparison with other advanced monocular 3D detection methods on KITTI validation and test sets, as demonstrated in Tables 1 and 2. The results indicate that our MonoTCM model achieved the best performance for most configurations in both $3D$ and BEV space without

Fig. 3. Qualitative results of MonoTCM. The first column exhibits the 3D detection outcomes. The second column showcases the attention map subsequent to the aggregation of the TCM module. The third column displays the clustering tokens following the TCM merging process.

additional data. In terms of AP_{bev}, MonoTCM outperformed the second-best models by +0.33%, +0.49%, and +0.78% under the easy, moderate and hard difficulty levels of test set, respectively. In the KITTI validation dataset, the performance of MonoTCM was further improved, with an average AP increase of 4.97% compared to the test dataset. Additionally, apart from the moderate difficulty level in 3D detection, MonoTCM consistently outperformed existing methods in other metrics.

Figure 3 displays the visualization of MonoTCM, showcasing qualitative results from three perspectives: the visualization of 3D detection bounding boxes, attention maps, and token aggregation. The figure reveals that the 3D bounding boxes and the attention maps with high scores were consistently aligned with the foreground objects, in close proximity to the ground truth. Furthermore, the clustering tokens, which were irregular in shape and represented by a single color for merged tokens, effectively grouped tokens with similar semantic representations. Even when tokens are not contiguous, such as two cars separated by a distance, the TCM module adeptly aggregated these long-range tokens with similar representations.

Table 3. Effectiveness of the proposed single TCM. The comparison is based on the 3D average precision for the car category on KITTI validation sets.

Module	Val, AP_{3D}		
	Easy	Mod.	Hard
None	27.21	20.24	17.20
Multi-Convs	27.70	19.60	16.27
Single-Conv	27.44	18.96	15.67
Multi-TCMs	26.62	19.06	15.88
Single-TCM	**28.92**	**20.59**	**17.31**

Table 4. Effectiveness of each components of TCM. C&M refers to clustering and merging operation. TB represents the transformer block.

C&M	**TB**	Val, AP_{3D}		
		Easy	Mod.	Hard
✗	✗	27.21	20.24	17.20
✓	✗	26.31	19.00	15.88
✗	✓	26.95	20.01	16.18
✓	✓	**28.92**	**20.59**	**17.31**

4.4   Ablation Study

Effect of Single TCM. To ascertain the efficacy of the proposed TCM module, we initiated an ablation study to replace the aggregation module with other blocks. As shown in Table 3, we replaced the **single-TCM** module with four alternative configurations: **None**, a baseline that employs multi-level features directly for 3D detection without any aggregation; **Single-Convs**, which adopts single convolution layer; **Multi-Convs**, which adopts multiple convolution layers with ReLU activation; and **Multi-TCMs**, where each TCM is assigned to a specific level of features. The comparison was based on the 3D average precision for the car category in the KITTI validation set. The results suggest that a poorly designed aggregation module yielded inferior outcomes compared to the baseline model, and in some cases, it might even hinder performance. When employing convolutional layers, an increase in the number of layers was correlated with better performance. However, the single-TCM setup proved to be the most effective. A plausible explanation for this phenomenon is that multiple aggregation layers with dedicated TCM modules process featured at each level in isolation, failing to synthesize semantic and depth information across different levels. Conversely, the single-TCM module is designed to process multi-scale features concurrently, resulting in the learning of both global and local image information across various dimensions. This holistic understanding of the data is particularly advantageous for subsequent detection tasks.

Effect of Each Component in TCM. As illustrated in Fig. 2, the TCM is predominantly constructed from three components: Clustering, Merging (C&M) Module, and Cross-attention Transformer Block (TB). To evaluate the effectiveness of these components, we conducted an ablation study, with the results detailed in Table 4. The integration of both modules compared to the baseline model (which lacks these components) produced an average improvement of 0.76% in the results. When only one module was employed, the TB module demonstrated a more pronounced impact on performance than the C&M module. This is attributed to the fact that clustering and merging operations without the TB's token interactions failed to integrate global and local information effectively, resulting in sub-optimal detection. Another intriguing phenomenon is that the detection results were enhanced when neither module is included. This suggests that a single module might be inadequate for the aggregation task, and the presence of extraneous components could contribute to detrimental effects on model optimization.

5   Conclusion

In this paper, we proposed MonoTCM, a DETR-based semantic-depth fusion transformer for monocular 3D object detection with token clustering and merging. The single-stage Token Clustering and Merging (TCM) module was introduced to effectively aggregate the insignificant background segments while preserving detailed representations in important regions. Furthermore, the aggregated representations were then decoupled into semantic and depth hints for respective feature extraction and fuse sequentially through a semantic-depth fusion decoder for accurate 3D and 2D properties estimation. The outstanding experimental results on the KITTI dataset demonstrated the efficacy of MonoTCM in the field of monocular 3D detection.

However, MonoTCM was subject to two primary limitations: (1) The computational complexity of the DPC-KNN algorithm used for clustering center selection is quadratic with respect to the number of tokens, which constrains the TCM module's applicability to high-resolution images. (2) MonoTCM is currently designed for monocular image input only; the potential for adaptation to a multi-view and multi-modal framework has not been investigated. These deficiencies will steer future research efforts towards enhancing the precision and efficiency of 3D detection.

Acknowledgement. This work received financial support from Jiangsu Industrial Technology Research Institute (JITRI), China, and Wuxi National Hi-Tech District (WND), China.

References

1. Brazil, G., Liu, X.: M3D-RPN: monocular 3D region proposal network for object detection. In: Proceedings of the IEEE/CVF International Conference on Computer Vision, pp. 9287–9296 (2019)
2. Brazil, G., Pons-Moll, G., Liu, X., Schiele, B.: Kinematic 3D object detection in monocular video. In: Vedaldi, A., Bischof, H., Brox, T., Frahm, J.-M. (eds.) Computer Vision – ECCV 2020: 16th European Conference, Glasgow, UK, August 23–28, 2020, Proceedings, Part XXIII, pp. 135–152. Springer International Publishing, Cham (2020). https://doi.org/10.1007/978-3-030-58592-1_9
3. Carion, N., Massa, F., Synnaeve, G., Usunier, N., Kirillov, A., Zagoruyko, S.: End-to-end object detection with transformers. In: Vedaldi, A., Bischof, H., Brox, T., Frahm, J.-M. (eds.) Computer Vision – ECCV 2020: 16th European Conference, Glasgow, UK, August 23–28, 2020, Proceedings, Part I, pp. 213–229. Springer International Publishing, Cham (2020). https://doi.org/10.1007/978-3-030-58452-8_13
4. Chen, H., Huang, Y., Tian, W., Gao, Z., Xiong, L.: MonoRUn: monocular 3D object detection by reconstruction and uncertainty propagation. In: Proceedings of the IEEE/CVF Conference on Computer Vision and Pattern Recognition, pp. 10379–10388 (2021)
5. Chen, T., Kornblith, S., Norouzi, M., Hinton, G.: A simple framework for contrastive learning of visual representations. In: International Conference on Machine Learning, pp. 1597–1607. PMLR (2020)
6. Chen, Y., Tai, L., Sun, K., Li, M.: MonoPair: monocular 3D object detection using pairwise spatial relationships. In: Proceedings of the IEEE/CVF Conference on Computer Vision and Pattern Recognition, pp. 12093–12102 (2020)
7. Ding, M., et al.: Learning depth-guided convolutions for monocular 3D object detection. In: Proceedings of the IEEE/CVF Conference on Computer Vision and Pattern Recognition Workshops, pp. 1000–1001 (2020)

8. Geiger, A., Lenz, P., Urtasun, R.: Are we ready for autonomous driving? The KITTI vision benchmark suite. In: 2012 IEEE Conference on Computer Vision and Pattern Recognition, pp. 3354–3361. IEEE (2012)
9. He, K., Zhang, X., Ren, S., Sun, J.: Deep residual learning for image recognition. In: Proceedings of the IEEE Conference on Computer Vision and Pattern Recognition, pp. 770–778 (2016)
10. Huang, K.C., Wu, T.H., Su, H.T., Hsu, W.H.: MonoDTR: monocular 3D object detection with depth-aware transformer. In: Proceedings of the IEEE/CVF Conference on Computer Vision and Pattern Recognition pp. 4012–4021 (2022)
11. Kuhn, H.W.: The Hungarian method for the assignment problem. Naval Res. Logistics Q. $\mathbf{2}$(1–2), 83–97 (1955)
12. Li, P., Zhao, H., Liu, P., Cao, F.: RTM3D: real-time monocular 3D detection from object keypoints for autonomous driving. In: Vedaldi, A., Bischof, H., Brox, T., Frahm, J.-M. (eds.) Computer Vision – ECCV 2020: 16th European Conference, Glasgow, UK, August 23–28, 2020, Proceedings, Part III, pp. 644–660. Springer International Publishing, Cham (2020). https://doi.org/10.1007/978-3-030-58580-8_38
13. Liang, T., et al.: BEVFusion: a simple and robust lidar-camera fusion framework. Adv. Neural. Inf. Process. Syst. $\mathbf{35}$, 10421–10434 (2022)
14. Lin, T.Y., Dollár, P., Girshick, R., He, K., Hariharan, B., Belongie, S.: Feature pyramid networks for object detection. In: Proceedings of the IEEE Conference on Computer Vision and Pattern Recognition, pp. 2117–2125 (2017)
15. Liu, S., Qi, L., Qin, H., Shi, J., Jia, J.: Path aggregation network for instance segmentation. In: Proceedings of the IEEE Conference on Computer Vision and Pattern Recognition, pp. 8759–8768 (2018)
16. Liu, Z., Wu, Z., Tóth, R.: SMOKE: single-stage monocular 3D object detection via keypoint estimation. In: Proceedings of the IEEE/CVF Conference on Computer Vision and Pattern Recognition Workshops, pp. 996–997 (2020)
17. Liu, Z., et al.: BEVFusion: multi-task multi-sensor fusion with unified bird's-eye view representation. In: 2023 IEEE International Conference on Robotics and Automation (ICRA), pp. 2774–2781. IEEE (2023)
18. Liu, Z., Zhou, D., Lu, F., Fang, J., Zhang, L.: AutoShape: real-time shape-aware monocular 3D object detection. In: Proceedings of the IEEE/CVF International Conference on Computer Vision, pp. 15641–15650 (2021)
19. Lu, Y., et al.: Geometry uncertainty projection network for monocular 3D object detection. In: Proceedings of the IEEE/CVF International Conference on Computer Vision, pp. 3111–3121 (2021)
20. Ma, X., Liu, S., Xia, Z., Zhang, H., Zeng, X., Ouyang, W.: Rethinking pseudo-LiDAR representation. In: Vedaldi, A., Bischof, H., Brox, T., Frahm, J.-M. (eds.) Computer Vision – ECCV 2020: 16th European Conference, Glasgow, UK, August 23–28, 2020, Proceedings, Part XIII, pp. 311–327. Springer International Publishing, Cham (2020). https://doi.org/10.1007/978-3-030-58601-0_19
21. Mousavian, A., Anguelov, D., Flynn, J., Kosecka, J.: 3D bounding box estimation using deep learning and geometry. In: Proceedings of the IEEE Conference on Computer Vision and Pattern Recognition, pp. 7074–7082 (2017)
22. Qi, C.R., Yi, L., Su, H., Guibas, L.J.: PointNet++: deep hierarchical feature learning on point sets in a metric space. In: Advances in Neural Information Processing Systems, vol. 30 (2017)
23. Qin, Z., Wang, J., Lu, Y.: MonoGRNet: a geometric reasoning network for monocular 3D object localization. In: Proceedings of the AAAI Conference on Artificial Intelligence, vol. 33, pp. 8851–8858 (2019)

24. Radford, A., et al.: Learning transferable visual models from natural language supervision. In: International Conference on Machine Learning, pp. 8748–8763. PMLR (2021)
25. Reading, C., Harakeh, A., Chae, J., Waslander, S.L.: Categorical depth distribution network for monocular 3D object detection. In: Proceedings of the IEEE/CVF Conference on Computer Vision and Pattern Recognition, pp. 8555–8564 (2021)
26. Simonelli, A., Bulo, S.R., Porzi, L., López-Antequera, M., Kontschieder, P.: Disentangling monocular 3D object detection. In: Proceedings of the IEEE/CVF International Conference on Computer Vision, pp. 1991–1999 (2019)
27. Simonelli, A., Buló, S.R., Porzi, L., Ricci, E., Kontschieder, P.: Towards generalization across depth for monocular 3D object detection. In: Vedaldi, A., Bischof, H., Brox, T., Frahm, J.-M. (eds.) Computer Vision – ECCV 2020: 16th European Conference, Glasgow, UK, August 23–28, 2020, Proceedings, Part XXII, pp. 767–782. Springer International Publishing, Cham (2020). https://doi.org/10.1007/978-3-030-58542-6_46
28. Wang, Y., Sun, Y., Liu, Z., Sarma, S.E., Bronstein, M.M., Solomon, J.M.: Dynamic graph CNN for learning on point clouds. ACM Trans. Graph. (ToG) **38**(5), 1–12 (2019)
29. Zeng, W., et al.: Not all tokens are equal: Human-centric visual analysis via token clustering transformer. In: Proceedings of the IEEE/CVF Conference on Computer Vision and Pattern Recognition, pp. 11101–11111 (2022)
30. Zhang, R., et al.: MonoDETR: depth-guided transformer for monocular 3D object detection. In: Proceedings of the IEEE/CVF International Conference on Computer Vision, pp. 9155–9166 (2023)
31. Zhang, Y., Lu, J., Zhou, J.: Objects are different: flexible monocular 3D object detection. In: Proceedings of the IEEE/CVF Conference on Computer Vision and Pattern Recognition, pp. 3289–3298 (2021)
32. Zhou, Y., Zhu, H., Liu, Q., Chang, S., Guo, M.: MonoATT: online monocular 3D object detection with adaptive token transformer. In: Proceedings of the IEEE/CVF Conference on Computer Vision and Pattern Recognition, pp. 17493–17503 (2023)

Fourier-Guided Illumination Difference and Reflection Component Prediction for Enhancing Low-Light Images

Zeyu Li[1,2(✉)]

[1] Institute of Software, Chinese Academy of Sciences, Beijing, China
lizeyu2023@iscas.ac.cn
[2] University of Chinese Academy of Sciences, Beijing, China

Abstract. Low-light image enhancement improves visibility and detail in poorly lit conditions, crucial for applications in security, medical imaging, and autonomous systems, but existing methods face many challenges, including noise and color distortion, complex multi-stage training process, and insufficient modeling of long-range dependencies. The noise and color distortion problems arise because existing methods fail to effectively suppress noise and maintain color consistency during the brightening process. The complex multi-stage training process not only increases the training time and computational cost, but also leads to error accumulation between different stages. Moreover, current techniques often struggle to effectively capture long-range interactions and the inherent self-similarity found in different regions of images, which affects the processing effect of complex low-light images. We propose an innovative method to address these problems through a decomposition and reconstruction framework of illumination and reflectance components, supplemented by Fourier-guided component prediction (FGIDRCP). This method simplifies the training process and improves efficiency, and uses a signal-to-noise ratio (SNR) dynamic weight adjustment method to adaptively handle noise and details in different regions. First, our model predicts the illumination difference and reflectance components to generate a preliminary normal-light image, which is then refined and denoised to finally generate the enhanced image. Through Fourier-guided component prediction, we are able to better extract and utilize the structure and texture information of the image, helping the network to decompose the reflectance component and thus reduce image distortion. The SNR-guided dynamic weight adjustment module optimizes denoising performance by adapting to varying noise levels and improving image clarity. Experimental results show that FGIDRCP performs well in low-light image enhancement tasks.

Keywords: Low-light image enhancement · Retinex · SNR · Fourier

1 Introduction

Low-light images enhancement is useful for improving quality in dimly lit conditions, as these images often face challenges like insufficient brightness, low contrast, color inaccuracies, and increased noise, impairing visual quality and subsequent image processing tasks like object recognition and classification. Enhancing these images is critical for applications in surveillance, photography, and autonomous driving.

Despite advances in image enhancement technologies, significant challenges remain. Many existing methods rely on the Retinex theory [3, 12, 28, 32, 33, 38, 46], decomposing images into reflectance and illumination components. However, these methods often struggle with noise and color distortion during enhancement. Additionally, the complex multi-stage training processes [32, 44, 45] increase computational cost and time, leading to error accumulation. Convolutional neural networks (CNNs) [12, 13, 16, 21, 25, 26, 34], which are frequently employed in these approaches, often fall short in effectively capturing extended contextual dependencies and global self-similarity patterns across images, making it difficult to effectively process complex low-light images. Moreover, current methods often fail to adaptively optimize enhancements across different image regions, resulting in uneven quality and inadequate denoising.

To address these challenges, we introduce an innovative method to address these problems through a decomposition and reconstruction framework of illumination and reflectance components, supplemented by Fourier-guided component prediction (FGIDRCP), a cutting-edge framework that integrates illumination and reflectance decomposition with Fourier-guided prediction. This approach enhances training efficiency and effectiveness by incorporating several key components. The Illumination Difference and Reflectance Component Estimation module generates an initial normal-light image by predicting variations in illumination and reflectance. The Fourier-Guided Component Prediction mechanism leverages the frequency domain to precisely model and decompose image components, enhancing detail fidelity and minimizing distortion. The Reflectance Component Guidance utilizes a pre-trained Retinex model to refine reflectance estimates, while the Denoising Module employs a dynamic weight adjustment method based on the SNR to adaptively process different regions of the image. Our method enhances low-light images by seamlessly integrating global context with local details, leading to exceptional improvement in image quality. Our main contributions are as follows:

- A model for predicting illumination difference, guiding the modeling of long-distance dependencies, simplifying the training process and improving efficiency.
- Fourier-guided illumination difference and reflection component prediction, extracting image structure and texture information, helping to decompose reflection components, thereby enhancing detail performance and reducing distortion.
- A reflection-guided method based on a pre-trained Retinex model ensures reflection consistency and reduces distortion by exchanging decomposition losses.

– A dynamic weight adjustment method based on signal-to-noise ratio (SNR) adaptively processes different regions, significantly improves denoising performance and image quality, and enhances overall visual consistency.

2   Related Works

2.1   Traditional Methods

Low-light image enhancement strategies are generally classified into basic and traditional methods. Basic approaches, like histogram equalization [1,15] and gamma adjustment, work by directly modifying image attributes to improve visibility and contrast in poorly lit conditions. While these methods are straightforward and computationally efficient, they often overlook illumination factors, leading to enhanced images that may not appear natural under normal lighting conditions [20]. On the other hand, traditional cognitive approaches are rooted in Retinex theory [19], which involves decomposing images into reflectance and illumination components to separately enhance reflectance and adjust illumination. For instance, Guo et al. [14] refined the estimated illumination map by incorporating a structural prior. Although these methods use more sophisticated models, they typically assume that low-light images are free from artifacts, which can result in noise and color distortion. Additionally, they heavily depend on manually crafted priors [23], which limits their generalization capabilities.

2.2   Deep Learning Methods

Initial deep learning methods for enhancing low-light images predominantly utilized convolutional neural networks (CNNs). For example, Wang et al.'s Deep-UPE [29] utilizes a straightforward CNN architecture to directly estimate the lighting map, thereby simplifying the enhancement procedure. However, this approach overlooks potential image degradations, leading to issues such as increased noise and color distortion. Some methods design spatial local filters for enhancement, such as Sean et al. [25], but fail to adaptively optimize low-light data, resulting in artifacts. Unsupervised methods such as Zero-DCE [13] build lightweight networks for dynamic range adjustment, Sparse [36] use sparse representation to enhance details and reduce noise, EnGAN [16] uses generative adversarial networks to balance image quality and noise suppression, RUAS [24] uses Retinex theory and unsupervised adversarial training for enhancement, FIDE [35] combines multi-frequency field perception and sparse networks for structural analysis, MF [11] applies multi-frequency analysis to capture different image scales, and DRBN [35] combines Retinex theory with deep learning for robust enhancement. Strong, KinD [45] uses knowledge distillation for enhancement, LPNet [22] processes different illumination levels separately to improve clarity, Restormer [41] uses transformer architecture to capture long-range dependencies, MIRNet [42] combines multi-scale features for comprehensive enhancement, SNR-Net [34] uses SNR-aware transformer and convolutional model for dynamic pixel enhancement, and Retinexformer [3] combines Retinex theory with single-stage transformer design to further optimize enhancement, SKF [40] utilizes semantic-aware knowledge guidance to improve the quality and clarity of images, EMNet [37] leverages external memory to improve image quality in dark environments, WaveNet [6] enhances images by effectively accounting for wave patterns, CIDNet [9] decouples color and intensity for more precise image restoration, LYT-Net [2] introduced a Multi-Scale State-Space Model.

Compared with existing methods, our method (FGIDRCP) innovatively uses the illumination difference and reflectance decomposition and reconstruction framework, supplemented by Fourier-guided prediction and Retinex decomposition network guidance. Our model better captures long-range dependencies and structural information, significantly enhances detail representation, and reduces color distortion.

3   Proposed Method

Our proposed method leverages a multi-stage network architecture to effectively decompose and reconstruct images, as depicted in Fig. 1. It integrates several key components, including Illumination Difference and Reflection Component Estimation (IDR), a Fourier-Guided Prediction module (FFT), a Retinex Decomposition Network-guided Reflectance Component Prediction (RDN) (depicted in Fig. 2), and a Signal-to-Noise Ratio-guided Denoising mechanism (SGD).

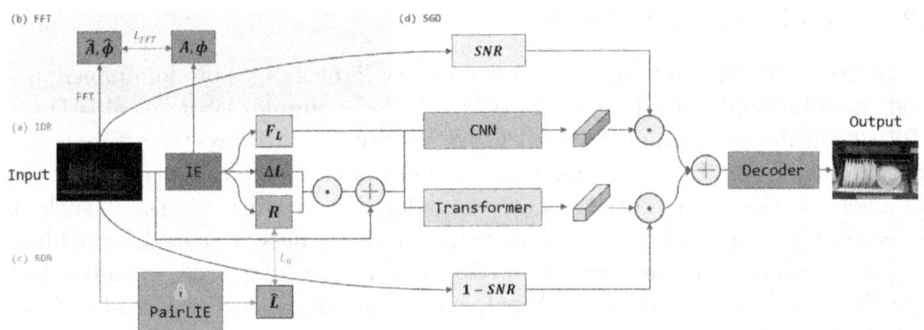

Fig. 1. Overview of the proposed network architecture. The key components of the network include the Illumination Difference and Reflection component estimation (IDR) module, the Fourier-Guided Prediction module (FFT), the Retinex decomposition network-guided reflectance component prediction module (RDN), and the SNR-guided denoising module (SGD). These components work together to output the illumination difference and reflectance component based on the input image for brightening the image, and dynamically denoise the image based on the signal-to-noise ratio.

3.1 Estimation of Illumination Difference and Reflectance Component

The Estimation of Illumination Difference and Reflectance Component (IDR) module is fundamental to our approach, focusing on the initial processing of images to separate illumination variations from reflectance, thereby creating a detailed low-light feature map. Building on Retinex theory [18,19], the decomposition of an image $\mathbf{I}$ into its fundamental parts is represented by:

$$\mathbf{I} = \mathbf{R} \cdot \mathbf{L} \tag{1}$$

The primary goal of low-light image enhancement is to derive a well-lit image $\mathbf{I}_{high}$ from its dim counterpart $\mathbf{I}_{low}$. This entails isolating the illumination $\mathbf{L}_{low}$ and reflectance $\mathbf{R}$ from $\mathbf{I}_{low}$ and similarly decomposing $\mathbf{I}_{high}$ into $\mathbf{L}_{high}$ and $\mathbf{R}$:

$$\mathbf{I}_{low} = \mathbf{R} \cdot \mathbf{L}_{low} \tag{2}$$

$$\mathbf{I}_{high} = \mathbf{R} \cdot \mathbf{L}_{high} \tag{3}$$

In these equations, $\mathbf{L}_{low}$ and $\mathbf{L}_{high}$ denote the illumination levels under dim and standard lighting conditions, respectively, while $\mathbf{R}$ signifies the scene's fundamental attributes, such as texture and color, which remain constant regardless of the lighting situation.

To estimate these components, the IDR module processes the low-light image $\mathbf{I}_{low}$ along with an illumination prior $\mathbf{L}_p$. This prior is derived by calculating the average pixel values across all color channels:

$$\mathbf{L}_p = \frac{1}{C} \sum_{c=1}^{C} \mathbf{I}_{low}^{c} \tag{4}$$

where C is the number of color channels, and the averaging operation mean_c is executed across these channels.

The IDR module then estimates the illumination difference $\Delta\mathbf{L}$ and the reflectance $\mathbf{R}$. Specifically, $\Delta\mathbf{L}$ captures the change between normal and low-light illumination conditions.

With these estimates, the predicted normal-light image $\mathbf{I}_{pred}$ is calculated as follows:

$$\mathbf{I}_{pred} = \mathbf{I}_{low} + \Delta\mathbf{L} \odot \mathbf{R} \tag{5}$$

The resulting brightened image $\mathbf{I}_{pred}$ and the associated feature map $\mathbf{F_L}$ [3] are then passed to the subsequent enhancement module SGD, producing the final enhanced image $\mathbf{I}_{enhanced} \in \mathbb{R}^{H \times W \times 3}$:

$$\mathbf{I}_{enhanced} = SGD(\mathbf{I}_{pred}, \mathbf{F_L}) \tag{6}$$

3.2 Fourier-Guided Prediction

Fourier-Guided prediction is employed to distinguish between illumination and reflectance components by analyzing their frequency characteristics. We transform the image into the frequency domain to facilitate separation, yielding:

$$\mathbf{F} = \mathcal{A} \cdot e^{j\Theta} \tag{7}$$

where $\mathcal{A}$ represents the amplitude spectrum and Θ is the phase spectrum. In this domain, the interaction between illumination and reflectance can be modeled as:

$$\mathbf{F} = \mathbf{L}_{\text{freq}} \cdot \mathbf{R}_{\text{freq}} \tag{8}$$

Here, $\mathbf{L}_{\text{freq}}$ and $\mathbf{R}_{\text{freq}}$ are the Fourier representations of illumination and reflectance, respectively. This relationship can be further detailed as:

$$\mathbf{F} = \mathcal{A}_{\mathbf{L}} \cdot \mathcal{A}_{\mathbf{R}} \cdot e^{j(\Theta_{\mathbf{L}} + \Theta_{\mathbf{R}})} \tag{9}$$

where $\mathcal{A}_{\mathbf{L}}$ and $\mathcal{A}_{\mathbf{R}}$ are the amplitude components for illumination and reflectance, and $\Theta_{\mathbf{L}}$ and $\Theta_{\mathbf{R}}$ are their respective phases.

To ensure precise separation, we use a loss function that optimizes both amplitude and phase predictions:

$$\mathcal{L}_{\text{Fourier}} = \|\mathcal{A} - \hat{\mathcal{A}}\|_1 + \|\Theta - \hat{\Theta}\|_1 \tag{10}$$

In this equation, $\mathcal{A}$ and $\hat{\mathcal{A}}$ denote the actual and predicted amplitude spectra, while Θ and $\hat{\Theta}$ represent the actual and predicted phase spectra. By focusing on accurate amplitude prediction to guide illumination extraction and phase prediction to guide reflectance extraction, this approach enhances the effectiveness of low-light image processing.

3.3 Retinex Decomposition Network-Guided Reflectance Component Prediction

Our methodology incorporates a pre-trained Retinex decomposition network to enhance the prediction accuracy of reflectance components for low-light image enhancement, as depicted in Fig. 2. This network, which remains fixed during the training phase, supplies estimated illumination components $\hat{\mathbf{L}}$, aiding in the precise forecasting of reflectance components.

In the training process, we input paired images of low-light and well-lit scenarios into the network to derive their respective illumination components, $\mathbf{L}_1^{\text{low}}$ and $\mathbf{L}_1^{\text{high}}$, using the Retinex decomposition model. The IDR module then estimates the reflectance components $\mathbf{R}_1^{\text{low}}$ and $\mathbf{R}_1^{\text{high}}$. Following this, we interchange the reflectance components between the low-light and well-lit images, as outlined in [8], to reconstruct new images: $\mathbf{I}_2^{\text{low}} = \mathbf{R}_1^{\text{high}} \odot \mathbf{L}_1^{\text{low}}$ and $\mathbf{I}_2^{\text{high}} = \mathbf{R}_1^{\text{low}} \odot \mathbf{L}_1^{\text{high}}$. These reconstructed images are then reprocessed with the IDR module to extract $\mathbf{R}_2^{\text{high}}$ and $\mathbf{R}_2^{\text{low}}$. A comparison with the initial reflectance components $\mathbf{R}_1^{\text{low}}$ and $\mathbf{R}_1^{\text{high}}$ is conducted to maintain consistency and ensure accuracy.

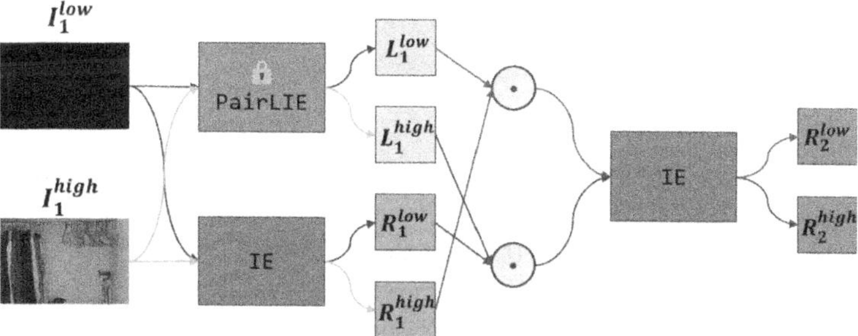

Fig. 2. Diagram of the Reflectance Component Prediction guided by the Retinex Decomposition Network.

The modified loss function used to enforce this consistency is:

$$
\begin{aligned}
\mathcal{L}_R =& \|\mathbf{R}_1^{\text{high}} - \mathbf{R}_2^{\text{high}}\|_1 + \|\mathbf{R}_1^{\text{low}} - \mathbf{R}_2^{\text{low}}\|_1 + \|\mathbf{R}_1^{\text{low}} - \mathbf{R}_1^{\text{high}}\|_1 \\
&+ \|\mathbf{R}_1^{\text{low}} \odot \mathbf{L}_1^{\text{low}} - \mathbf{I}_1^{\text{low}}\|_1 + \|\mathbf{R}_1^{\text{high}} \odot \mathbf{L}_1^{\text{high}} - \mathbf{I}_1^{\text{high}}\|_1 \\
&+ \|\nabla \mathbf{R}_1^{\text{low}}\|_1 + \|\nabla \mathbf{L}_1^{\text{low}}\|_1 + \|\nabla \mathbf{R}_1^{\text{high}}\|_1 + \|\nabla \mathbf{L}_1^{\text{high}}\|_1
\end{aligned}
\tag{11}
$$

This approach ensures that our model effectively learns to predict reflectance components that remain consistent even after illumination variations are accounted for, thereby improving the overall quality of low-light image enhancement.

3.4 SNR-Guided Denoising

The SNR-Guided Denoising module adjusts weights dynamically based on the Signal-to-Noise Ratio (SNR). By utilizing CNNs for local detail enhancement and Transformers for capturing global context, this method effectively balances noise reduction with detail preservation. The SNR is computed using a mean filter to estimate the noise level, guiding the denoising process to optimize image quality. To compute SNR, we first apply a mean filter $\mathbf{H}$ to the original low-light image $\mathbf{I}_g$, defined as:

$$
\mathbf{H} = \frac{1}{k^2} \mathbf{1}_{k \times k}
\tag{12}
$$

where $\mathbf{1}_{k \times k}$ is an all-one matrix. Mean filtering is represented as the convolution of the image $\mathbf{I}_g$ with the kernel $\mathbf{H}$:

$$
\widehat{\mathbf{I}}_g = \mathbf{I}_g * \mathbf{H}
\tag{13}
$$

Next, we calculate the difference between the processed image and the original image, representing the noise component $\mathbf{N}$:

$$
\mathbf{N} = \text{abs}(\mathbf{I}_g - \widehat{\mathbf{I}}_g)
\tag{14}
$$

The SNR S is defined as the ratio of the denoised image to the noise:

$$S = \frac{\widehat{\mathbf{I}}_g}{\mathbf{N}} \tag{15}$$

Based on the value of SNR S, we dynamically adjust the contributions of the CNN and the Transformer in the feature space. In areas with low SNR, where more global information is needed, the weight of the Transformer is increased; conversely, in areas with high SNR, where more local detail is necessary, the weight of the CNN is increased. This adjustment of weights is performed as follows:

$$\alpha(x, y) = \frac{S(x, y)}{\max(S)} \tag{16}$$

$$\mathbf{F}(x, y) = \alpha(x, y) \cdot \mathbf{C}(x, y) + (1 - \alpha(x, y)) \cdot \mathbf{T}(x, y) \tag{17}$$

Finally, the processed features $\mathbf{F}$ are used to generate the final output image. This method combines the CNN's strength in local detail enhancement with the Transformer's ability to integrate global context, achieving a balanced refinement that adjusts to the image's content needs.

4 Experiment

4.1 Datasets and Implementation Details

Datasets. Our evaluation utilizes three datasets to assess the framework's performance: LOL-v1 [32], LOL-v2-real [36], and LOL-v2-synthetic [36].

Implementation Details. Our framework was implemented in PyTorch and tested on an NVIDIA RTX 3090 GPU. Training used a batch size of 8 and 8 data loading processes, with geometric augmentations such as random rotations, scaling, and flips to improve generalization. Training involved 150,000 iterations with the Adam optimizer, starting at a learning rate of 2×10^{-4}. Mixup augmentation was applied with a parameter of 1.2.

4.2 Comparison with Current Methods

We compare FGIDRCP with serveral methods for low-light enhancement, including SRIE [10], BIMEF [39], RRM [23], SID [4], 3DLUT [43], DeepUPE [29], A3DLUT [30], RF [17], DeepLPF [27], IPT [5], UFormer [31], Dong [7], LIME [14], RetinexNet [32], Sparse [36], EnGAN [16], RUAS [24], FIDE [35], MF [11], DRBN [35], KinD [45], LPNet [22], Restormer [41], MIRNet [42], Retinexformer [3], and SNR-Net [34].

Table 1. Quantitative comparison on LOL-v1, LOL-v2-real, and LOL-v2-synthetic datasets.

Methods	LOL-v1		LOL-v2-real		LOL-v2-syn	
	PSNR↑	SSIM↑	PSNR↑	SSIM↑	PSNR↑	SSIM↑
SRIE [10]	11.86	0.500	17.34	0.686	14.50	0.616
BIMEF [39]	13.86	0.580	17.85	0.653	17.20	0.713
RRM [23]	13.88	0.660	17.34	0.686	17.15	0.727
SID [4]	14.35	0.436	13.24	0.442	15.04	0.610
3DLUT [43]	14.35	0.445	17.59	0.721	18.04	0.800
DeepUPE [29]	14.38	0.446	13.27	0.452	15.08	0.623
A3DLUT [30]	14.77	0.458	18.19	0.745	18.92	0.838
RF [17]	15.23	0.452	14.05	0.458	15.97	0.632
DeepLPF [27]	15.28	0.473	14.10	0.480	16.02	0.587
IPT [5]	16.27	0.504	19.80	0.813	18.30	0.811
UFormer [31]	16.36	0.771	18.82	0.771	19.66	0.871
Dong [7]	16.72	0.580	17.26	0.527	16.90	0.749
LIME [14]	16.76	0.560	15.24	0.470	16.88	0.776
RetinexNet [32]	16.77	0.560	15.47	0.567	17.13	0.798
Sparse [36]	17.20	0.640	20.06	0.816	22.05	0.905
EnGAN [16]	17.48	0.650	18.23	0.617	16.57	0.734
RUAS [24]	18.23	0.720	18.37	0.723	16.55	0.652
FIDE [35]	18.27	0.665	16.85	0.678	15.20	0.612
MF [11]	18.79	0.640	18.73	0.559	17.50	0.751
DRBN [35]	20.13	0.830	20.29	0.831	23.22	0.927
KinD [45]	20.86	0.790	14.74	0.641	13.29	0.578
LPNet [22]	21.46	0.802	17.80	0.792	19.51	0.846
Restormer [41]	22.43	0.823	19.94	0.827	21.41	0.830
MIRNet [42]	24.14	0.830	20.02	0.820	21.94	0.876
Retinexformer [3]	23.43	0.881	21.74	0.854	25.67	0.930
SNR-Net [34]	24.02	0.884	20.62	0.856	24.14	0.928
Ours	**24.33**	**0.904**	**22.69**	**0.884**	**25.96**	**0.956**

Quantitative Results. Table 1 displays the performance of FGIDRCP on the LOL-v1, LOL-v2-real, and LOL-v2-synthetic datasets. FGIDRCP achieves the highest PSNR and SSIM scores across all datasets, demonstrating its superior capability in enhancing low-light images with exceptional detail and visual quality.

Qualitative Results.. In this section, we qualitatively analyze the performance of FGIDRCP on three datasets. Figures 3, 4, and 5 showcase visual comparisons across these datasets, demonstrating the effectiveness of our approach in enhancing low-light images. FGIDRCP significantly improves visibility, reduces noise, and enhances contrast on the LOL-v1 dataset (Fig. 3) For LOL-v2-real (Fig. 4), it robustly handles real-world scenarios, delivering better illumination and detail preservation. On the LOL-v2 synthetic dataset (Fig. 5), the method generalizes well across controlled low-light conditions, consistently producing superior results. Overall, the qualitative results indicate FGIDRCP's clear advantage in detail preservation, noise reduction, and visual quality over existing techniques.

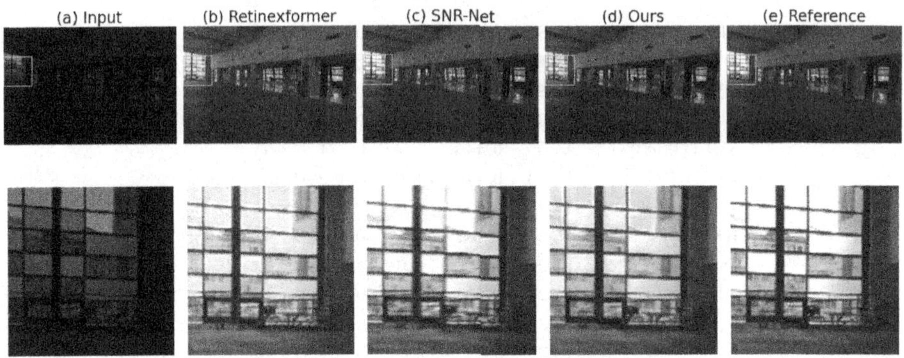

Fig. 3. Visual analysis of the outcomes on the LOL-v1 dataset.

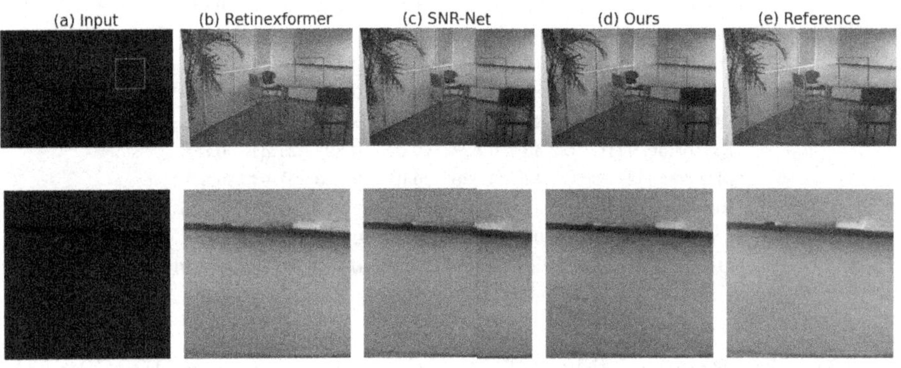

Fig. 4. Visual analysis of the outcomes on the LOL-v2-real dataset.

4.3   Ablation Study

To evaluate the contribution of each component in our framework, we performed ablation studies on the LOL-v1, LOL-v2-real, and LOL-v2-synthetic datasets. The study involved testing various configurations, including the following:

- **Ours w/o FFT**: This setting removes the Fourier-Guided Prediction component.
- **Ours w/o RDN**: This setting excludes the Retinex Decomposition Network-Guided Reflectance Component Prediction.

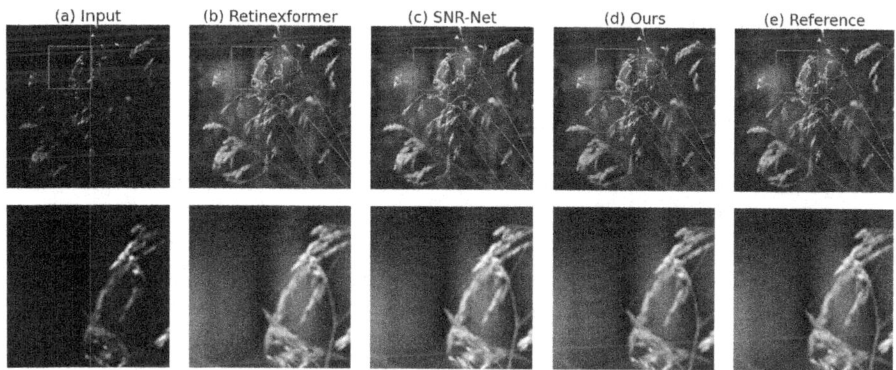

Fig. 5. Visual analysis of the outcomes on the LOL-v2-synthetic dataset.

- **Ours w/o SGD**: This setting omits the SNR-Guided Denoising component.
- **Ours**: The complete framework with all components integrated.

Table 2 presents the results of our ablation studies. By comparing the performance metrics across all settings, it is evident that our full framework consistently achieves the highest PSNR and SSIM scores, and the lowest LPIPS values. This validates the effectiveness of integrating all components. The individual removal of FFT, RDN, and SGD components illustrates their respective contributions to the overall performance, as each removal results in a notable decline in performance metrics.

Table 2. Ablation study results on LOL-v1, LOL-v2-real, and LOL-v2-synthetic datasets.

Method	LOL-v1			LOL-v2-real			LOL-v2-synthetic		
	PSNR↑	SSIM↑	LPIPS↓	PSNR↑	SSIM↑	LPIPS↓	PSNR↑	SSIM↑	LPIPS↓
Ours w/o FFT	23.00	0.877	0.133	22.33	0.883	0.172	25.52	0.937	0.063
Ours w/o RDN	22.89	0.871	0.138	22.15	0.879	0.176	25.03	0.929	0.066
Ours w/o SGD	22.45	0.865	0.142	21.93	0.872	0.179	24.48	0.923	0.073
Ours	**24.33**	**0.904**	**0.126**	**22.69**	**0.884**	**0.160**	**25.96**	**0.956**	**0.054**

These results highlight the importance of each component in our framework, demonstrating that the combined use of Fourier-Guided Prediction, Retinex Decomposition Network-Guided Reflectance Component Prediction, and SNR-Guided Denoising is crucial for achieving superior image enhancement performance. The effectiveness of FGIDRCP is thus confirmed across different datasets, underscoring its robustness and generalizability.

5   Conclusion

In this paper, we proposed a novel approach for low-light image enhancement, integrating several key components to address existing challenges. Our model includes an Illumination Difference and Reflection Component Estimation Module for predicting illumination differences and reflectance components, a Fourier-Guided Component Prediction mechanism to capture structural information and reduce distortion, a Retinex Decomposition Network-Guided Reflectance Component Prediction to ensure consistency and minimize distortion, and an SNR Dynamic Weight Adjustment for Denoising to improve image quality. Experimental results show that FGIDRCP achieves the highest PSNR and SSIM across multiple datasets, demonstrating superior enhancement capabilities while preserving details and improving visual quality. Visual comparisons reveal significant improvements in detail visibility, noise reduction, and contrast. Ablation studies confirm the importance of each component. For future work, we plan to consider additional factors such as varying lighting conditions and different types of noise, optimize the model architecture, and explore applications in video enhancement and multi-exposure image fusion.

Acknowledgment. This work was supported by the innovation fund (No.2021YFB 3601400). I would like to express my sincere gratitude to my advisors, Prof. Fanjiang Xu and Prof. Xiongxin Tang, for their invaluable guidance and support throughout this research. I also extend my thanks to Prof. Hanxiang Yang, and Prof. Qiao Chen for their insightful feedback and assistance in the development of this work. Furthermore, I am deeply grateful to the development team of the SEE series software for providing the essential simulation support that greatly contributed to the completion of this research.

References

1. Abdullah-Al-Wadud, M., Kabir, M.H., Dewan, M., Chae, O.: A dynamic histogram equalization for image contrast enhancement. IEEE Trans. Consum. Electron. **53**(2), 593–600 (2007)
2. Brateanu, A., Balmez, R., Avram, A., Orhei, C.: LYT-NET: Lightweight YUV transformer-based network for low-light image enhancement. arXiv preprint arXiv:2401.15204 (2024)
3. Cai, Y., Bian, H., In, J., Wang, H., Timofte, R., Zhang, Y.: Retinexformer: one-stage retinex-based transformer for low-light image enhancement. In: ICCV (2023)
4. Chen, C., Chen, Q., Do, M.N., Koltun, V.: Seeing motion in the dark. In: ICCV (2019)
5. Chen, H., et al.: Pre-trained image processing transformer. In: Proceedings of the IEEE Conference on Computer Vision and Pattern Recognition (CVPR) (2021)
6. Dang, J., Li, Z., Zhong, Y., Wang, L.: WaveNet: wave-aware image enhancement. In: Chaine, R., Deng, Z., Kim, M.H. (eds.) Pacific Graphics Short Papers and Posters. The Eurographics Association (2023). https://doi.org/10.2312/pg. 20231267

7. Dong, X., et al.: Fast efficient algorithm for enhancement of low lighting video. In: Proceedings of the International Conference on Multimedia and Expo (ICME), pp. 5, 6 (2011)

8. Du, Z., Shi, M., Deng, J.: Boosting object detection with zero-shot day-night domain adaptation. In: Proceedings of the IEEE/CVF Conference on Computer Vision and Pattern Recognition, pp. 12666–12676 (2024)

9. Feng, Y., Zhang, C., Wang, P., Wu, P., Yan, Q., Zhang, Y.: You only need one color space: an efficient network for low-light image enhancement (2024)

10. Fu, X., et al.: A weighted variational model for simultaneous reflectance and illumination estimation. In: Proceedings of the IEEE Conference on Computer Vision and Pattern Recognition, pp. 2782–2790 (2016)

11. Fu, X., Zeng, D., Huang, Y., Liao, Y., Ding, X., Paisley, J.W.: A fusion-based enhancing method for weakly illuminated images. Signal Process. **129**, 82–96 (2016)

12. Fu, Z., Yang, Y., Tu, X., Huang, Y., Ding, X., Ma, K.: Learning a simple low-light image enhancer from paired low-light instances. In: CVPR (2023)

13. Guo, C.G., Li, C., Guo, J., Loy, C.C., Hou, J., Kwong, S., Cong, R.: Zero-reference deep curve estimation for low-light image enhancement. In: Proceedings of the IEEE Conference on Computer Vision and Pattern Recognition (CVPR), pp. 1780–1789 (2020)

14. Guo, X., Li, Y., Ling, H.: LIME: low-light image enhancement via illumination map estimation. IEEE Trans. Image Process. **26**(2), 982–993 (2016)

15. Ibrahim, H., Kong, N.: Brightness preserving dynamic histogram equalization for image contrast enhancement. IEEE Trans. Consum. Electron. **53**(4), 1752–1758 (2007)

16. Jiang, Y., et al.: EnlightenGAN: deep light enhancement without paired supervision. IEEE TIP **30**, 2340–2349 (2021)

17. Kosugi, S., Yamasaki, T.: Unpaired image enhancement featuring reinforcement-learning-controlled image editing software. In: Proceedings of the AAAI Conference on Artificial Intelligence (AAAI) (2020)

18. Land, E.H.: The retinex theory of color vision. Sci. Am. **237**(6), 108–129 (1977)

19. Land, E.H., McCann, J.J.: Lightness and retinex theory. JOSA **61**(1), 1–11 (1971)

20. Land, E.H.: An alternative technique for the computation of the designator in the retinex theory of color vision. Proc. Natl. Acad. Sci. **83**(10), 3078–3080 (1986)

21. Li, C., Guo, C., Loy, C.C.: Learning to enhance low-light image via zero-reference deep curve estimation. IEEE TPAMI **44**(8), 4225–4238 (2021)

22. Li, J., Li, J., Fang, F., Li, F., Zhang, G.: Luminance-aware pyramid network for low-light image enhancement. IEEE Trans. Multimedia (2020)

23. Li, M., Liu, J., Yang, W., Sun, X., Guo, Z.: Structure-revealing low-light image enhancement via robust retinex model. IEEE Trans. Image Process. **27**(6), 2828–2841 (2018)

24. Liu, R., Ma, T., Zhang, J., Fan, X., Luo, Z.: Retinex-inspired unrolling with cooperative prior architecture search for low-light image enhancement. In: CVPR (2021)

25. Lore, K.G., Akintayo, A., Sarkar, S.: LLNet: a deep autoencoder approach to natural low-light image enhancement. Pattern Recogn. **61**, 650–662 (2017)

26. Lv, F., Lu, F., Wu, J., Lim, C.: MBLLEN: low-light image/video enhancement using CNNs. In: BMVC, vol. 220, p. 4 (2018)

27. Moran, S., Marza, P., McDonagh, S., Parisot, S., Slabaugh, G.: DeepLPF: deep local parametric filters for image enhancement. In: Proceedings of the IEEE Conference on Computer Vision and Pattern Recognition (CVPR) (2020)

28. Stutz, D.: Retinex theory and algorithm. https://davidstutz.de/retinex-theory-and-algorithm (2021)
29. Wang, R., Zhang, Q., Fu, C.W., Shen, X., Zheng, W.S., Jia, J.: Underexposed photo enhancement using deep illumination estimation. In: Proceedings of the IEEE Conference on Computer Vision and Pattern Recognition (CVPR), pp. 2, 5, 6, 7 (2019)
30. Wang, T., Li, Y., Peng, J., Ma, Y., Wang, X., Song, F., Yan, Y.: Real-time image enhancer via learnable spatial-aware 3D lookup tables. In: Proceedings of the International Conference on Computer Vision (ICCV), pp. 2, 5, 6, 7 (2021)
31. Wang, Z., Cun, X., Bao, J., Liu, J.: Uformer: a general u-shaped transformer for image restoration. In: Proceedings of the IEEE Conference on Computer Vision and Pattern Recognition (CVPR) (2022)
32. Wei, C., Wang, W., Yang, W., Liu, J.: Deep retinex decomposition for low-light enhancement. In: BMVC (2018)
33. Wu, W., Weng, J., Zhang, P., Wang, X., Yang, W., Jiang, J.: URetinex-net: retinex-based deep unfolding network for low-light image enhancement. In: CVPR (2022)
34. Xu, X., Wang, R., Fu, C.W., Jia, J.: SNR-aware low-light image enhancement. In: Proceedings of the IEEE/CVF Conference on Computer Vision and Pattern Recognition (CVPR) (2022)
35. Yang, W., Wang, S., Fang, Y., Wang, Y., Liu, J.: From fidelity to perceptual quality: a semi-supervised approach for low-light image enhancement. In: CVPR (2020)
36. Yang, W., Wang, W., Huang, H., Wang, S., Liu, J.: Sparse gradient regularized deep retinex network for robust low-light image enhancement. IEEE Trans. Image Process. 30, 2072–2086 (2021)
37. Ye, D., Ni, Z., Yang, W., Wang, H., Wang, S., Kwong, S.: Glow in the dark: low-light image enhancement with external memory. IEEE Trans. Multimedia, 1–16 (2023). https://doi.org/10.1109/TMM.2023.3293736
38. Yi, X., Xu, H., Zhang, H., Tang, L., Ma, J.: Diff-Retinex: rethinking low-light image enhancement with a generative diffusion model. In: ICCV (2023)
39. Ying, Z., Li, G., Gao, W.: A bio-inspired multi-exposure fusion framework for low-light image enhancement. arXiv preprint arXiv:1711.00591 p. 5 (2017)
40. Yuhui, W., et al.: Learning semantic-aware knowledge guidance for low-light image enhancement. In: Proceedings of the IEEE/CVF Conference on Computer Vision and Pattern Recognition (2023)
41. Zamir, S.W., Arora, A., Khan, S., Hayat, M., Khan, F.S., Yang, M.: Restormer: efficient transformer for high-resolution image restoration. In: CVPR (2022)
42. Zamir, S.W., et al.: Learning enriched features for real image restoration and enhancement. In: Proceedings of the European Conference on Computer Vision (ECCV) (2020)
43. Zeng, H., Cai, J., Li, L., Cao, Z., Zhang, L.: Learning image-adaptive 3D lookup tables for high performance photo enhancement in real-time. IEEE Trans. Pattern Anal. Mach. Intell., 2, 5, 6, 7 (2020)
44. Zhang, Y., Guo, X., Ma, J., Liu, W., Zhang, J.: Beyond brightening low-light images. IJCV 129(4), 1013–1037 (2021)
45. Zhang, Y., Zhang, J., Guo, X.: Kindling the darkness: a practical low-light image enhancer. In: ACM MM (2019)
46. Zhao, Z., Xiong, B., Wang, L., Ou, Q., Yu, L., Kuang, F.: RetinexDIP: a unified deep framework for low-light image enhancement. IEEE Trans. Circuits Syst. Video Technol. 32(3), 1076–1088 (2022). https://doi.org/10.1109/TCSVT.2021.3073371

Efficient Visual Object Tracking with Temporal Context-Aware Token Learning and Scale Adaptive Token Pruning

Yan Gui[1(✉)], Yiru Ou[1], Ruojun Guo[1], Jianming Zhang[1], and Zhihua Chen[2]

[1] School of Computer and Communication Engineering, Changsha University of Science and Technology, Changsha 410076, Hunan, China
`guiyan@csust.edu.cn`
[2] Department of Computer Science and Engineering, East China University of Science and Technology, Shanghai 200237, Shanghai, China

Abstract. Transformer-based trackers have recently demonstrated remarkable performance in the visual tracking community. However, leveraging rich information across temporal frames is difficult for conventional visual Transformers due to the quadratically increasing complexity of the self-attention computation, thus limiting the performance potential of Transformer-based trackers. In this paper, we propose a more efficient one-stream tracking framework by integrating temporal context-aware token learning and scale-adaptive token pruning. Specifically, we employ a temporal context-aware encoder that simultaneously enhances feature learning and relation modeling by mutually interacting the inherent template, dynamic templates, and the search region solely through self-attention. To balance the tracking accuracy and speed, a scale-adaptive token pruning module is proposed based on learnable scale attention, thus making the search token pruning more adaptable to different target scales. Furthermore, to handle multiple templates during inference and improve tracking robustness, we design a temporal template updating strategy that dynamically selects the templates capturing appearance variations of target objects. The results on five benchmarks show that our tracker obtains superior tracking performance while maintaining fast inference speed.

Keywords: Temporal contexts · Token pruning · Template updating · One-stream framework · Transformer-based trackers

1 Introduction

Visual object tracking (VOT) [1,2] is a hot topic in the field of computer vision, aiming to predict the states of a target object in a video based on the ground truth provided for the target in the first frame. It has a variety of applications in diverse fields including surveillance [3], robotic vision [4], autonomous driving [5], and augmented reality [6]. Despite significant progress, accurate and robust tracking remains challenging for real-world video scenarios of complexity. Large variations in appearance caused by illumination, deformation, motion,

M. Mahmud et al. (Eds.): ICONIP 2024, CCIS 2297, pp. 361–375, 2026.
https://doi.org/10.1007/978-981-96-7036-9_24

and occlusion introduce ambiguities that in turn induce inaccuracies and instabilities in target localization. Besides, the real-time tracking is critical in practical applications. Recently, numerous methods [2,7–13] have been proposed to enhance the performance of trackers. In this work, we consider Siamese-based and Transformer-based tracking methods.

Siamese-based tracking methods [1,2,7,14–16] typically cast VOT as a matching problem. Specifically, they first use a convolutional Siamese backbone to extract deep visual features from the inputs. Then, they are fused by performing a cross-correlation for classification and regression tasks. However, this method only utilizes local CNN (Convolutional Neural Network) features from convolutions for matching and prediction, which limits the ability of Siamese trackers to fully exploit global contextual information. This deficiency hinders the further improvement of location accuracy. Transformer-based methods [12,13,17,18] utilize a self-attention mechanism [19] to effectively capture long-range dependencies of features and learn global features. Trackers [12,13] introduce transformer-based fusion modules to capture non-linear interactions between the template and search region features. However, they still follow a two-stream, two-stage pipeline that involves separate steps for feature learning and interaction. Trackers [20,21] propose one-stream, one-stage tracking architectures that simultaneously extract strong discriminative features and perform comprehensive feature interactions between the template and the search region. However, the aforementioned transformer-based methods incur significantly expensive computational and memory costs due to the attention mechanism, which limits their input resolution and practicality and slows down the tracking speed.

To address this limitation, OSTrack [20] incorporates a similarity prior-based candidate elimination module to remove noisy background candidates. With the candidate elimination module, OSTrack achieves a more stable feature interaction between the template and search region, while enhancing the inference efficiency. Despite its advantages, the fixed token keeping ratio used for the module easily leads to the incorrect elimination of target candidates. This motivates us to explore a flexible candidate elimination mechanism. Thus, we propose a scale adaptive token pruning module that combines a learnable token keeping ratio to dynamically vary the number of tokens. The proposed module not only effectively captures target appearance changes in various video scenes, but also strongly influences the accuracy and speed of tracking.

Additionally, the incorporation of both spatial and temporal contexts is crucial for successful visual tracking. Trackers in [13,17,22] use multiple target templates and explore the spatio-temporal relationships among them through a transformer. STARK [13] feeds a dynamically updated template into the transformer, enabling it to capture temporal information of the target. ROMTrack [23] exploits the temporal contexts by utilizing the encoded features from the previous frame. However, the usage of fixed templates and a single dynamic template inadequately harnesses temporal contextual information within video sequences. In contrast, we propose a temporal template updating strategy to maintain multiple dynamic templates during inference, capturing the appearance changes of the targets and thus improving the tracker's robustness.

In summary, our work has the following three contributions: (1) We are the first to introduce temporally ordered templates into a one-stream tracking framework and exploit rich temporal information across them for discriminative feature representation learning. We also apply a temporal template updating strategy to accommodate the situation of multiple templates. (2) We propose a scale adaptive token pruning module to effectively eliminate noisy tokens by capturing target scale information in videos. (3) The proposed trackers achieve very excellent performance on five benchmarks, including GOT-10k, LaSOT, TrackingNet, OTB100, and NFS30, while reaching the inference speed of 95 FPS. The source codes and models are available at https://github.com/samT0/TCSA.

2   Method

2.1   Overview

We present a more efficient one-stream tracking framework by integrating temporal context-aware token learning and scale-adaptive token pruning, dubbed as TCSA. First, we introduce a temporal context-aware (TCA) encoder to unify the feature extraction and relation modeling, capturing rich spatio-temporal context information among the inherent and dynamic templates. And, it combines a scale adaptive token pruning (SA-TP) module to selectively filter useful tokens, enabling our TCSA to be more adaptive to deformation and large appearance changes. Then, we utilize the output search region features for final target classification and regression. Additionally, we describe the training and inference of TCSA by applying a temporal template updating (TTU) strategy. The architecture of the TCSA is illustrated in Fig. 1.

2.2   Network Architecture

Temporal Context-Aware Encoder. As shown in the right part of Fig. 1, the TCA encoder is made up of several transformer blocks. It can capture a richer temporal context of the target and produce more discriminative token learning and relation modeling by considering additional dynamic templates as input. Specifically, given a template set $(z, t) \in \mathbb{R}^{(1+n) \times 3 \times h \times w}$ and a search region $x \in \mathbb{R}^{3 \times H \times W}$, where z is the inherent template, $t = \{t^i \mid i = 1, \ldots, \}$ represents the dynamic templates, and n is the total number of dynamic templates. We first divide them into non-overlapped patches, which are then flattened into a sequence of patches $z_p \in \mathbb{R}^{N_z \times (3 \times p \times p)}$, $t_p \subset \mathbb{R}^{N_t \times (3 \times p \times p)}$ and $x_p \subset \mathbb{R}^{N_x \times (3 \times p \times p)}$, where $p \times p$ is the spatial resolution of the patch. $N_z = hw/p \times p$, $N_t = nhw/p \times p$, and $N_x = HW/p \times p$ are the number of patches corresponding to the multiple templates and the search region. They are encoded by a patch embedding layer. Then, we add learnable position embeddings to the patch embeddings and get tokens F_z, F_t and F_x, respectively. These tokens are concatenated as $[F_z; F_t; F_x]$ and used as the input for the subsequent self-attention module.

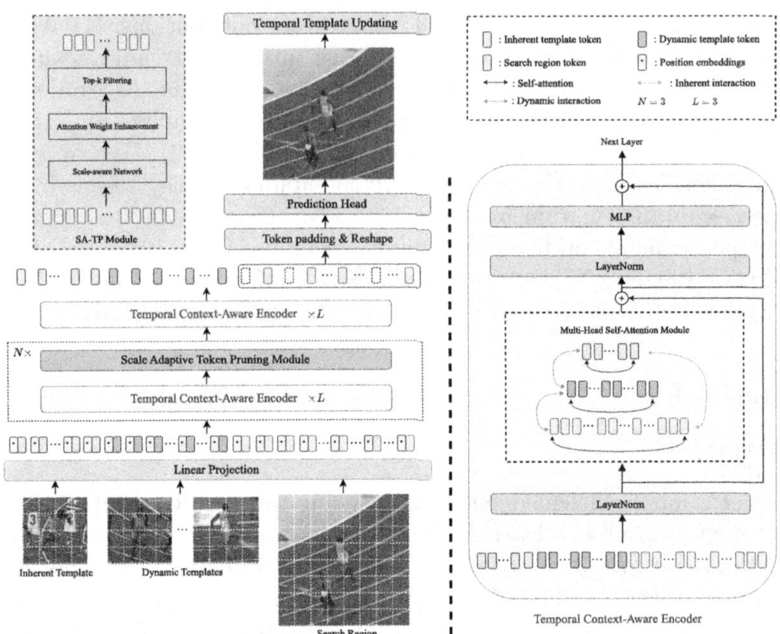

Fig. 1. The overall framework of the proposed TCSA. It takes multiple templates and one search region as inputs. The TCA encoder conducts feature extraction and interaction. Given the learned tokens from the TCA encoder, the SA-TP module effectively eliminates noisy background tokens. And, the TTU module generates new dynamic templates during inference.

For attention computation, we employ three different linear projection layers on $[F_z; F_t; F_x]$ to produce the features $[Q_z; Q_t; Q_x]$, $[K_z; K_t; K_x]$, and $[V_z; V_t; V_x]$. We then conduct self-attention on these features to learn globally enhanced spatio-temporal features A:

$$A = \text{Softmax}\left(\frac{[Q_z; Q_t; Q_x][K_z; K_t; K_x]^T}{\sqrt{d_k}}\right)[V_z; V_t; V_x] \tag{1}$$

$$\text{Softmax}\left(\frac{[Q_z; Q_t; Q_x][K_z; K_t; K_x]^T}{\sqrt{d_k}}\right) = \begin{bmatrix} W_{zz} & W_{zt} & W_{zx} \\ W_{tz} & W_{tt} & W_{tx} \\ W_{xz} & W_{xt} & W_{xx} \end{bmatrix} \tag{2}$$

where W_{zx} (W_{tx}) represents the weight matrix between the inherent template (dynamic templates) and the search region. W_{xx} is the weight matrix measured within the search region, while W_{zz} and W_{tt} are similar. Based on Eq. 2, the output features A can be rewritten as:

$$A = \begin{bmatrix} W_{zz} & W_{zt} & W_{zx} \\ W_{tz} & W_{tt} & W_{tx} \\ W_{xz} & W_{xt} & W_{xx} \end{bmatrix} \begin{bmatrix} V_z \\ V_t \\ V_x \end{bmatrix} = \begin{bmatrix} W_{zz}V_z + W_{zt}V_t + W_{zx}V_x \\ W_{tz}V_z + W_{tt}V_t + W_{tx}V_x \\ W_{xz}V_z + W_{xt}V_t + W_{xx}V_x \end{bmatrix} \triangleq \begin{bmatrix} A_z \\ A_t \\ A_x \end{bmatrix} \tag{3}$$

where A_z, A_t, and A_x represent the enhanced inherent template, dynamic templates, and search region features, respectively. A_z has aggregated informative features from the dynamic template (e.g., $W_{zt} V_t$) and search region (e.g., $W_{zx} V_x$). The $W_{zt} V_t$ and $W_{zx} V_x$ terms in A_z have respectively transferred the features of dynamic templates and search region into the inherent template, helping the output inherent template tokens to leverage contextual appearance information of the dynamic templates and eliminate inference from the search region. Similarly, A_t has aggregated features from the inherent template and search region (e.g., $W_{tz} V_z$, and $W_{tx} V_x$). A_x has aggregated features from the inherent template and dynamic templates (e.g., $W_{xz} V_z$, and $W_{xt} V_t$). This further enhances feature extraction and relation modeling, resulting in more discriminative features. In our implementation, we have set up 12 TCA encoders and 3 dynamic templates.

Scale Adaptive Token Pruning Module. We use the SA-TP module to capture the scale information of targets, enhance the discrimination of tokens in the search region, and effectively filter out tokens of the background region. Given the search region tokens $A_x \in \mathbb{R}^{N_x \times C}$ (see details in Eq. 3), we first employ a scale-aware sub-network, depicted in Fig. 2, to compute a scale attention $\alpha \in [0, 1]^{1 \times 1}$:

$$\alpha = \mathrm{Sigmoid}\left(\delta\left(\mathrm{FC}\left(\mathrm{MP}\left(A_x\right)\right)\right)\right), \tag{4}$$

where MP(.) and FC(.) denote the maximum pooling layer and the fully connected layer. $\delta(.)$ and Sigmoid(.) represent the ReLU and the sigmoid function. The α varies in each transformer block, allowing it to adapt to targets of different scales. Then, we apply the scale attention α to the attention weights W_{zx} and W_{tx} (or W_{xz} and W_{xt}, see details in Eq. 2) by division operation, to obtain enhanced attention weights W'_{zx} and W'_{tx}:

$$[W'_{zx}; W'_{tx}] = [W_{zx}; W_{tx}]/\alpha, \tag{5}$$

where the values of elements in W'_{zx} and W'_{tx} increase if the template tokens are highly similar to the search region tokens, and decrease otherwise. Thus, the discrimination between the token pairs in the templates and search region is further enhanced.

Furthermore, to mitigate the adverse effects caused by noisy background regions in the templates, we use W'_{zx} and W'_{tx}, and average them by $\bar{W}'_{z't'x} = \mathrm{avg}(W'_{z'x} + W'_{t'x})$, where the subscripts z' and t' represent tokens corresponding to the center parts of the inherent and dynamic templates, and avg(.) denotes the averaging operation. Finally, to reduce the number of tokens in A_x, we can apply top-k filtering and keep only the k largest elements in $\bar{W}'_{z't'x}$, producing the pruned search region tokens $A_{x'} \in \mathbb{R}^{k \times C}$, where the hyperparameter $k = \alpha \times N_x$ is the token keeping length. The background tokens of the search region are thus effectively identified and eliminated. Note that the above description is based on having a single attention head, but it remains similar in the case of multiple attention heads. In our experiment, we insert the proposed SA-TP module after

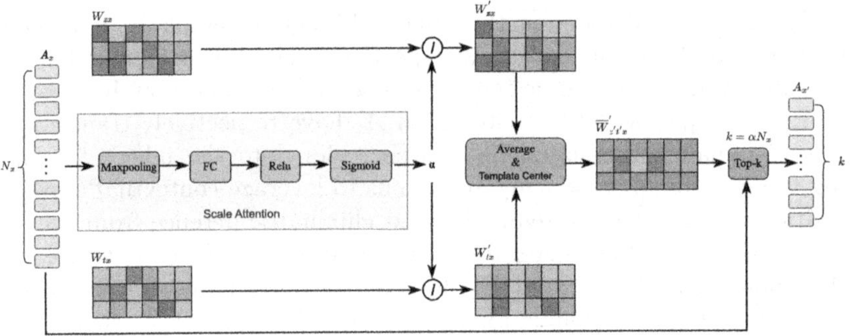

Fig. 2. Illustration for scale adaptive token pruning module.

the 3rd, 6th and 9th TCA encoder layers for better token pruning. In addition, by zero-padding the positions of the discarded tokens, we obtain the padded sequence of search region tokens. This sequence is then reshaped into a feature map and sent to a center-based localization head [24] for target box predictions.

In Fig. 3, we present the token pruning results of three TCA encoder layers on three practical application examples. The input images in the top row, which are selected from the same video sequence, contain a target object with different scales. The input images in the next row are the remaining two examples. The results show that the proposed scale-adaptive token pruning can effectively capture the scale information of the target objects and dynamically remove irrelevant background tokens based on the learnable token keeping ratios. This facilitates enhancing the efficiency and accuracy of our TCSA.

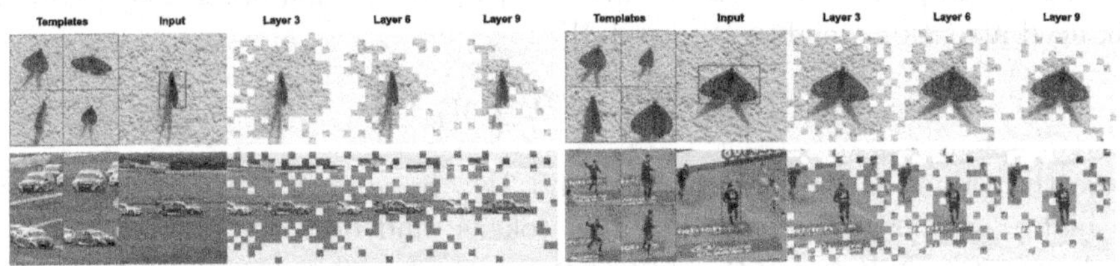

Fig. 3. Results of the scale-adaptive token pruning module on three practical application examples. The templates consist of one inherent template and three dynamic templates. The input image is the search region. The templates and the target object are highlighted with red boxes. The masked regions in the results indicate the discarded background tokens. (Color figure online)

2.3   Training and Inference

Training. During offline training, we randomly select 5 temporally ordered frames from a pre-defined time interval (e.g., 200) in a video as a training sample. Note that, there is no restriction on the time interval between two sequentially ordered frames. The first and last frames respectively serve as the inherent template and the search frames, while the middle three frames are used as the dynamic template frames. For network training, the Gaussian-weighted focal loss [25] is employed for training the classification branch, while the generalized IoU (Intersection-over-Union) loss [26] and L_1 loss are utilized for training the regression branches. The formulation of our loss is as follows:

$$L_{\text{total}} = \lambda_1 L_{\text{cls}} + \lambda_2 L_{\text{giou}} + \lambda_3 L_1 \tag{6}$$

In our experiment, we set $\lambda_1{=}1$, $\lambda_2{=}2$, and $\lambda_3{=}5$.

Temporal Template Updating. We dynamically maintain multiple temporal templates to better adapt to appearance changes. Initially, the dynamic template set t is constructed by duplicating the inherent template from the first frame n times. Considering that high-quality templates contribute to superior tracking performance, the given predicted target object b_i is treated as a candidate template when its confidence score r_i is greater than 0.7. Moreover, when r_i is less than the highest confidence score of t, i.e., $r_i < max(r_i)$, indicating a large appearance change, our TTU will generate a new template from b_i and use it to update the dynamic template set, where r_i represents a set of confidence score of dynamic templates in t. We replace the template in t with the highest confidence score using the new template, ensuring that the number of templates in t is always equal to n. In our experiment, we set $n = 3$ for the best performance. The main steps of temporal templates updating are summarized in Algorithm 1.

Algorithm 1 Inference with temporal template updating

Input: *video frames* $V = [I_1, .., I_n]$, *first box* $b_{initial}$, *and* $TCSA(\cdot)$;
Output: *tracking result for each frame* $B = [b_1, ..., b_n]$;
  1: **for** *each frame* I_i *in* V **do**
  2:    **if** $i == 1$ **then**
  3:      *initial templates* $z = crop(I_i, b_{inital})$, $t = [t_1, ..., t_m] = z$;
  4:      *predict box* $b_i = b_{initial}$, $rt = [rt_1, ..., rt_m] = 1$;
  5:    **else**
  6:      *Search Region* $x = crop(I_i, b_{i-1})$
  7:      $b_i, r_i = TCSA(x, t, z)$;
  8:      **if** $r_i > 0.7$ **then**
  9:        $j = argmax(rt)$
10:        **if** $r_i < max(rt)$ **then**
11:          $rt[j] = r_i$
12:          $t[j] = crop(I_i, b_i)$
13:        **end if**
14:      **end if**
15:    **end if**
16:    *Predictions* $B[i] = b_i$;
17: **end for**
18: *return* $B = [b_1, ..., b_n]$;

Inference. During inference, we feed one inherent template, n dynamic templates, and a search region into the network to generate the confidence scores and target bounding box. The dynamic templates are updated only when the new templates are generated by the temporal template updating strategy.

3   Experiments

3.1   Implementation Details

We take the ViT-Base [27] model pre-trained on ImageNet [28] as the main architecture. By considering two different input resolutions, we have TCSA-256 and TCSA-384. The training is performed on two NVIDIA RTX 3090 GPUs with 24 GB memory. We use COCO [29], GOT-10k [30], LaSOT [31], and TrackingNet [32] as the training datasets. During training, we consider horizontal flipping and brightness jittering for data augmentations. AdamW [33] is used as the optimizer, and the weight decay is set to 10^{-4}. We set initial learning rates of the backbone and other modules to 4×10^{-5} and 4×10^{-4}, which are reduced by 10 times after 100 epochs. We conduct 300 training epochs, with each epoch containing 60,000 training samples.

3.2   Comparison with State-of-the-Arts

GOT-10k. GOT-10k [30] contains more than 1000 videos and has 180 videos for testing. The results are evaluated using the official server. As shown in Table 1, TCSA-256 outperforms all other trackers in all three metrics, e.g., 3.5% in $SR_{0.75}$ compared to SeqTrack [34], and 2.6% in AO (Average Overlap) compared to ROMTrack [23]. Both SeqTrack and ROMTrack only employ an additional dynamic template, which has a limited ability to capture various states of target objects in complex video scenarios. As a result, they fail to fully leverage temporal context information. Furthermore, our TCSA-384 further improves the performance on the GOT-10k test set. This demonstrates that our tracker can accurately identify and locate target objects in videos with complex scenarios through discriminative feature extraction and robust relation modeling.

LaSOT. LaSOT [31] includes 1120 long-term videos for training and 280 videos for testing. Table 1 shows that our TCSA-384 and TCSA-256 achieve AUC (Area Under the Curve) scores of 71.3% and 69.7%, respectively, as well as P_{norm} (normalized Precision) scores of 80.8% and 79.6%. The results are highly comparable to SeqTrack and ROMTrack. Besides, TCSA-256 runs at 95 FPS, being 2.4 times faster than SeqTrack. With TCA encoder integrating spatio-temporal context information and SA-TP module eliminating noisy tokens, the performance of our tracker strikes a good balance, making it suited for long-term tracking scenarios.

TrackingNet. TrackingNet [32] contains over 30,000 short-term video sequences. The evaluation of our TCSA is conducted on the test set, which includes 511 videos. The tracking results are evaluated on the office server for performance evaluation. From Table 1, our TCSA-384 obtains a good performance, with an AUC score of 84.4% and a P_{norm} score of 89.0%, respectively, while running at 51 FPS. This demonstrates the advantages of our tracker in short-term visual tracking.

Table 1. Comparison results on GOT-10k, LaSOT, and TrackingNet. The best result is highlighted in bold and the second best results is highlighted in underline. * denotes the tracker trained using the GOT-10k training set.

Method	GOT-10k*			LaSOT			TrackingNet			FPS
	AO	$SR_{0.5}$	$SR_{0.75}$	AUC	P_{Norm}	P	AUC	P_{Norm}	P	
TCSA-256	75.5	85.9	75.3	69.7	79.6	75.7	84.1	88.8	82.5	95
SeqTrack-256 [34]	74.7	84.7	71.8	69.9	79.7	76.3	83.3	88.3	82.2	40
ROMTrack-256 [23]	72.9	82.9	70.2	69.3	78.8	75.6	83.6	88.4	82.7	62
MixFormer-22k [22]	72.6	82.2	68.8	69.2	78.7	74.7	83.1	88.1	81.6	92
MixFormer-11k [22]	71.2	79.9	65.8	67.9	77.3	73.9	82.6	87.7	81.2	92
SimTrack-B/16 [21]	68.6	78.9	62.4	69.3	78.5	74.0	-	-	-	-
SwinTrack-T224 [35]	71.3	81.9	64.5	67.2	-	70.8	81.1	-	78.4	98
OSTrack-256 [20]	71.0	80.4	68.2	69.1	78.7	75.2	83.1	87.8	82.0	105
AiATrack [36]	69.6	80.0	67.8	69.0	79.4	73.8	82.7	87.8	80.4	38
STARK [13]	68.8	78.1	64.1	67.1	77.0	-	82.0	86.9	-	41
TransT [12]	67.1	76.8	60.9	64.9	73.8	69.0	81.4	86.7	80.3	30
TrDiMP [17]	67.1	77.7	58.3	63.9	-	61.4	78.4	83.3	73.1	35
SiamR-CNN [41]	64.9	72.8	59.7	64.8	72.2	-	81.2	85.4	80.0	5
SiamAttn [42]	-	-	-	56.0	64.8	-	75.2	71.7	-	45
Ocean [8]	61.1	72.1	47.3	56.0	65.1	56.6	-	-	-	58
SiamRPN++ [9]	51.7	61.6	32.5	49.6	56.9	49.1	73.3	80.0	69.4	35
SiamFC [2]	34.8	35.3	7.8	33.6	42.0	33.9	57.1	66.3	53.3	58
TCSA-384	78.6	88.2	78.7	71.3	80.8	77.7	84.4	89.0	83.6	51
SeqTrack-384 [34]	74.5	84.3	71.4	71.5	81.1	77.8	83.9	88.8	83.6	15
ROMTrack-384 [23]	74.2	84.3	72.4	71.4	81.4	78.2	84.1	89.0	83.7	-
MixFormer-L [22]	-	-	-	70.1	79.9	76.3	83.9	88.9	83.1	18
SimTrack-L [21]	69.8	78.8	66.0	69.3	78.5	74.0	83.4	87.4	-	-
OStrack-384 [20]	73.7	83.2	70.8	71.1	81.1	77.6	83.9	88.5	83.2	58

Table 2. Comparison results on OTB100 and NFS30. The results are compared based on the AUC (%) score. The best result is highlighted in bold and the second best results is highlighted in underline.

	TransT [12]	STARK [13]	KeepTrack [37]	ToMP [38]	MixFormer −L [22]	OSTrack −384 [20]	ROMTrack −256 [23]	SeqTrack −256 [34]	**TCSA −256**	**TCSA −384**
OTB100	69.4	68.5	70.9	70.1	70.4	70.1	71.4	-	72.6	72.3
NFS30	65.7	65.2	66.4	66.7	-	66.5	68.0	67.6	68.5	68.9

OTB100 and NFS30. We also verify the performance of our TCSA on two other popular benchmarks, including OTB100 [39] and NFS30 [40]. The results are shown in Table 2. Our TCSA-256 and TCSA-384 achieve the best performance on both benchmarks, further indicating their generalizability.

3.3 Ablation Study and Analysis

Study on Key Components. We present the performance of our models on the LaSOT in Table 3. With the use of all proposed components, our TCSA-384 and TCSA-256 obtain the highest AUC score and P_{Norm} score. For the TCSA-256, when we disable the SA-TP module in the model, the performance degrades to 69.1% and 78.6%, respectively. This is because the SA-TP module

Table 3. Ablation study on key components of our models. Bold indicates the best results.

Model	SA-TP	TTU	LaSOT		
			AUC	P_{Norm}	P
TCSA-384	✓	✓	**71.3**	**80.8**	**77.7**
TCSA-256	✓	✓	**69.7**	**79.6**	**75.7**
	✓		68.7	78.0	74.2
		✓	69.1	78.6	75.1
			66.8	76.1	73.2

Table 4. Ablation study on dynamic template number. Bold indicates the best results.

# of dynamic template	LaSOT			FPS
	AUC	P_{Norm}	P	
0	67.4	76.7	73.5	**148**
1	68.3	77.5	74.2	129
2	69.1	78.9	75.1	112
3	**69.7**	**79.6**	**75.7**	95
4	69.3	79.1	75.2	77

can effectively mitigate the influence of background tokens during token learning. Additionally, removing the TTU module in TCSA-256 decreases the AUC score by 1.0% and the P_{Norm} score by 1.6%. The main reason is that the TTU module assists the tracker in adapting large appearance variations. Thus, both SA-TP and TTU are essential components for achieving optimal performance.

Study on Dynamic Template Number. We test the effect of the number of dynamic templates on the model. Table 4 shows the results on LaSOT. Our TCSA-256 can get much better performance with the use of multiple dynamic templates. When we input an additional dynamic template, TCSA-256 achieves improved performance. This is because it combines an inherent template to provide the target object with different states and rich temporal information, enhancing the target feature representation. When the number of dynamic templates is set to 3, TCSA-256 obtains the highest AUC score of 69.7% on LaSOT, demonstrating its strong adaptability to the target appearance changes through temporal information modeling. However, having more than 3 dynamic templates leads to worse performance, as it introduces redundant information and unwanted noise into the model, which negatively affects tracking.

Study on Different Token Pruning Modules. The number of tokens can dynamically vary when the token pruning module is used during feature learning. In Table 5, we present the effect of pre-defined and learnable token keeping rates α on model performance. Overall, integrating the token pruning module in

Table 5. Ablation study on different token pruning modules. Bold indicates the best results.

Method	α	LaSOT			MACs (G)	FPS
		AUC	P_{Norm}	P		
Baseline	-	69.1	78.6	75.1	29.0	84
Pre-defined value	0.6	68.6	78.4	74.7	19.6	**106**
	0.7	69.2	79.0	75.1	21.9	101
	0.8	68.9	78.7	74.9	23.6	93
	0.9	68.8	78.6	74.8	26.2	90
Scale adaptive	[0.6,0.8]	**69.7**	**79.6**	**75.7**	**21.5**	95

model can significantly improve the tracking performance. For the token pruning module with a pre-defined α, the performance changes less. Obviously, using an unreasonable α easily leads to either under-pruning or over-pruning. When we use the learned α ranging from 0.6 to 0.8, the TCSA-256 can boost the AUC score on LaSOT by 0.5%, while achieving comparable inference speed and decreasing the MACs by 0.4%. This verifies that the SA-TP module can effectively eliminate noisy tokens by capturing target scales.

3.4 Qualitative Comparison and Application

We show the tracking results of the eight challenging image sequences of GOT-10k in Fig. 4, generated by our TCSA and three other trackers from OSTrack [20], Mixformer [22], and ROMTrack [23]. Overall, our tracker achieves favorable results in most of these image sequences.

Fig. 4. The bounding boxes results of our TCSA and other three trackers on eight challenging video sequences from GOT-10k.

Especially for scale variation, occlusion, and fast motion, as shown in the first three rows of Fig. 4, our TCSA can accurately locate the target; however, the compared trackers exhibit limited performance in handling these challenging scenarios. For other examples shown in the last two rows of Fig. 4, the target object in each of the image sequences has large appearance changes due to scale variation. Our TCSA generates relatively accurate bounding boxes for the target, as we introduce scale-adaptive token pruning and temporal template updating into our method to improve tracking accuracy and robustness.

We also apply our TCSA to video surveillance and autonomous driving. In Fig. 5, we show the tracking bounding boxes of OSTrack [20], ROMTrack [23], and our TCSA on four challenging video sequences. The first two examples are indoor and outdoor surveillance videos from the OTB100 dataset, while the last two examples are aerial videos from UAV123. It can be seen that ROMTrack suffers from significant drifts in situations involving similar and small objects, due to a lack of an efficient template updating mechanism. OSTrack obtains improved prediction boxes in the face of these challenges by utilizing dynamic templates to handle object appearance changes. Apparently, our TCSA produces tracking results that are sufficiently accurate compared to the ground truth.

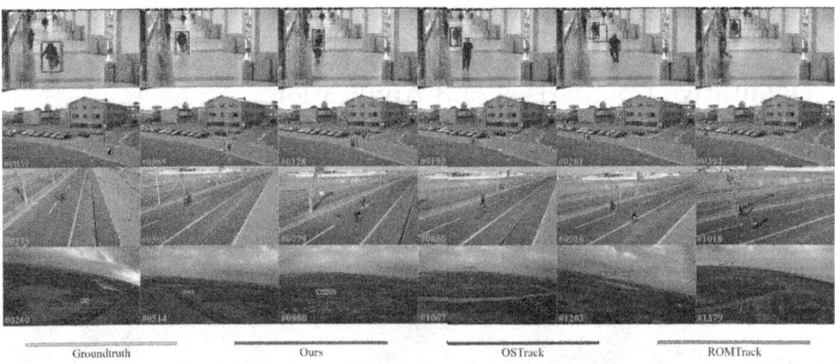

Fig. 5. Results of OSTrack, ROMTrack, and ours on the OTB100 and UAV123.

4   Conclusion

This paper presents TCSA, an efficient one-stream framework for visual object tracking. It incorporates temporal context-aware feature representation enhancement and scale-adaptive token pruning, aiming to fully exploit temporal contexts across multiple templates and achieve a good speed and tracking accuracy. In addition, the proposed temporal template updating strategy is introduced to maintain dynamically evolving templates during inference. The results demonstrate that our trackers surpass recent trackers in terms of tracking performance.

One major limitation of the TCSA is that the SA-TP module mistakenly removes tokens that are useful for target localization, resulting in the loss of information and thus decreasing tracking performance. To mitigate this, we perform token pruning at only a few layers of the Transformer encoder using the SA-TP module. Another limitation is that the use of temporal template updating during inference results in heavy computation, which limits the trackers' ability to achieve fast tracking speed. Therefore, we plan to explore lightweight transformer-based tracking architectures to reduce the computational complexity and develop an efficient multimodal visual Transformer based on our TCSA for RGB-T tracking.

Acknowledgement. This work was supported by National Natural Science Foundation of China under Grants 62272164 and 61972056, and Hunan Provincial Natural Science Foundation of China under Grant 2023JJ30050.

References

1. Tao, R., Gavves, E., Smeulders, A.W.: Siamese instance search for tracking. In: Proceedings of the IEEE Conference on Computer Vision and Pattern Recognition, pp. 1420–1429 (2016)
2. Bertinetto, L., Valmadre, J., Henriques, J.F., Vedaldi, A., Torr, P.H.: Fully-convolutional Siamese networks for object tracking. In: Proceedings of European Conference on Computer Vision, pp. 850–865 (2016)
3. Emami, A., Dadgostar, F., Bigdeli, A., Lovell, B.C.: Role of spatiotemporal oriented energy features for robust visual tracking in video surveillance. In: IEEE Ninth International Conference on Advanced Video and Signal-Based Surveillance, pp. 349–354 (2012)
4. Liu, L., Xing, J., Ai, H., Ruan, X.: Hand posture recognition using finger geometric feature. In: Proceedings of the 21st International Conference on Pattern Recognition, pp. 565–568 (2012)
5. Lee, K.H., Hwang, J.N.: On-road pedestrian tracking across multiple driving recorders. In: IEEE Trans. Multimedia, pp. 1429–1438 (2015)
6. Zhang, G., Vela, P.A.: Good features to track for visual slam. In: Proceedings of the IEEE Conference on Computer Vision and Pattern Recognition, pp. 1373–1382 (2015)
7. Li, B., Yan, J., Wu, W., Zhu, Z., Hu, X.: High performance visual tracking with Siamese region proposal network. In: Proceedings of the IEEE Conference on Computer Vision and Pattern Recognition, pp. 8971–8980 (2018)
8. Zhang, Z., Peng, H., Fu, J., Li, B., Hu, W.: Ocean: object-aware anchor-free tracking. In: Proceedings of European Conference on Computer Vision, pp. 771–787 (2020)
9. Li, B., Wu, W., Wang, Q., Zhang, F., Xing, J. and Yan, J.: SiamRPN++: evolution of Siamese visual tracking with very deep networks. In: Proceedings of the IEEE/CVF Conference on Computer Vision and Pattern Recognition, pp. 4282–4291 (2019)
10. Guo, D., Wang, J., Cui, Y., Wang, Z., Chen, S.: SiamCAR: Siamese fully convolutional classification and regression for visual tracking. In: Proceedings of the IEEE/CVF Conference on Computer Vision and Pattern Recognition, pp. 6269–6277 (2020)

11. Wang, Q., Teng, Z., Xing, J., Gao, J., Hu, W., Maybank, S.: Learning attentions: residual attentional Siamese network for high performance online visual tracking. In: Proceedings of the IEEE Conference on Computer Vision and Pattern Recognition, pp. 4854–4863 (2018)
12. Chen, X., Yan, B., Zhu, J., Wang, D., Yang, X., Lu, H.: Transformer tracking. In: Proceedings of the IEEE/CVF Conference on Computer Vision and Pattern Recognition, pp. 8126–8135 (2021)
13. Yan, B., Peng, H., Fu, J., Wang, D., Lu, H.: Learning spatio-temporal transformer for visual tracking. In: Proceedings of the IEEE/CVF International Conference on Computer Vision, pp. 10448–10457 (2021)
14. Zhu, Z., Wang, Q., Li, B., Wu, W., Yan, J.M., Hu, W.: Distractor-aware Siamese networks for visual object tracking. In: Proceedings of the European Conference on Computer Vision, pp. 101–117 (2018)
15. Fan, H., Ling, H.: Siamese cascaded region proposal networks for real-time visual tracking. In: Proceedings of the IEEE/CVF Conference on Computer Vision and Pattern Recognition, pp. 7952–7961 (2019)
16. Du, F., Liu, P., Zhao, W., Tang, X.: Correlation-guided attention for corner detection based visual tracking. In: Proceedings of the IEEE/CVF Conference on Computer Vision and Pattern Recognition, pp. 6836–6845 (2020)
17. Wang, N., Zhou, W., Wang, J., Li, H.: Transformer meets tracker: exploiting temporal context for robust visual tracking. In: Proceedings of the IEEE/CVF Conference on Computer Vision and Pattern Recognition, pp. 1571–1580 (2021)
18. Zhang, J., He, Y., Chen, W., Kuang, L.D., Zheng, B.: CorrFormer: context-aware tracking with cross-correlation and transformer. In: Proceedings of Computers and Electrical Engineering, p.109075 (2024)
19. Vaswani, A., et al.: Attention is all you need. In: Advances in Neural Information Processing Systems, vol. 30 (2017)
20. Ye, B., Chang, H., Ma, B., Shan, S, Chen, X.: Joint feature learning and relation modeling for tracking: a one-stream framework. In: European Conference on Computer Vision, pp. 341–357 (2022)
21. Chen, B., et al.: Backbone is all your need: a simplified architecture for visual object tracking. In: European Conference on Computer Vision, pp. 375–392 (2022)
22. Cui, Y., Jiang, C., Wang, L., Wu, G.: MixFormer: end-to-end tracking with iterative mixed attention. In: Proceedings of the IEEE/CVF Conference on Computer Vision and Pattern Recognition, pp. 13608–13618 (2022)
23. Cai, Y., Liu, J., Tang, J., Wu, G.: Robust object modeling for visual tracking. In: Proceedings of the IEEE/CVF International Conference on Computer Vision, pp. 9589–9600 (2023)
24. Zhou, X., Wang, D., Krähenbühl, P.: Objects as points. arxiv preprint arxiv:1904.07850 (2019)
25. Law, H., Deng, J.: CornerNet: detecting objects as paired keypoints. In: Proceedings of the European Conference on Computer Vision, pp. 734–750 (2018)
26. Rezatofighi, H., Tsoi, N., Gwak, J., Sadeghian, A., Reid, I., Savarese, S.: Generalized intersection over union: a metric and a loss for bounding box regression. In: Proceedings of the IEEE/CVF Conference on Computer Vision and Pattern Recognition, pp. 658–666 (2019)
27. Dosovitskiy, A., et al.: An image is worth 16×16 words: transformers for image recognition at scale. arxiv preprint arxiv:2010.11929 (2020)
28. Deng, J., Dong, W., Socher, R., Li, L.J., Li, K., Fei-Fei, L.: ImageNet: a large-scale hierarchical image database. In: IEEE Conference on Computer Vision and Pattern Recognition, pp. 248–255 (2009)

29. Lin, T.Y., et al: Microsoft COCO: common objects in context. In: Proceedings of European Conference on Computer Vision, pp. 740–755 (2014)
30. Huang, L., Zhao, X., Huang, K.: GOT-10k: a large high-diversity benchmark for generic object tracking in the wild. In: Proceedings of IEEE Transactions on Pattern Analysis and Machine Intelligence, pp. 1562–1577 (2019)
31. Fan, H., et al.: A high-quality benchmark for large-scale single object tracking. In: Proceedings of the IEEE/CVF Conference on Computer Vision and Pattern Recognition, pp. 5374–5383 (2019)
32. Muller, M., Bibi, A., Giancola, S., Alsubaihi, S., Ghanem, B.: TrackingNet: a large-scale dataset and benchmark for object tracking in the wild. In: Proceedings of the European Conference on Computer Vision, pp. 300–317 (2018)
33. Loshchilov, I., Hutter, F.: Decoupled weight decay regularization. arxiv preprint arxiv:1711.05101 (2017)
34. Chen, X., Peng, H., Wang, D., Lu, H., Hu, H.: SeqTrack: sequence to sequence learning for visual object tracking. In: Proceedings of the IEEE/CVF Conference on Computer Vision and Pattern Recognition, pp. 14572–14581 (2023)
35. Lin, L., Fan, H., Zhang, Z., Xu, Y., Ling, H.: SwinTrack: a simple and strong baseline for transformer tracking. In: Advances in Neural Information Processing Systems, pp. 16743–16754 (2022)
36. Gao, S., Zhou, C., Ma, C., Wang, X., Yuan, J.: AiATrack: attention in attention for transformer visual tracking. In: European Conference on Computer Vision, pp. 146-164 (2022)
37. Mayer, C., Danelljan, M., Paudel, D.P., Van Gool, L.: Learning target candidate association to keep track of what not to track. In: Proceedings of the IEEE/CVF International Conference on Computer Vision, pp. 13444–13454 (2021)
38. Mayer, C., et al.: Transforming model prediction for tracking. In: Proceedings of the IEEE/CVF Conference on Computer Vision and Pattern Recognition, pp. 8731–8740 (2022)
39. Wu, Yi., Lim, J., Yang, Mi.: Object tracking benchmark. IEEE Trans. Pattern Anal. Mach. Intell. **37**(9), 1834–1848 (2015)
40. Kiani Galoogahi, H., Fagg, A., Huang, C., Ramanan, D., Lucey, S.: Need for speed: a benchmark for higher frame rate object tracking. In: Proceedings of the IEEE International Conference on Computer Vision, pp. 1125–1134 (2017)
41. Voigtlaender, P., Luiten, J., Torr, P.H., Leibe, B.: Siam R-CNN: visual tracking by re-detection. In: Proceedings of the IEEE/CVF Conference on Computer Vision and Pattern Recognition, pp. 6578–6588 (2020)
42. Yu, Y., xiong, Y., Huang, W., Scott, M.R.: Deformable Siamese attention networks for visual object tracking. In: Proceedings of the IEEE/CVF Conference on Computer Vision and Pattern Recognition, pp. 6728–6737 (2020)

Towards Unveiling the Potential of Fuzzy Values as Features: A Comparative Study in Cybercrime Text Analysis

Faizad Ullah[(✉)] [iD], Muhammad Sohaib Ayub [iD], Ali Faheem [iD], Mian Muhammad Awais [iD], and Asim Karim [iD]

Department of Computer Science, Lahore University of Management Sciences, Lahore, Pakistan
`{20030057,15030039,ali.faheem,awais,akarim}@lums.edu.pk`

Abstract. Accurate detection and classification of cybercrime text present significant challenges for machine learning models, primarily due to the data's complex boundaries and overlapping characteristics. In this context, the role of data features becomes critical, as they provide crucial insights and prejudiced strength necessary to devastate the inherent complexities and enhance the model's accuracy. This paper proposes a novel approach incorporating fuzzy values as features with standard feature extraction techniques to overcome issues arising from unclear boundaries in cybercrime and hate speech texts. By assigning fuzzy values to individual tweets, we capture the degree of relatedness to different cybercrime classes, providing valuable insights into their associations. Additionally, we explore the potential of feature fusion by combining fuzzy values with Bag-of-Words (BoW) and Term Frequency-Inverse Document Frequency (TF-IDF) representations. This fusion results in a more discriminative and informative feature set that captures semantic relevance and contextual significance. Through extensive experimental evaluations, we demonstrate the potential of our proposed approach compared to standard feature extraction techniques, highlighting its effectiveness in handling the complexities of cybercrime boundaries. We present the evaluation of the RUHSOLD and state-of-the-art Cybercrimes in Roman Urdu (CRU) dataset, contribute to advancing cybercrime detection methodologies, and encourage further investigations in multi-class classification challenges within cybersecurity.

Keywords: Cybercrime Detection · Fuzzy Values · Intelligent Computing · Multi-class Classification · Semantic Analysis

1 Introduction

The steadily increasing online communication and the proliferation of social networking platforms have created new avenues for cybercrime. Such crimes cause issues for individuals, organizations, and, occasionally, national security. Therefore, the timely detection of these crimes is crucial. With the increasing amount

M. Mahmud et al. (Eds.): ICONIP 2024, CCIS 2297, pp. 376–387, 2026.
https://doi.org/10.1007/978-981-96-7036-9_25

of text data generated from online activities, analyzing and classifying cybercrime text data has become a crucial challenge in cybercrime detection and prevention. Natural Language Processing (NLP) techniques have shown promise in tackling this challenge by enabling the automated analysis of large volumes of text data. However, automatically classifying such data is a complex task, as these crimes have a thin border in differentiating cybercrimes text from other offensive (non-cybercrimes) text [16]. To address the challenges posed by cybercrime text classification, our research aims to develop an advanced classification framework that leverages feature extraction and NLP techniques. By incorporating machine learning algorithms and standard feature extraction methods, we aim to identify cybercrime and hate speech (a type of cybercrime) related text data accurately and efficiently. Our approach also focuses on differentiating cybercrimes from other offensive text, facilitating timely detection and prevention of cybercrime activities in online platforms.

In recent years, Machine Learning (ML) models and advanced pre-trained multilingual transformers have demonstrated impressive results in various NLP tasks, including text classification [9]. Machine learning algorithms can automatically learn from data and identify patterns indicative of cybercrime activities. However, machine learning algorithms' effectiveness heavily depends on the quality of the features used to represent the textual data. The accurate classification of cybercrime text remains challenging for machine learning models, primarily attributed to the intricate boundaries and overlapping characteristics inherent in the data. These challenges are further compounded by the confusion experienced by human annotators during the annotation process, stemming from the intricate and overlapping nature of cybercrime-related content. In this context, the significance of data features becomes increasingly apparent, as they serve as vital sources of insight and discriminatory power necessary to navigate the complexities and enhance the precision of text classification in cybercrime detection.

Traditional machine learning models often struggle to accurately classify cybercrime instances due to the complex and dynamic nature of the problem. Cybercrime tweets can exhibit various levels of relatedness to multiple categories, blurring the lines between different types of cybercrimes [16]. Assigning a crisp label to such instances fails to capture the data's subtle nuances and shades of grey. One option could be to turn it into a multi-label classification problem, but the utilized datasets, i.e., CRU and RUHSOLD are designed for multi-class classification. To address this limitation, we introduce the concept of fuzzy values, which serve as a measure of relatedness or membership to each cybercrime class. For this purpose, we utilizes six function such as Quick Ratio (QR), Partial Ratio (PR), Partial Token Set Ratio (PTSER), Partial Token Sort Ratio (PTSOR), Token Set Ratio (TSER), and Token Sort Ratio (TSOR) from FuzzyWuzzy[1] library. We enable a more flexible and granular representation of cybercrime instances by assigning fuzzy values produced by these functions to each tweet. These fuzzy values reflect the similarity between a tweet and each cybercrime class, accounting for the data's inherent uncertainties and overlapping

---

[1] https://pypi.org/project/fuzzywuzzy/.

characteristics. Rather than forcing a tweet into a single, discrete category, our approach adopts the intrinsic complexity of cybercrime classification and provides a more nuanced understanding of the data.

Incorporating fuzzy values as features empowers machine learning models to make informed decisions based on the degrees of relatedness to different cybercrime classes. This allows for more accurate and robust classification, better capturing the fine boundaries between various cybercrimes. The fuzzy values serve as a mechanism to quantify the uncertainty and ambiguity in the data, providing a richer representation that aligns with the complexities of real-world cybercrime instances. To evaluate the effectiveness of our proposed approach, we compare its performance with classical machine learning models on the benchmark cybercrime dataset for Roman Urdu (CRU) [16] and Roman Urdu Hate Speech and Offensive Language Dataset (RUHSOLD) [13]. We also compare the results with advanced pre-trained multilingual transformers to assess the impact of our fuzzy value-based approach. Through extensive experimentation and analysis, we demonstrate the admirable performance of our method in accurately classifying cybercrime tweets and overcoming the challenges posed by the thin boundaries between different classes. We present a novel approach that leverages advanced feature extraction techniques, fuzzy values, and fusion methodologies to improve the accuracy of cybercrime tweet classification. By addressing existing methods' limitations and enhancing the classification models' robustness, our research contributes to the advancement of cybercrime detection and classification methodologies in cybersecurity.

1.1 Problem Formulation

Let $\mathcal{T} = \{t_1, t_2, t_3, \ldots, t_n\}$ denote the set of cybercrime tweets in the CRU dataset. $C = \{hate\ speech, cyber\ terrorism, cyber\ harassment, offensive, normal\}$ is a set of labels, where $\forall t_i \in C$. However, it is essential to note that the characteristics of cybercrimes often exhibit overlapping properties. Therefore, although the CRU dataset assigns a single class label to each tweet, it is crucial to acknowledge cybercrimes' inherent ambiguity and overlapping nature during the classification process. To address the challenges arising from the overlapping characteristics of cybercrimes, we propose a novel approach that integrates fuzzy values as a measure of relatedness or membership to each cybercrime class. The following function $f(.)$ calculates the fuzzy values as:

$$\mu_{ij} = f(t_i, C_j) \tag{1}$$

where $f(.)$ can be any string matching function such as QR, PR, PTSER, PTSOR, TSER, or TSOR. This equation quantifies the degree of association between a tweet t_i and a specific cybercrime class C_j, considering the overlapping nature of cybercrime. μ_{ij} represent the fuzzy value assigned to an individual tweet t_i for cybercrime class C_j, by capturing its degree of association with that class using $f(.)$. By giving fuzzy values, we aim to capture the degrees of relatedness to different cybercrime classes, accounting for the complexities and uncertainties present in the data. This approach allows for a more nuanced representation of cybercrime instances beyond the traditional binary class labels.

In the subsequent sections of this paper, we provide a comprehensive overview of related work, discuss the methodology, present our proposed fuzzy value-based approach in detail, report experimental results and analysis, and conclude with insights and directions for future research.

2 Related Work

The detection and prevention of cybercrime is a growing concern in cybersecurity. With the increasing use of digital technology, the amount of data related to cybercrime is also increasing rapidly [19]. Analyzing this data to identify patterns and trends is essential to understanding and combating cybercrime. Significant work has been done so far to combat cybercrimes. Law enforcement agencies educate people using different awareness measures, and social networking sites/apps such as Facebook and Yahoo! update their Terms of Service (ToS) to prevent cybercrimes. Facebook ToS prohibits posting content that is: pornographic, hateful, threatening, or inciting violence. Yahoo! ToS are similar to forbidding hateful, abusive, defamatory, vulgar, obscene, invasive of another's privacy, ethnically, racially, or otherwise objectionable content. The research community also plays a vital role in detecting and preventing cybercrimes.

Much of the existing research consists of diverse yet related tasks such as the detection of racism and sexism on X (formerly Twitter) [18], differentiating offensive language from hate speech [2], hateful and non-hateful speech in a white supremacy forum [3]. Generally, researchers identify/classify cybercrimes by extracting specific crime-related keywords from the text. Other features used in the literature include (1) bag-of-words, n-grams [12], (2) sentiment analysis, (3) lexical resources, (4) linguistic features, (5) knowledge-based features, (6) meta-information, (7) word embeddings.

Researchers used various techniques for detecting hate speech and other offensive language in literature. Kwok and Wang [8] merged a Naive Bayes classifier with the Bag-of-Words (BoW) method for hate speech detection. While Grevy et al. [7] utilized Support Vector Machines (SVM) and BoW. Traditionally, researchers used various features for machine learning methods such as SVM, Naive Bayes, and Logistic Regression [2]. However, these models are based on features, and with the emergence of data-driven models, people shifted from features-based models to data-based models.

Recently, advanced pre-trained multilingual transformer models, such as Bidirectional Encoder Representations from Transformers (BERT) and Generative Pre-trained Transformer (GPT), have achieved state-of-the-art results in various natural language processing tasks [5,10]. These models can be fine-tuned on specific tasks, such as text classification or sentiment analysis [11,15], and have been shown to outperform traditional machine learning techniques.

In addition to machine learning and fuzzy logic techniques, data augmentation has also been proposed to improve the performance of cybercrime text data analysis models [1]. EDA techniques, such as synonym replacement and random insertion, effectively improve the accuracy of text classification models. Overall, the literature suggests that machine learning, fuzzy logic, advanced transformer models, and data augmentation techniques are all promising approaches to cybercrime text data analysis [6]. However, there is still room for further research and experimentation to identify the most practical combination of these techniques for different types of cybercrime data and applications. Researchers have explored fuzzy logic in cybercrime text analysis to address these limitations. Fuzzy logic provides a flexible framework for dealing with imprecise or uncertain data, often in cybercrime investigations. Several studies have demonstrated the effectiveness of fuzzy logic in classifying cybercrime-related text data. In particular, fuzzy logic-based feature extraction methods effectively improve the performance of machine learning algorithms [20].

Overall, the literature suggests that combining machine learning, fuzzy logic, advanced pre-trained transformers, and data augmentation techniques can improve the accuracy and robustness of cybercrime text analysis. Using fuzzy values as features with advanced pre-trained multilingual transformers and machine learning techniques for cybercrime text analysis represents a promising avenue for future research which is presented in this paper.

3   Proposed Methodology

This section discusses the proposed methods to improve cybercrime text classification accuracy by combining fuzzy values with other standard feature extraction techniques, as shown in Fig. 1. We utilized RUHSOLD [13] and CRU [16] datasets for evaluating our proposed methodology. These datasets are relevant to our work, and both of the datasets are in Roman Urdu.

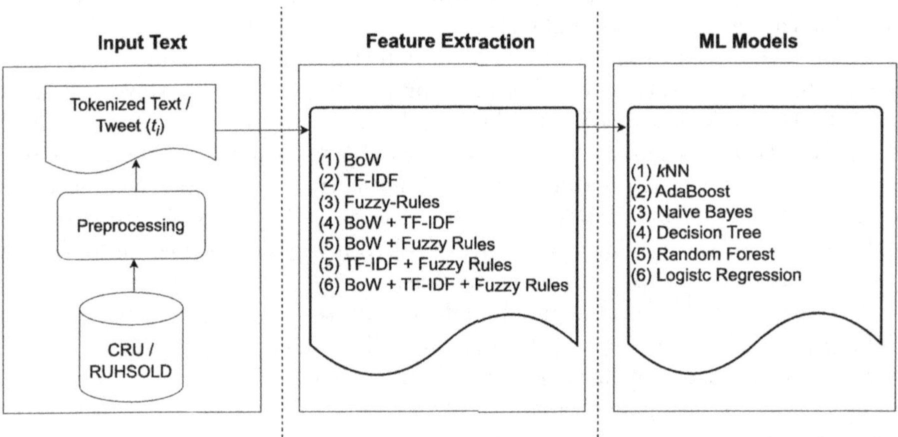

Fig. 1. Proposed methodology

3.1   Fuzzy Value Calculation

Calculating fuzzy values μ_{ij} allows us to quantify the degree of relatedness of each tweet t_i to the different cybercrime classes C_j, accounting for the inherent complexities and boundaries within the dataset. We first divided the dataset into five subsets to calculate the fuzzy values, each containing tweets specific to one of the cybercrime classes C_j. After stop-word removal, we performed tokenization and lemmatization within each set to extract the keywords and their base forms. Next, we created a list of the top-k words for each C_j based on their frequency and relevance within the subset. The utilized string-matching techniques generate values between 0 and 100, indicating the degree of matching or relatedness. We then normalize the values between 0 and 1. Next, we calculate the fuzzy values for t_i and l_{c_j} (list of top-k words). The QR, PR, $PTSER$, $PTSOR$, $TSER$, $TSOR$ for t_i and l_{c_j} is defined as follows, respectively:

$$QR(t,l) = \frac{2 \cdot |\mathcal{I}(t,l)|}{|t| + |l|} \tag{2}$$

$$PR(t,l) = \frac{2 \cdot |\mathcal{M}(t,l)|}{|t| + |l|} \tag{3}$$

$$PTSER(t,l) = \frac{|\omega(t,l)|}{|\Omega(t,l)|} \tag{4}$$

$$PTSOR(t,l) = \frac{|\omega_{\text{sorted}}(t,l)|}{|\Omega(t,l)|} \tag{5}$$

$$TSER(t,l) = \frac{|\omega(t,l)|}{|\Omega(t,l)|} \tag{6}$$

$$TSOR(t,l) = \frac{|\omega_{\text{sorted}}(t,l)|}{|\Omega(t,l)|} \tag{7}$$

where $|t|$ denotes the length of the string t_i, $\mathcal{I}(t,l)$ represents the set of common characters between strings t and l. $\mathcal{M}(t,l)$ returns the set of characters common to strings t and l. $\omega(t,l)$ and $\Omega(t,l)$ denote the number of matching tokens and the set of unique tokens present in both strings t and l, respectively. By applying these fuzzy matching techniques to each t_i and l_{cj}, we obtain six fuzzy values μ_{ij}. Hence, we obtain 36 fuzzy values for each tweet (6 classes $\times$ 6 fuzzy matching techniques). These fuzzy values serve as additional features, enhancing our classification models' robustness and discriminative power.

Symbolically, the feature vector $\mathbf{x}_{\text{fuzzy}}(t)$ can be represented as

$$\mathbf{x}_{\text{fuzzy}}(t) = [\mu_1, \mu_2, \ldots, \mu_{36}]$$

Next, we integrate fuzzy values with other classical feature extraction techniques to enhance classification performance. This feature fusion step aims to leverage the complementary information provided by different feature sets, including BoW and TF-IDF, in conjunction with the fuzzy values. The feature vector $\mathbf{x}_{\text{fusion}}(t)$ after fusion includes both the BoW or TF-IDF features, along with the fuzzy values.

For the fusion of fuzzy values with BoW, we concatenate the BoW feature vector $\mathbf{x}_{\text{bow}}(t)$ with the fuzzy values feature vector $\mathbf{x}_{\text{fuzzy}}(t)$, resulting in a combined feature vector denoted as $\mathbf{x}_{\text{fusion-bow}}(t)$. Mathematically, we have:

$$\mathbf{x}_{\text{fusion-bow}}(t) = [\mathbf{x}_{\text{bow}}(t) \oplus \mathbf{x}_{\text{fuzzy}}(t)]$$

Similarly, for the fusion of fuzzy values with TF-IDF, we concatenate the TF-IDF feature vector $\mathbf{x}_{\text{tfidf}}(t)$ with the fuzzy values feature vector $\mathbf{x}_{\text{fuzzy}}(t)$, resulting in a combined feature vector denoted as $\mathbf{x}_{\text{fusion-tfidf}}(t)$. Mathematically, we have:

$$\mathbf{x}_{\text{fusion-tfidf}}(t) = [\mathbf{x}_{\text{tfidf}}(t) \oplus \mathbf{x}_{\text{fuzzy}}(t)]$$

We explore the fusion of fuzzy values with both BoW and TF-IDF. We concatenate the BoW feature vector $\mathbf{x}_{\text{bow}}(t)$, the TF-IDF feature vector $\mathbf{x}_{\text{tfidf}}(t)$, and the fuzzy values feature vector $\mathbf{x}_{\text{fuzzy}}(t)$, resulting in a combined feature vector denoted as $\mathbf{x}_{\text{fusion-bow-tfidf}}(t)$. Mathematically, we have:

$$\mathbf{x}_{\text{fusion-bow-tfidf}}(t) = [\mathbf{x}_{\text{bow}}(t) \oplus \mathbf{x}_{\text{tfidf}}(t) \oplus \mathbf{x}_{\text{fuzzy}}(t)]$$

Finally, for the fusion of BoW and TF-IDF, we concatenate the BoW feature vector $\mathbf{x}_{\text{bow}}(t)$ with the TF-IDF feature vector $\mathbf{x}_{\text{tfidf}}(t)$, resulting in a combined feature vector denoted as $\mathbf{x}_{\text{bow-tfidf}}(t)$. Mathematically, we have:

$$\mathbf{x}_{\text{bow-tfidf}}(t) = [\mathbf{x}_{\text{bow}}(t) \oplus \mathbf{x}_{\text{tfidf}}(t)]$$

The fusion of features in our approach leverages the complementary strengths of fuzzy values, leading to a more discriminative and informative feature set. We combine fuzzy values with other features to capture the tweets' semantic relevance and contextual significance, improving accuracy and enhancing cybercrime detection capabilities. This feature fusion approach enhances the overall performance of the classification system, allowing for more effective identification and classification of cybercrime-related tweets.

3.2 Machine Learning Models

We describe the selection and application of various machine learning algorithms to classify cybercrime tweets based on the fused feature set. We consider six machine learning models: AdaBoost, Naive Bayes, K-Nearest Neighbors (KNN), Logistic Regression, Decision Tree, and Random Forest. Our goal is to compare the performance of different models and identify the most suitable approach for cybercrime classification. We train and evaluate each model's performance using different combinations of feature sets. We apply the BoW feature set to each model individually. This approach leverages the frequency-based representation of words in the tweets. We analyze the performance of each model when trained on BoW features alone. Next, we extend the feature set by incorporating the fuzzy values. We concatenate the $\mathbf{x}_{\text{fuzzy}}(t)$ with the BoW representation, creating an augmented feature set $\mathbf{x}_{\text{fusion-bow}}(t)$. This feature fusion enables the models to leverage the fuzzy matching-based relatedness information in addition to the lexical characteristics captured by BoW. Additionally, we explore the performance of the models when incorporating the term TF-IDF features. We apply TF-IDF as a standalone feature set and combine it with the fuzzy values, creating a hybrid feature set $\mathbf{x}_{\text{fusion-tfidf}}(t)$ that captures both the importance of words in tweet t_i and the relatedness information provided by the fuzzy values. We aim to identify the best-performing approach for cybercrime classification by evaluating the models across different feature sets.

To conduct the comparative analysis, we also selected three advanced pre-trained multilingual transformers for comparison. By outperforming existing state-of-the-art methods, specifically, pre-trained multilingual transformers such as BERT-Multi [4], BERT-Distill [14], and Mini-LM [17], our approach opens up new possibilities for improving the accuracy and effectiveness of cybercrime classification systems using classical machine learning techniques. Overall, the comparative analysis serves as solid evidence of the effectiveness of our proposed approach in addressing the challenges of cybercrime classification. The results also validate the effectiveness of incorporating fuzzy values as features, establishing our approach as a significant contribution to cybercrime detection and classification. To ensure reproducibility, we have made our code publicly available[2].

4 Result and Discussion

In this section, we present the results of our comprehensive experiments conducted to evaluate the performance of the proposed approach. To establish a baseline for comparison, we first evaluate the performance of the machine learning models when trained solely on the BoW features. This baseline helps us assess the effectiveness of the additional feature sets and their impact on classification accuracy.

Next, we analyze the performance of the models when incorporating fuzzy values as a feature. By examining the accuracy achieved using the fuzzy values alone, we gain insights into their effectiveness in capturing the relatedness to different cybercrime classes. We then evaluate the models' performance when using the feature fusion approach, combining the fuzzy values with the BoW and TF-IDF representations. This analysis helps us assess the impact of feature fusion on classification accuracy and determine if it leads to improved performance compared to individual feature sets. We track and compare the accuracy of each model and feature set combination to provide valuable insights into its practical applicability in real-world scenarios, as shown in Table 1.

Our proposed method, "TF-IDF + Fuzzy Values", demonstrates robust performance and improved accuracy compared to other feature extraction techniques for text classification. For both datasets (CRU and RUHSOLD), our

---

[2] https://github.com/sohaibayub/iconip.

Table 1. Accuracy comparison of various embedding for CRU and RUHSOLD datasets using various machine learning models. In this table, AB stands for AdaBoost, NB for Naive Bayes, kNN for K-Nearest Neighbors, LR for Logistic Regression, DT for Decision Tree, and RF for Random Forest. Best values are shown in bold.

	CRU						RUHSOLD					
	AB	NB	kNN	LR	DT	RF	AB	NB	kNN	LR	DT	RF
Bag of Words	53	57	65	76	71	78	54	**47**	71	80	81	82
BoW + Fuzzy Values	56	57	66	79	71	81	57	**47**	72	84	86	84
BoW + TF-IDF	53	**58**	65	76	70	76	54	**47**	71	80	80	83
BoW + TF-IDF + Fuzzy	61	**58**	66	79	71	82	54	**47**	73	**84**	86	85
Fuzzy Values	49	49	54	57	58	67	64	44	**78**	63	79	84
TF-IDF	53	57	65	77	70	77	54	46	72	80	81	82
TF-IDF + Fuzzy	**71**	57	**67**	**81**	**87**	**88**	**66**	46	74	83	**90**	**89**

method achieves the highest accuracy on all the utilized ML models except for Naive Bayes. Furthermore, our method outperforms pre-trained multilingual transformer models as shown in Fig. 2, including BERT-Distill, BERT-Multi, and Mini-LM, achieving the highest accuracy of 90% for the RUHSOLD dataset and 88% for the CRU dataset. Overall, the experimental results highlight the effectiveness of our proposed method, "TF-IDF + Fuzzy Values", in improving text classification accuracy, outperforming traditional embedding techniques.

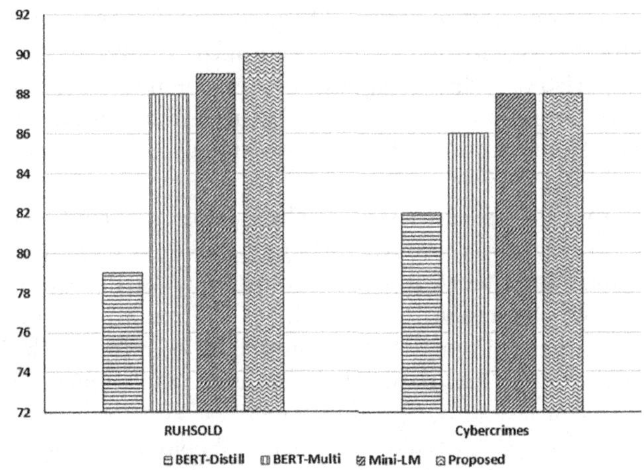

Fig. 2. Accuracy comparison of various transformers and our proposed approach "TF-IDF + Fuzzy Values" on CRU and RUHSOLD datasets.

We also observed that the Random Forest and Decision Tree models consistently achieved high accuracy in classifying the cybercrime tweets across different feature sets, as shown in Fig. 3. These models demonstrated robust performance and outperformed the other models in most cases. The high accuracy achieved by Random Forest (RF) and Decision Tree (DT) models in our study can be attributed to their ability to handle the unique complexities of the cybercrime datasets, particularly with the fusion of TF-IDF and fuzzy values. Specifically, in the CRU dataset, cybercrime tweets exhibit significant overlap between categories such as hate speech, cyber terrorism, and offensive language. The introduction of fuzzy values allowed us to represent the degree of relatedness between tweets and multiple cybercrime classes. Both RF and DT models leveraged this nuanced feature set to better navigate these overlapping boundaries. RF, with its ensemble of decision trees, performed particularly well by considering different feature combinations, reducing overfitting, and enhancing generalization across the complex feature spaces created by combining TF-IDF and Fuzzy Values. DT's hierarchical structure further enabled precise splitting based on these fuzzy values features, improving classification performance.

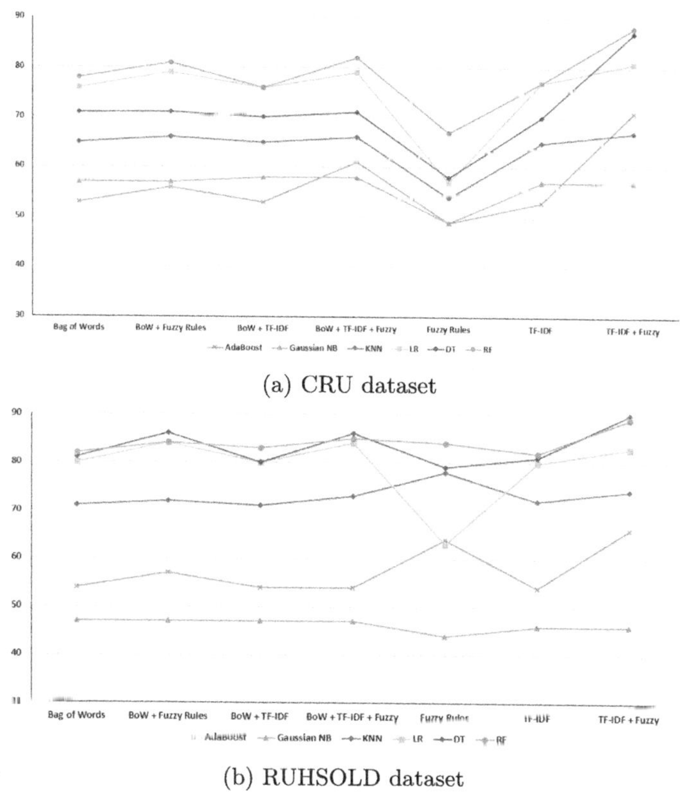

(a) CRU dataset

(b) RUHSOLD dataset

Fig. 3. Accuracy comparison of various feature extraction techniques for classifying CRU and RUHSOLD datasets.

5　Conclusion

Cybercrimes and other hate speech data pose thin boundaries in text. Classification models (classical machine learning models) underperform in such scenarios. To overcome this issue, we proposed a novel approach that combines fuzzy values with TF-IDF for improved performance in text classification tasks. We compared our method with various classical feature extraction techniques and machine learning models on CRU and RUHSOLD datasets and observed significant improvements in accuracy. These results indicate the effectiveness of our proposed approach in handling complex and diverse textual data.

Furthermore, we compared our approach with pre-trained multilingual transformers, including BERT-Distill, BERT-Multi, and Mini-LM. Surprisingly, our "TF-IDF + Fuzzy Values" method surpassed the transformer models, achieving the highest accuracy of 88% on the CRU dataset and 90% on the RUHSOLD dataset. This finding suggests that our approach can be a viable alternative to expensive and computationally intensive transformer models.

The successful application of fuzzy values with TF-IDF in text classification opens up new possibilities for enhancing various natural language processing tasks. Future research can explore the integration of additional linguistic features and fine-tuning techniques to improve the performance of our approach further. We also plan to explore this approach in other languages, including Arabic, Chinese, and English. This research contributes to text classification by providing a promising, effective and computationally efficient alternative.

References

1. Dabare, R., Wong, K.W., Shiratuddin, M.F., Koutsakis, P.: Fuzzy data augmentation for handling overlapped and imbalanced data. In: Mantoro, T., Lee, M., Ayu, M.A., Wong, K.W., Hidayanto, A.N. (eds.) ICONIP 2021. CCIS, vol. 1516, pp. 625–633. Springer, Cham (2021). https://doi.org/10.1007/978-3-030-92307-5_73
2. Davidson, T., Warmsley, D., Macy, M., Weber, I.: Automated hate speech detection and the problem of offensive language. In: International AAAI Conference on Web and Social Media, vol. 11, pp. 512–515 (2017)
3. De Gibert, O., Perez, N., García-Pablos, A., Cuadros, M.: Hate speech dataset from a white supremacy forum. arXiv preprint arXiv:1809.04444 (2018)
4. Devlin, J., Chang, M.W., Lee, K., Toutanova, K.: BERT: pre-training of deep bidirectional transformers for language understanding. North American Chapter of the Association for Computational Linguistics (2019)
5. Faheem, A., Ullah, F., Ayub, M.S., Karim, A.: UrduMASD: a multimodal abstractive summarization dataset for Urdu. In: Proceedings of the 2024 Joint International Conference on Computational Linguistics, Language Resources and Evaluation (LREC-COLING 2024), pp. 17245–17253. ELRA and ICCL, Torino, Italia (2024). https://aclanthology.org/2024.lrec-main.1498
6. Ge, Y., Hu, X., Li, P., Wang, H., Zhao, J., Li, J.: Multi-label learning with data self-augmentation. In: International Conference on Neural Information Processing, pp. 336–347 (2022)

7. Greevy, E., Smeaton, A.F.: Classifying racist texts using a support vector machine. In: SIGIR Conference on Research and Development in Information Retrieval, pp. 468–469 (2004)
8. Kwok, I., Wang, Y.: Locate the hate: detecting tweets against blacks. In: AAAI Conference on Artificial Intelligence (2013)
9. Li, L., Xiao, L., Jin, W., Zhu, H., Yang, G.: Text classification based on word2vec and convolutional neural network. In: Cheng, L., Leung, A., Ozawa, S. (eds.) ICONIP 2018. LNCS, vol. 11305, pp. 450–460. Springer, Cham (2018). https://doi.org/10.1007/978-3-030-04221-9_40
10. Maity, K., Bhattacharya, S., Saha, S., Janoai, S., Pasupa, K.: FastThaiCaps: a transformer based capsule network for hate speech detection in Thai language. In: International Conference of Neural Information Processing, pp. 425–437 (2023)
11. Maity, K., Saha, S.: A multi-task model for sentiment aided cyberbullying detection in code-mixed indian languages. In: International Conference of Neural Information Processing, pp. 440–451 (2021)
12. Nobata, C., Tetreault, J., Thomas, A., Mehdad, Y., Chang, Y.: Abusive language detection in online user content. In: International Conference on World Wide Web, pp. 145–153 (2016)
13. Rizwan, H., Shakeel, M.H., Karim, A.: Hate-speech and offensive language detection in Roman Urdu. In: Empirical Methods in Natural Language Processing (EMNLP), pp. 2512–2522 (2020)
14. Sanh, V., Debut, L., Chaumond, J., Wolf, T.: DistilBERT, a distilled version of BERT: smaller, faster, cheaper and lighter. arXiv preprint arXiv:1910.01108 (2019)
15. Ullah, F., Azam, U., Faheem, A., Kamiran, F., Karim, A.: Comparing prompt-based and standard fine-tuning for Urdu text classification. In: Findings of the Association for Computational Linguistics: EMNLP 2023, pp. 6747–6754. Association for Computational Linguistics, Singapore (2023). https://aclanthology.org/2023.findings-emnlp.449
16. Ullah, F., Faheem, A., Azam, U., Ayub, M.S., Kamiran, F., Karim, A.: Detecting cybercrimes in accordance with pakistani law: dataset and evaluation using PLMs. In: Proceedings of the 2024 Joint International Conference on Computational Linguistics, Language Resources and Evaluation (LREC-COLING 2024), pp. 4717–4728 (2024)
17. Wang, W., Wei, F., Dong, L., Bao, H., Yang, N., Zhou, M.: MiniLM: deep self-attention distillation for task-agnostic compression of pre-trained transformers. In: Advances in Neural Information Processing Systems, vol. 33, pp. 5776–5788 (2020)
18. Waseem, Z.: Are you a racist or am I seeing things? Annotator influence on hate speech detection on Twitter. In: Workshop on NLP and Computational Social Science, pp. 138–142 (2016)
19. Zhang, M., Xu, B., Bai, S., Lu, S., Lin, Z.: A deep learning method to detect web attacks using a specially designed CNN. In: Liu, D., Xie, S., Li, Y., Zhao, D., El-Alfy, E.-S.M. (eds.) ICONIP 2017. LNCS, vol. 10638, pp. 828–836. Springer, Cham (2017). https://doi.org/10.1007/978-3-319-70139-4_84
20. Zhou, J., Rajapakse, J.C.: Extraction of fuzzy features for detecting brain activation from functional MR time-series. In: King, I., Wang, J., Chan, L.-W., Wang, D.L. (eds.) ICONIP 2006. LNCS, vol. 4234, pp. 983–992. Springer, Heidelberg (2006). https://doi.org/10.1007/11893295_108

Hybrid Niching Differential Evolution with Restart Strategy for Multimodal Optimization

Xinyu Zhou, Jingjin Yin, Hua Chen, Wenlong Ni, and Mingwen Wang[✉]

School of Computer and Information Engineering, Jiangxi Normal University,
Nanchang 330022, China
{xyzhou,jjyin,hua.chen,wni,mwwang}@jxnu.edu.cn

Abstract. As an effective diversity preservation technique, niching technique has been extensively used to assist evolutionary algorithms (EAs) for multimodal optimization problems (MMOPs). However, different niching techniques have different preferences for exploration and exploitation, and most existing niching techniques are used alone, which influence the performance of multimodal optimization algorithms to some extent. Therefore, an enhanced differential evolution (DE) variant, called HNDE-RS, is proposed by designing a hybrid niching strategy. In this strategy, two niching techniques, i.e., crowding and speciation, are utilized simultaneously and integrated with DE to evolve the population. In addition, HNDE-RS has another strategy, i.e., a novel restart strategy, which utilizes the DBSCAN clustering technique and an external archive to construct a new population when stagnation or premature convergence occurs. In the experiments, HNDE-RS is verified on CEC2013 multimodal test suit and compared with 6 state-of-the-art multimodal optimization algorithms. The experimental results demonstrate the competitive performance of HNDE-RS, especially on the functions with a large number of global optimal solutions.

Keywords: Evolutionary algorithm · Multimodal optimization ·
Hybrid niching · Restart strategy

1 Introduction

Multimodal optimization problems (MMOPs) are ubiquitous in real life. In general, they have more than one global optimal solution and require the algorithm to find all the global optimal solutions. For example, the electromagnetic design [18], multiple ellipses detection [19], and protein structure prediction [17] are some representative MMOPs that contain multiple global optimal solutions. For MMOPs, the optimization algorithms are required not only to locate as many global optimal solutions as possible in a single run, but also to converge on the global optimal solutions as accurately as possible, which means that the optimization algorithms must have both a good exploration capability and a good exploitation capability.

Traditional optimization methods, such as Newton's method and gradient descent method, are suitable to solve the problems with some mathematical

M. Mahmud et al. (Eds.): ICONIP 2024, CCIS 2297, pp. 388–403, 2026.
https://doi.org/10.1007/978-981-96-7036-9_26

properties, such as continuous and differentiable. However, in the real world, not all the problems have these properties. In this scenario, evolutionary algorithms (EAs) and swarm intelligence algorithms (SIs), such as genetic algorithm (GA) [4], differential evolution (DE) [3], and artificial bee colony (ABC) algorithm [23], have been widely used to deal with many practical problems, since they almost have no requirements on the mathematical properties. But it should be noted that the EAs and SIs are originally designed to identify only one global optimal solution and cannot be directly employed to solve the MMOPs due to the lack of population diversity maintenance mechanism.

To address this issue, numerous efforts have been made. One of the most common method is the niching technique [9]. Niche is a biological concept that refers to a specific part of a habitat or the space and environment in which an individual is located at a specific time. In the community of multimodal optimization, a niche is defined as the subspace around peaks in the search space, and a species is defined as the individuals within the niche.

The key idea of the niching technique is to divide the population into several niches (so-called subpopulations), and each niche is responsible for locating one or more global optima. Crowding [13], speciation [6], clustering [5], and fitness sharing [7] are some well-known niching techniques, and by combining with EAs or SIs, some classic multimodal optimization algorithms were proposed. For example, Thomsen [13] proposed a crowding-based DE (CDE) by combining DE with the crowding technique. Li [6] proposed a speciation-based DE (SDE) by combining DE with the speciation technique. Qu et al. [12] introduced the concept of neighbourhood-based mutation into CDE and SDE, and two well-known multimodal optimization algorithms, i.e., NCDE and NSDE, were developed.

So far, many different multimodal optimization algorithms have been designed. Although these algorithms have been shown impressive performance, almost all of them only use a single niching technique. In fact, although the core role is to preserve population diversity, some niching techniques inherently favor exploration, e.g., crowding, while others inherently favor exploitation, e.g., speciation and clustering. When used alone, these niching techniques affect the efficiency of algorithms to some extent. Motivated by this, we attempt to make a better tradeoff between exploration and exploitation by combining different niching techniques that can complement one another perfectly. In this work, we propose a multimodal optimization algorithm, named as HNDE-RS. The main contributions of HNDE-RS can be summarized as follows:

1) A hybrid niching strategy is designed by combining the crowding technique and the speciation technique. In this strategy, the crowding technique and the speciation technique are integrated with DE separately, and then the resulting DE variants are used simultaneously to evolve the population. According to this strategy, the crowding and speciation techniques can complement and reinforce each other, and finally tackle the problem where onefold niching technique favors exploration or exploitation during the evolutionary process.

2) A novel restart strategy is designed to address the issue of stagnation and premature convergence. Specifically, once the population is recognized as stag-

nation or premature convergence, a new population is constructed by the DBSCAN clustering and replace the stagnant population. Furthermore, to reduce the unnecessary consumption of fitness evaluations in the reconstruction process, an archive is constructed based on the discarded individuals.

To evaluate the effectiveness and efficiency of HNDE-RS, some experiments are conducted on the CEC2013 multimodal test suit [8], and 6 state-of-the-art multimodal optimization algorithms are included in the performance comparison. The experimental results show that the performance of HNDE-RS is competitive. The rest of this paper is organized as follows: Sect. 2 reviews the related works about multimodal optimization algorithms. Section 3 describes HNDE-RS in detail. Section 4 gives the experimental results and related discussions. Section 5 presents the conclusions and future work.

2   Related Works

2.1   Multimodal Optimization Algorithms

Over the past few decades, a variety of multimodal optimization algorithms have been designed by researchers, and most of them can be classified into two parts. The first part is niching technique-based algorithms, and the second part is multiobjective technique-based algorithms.

Niching Technique-Based Algorithms. As an effective diversity preservation technique for solving MMOPs, niching technique with EAs and SIs can conduct parallel searching to locate multiple global optimal solutions. The two most representative niching techniques are crowding and speciation, and when these two niching techniques are integrated with DE, the crowding-based DE (CDE) and speciation-based DE (SDE) were proposed by Thomsen [13] and Li [6], respectively. Qu et al. [12] proposed a neighborhood mutation-based DE (NSDE and NCDE), which applies the neighborhood search strategy in CDE and SDE. Similarly, Gao et al. [5] presented two adaptive clustering analysis methods (self-CCDE and self-CSDE) by applying a self-adaptive parameter control technique and a clustering method in CDE and SDE, respectively.

In addition to the crowding and speciation, clustering is another hot niching technique. For instances, Wang et al. [15] proposed an automatic niching DE (ANDE) by employing the affinity propagation clustering to adaptively construct species. Luo et al. [11] proposed a neighbor clustering method (NBNC) and a novel PSO algorithm based on NBNC (NBNC-PSO-ES), which combines PSO and covariance matrix adaptive evolution strategy (CMAES).

Recently, some novel niching techniques have been designed. For example, Zhang et al. [21] proposed a Voronoi neighborhood-based CDE (VNCDE), which uses Voronoi diagram to determine the neighbors of each individual. Zhao et al. [22] proposed LBPADE to solve MMOPs, which borrowed the idea of local binary pattern (LBP) in image processing and developed a niching technique based on LBP. Wang et al. [16] proposed a distributed differential evolution algorithm with an adaptive estimation distribution (AED-DDE), in which each individual finds its own appropriate niche size to form a niche. Li et al. [10] proposed a novel differential evolution variant (ESPDE) based on the Hill-Valley technique for multimodal optimization, which uses history information to classify individuals on the same peak as one species.

Multiobjective Technique-Based Algorithms. To solve MMOPs, some researchers have emphasized the multiobjective techniques in addition to the niching techniques. For instance, Yao et al. [20] developed a bi-objective multi-population genetic algorithm (BMPGA), in which MMOP itself acted as one objective, and the gradient information of the objective function is used as another objective. Basak et al. [1] proposed a bi-objective DE (MOBiDE), where the two objectives are the objective function and the average distance of a solution from all other solutions, respectively. Wang et al. [14] proposed an improved construction method (MOMMOP) by considering the conflict between objectives. Cheng et al. [2] proposed a novel method (EMO-MMO) by designing a grid-based diversity indicator as the second optimization objective, so that the MMOPs are transformed into a multiobjective optimization problem.

3   The Proposed Method

3.1   Motivation

As briefly reviewed above, in the last few years, a variety of niching technique-based multimodal optimization algorithms have been proposed. However, almost without exception, these algorithms use onefold niching technique, while different niching techniques have different preferences for exploitation and exploration, which results in the efficiency of these algorithms being limited to some extent. Taking CDE [13] and NSDE [12] as a simple example, Fig. 1 shows the final population distribution of CDE and NSDE on F6 and F13 in the CEC2013 multimodal test suit, respectively. As can be seen from Fig. 1(a) and Fig. 1(c), since the crowding technique is used alone and inherently favors exploration, all the optimal peaks have been located, but they have not been fully exploited due to most individuals are trapped in the local optima. In contrast, in Fig. 1(b) and Fig. 1(d), since the speciation technique is used alone and inherently favors exploitation, some optimal peaks are missed. However, those optimal peaks that have been located are fully exploited due to most individuals are attracted to the individuals that close to them. Hence, both of CDE and NSDE have not found all the optimal solutions on F6 and F13.

Furthermore, during the evolutionary process, the population easily suffers from stagnation and premature convergence. To address this issues, reinitializing the current population and combining the current population with an external archive are the two commonly used methods to reconstruct a new population. However, they still have some shortcomings. Specifically, the first method causes the unnecessary waste of fitness evaluations, since it both generates scores of new

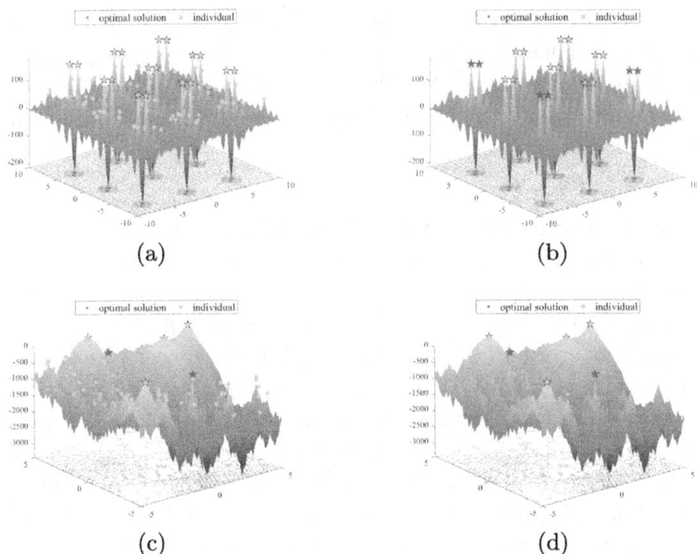

Fig. 1. Final population distribution of CDE and NSDE on F6 and F13 in CEC2013 multimodal test suit. (a) final population distribution of CDE on F6. (b) final population distribution of NSDE on F6. (c) final population distribution of CDE on F13. (d) final population distribution of NSDE on F13.

individuals and does not make full use of the promising individuals. As for the second method, although it does not consume the number of fitness evaluations, it doubles the population size several times and causes the individuals trapped in local optima to remain in the new population.

Considering the above observations, we are motivated to propose a more effective multimodal optimization algorithm, i.e., HNDE-RS, by designing a hybrid niching strategy and a novel restart strategy.

3.2 General Framework of HNDE-RS

Figure 2 shows the general framework of HNDE-RS, which illustrates how the hybrid niching strategy and restart strategy are employed in HNDE-RS. As can be seen from Fig. 2, after the initialization, the population continues to evolve based on the hybrid niching strategy until it reaches stagnation or satisfies the termination condition. During this period, the crowding-based evolutionary process and the speciation-based evolutionary process are sequentially used to population evolution. In addition, to refine the individuals and avoid promising individuals being attracted to other peaks, a local search process is employed between these two evolutionary processes. The crowding-based evolutionary process, local search process, and speciation-based evolutionary process together constitute the three phases of the hybrid niching strategy. It should be noted that all the discarded individuals are placed in an external archive

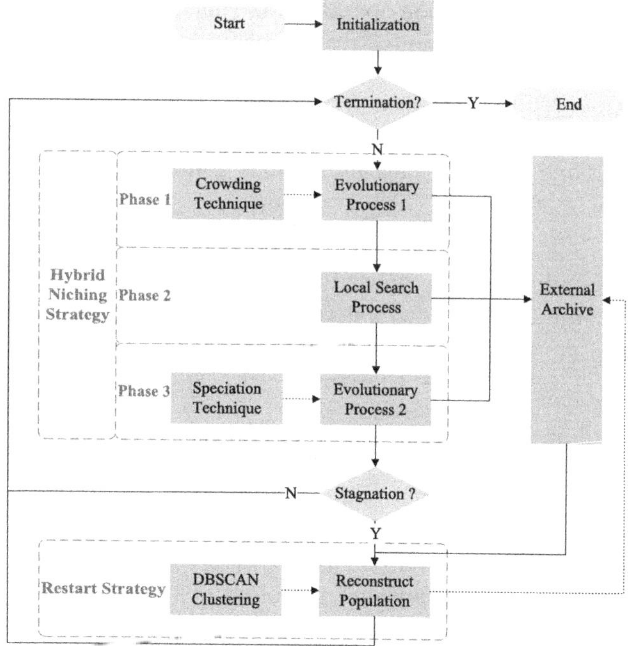

Fig. 2. General framework of HNDE-RS.

during the evolution. Subsequently, the restart strategy is triggered only when the population is stagnant. In this strategy, the DBSCAN clustering technique and the external archive are utilized to reconstruct the population. Once the new population is obtained, it will replace the original population and start a new evolution, and the external archive is emptied. More details about these two strategies are given in the following sections.

3.3 Hybrid Niching Strategy

In the crowding-based evolutionary process of hybrid niching strategy, for each individual in the population, there are two ways to generate its corresponding offspring. The first way is to randomly select 3 different individuals from the population and then execute the mutation and crossover operations of DE, while the other way is to select 5 nearest (measured by the Euclidean distance) individuals from the neighborhood and then execute the mutation and crossover operations. These two ways are chosen randomly. After the offspring are produced, each offspring will find its most similar individual (measured by the minimum Euclidean distance) in the population for comparison, and then the better one will survive into the next generation while the inferior one is discarded into the external archive. It should be noted that compared with CDE [13] and NCDE [12], this evolutionary process possesses their characteristics to generate offspring, while have no difference in the selection operation. In this phase, the two parameters of DE, F and Cr are set to 0.9 and 0.3 respectively, and the $DE/rand/1$ mutation strategy of DE is used.

In the local search process of hybrid niching strategy, an adaptive Gaussian distribution is directly utilized as the local search operation, which has been widely used to refine the promising individuals and can be written as:

$$X_i^{new} = \text{Gaussian}\ (X_i,\ \sigma) \tag{1}$$

where X_i is the individual on which to perform the local search, X_i^{new} is the new individual sampled around X_i. σ is the standard deviation and is set as:

$$\sigma = 10^{-1-(\frac{10}{D}+3)\cdot\frac{FEs}{MaxFEs}} \tag{2}$$

where D is the dimension of the search space, FEs is the number of currently consumed fitness evaluations, and $MaxFEs$ is the maximum number of fitness evaluations. Specifically, for each individual, 2 new individuals are sampled by using Eq. (1) and Eq. (2), and of these 3 individuals, the best one will be kept in population, while the others are placed into the external archive.

In the speciation-based evolutionary process of hybrid niching strategy, the population is divided into several different niches by using the speciation technique, and within each niche, the offspring are generated by executing the mutation and crossover operations of DE. After that, each offspring is compared with its most similar individual in the corresponding niche. It should be noticed that the steps of population division are the same as that in NSDE, while the selection operation is the same as CDE. As in previous phases, the discarded individuals in this phase are also placed into the archive. In this pahse, the two parameters of DE, F and Cr, are both set to 0.5, and the $DE/best/1$ mutation strategy of DE is used.

3.4   Restart Strategy

To deal with the problem of stagnation and premature convergence, restart strategies have been widely applied. However, as described in Sect. 3.1, most existing restart strategies still have some limitations. Therefore, we develop a novel restart strategy based on DBSCAN clustering technique and external archive. The description of when the restart strategy is triggered, how to prepare external archive for the restart strategy, and how to construct new population using DBSCAN clustering is as follows.

Identification of Population Stagnation. To identify the stagnant state, in this paper, the distribution information of the population is utilized. The mean μ_P and the std σ_P of the population P in generation G are calculated as follows:

$$\mu_P^G = \frac{1}{NP}\sum_{i=1}^{NP} X_i^G \tag{3}$$

$$\sigma_P^G = \sqrt{\frac{1}{NP} \sum_{i=1}^{NP} \left(X_i^G - \mu_P^G\right)^2} \tag{4}$$

where NP represents the population size, and X_i^G represents the ith individual in generation G. If $\max \left|\sigma_P^G - \sigma_P^{G-1}\right|$ is less than 0.001, the population stagnation counter (psc) plus 1, otherwise resets to 0. If psc exceeds the threshold rho_{psc}, the restart strategy is triggered.

Preparation of External Archive. During the reconstruction, to reduce the unnecessary consumption of fitness evaluations, an external archive is constructed based on the discarded individuals and continuously updated. When stagnation or premature convergence occurs, the individuals in this external archive are merged into the stagnant population to form a new population. The steps to construct and update the external archive are shown in Fig. 3.

As can be seen from Fig. 3, whenever a discarded individual X_d is obtained, the size of external archive A is checked. If A is empty, X_d will be directly added to A. Otherwise, the most similar (measured by the minimum Euclidean distance) individual X_s in A to X_d will be identified for further comparison. If the distance between them is less than the threshold rho_d, the one with better fitness value will remain in A. On the contrary, if the distance between them exceeds the threshold, A will be checked again to see if it is full. If there is remaining capacity, X_d will be directly added to A. Otherwise, X_d will be compared with the worst individual X_w in A and replaces X_w if X_d is better than X_w. In this paper, rho_d is set 0.05, and the archive size A_{max} is set $2NP$.

Clustering-Assisted Population Reconstruction. When the population is stagnant, it is common to form a new population by directly merging the current population with an external archive. Although this method is simple and indeed works to some extent, it changes the population size and does not eliminate the individuals that are trapped in local optima, which accelerates the consumption of fitness evaluations. Therefore, we utilize the DBSCAN clustering technique to reconstruct a population. The main process of population reconstruction is shown in Fig. 4.

First, a combined set U is obtained by combining the population P and external archive A. Subsequently, P and A are set to empty, and the DBSCAN clustering is utilized to partition the set U into several clusters. If no cluster is identified, it means that the individuals in this population are distant from each other, and then the NP fitter individuals in U are kept as the new population. Otherwise, for each cluster C_i, the seed (best individual) is compared with the best individual in U. If the difference in their fitness values is less than 0.0001, the seed is retained and placed into P. If the difference exceeds 0.0001, and the number of individuals in C_i is more than 10, the 10 fitter individuals are retained and placed into P. If neither of the above two conditions is met, all the individuals in C_i are retained and placed into P. After this operation, the

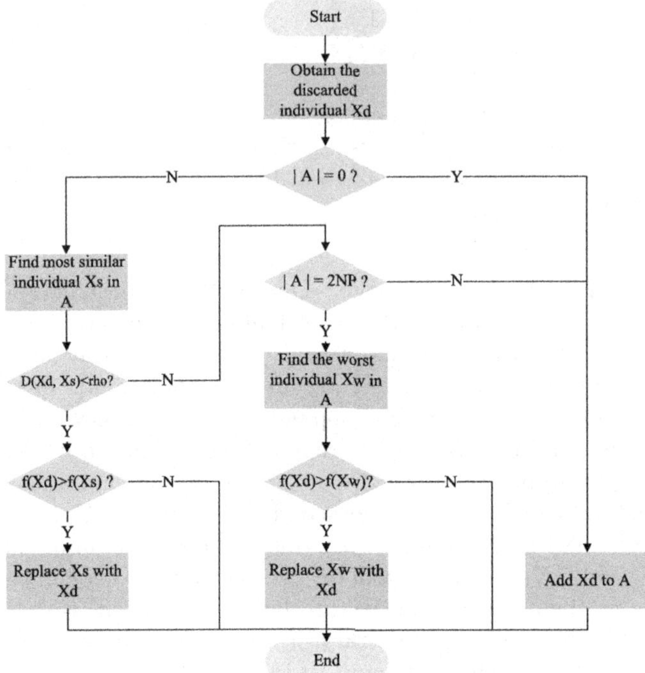

Fig. 3. The flow chart of preparing external archive.

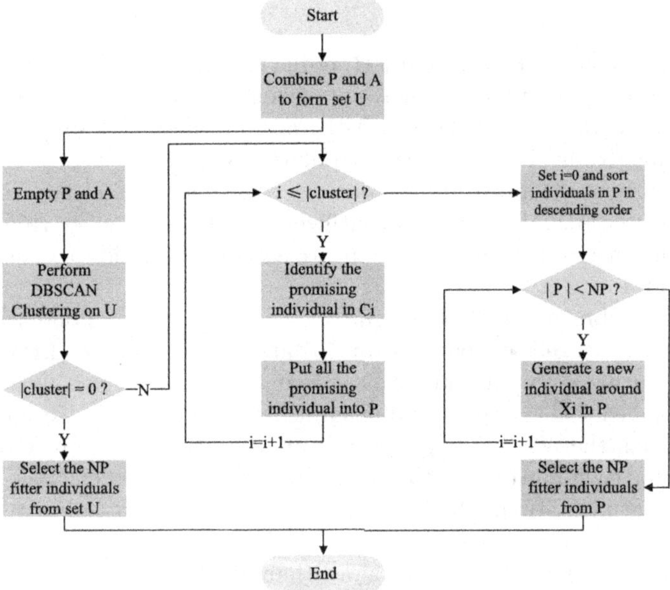

Fig. 4. The flow chart of population reconstruction.

redundant individuals are removed, and the promising individuals are selected as new individuals. Next, all the new individuals in P are sorted in descending order, and the size of P is checked. If the size of P is less than NP, new individual is generated by using Eq. (1) and Eq. (2) until the size of P is equal to NP. Otherwise, the NP fitter new individuals are retained. At this point, the construction of the new population P is completed. In this paper, the two parameters of DBSCAN clustering, i.e., *minpoints* and *eps*, are set to 2 and 0.1, respectively.

Complete Procedure of HNDE-RS. Based on the above description, the pseudocode of HNDE-RS is shown in Algorithm 1. In the initialization (lines

---

Algorithm 1. HNDE-RS

Input: population size NP, maximum number of fitness evaluations $MaxFEs$
Output: population P, external archive A
1: Set $FEs = 0$, $G = 1$, $psc = 0$, $rho_{psc} = 15$, $rho_d = 0.05$, $A_{max} = 2NP$, $minpoints = 2$, $eps = 0.1$;
2: Randomly initialize the population P in the entire search space;
3: According to Eqs. (3) and (4), calculate the initial distribution information of P;
4: **while** $FEs \leq MaxFEs$ **do**
5:    /* **Hybrid niching strategy** */
6:    Execute the crowding-based evolutionary process;
7:    According to Eq. (1) and Eq. (2), execute the adaptive local search process;
8:    Execute the speciation-based evolutionary process;
9:    Update the external archive A;
10:    According to Eq. (3) and Eq. (4), calculate the μ_P^G and σ_P^G of P;
11:    **if** $\max \left| \sigma_P^G - \sigma_P^{G-1} \right| < 0.001$ **then**
12:      $psc = psc + 1$;
13:    **else**
14:      $psc = 0$;
15:    **end if**
16:    **if** $psc > rho_{psc}$ **then**
17:      /* **Restart strategy** */
18:      $U = P \cup A$;
19:      Set $P = \emptyset$, $A = \emptyset$, and $psc = 0$;
20:      Perform DBSCAN clustering operation on individuals in U;
21:      Store the promising individuals in each cluster into P;
22:      Sort individuals in P in descending order;
23:      **if** $|P| < NP$ **then**
24:        According to Eq. (1) and Eq. (2), generate $NP - |P|$ individuals;
25:      **else**
26:        Keep the NP fitter individuals in P;
27:      **end if**
28:      According to (3) and (4), calculate the new distribution information of P;
29:    **end if**
30:    $G = G + 1$;
31: **end while**

1–2), the parameters of HNDE-RS are set, and the population P is generated. Then, the initial distribution information of population P is calculated according to Eq. (3) and Eq. (4) (line 3). Next, the hybrid niching strategy is executed (lines 6–9) to address the issue of the crowding technique favors exploration and the speciation technique favors exploitation. Subsequently, the distribution information of population P in this generation is calculated (line 10), and the counter psc is updated (lines 11–15). If the population stagnation or premature convergence occurs, the restart strategy is triggered to construct a new population and start a new evolution (lines 16–29). The procedure of hybrid niching strategy and restart strategy will repeat until termination condition is met.

4   Experiments

4.1   Benchmark Functions

The experiments are conducted on the CEC2013 multimodal test suite, which is widely used in the community of multimodal optimization. To evaluate the performance of algorithms, two indicators, peak ratio (PR) and success rate (SR) are used. More details about the CEC2013 multimodal test suit and the evaluation criteria can be found in the literature [8]. PR and SR are calculated as follows:

$$PR = \frac{1}{N_o \cdot N_r} \sum_{i=1}^{N_r} N_f^i \tag{5}$$

$$SR = \frac{N_s}{N_r} \tag{6}$$

where N_r denotes the number of runs, N_f^i denotes the number of global optimal solutions discovered in the ith run, N_o denotes the number of all the optimal solutions in the test function, and N_s denotes the number of successful runs.

For a fair comparison, it is worth nothing that the population size and the maximum number of fitness evaluations on each function are set to the same as the recommendations of CEC2013 multimodal competition. Furthermore, to eliminate the error of random factors, all the algorithms are independently executed for 51 times per test function and the average results are recorded.

4.2   Experimental Results

Comparison with Other Algorithms. With the accuracy level of $\epsilon = 1.0E - 04$, HNDE-RS is compared with 6 state-of-the-art multimodal algorithms, including ESPDE [10], AED-DDE [16], ANDE [15], MOMMOP [14], LBPADE [22], and VNCDE [21], which are some representative multimodal optimization algorithms in the past three years. The detailed results of PR and SR are shown in Table 1. In Table 1, the best PR results are marked in **boldface**. The average of PR values (Avg PR) and the number of best PR results (Nbr PR) obtained by each algorithm are shown in the last two row, respectively. It should be noticed that all the results are obtained from their respective papers.

From Table 1, we can find that HNDE-RS outperforms the other comparative algorithms, with the best average PR value (0.846) and 15 highest PR values out of 20 test problems. Specifically, on F01-F06, most of the algorithms including HNDE-RS can find all the global optimal solutions due to these functions are low-dimensional and contain a small number of optimal solutions. On F07-F10, only MOMMOP obtains the best results on all these functions. Although HNDE-RS is worse than MOMMOP, the difference between HNDE-RS and MOMMOP is insignificant (less than 1.2%), and the performance of HNDE-RS is still significantly better than the other 5 compared algorithms. On F11–F15, HNDE-RS provides the best results among all algorithms, except for F15. On F15, HNDE-RS is slightly worse than ESPDE and VNCDE. Nevertheless, HNDE-RS still achieves a better performance to other compared algorithms. On F16-F18, HNDE-RS is only outperformed by ESPDE on F17 but is equivalent to these state-of-the-art algorithms on other functions. On F19-F20, although the performance of HNDE-RS is slightly worse than VNCDE, the difference in performance is insignificant (less than 3%).

Table 1. Results in PR and SR of all comparison algorithms on CEC2013 at the accuracy level $\epsilon = 1.0E - 04$

Func.	HNDE-RS		ESPDE		AED-DDE		ANDE		MOMMOP		LBPADE		VNCDE	
	PR	SR	PR	SR	PR	SR	PR	SR	PR	SR	PR	SR	PR	SR
F01	**1.000**	1.000	**1.000**	1.000	**1.000**	1.000	**1.000**	1.000	**1.000**	1.000	**1.000**	1.000	**1.000**	1.000
F02	**1.000**	1.000	**1.000**	1.000	**1.000**	1.000	**1.000**	1.000	**1.000**	1.000	**1.000**	1.000	**1.000**	1.000
F03	**1.000**	1.000	**1.000**	1.000	**1.000**	1.000	**1.000**	1.000	**1.000**	1.000	**1.000**	1.000	**1.000**	1.000
F04	**1.000**	1.000	**1.000**	1.000	**1.000**	1.000	**1.000**	1.000	**1.000**	1.000	**1.000**	1.000	**1.000**	1.000
F05	**1.000**	1.000	**1.000**	1.000	**1.000**	1.000	**1.000**	1.000	**1.000**	1.000	**1.000**	1.000	**1.000**	1.000
F06	**1.000**	1.000	**1.000**	1.000	**1.000**	1.000	**1.000**	1.000	**1.000**	1.000	**1.000**	1.000	0.967	0.540
F07	**1.000**	1.000	0.963	0.360	0.838	0.039	0.933	0.176	**1.000**	1.000	0.889	0.000	0.746	0.000
F08	**1.000**	1.000	0.880	0.000	0.747	0.000	0.944	0.078	**1.000**	1.000	0.575	0.000	0.558	0.000
F09	0.988	0.843	0.729	0.000	0.384	0.000	0.512	0.000	**1.000**	0.940	0.476	0.000	0.298	0.000
F10	**1.000**	1.000	**1.000**	1.000	**1.000**	1.000	**1.000**	1.000	**1.000**	1.000	**1.000**	1.000	**1.000**	1.000
F11	**1.000**	1.000	**1.000**	1.000	**1.000**	1.000	**1.000**	1.000	0.717	0.020	0.674	0.000	**1.000**	1.000
F12	**1.000**	1.000	0.930	0.560	**1.000**	1.000	**1.000**	1.000	0.960	0.700	0.750	0.000	0.993	0.940
F13	**0.948**	0.765	0.793	0.008	0.686	0.000	0.686	0.000	0.667	0.000	0.667	0.000	0.687	0.000
F14	**0.752**	0.039	0.727	0.000	0.667	0.000	0.667	0.000	0.667	0.000	0.667	0.000	0.667	0.000
F15	0.691	0.000	0.730	0.000	0.637	0.000	0.632	0.000	0.605	0.000	0.654	0.000	**0.748**	0.000
F16	**0.667**	0.000	**0.667**	0.000	**0.667**	0.000	**0.667**	0.000	**0.667**	0.000	**0.667**	0.000	**0.667**	0.000
F17	0.495	0.000	**0.685**	0.000	0.375	0.000	0.397	0.000	0.518	0.000	0.532	0.000	0.593	0.000
F18	**0.667**	0.000	0.660	0.000	0.654	0.000	0.654	0.000	0.500	0.000	**0.667**	0.000	**0.667**	0.000
F19	0.460	0.000	0.445	0.000	0.375	0.000	0.363	0.000	0.250	0.000	**0.475**	0.000	**0.475**	0.000
F20	0.260	0.000	0.265	0.000	0.250	0.000	0.248	0.000	0.125	0.000	0.275	0.000	**0.290**	0.000
Avg PR	**0.846**		0.824		0.764		0.785		0.784		0.748		0.768	
Nbr PR	**15**		10		10		10		11		10		12	

Ablation Experiment. HNDE-RS's competitive performance is evident from the above experiment results. To further investigate the effectiveness and influence of the two proposed strategies, three variants based on HNDE-RS are designed as the competitors, which are listed as follows:

1) HNDE-RS-V1: HNDE-RS without the crowding technique, and only the speciation technique and restart strategy are used.
2) HNDE-RS-V2: HNDE-RS without the speciation technique, and only the crowding technique and restart strategy are used.
3) HNDE-RS-V3: HNDE-RS without the clustering and external archive assisted restart strategy, and only hybrid niching strategy is used.

The comparison PR results are shown in Fig. 5. From Fig. 5, it can be observed that HNDE-RS can outperform three variants on almost all test functions. Specifically, compared with HNDE-RS-V1, HNDE-RS obtains better results on all of 20 functions, which verifies that the use of crowding technique can indeed provide exploration. Similarly, compared with HNDE-RS-V2, although HNDE-RS-V2 has a better PR value on F9, HNDE-RS surpasses HNDE-RS-V2 on any other functions, which verifies the use of speciation technique can indeed provide exploitation. By comparing HNDE-RS, HNDE-RS-V1 and HNDE-RS-V2, we can draw the conclusion that the performance of the algorithm can indeed be improved by hybridizing crowding and speciation. Finally, compared with HNDE-RS-V3, HNDE-RS outperforms HNDE-RS-V3 on all of 20 functions, which implies that the restart strategy based on DBSCAN clustering and external archive is effective.

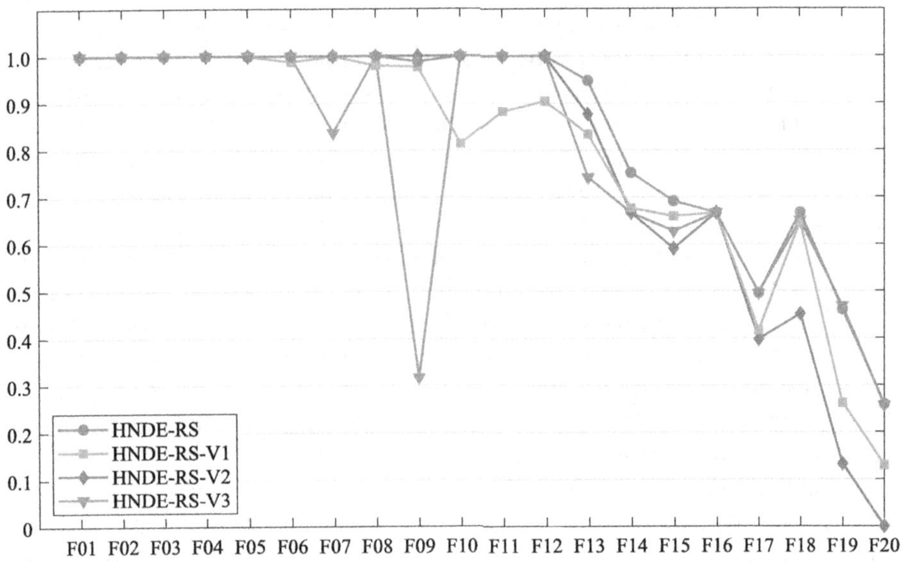

Fig. 5. Results in PR of HNDE-RS and its variants on CEC2013 at the accuracy level $\epsilon = 1.0E - 04$

5   Conclusion

In this paper, we propose a hybrid niching strategy and a restart strategy to integrate with DE algorithm, and the resulting algorithm is called HNDE-RS. In contrast to existing multimodal optimization algorithms, the most salient feature of HNDE-RS is that the exploration and exploitation can be better balanced according to the different preferences of niching techniques. To this aim, the crowding technique and speciation technique are hybridized to complement and reinforce each other. As another contribution of HNDE-RS, a restart strategy based on DBSCAN clustering and external archive is designed to reconstruct a high-quality population, with the aim of solving stagnation and premature convergence.

To evaluate the performance of HNDE-RS, computational experiments have been made, including the efficiency verification on CEC2013 multimodal test suite at three different accuracy levels and performance comparison with 6 related multimodal optimization algorithms. Experimental results indicate that HNDE-RS has a competitive performance when compared with other algorithms.

However, like other multimodal optimization algorithms, HNDE-RS cannot locate all the global optima on high-dimensional test functions. Thus, to improve the performance of HNDE-RS in dealing with complex MMOPs with high-dimensional characteristics, in the future, we will focus on how to improve the effectiveness and efficiency of the hybrid niching technique and the restart strategy, such as from the side of fitness landscape analysis techniques and reinforcement learning.

Acknowledgement. This work is supported by the National Natural Science Foundation of China (No. 62366022) and the Jiangxi Provincial Natural Science Foundation (No. 20232BAB202048).

References

1. Basak, A., Das, S., Tan, K.C.: Multimodal optimization using a biobjective differential evolution algorithm enhanced with mean distance-based selection. IEEE Trans. Evol. Comput. **17**(5), 666–685 (2013)
2. Cheng, R., Li, M.Q., Li, K., Yao, X.: Evolutionary multiobjective optimization-based multimodal optimization: Fitness landscape approximation and peak detection. IEEE Trans. Evol. Comput. **22**(5), 692–706 (2018)
3. Das, S., Suganthan, P.N.: Differential evolution: a survey of the state-of-the-art. IEEE Trans. Evol. Comput. **15**(1), 4–31 (2011)
4. Deb, K., Pratap, A., Agarwal, S., Meyarivan, T.A.M.T.: A fast and elitist multiobjective genetic algorithm: NSGA-II. IEEE Trans. Evol. Comput. **6**(2), 182–197 (2002)
5. Gao, W.F., Yen, G.G., Liu, S.Y.: A cluster-based differential evolution with self-adaptive strategy for multimodal optimization. IEEE Trans. Cybern. **44**(8), 1314–1327 (2014)

6. Li, X.D.: Efficient differential evolution using speciation for multimodal function optimization. In: Proceedings of the 7th Annual Conference on Genetic and Evolutionary Computation, pp. 873–880. Washington, DC, USA (2005)

7. Li, X.D.: A multimodal particle swarm optimizer based on fitness Euclidean-distance ratio. In: Proceedings of the 9th Annual Conference on Genetic and Evolutionary Computation, pp. 78–85. London, England (2007)

8. Li, X.D., Engelbrecht, A., Epitropakis, M.G.: Benchmark functions for cec'2013 special session and competition on niching methods for multimodal function optimization. Technical report, Evol. Comput. Mach. Learn. Group, RMIT Univ., Melbourne, VIC, Australia (2013)

9. Li, X.D., Epitropakis, M.G., Deb, K., Engelbrecht, A.: Seeking multiple solutions: an updated survey on niching methods and their applications. IEEE Trans. Evol. Comput. $21(4)$, 518–538 (2017)

10. Li, Y., et al.: History information-based hill-valley technique for multimodal optimization problems. Inf. Sci. 631, 15–30 (2023)

11. Luo, W.J., Qiao, Y.Y., Lin, X., Xu, P.L., Preuss, M.: Hybridizing niching, particle swarm optimization, and evolution strategy for multimodal optimization. IEEE Trans. Cybern. $52(7)$, 6707–6720 (2022)

12. Qu, B.Y., Suganthan, P.N., Liang, J.J.: Differential evolution with neighborhood mutation for multimodal optimization. IEEE Trans. Evol. Comput. $16(5)$, 601–614 (2012)

13. Thomsen, R.: Multimodal optimization using crowding-based differential evolution. In: Proceedings of the 2004 IEEE Congress on Evolutionary Computation, vol. 2, pp. 1382–1389. Portland, OR, USA (2004)

14. Wang, Y., Li, H.X., Yen, G.G., Song, W.: MOMMOP: multiobjective optimization for locating multiple optimal solutions of multimodal optimization problems. IEEE Trans. Cybern. $45(4)$, 830–843 (2015)

15. Wang, Z.J., et al.: Automatic niching differential evolution with contour prediction approach for multimodal optimization problems. IEEE Trans. Evol. Comput. $24(1)$, 114–128 (2020)

16. Wang, Z.J., Zhou, Y.R., Zhang, J.: Adaptive estimation distribution distributed differential evolution for multimodal optimization problems. IEEE Trans. Cybern. $52(7)$, 6059–6070 (2022)

17. Wong, K.C., Leung, K.S., Wong, M.H.: Protein structure prediction on a lattice model via multimodal optimization techniques. In: Proceedings of the 12th annual Conference on Genetic and Evolutionary Computation, pp. 155–162. Portland, OR, USA (2010)

18. Woo, D.K., Choi, J.H., Ali, M., Jung, H.K.: A novel multimodal optimization algorithm applied to electromagnetic optimization. IEEE Trans. Magn. $47(6)$, 1667–1673 (2011)

19. Yao, J., Kharma, N., Grogono, P.: A multi-population genetic algorithm for robust and fast ellipse detection. Pattern Anal. Appl. $8(1)$, 149–162 (2005)

20. Yao, J., Kharma, N., Grogono, P.: Bi-objective multipopulation genetic algorithm for multimodal function optimization. IEEE Trans. Evol. Comput. $14(1)$, 80–102 (2010)

21. Zhang, Y.H., Gong, Y.J., Gao, Y., Wang, H., Zhang, J.: Parameter-free voronoi neighborhood for evolutionary multimodal optimization. IEEE Trans. Evol. Comput. $24(2)$, 335–349 (2020)

22. Zhao, H., et al.: Local binary pattern-based adaptive differential evolution for multimodal optimization problems. IEEE Trans. Cybern. $50(7)$, 3343–3357 (2020)

23. Zhou, X.Y., Tan, G.S., Wu, Y.L., Wu, S.X.: Neighborhood learning for artificial bee colony algorithm: a mini-survey. In: Luo, B., Cheng, L., Wu, Z.G., Li, H.Y., Li, C.J. (eds.) ICONIP 2023. LNCS, vol. 14449, pp. 370–381. Springer, Singapore (2023). https://doi.org/10.1007/978-981-99-8067-3_28

StreetSyn: A Full Radiance Field Solution for Street and Vehicle Free-View Synthesis

Shenhao Zhu[1]($\boxtimes$), Li Wang[1], Xun Cao[1], Ruigang Yang[2], Xinxin Zuo[3], and Hao Zhu[1]

[1] Nanjing University, Nanjing & Suzhou, China
shenhaozhu@smail.nju.edu.cn
[2] Inceptio, Shanghai, China
[3] University of Alberta, Edmonton, Canada

Abstract. Starting from sparse views of real-captured scenes and a synthetic dataset of 3D vehicles, we aim to synthesize photo-realistic street views with moving vehicles, editable illumination, and controllable viewpoints which is a significant task for autonomous driving simulation. The problem is very challenging as only sparse views are available for recovering such a complex street environment. In this paper, we propose a full radiance field scheme for free-view synthesis of street scenes and vehicles. Benefiting from the scheme that both the scene and the vehicle are represented as radiance fields, illumination can be directly extracted from the real-captured scenes and transferred to the synthesized vehicle. The ambient illumination is modeled as a mixture of Spherical Gaussians (SGs) with different frequencies, which turns out to be effective in recovering the low-frequency sky illumination and high-frequency sun illumination. Experiments show that our model can synthesize street view and vehicle images in free views, and significantly outperforms previous works in photo-realism and lighting modeling accuracy.

Keywords: Scene synthesis · Driving simulation · Neural Radiance Field

1 Introduction

Humans can envision unseen scenarios in their minds, significantly enhancing the effectiveness of learning. Drawing inspiration from this capacity, researchers in the field of autonomous driving have focused on creating virtual environments with synthesized vehicles to train perception models. Extensive studies demonstrate that this approach is crucial in addressing challenges related to 3D visual-based decision-making problems [21,36].

The traditional 3D simulation involves the utilization of Computer-Aided Design (CAD) models. Nevertheless, generating extensive and varied street

Supplementary Information The online version contains supplementary material available at https://doi.org/10.1007/978-981-96-7036-9_27.

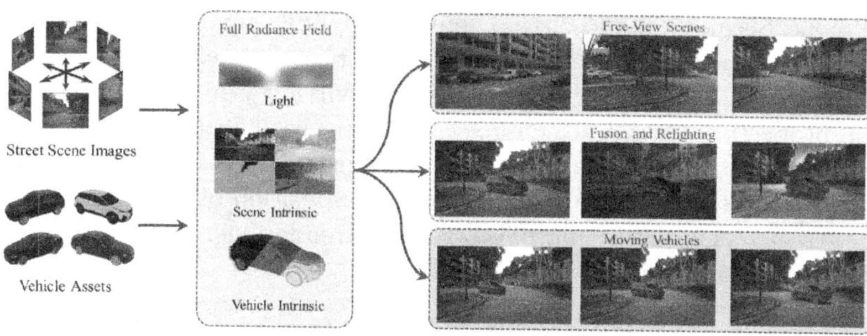

Fig. 1. We take sparse videos of real-captured street scenes as well as synthetic vehicles as input. Full neural radiance fields of both geometric and photometric properties such as lighting, materials are optimized for both the background scenes and synthetic vehicles to support flexible street scene simulation, such as free-view rendering, relighting, vehicle composition and manipulation in 3D.

scenes with purely CAD models proves highly inefficient and costly in practical terms. On the other hand, there are extensive street view datasets of real-captured scenes, which contain extremely diverse street scenes under real-world lighting information. Thus, in this paper, we propose a hybrid solution that models the street scenes with real-captured sparse videos and composites of synthetic 3D vehicles to synthesize photo-realistic free-view rendering in diverse scenarios.

Pioneering works try to synthesize street views by modeling vehicles as 3D objects and representing the background scenes as static images [5,20,32]. These methods exhibit limited control over the overall appearance of both scenes and vehicles, and the viewpoint remains fixed. Now benefiting from the emergence of neural radiance field (NeRF) [24] for a 3D scene representation, we can model background scenes and integrated vehicles separately and then composite them into a single radiance field to produce high rendering quality in free-view. Some recent works [2,33,35] exploited NeRF representation for street scene rendering and proposed to reconstruct scene geometry and recover intrinsic properties of the scene for relighting purpose. Nevertheless, the rendering quality is significantly constrained by the simplified lighting and material modeling employed to depict intricate street scenes.

As shown in Fig. 1, in this paper, we propose a full neural radiation field solution to build a street scene simulator, which supports free-view rendering, 3D vehicle composition, and relighting. The key idea is that by modeling both real-captured street view scenes and virtual vehicles as neural radiance fields we can integrate them in 3D space by considering both geometric constraints and the illumination effects.

First, to attain photo-realistic rendering and relighting of integrated background scenes and virtual vehicles, we put lots of effort into decomposing the intrinsic scene properties. Basically, we propose a novel scheme to effectively decompose the environmental lighting while optimizing the NeRF for captured street scenes. Specifically, we model the ambient illumination as a mixture of SGs with different frequencies, which turns out to be effective in recovering the low-frequency sky illumination and high-frequency sun illumination in a real-world environment. Second, for the virtual vehicles, we build up a neural radiance field library with their decomposed intrinsic properties as well. By leveraging the radiance fields of both background scenes and virtual vehicles, along with their intrinsic properties, we can execute scene composition in accordance with spatially varying rendering equations. Finally, a FusionNet is introduced to further eliminate the synthetic-to-real gap after the composition of the inserted vehicles and the street scene, producing photo-realistic renderings.

Our contribution can be summarized as:

- To the best of our knowledge, our method is the first full radiance field solution for street and vehicle free-view synthesis for autonomous driving simulation, which supports free-view rendering of the scene and greatly expands the simulation scene.
- To achieve realistic relighting and composition, we propose a method to decompose intrinsic properties of both street scenes and virtual vehicles while optimizing NeRF.
- Inverse rendering and a generative refinement are introduced to eliminate the bias between real-captured data and synthetic data, yielding photo-realistic fusing images.

2   Related Work

We first introduce the previous works of simulation for autonomous driving, then review the relighting and composition for synthesized objects, which are two key modules in generating photo-realistic scenes.

Driving View Simulation: Simulating driving views has been a key technique in training an autonomous driving system and is catching wide attention. The traditional simulator [7,27] leverages manually designed 3D models to render street scenes with moving vehicles. These renderings contain a synthetic-to-real gap, and a limited amount of 3D models makes it very difficult to build large-scale and diverse scenes. In the follow-up research, the performance boost of 2D conditioned generation models [4,15,26,31] provides an alternative to synthesize novel scenes with editable contents in 2D generated images [8,21]. These methods struggle in simulating 3D-related properties like rotations, shadows, and occlusions. Very recently, Neural Radiance Field (NeRF) [24] was introduced for its high-quality rendering and free-view synthesis performance [1,5,34,37]. These works represent vehicles with a NeRF and then fuse the renderings into scene images. Though 3D properties for vehicles are learned, the rendering views are limited as the scenes are still in a 2D representation. More importantly, the vehicles and scenes are represented in a radiance field and an image separately, so the lighting conditions cannot be accurately estimated and transferred, leading to a less realistic synthesis in complex and diverse lighting situations. In contrast, our method is a full radiance field solution merging street scenes and vehicles, going further in synthesizing relightable and free-view renderings.

Merging Objects and Scenes: Merging synthetic objects (vehicles) into reconstructed outdoor street scenes is a valuable task and of great significance to solving the long-tail scenario problem of autonomous driving. To obtain photo-realistic fusion results, some works [9,12–14,19] start from the background environmental images, trying to collect information from an outdoor background image to obtain the surrounding environmental lighting conditions, and represent the scene lighting as spherical harmonic coefficients or HDR maps of sun and sky. However, these methods cannot cover the impact of objects in the scene on lighting and the effect changes that occur after foreground objects are inserted. To solve this problem, some methods [29,38] decompose lighting into global and local parts for prediction respectively. However, due to the limitation that the background is a 2D image, the predicted illumination cannot guarantee the spatial consistency of the same scene.

After NeRF was proposed, some methods [22,28,33] used intrinsic decomposition and inverse rendering images to optimize scene geometry while estimating illumination to ensure spatial consistency. For example, NeRF-OSR [28] proposes a method that combines spherical harmonic coefficient illumination and diffuse reflection effects, but it cannot restore the high-frequency highlight effect. Based on the characteristics of the natural environment, we choose SG mixtures to represent environmental lighting, use a series of regularization terms to constrain and optimize the decomposition of the scene's intrinsic attributes, and use an inverse renderer to restore the color and shadow effects of foreground objects to the greatest extent. In addition, we use FusionNet to further enhance the realism after fusion, thereby proposing a complete solution to this problem.

3   Method

3.1   Overview

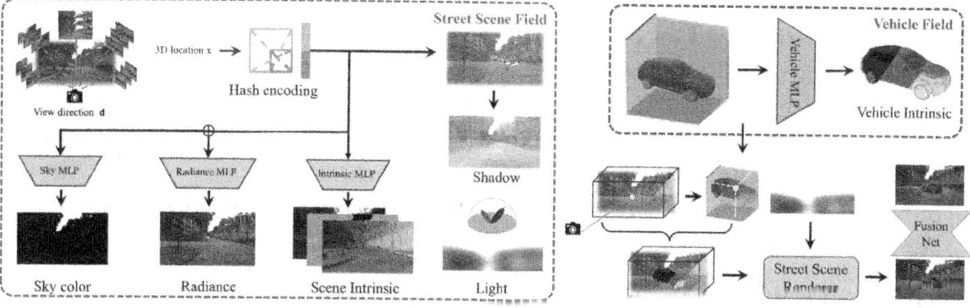

Fig. 2. Overall Pipeline. Our approach enables moving foreground objects to be inserted into real-world scenes. First, two neural radiance fields are generated for street scenes and vehicles separately, with the intrinsic properties decomposed. Then the two fields are fused to synthesize a lighting-uniform street scene with vehicles inserted.

As shown in Fig. 2, our approach enables moving foreground objects to be inserted into real-world scenes and is capable of generating a high-quality video of the composition in novel views. For the background scene, we take as input multiple frames of images extracted from a street scene video, the camera poses can be obtained from the IMU/GPS information, denoted as $\{I_i, C_i\}_{i=1}^N$, where $I_i \in \mathbb{R}^{H*W*3}$ is an image, $C_i \in SE(3)$ is its camera pose and N is the number of images. A neural radiance field for a street scene is built to decompose the intrinsic scene properties (Sect. 3.2).

We introduce a mixture of Spherical Gaussians (SGs) [30] with different frequencies to model ambient illuminations, which is sufficient to cover the low-frequency sky illumination and relatively high-frequency sun illumination in a real-world environment. A Street Scene Renderer is leveraged to jointly optimize the parameters of the neural field and lighting conditions in the inference phase (Sect. 3.3). During training, we model the sky and the scene separately and design their respective loss functions. A set of regularization terms optimizes this highly underdetermined problem (Sect. 3.4). For foreground objects, we take advantage of a virtual vehicle library and build a network to decompose the intrinsic properties of these vehicles. The vehicle information can be queried through the library, and the vehicle can be directly inserted into the street scene with a modified rendering equation. FusionNet is used to handle inconsistent color saturation and harsh boundaries and to eliminate the synthetic-to-real gap (Sect. 3.5). It is worth noting that our approach represents both foreground objects and background scenes in a unified field with their intrinsic properties decomposed, which enables the scene to be rendered in free views.

3.2 NeRF for Street Scenes

NeRF [24] represents a scene with a neural field $F : (\boldsymbol{x}, \boldsymbol{d}) \mapsto (\sigma, \boldsymbol{c})$ which maps 3D location $\boldsymbol{x}$ and view direction $\boldsymbol{d}$ to corresponding density $\sigma \in \mathbb{R}$ and color $\boldsymbol{c} \in \mathbb{R}^3$. To model a large-scale street scene, hash-based NeRF representation [25] is leveraged to improve the efficiency of training and inference.

Intrinsic Property Decomposition: A neural intrinsic field $F_\phi : \boldsymbol{x} \mapsto (\sigma, \boldsymbol{n}, \boldsymbol{k})$ is learned to decompose the intrinsic properties related to spatial location. Specifically, we encode input 3D location $\boldsymbol{x}$ with a hash table, then use three separated modules to decode the hash feature: the geometry module $\sigma = F_{geo}(\boldsymbol{x}; \theta_g)$ outputs the density $\sigma \in \mathbb{R}$; the normal module $\boldsymbol{n} = F_{norm}(\boldsymbol{x}; \theta_n)$ outputs the surface normal $\boldsymbol{n} \in \mathbb{R}^3$; and the appearance module $\boldsymbol{k} = (\boldsymbol{k}_d, \boldsymbol{k}_s) = F_{app}(\boldsymbol{x}; \theta_a)$ outputs the material properties of the surface point.

Virtual Vehicle Library: To enhance the diversity of our model's synthetic scenes, we have collected a series of virtual vehicle models and established a Virtual Vehicle Library to complement our street scene field. This enables the synthesis of various foreground vehicles, enhancing the diversity of the scenes. Each vehicle is represented as an implicit representation with triplanes and a small MLP decoder. We obtain the intrinsic decomposition of the vehicles, maintaining consistency with the scene, yielding a vehicle module $(\sigma_v, \boldsymbol{k}_v) = F_v(\boldsymbol{x}; \theta_v)$. This allows for direct integration of the vehicles into the scene at intrinsic level.

3.3   Street Scene Renderer

The Rendering Equation: The rendering equation [17] describes how light propagates in a 3D scene and determines the irradiance observed at a particular point. It provides a common formulation for rendering in computer graphics:

$$L_o(\boldsymbol{p}, \boldsymbol{w}_o) = \int_{\Omega} f_r(\boldsymbol{p}; \boldsymbol{w}_i, \boldsymbol{w}_o) L_i(\boldsymbol{p}, \boldsymbol{w}_i)(\boldsymbol{w}_i \cdot \boldsymbol{n}) d\boldsymbol{w}_i \tag{1}$$

Following the rendering equation, given a camera ray $r(t) = \boldsymbol{o} + t\boldsymbol{d}$, we first compute the location of $\boldsymbol{p}$ by the volume render depth $\hat{D}$ as $\boldsymbol{p} = \boldsymbol{o} + \hat{D}\boldsymbol{d}$, then use an inverse render workflow to synthesize photorealistic street scenes.

Lighting Representation: As the rendering equation is an integral equation without an analytical solution, existing methods either simplify the ambient lighting [22, 28] or calculate it through expensive Monte Carlo integration [10, 33], which reduces the fidelity and efficiency of rendering.

To address the problem, We represent the ambient lighting as a sum of spherical Gaussian lobes (SGs) [30]:

$$L_i(\boldsymbol{\omega}_i) = \sum_{k=1}^{M} G(\boldsymbol{\omega}_i; \boldsymbol{\phi}, \lambda, \boldsymbol{\mu}) \tag{2}$$

where $\boldsymbol{\phi} \in \mathbb{R}^3$ is the direction, $\lambda \in \mathbb{R}_+$ is the sharpness and $\boldsymbol{\mu} \in \mathbb{R}_+^3$ is the amplitude, for a particular SG lobe. We regard the SGs as a part of model parameters and jointly optimize them during the training stage.

Metal/Roughness Workflow: Street scenes encompass a wide variety of objects, each with distinct materials. In such scenes, many objects, such as bricks, roads, and plants, typically have a high roughness, resulting in diffuse reflection effects. However, for vehicles, the car paint often exhibits metallic properties and characteristics like lacquer, which cannot be accurately simulated by considering roughness alone. Therefore, to precisely decompose and render the intrinsic properties of street scenes, we introduce the metal/roughness workflow to model the materials in street scenes.

In our workflow, the material properties $\boldsymbol{k}_d$ and $\boldsymbol{k}_s$ control the diffuse appearance and the specular appearance respectively. $\boldsymbol{k}_d \in \mathbb{R}^3$ represents the base color of the object's surface. In our workflow, $\boldsymbol{k}_s$ is composed of roughness α and metalness m from the material properties.

Based on the aforementioned material properties, we define the BRDF in the Metal/roughness workflow as the sum of the diffuse component and the specular component: $f_r = f_d + f_o$. The diffuse part is determined by the base color $\boldsymbol{k}_d$ and the metalness m: $f_d = \frac{\boldsymbol{k}_d}{\pi} * (1 - m)$. The metalness here is used to control the intensity of the diffuse reflection. The specular component of BRDF consists of normal distribution function, Fresnel term and shadowing term. Metalness m serves here to align the color of the specular highlights with the base color, thereby enhancing the metallic sensation. Similar to [41], We represent the specular component of BRDF as a single SG, where h is exactly halfway between the light direction vector and the view direction vector:

$$f_s = G(\boldsymbol{h}; \boldsymbol{n}, \frac{1}{2\alpha^2(\boldsymbol{h} \cdot \boldsymbol{\omega}_o)}, \frac{\mathcal{M}}{\pi\alpha^2}) \tag{3}$$

As both lighting and BRDF are represented by SGs, we can first calculate the lighting and BRDF, then obtain the emitted irradiance C_{render} by closed-form hemispherical integration of the SGs [23].

Shadow Map: Owing to the intricate nature of street scenes, the presence of various objects obstructs light, leading to the formation of shadows throughout the scene. We compute a shadow map by uniformly sampling N_l incident rays on the upper hemisphere, the visibility V of an incident ray with direction $\boldsymbol{l}_k$ at surface point $\boldsymbol{p}$ can be denoted as the volume opacity of the ray: $r(t) = \boldsymbol{p} + t\boldsymbol{l}_k$. The ray is traced through the scene to obtain the visibility by volume rendering:

$$V = \exp(-\sum_i \sigma_i(t_i - t_{i-1})), \tag{4}$$

Then, we generate the shadow using a ratio between the occluded and unoccluded irradiance.

$$S(\boldsymbol{p}) = \frac{\sum_{k=1}^{N_l} f_r(\boldsymbol{p}; \boldsymbol{\omega}_o, \boldsymbol{l}_k) L_i(\boldsymbol{p}; \boldsymbol{l}_k) V(\boldsymbol{p}; \boldsymbol{l}_k)(\boldsymbol{l}_k \cdot \boldsymbol{n})}{\sum_{k=1}^{N_l} f_r(\boldsymbol{p}; \boldsymbol{\omega}_o, \boldsymbol{l}_k) L_i(\boldsymbol{p}; \boldsymbol{l}_k)(\boldsymbol{l}_k \cdot \boldsymbol{n})} \tag{5}$$

A shadow map can be added to the scene by a pixel-wise product: $L'_o = S \odot L_o$

3.4   Optimization and Regularization

The sparse training image perspective and unknown lighting conditions make the scene reconstruction an extremely ill-posed problem. Therefore, several data augmentation and regularization terms are leveraged to constrain the problem.

Rendering Loss: We use the consistency between the predicted color and ground-truth color from input images as our main supervision, which is formulated as:

$$\mathcal{L}_{\text{render}} = \sum_{r \in \mathcal{R}} |C_{\text{gt}} - C_{\text{render}}|^2 \tag{6}$$

where C_{render} is the RGB value calculated by the rendering pass of each camera ray $r \in \mathcal{R}$, and C_{gt} is the corresponding ground-truth color. An end-to-end training scheme is achieved with attribute decomposition and lighting parameters jointly optimized.

Sky Mask Loss: We use a binary cross-entropy loss term between the volume rendered alpha channel mask M_α and sky mask M_textgt, which is obtained from an off-the-shelf semantic segmentation network [16]. The sky mask loss term $\mathcal{L}_{\text{skymask}}$ is formulated as:

$$\mathcal{L}_{\text{skymask}} = \sum_{r \in \mathcal{R}} \text{BCE}(M_\alpha, M_{\text{gt}}) \tag{7}$$

Sky Modeling Loss: Because performing physically-based rendering on the pixels of the sky is meaningless, we used an off-the-shelf semantic segmentation network [16] to mask out the sky pixels. We first use a binary cross-entropy to find sky pixels, then the sky network can give the corresponding sky color according to the viewing direction. An MSE loss is used to evaluate the difference between the predicted sky color and the ground-truth sky color, formulated as:

$$\mathcal{L}_{\text{sky}} = \sum_{r \in \text{sky}} |C_{\text{gt}} - C_{\text{sky}}|^2 \tag{8}$$

Normal Loss: Normal extracted from the density field $\hat{n} = -\frac{\nabla\sigma(x)}{|\nabla\sigma(x)|}$ is used as a supervisory term for normal predicted from MLP. Additionally, we use an off-the-shelf normal estimator [18] to provide a reference normal n_r, as in:

$$\mathcal{L}_{\text{norm.}} = \sum_{r \in \mathcal{R}} \left(|n_r - n|^2 + |\hat{n} - n|^2\right) \tag{9}$$

Base Color Loss: We observed that without any constraints, the shadows in the scene tend to be incorrectly merged into the base color. Noting that shadows often appear on roads, we utilize a shadow detector [6] and a segmentation network [16]. The pixels identified by the shadow detector as being in shadow, and the road pixels, are combined to define the domain $\mathcal{S}$ where constraints should be applied to the base color. Then we mask out the shadow pixels in the image and calculate the average color of each semantic segment after removing shadows to serve as the reference color $C_{\text{ref.}}$. We encourage the base color to be consistent with the pixels without shadow, which is formulated as:

$$\mathcal{L}_{\text{base.}} = \sum_{r \in \mathcal{S}} |C_{\text{ref.}} - C_{\text{base.}}|^2 \tag{10}$$

Combining all supervision and regularization items, our overall loss function is denoted as:

$$\begin{aligned}
\mathcal{L} = {} & \mathcal{L}_{\text{render}} + \lambda_{\text{skymask}}\mathcal{L}_{\text{skymask}} \\
& + \lambda_{\text{sky}}\mathcal{L}_{\text{sky}} + \lambda_{\text{norm.}}\mathcal{L}_{\text{norm.}} + \lambda_{\text{base.}}\mathcal{L}_{\text{base.}}
\end{aligned} \tag{11}$$

In our experiments, we set $\lambda_{\text{rad.}} = \lambda_{\text{sky}} = 1$, $\lambda_{\text{skymask}} = \lambda_{\text{norm.}} = \lambda_{\text{base.}} = 0.1$, $\lambda_{\text{smooth}} = 0.01$. λ_{depth} is set to 0.1 on nuScenes [3] dataset.

3.5   Street Scene Fusion

Intrinsic Properties Merging: As shown in Fig. 3, a hybrid sampling method is used to integrate vehicles and street scenes together. Beyond the original sampling points within the scene, we introduce an additional set of sampling points

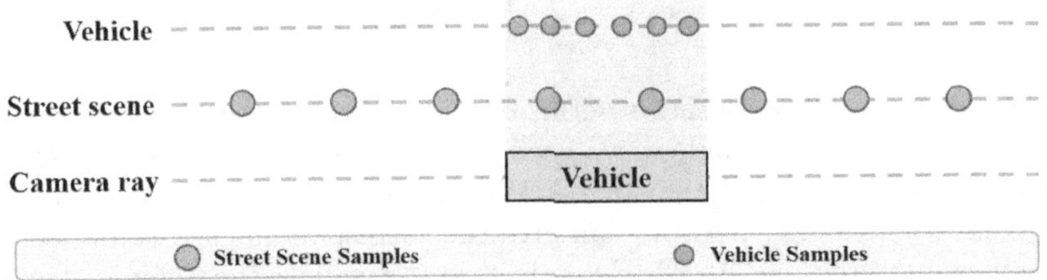

Fig. 3. We employ a hybrid sampling approach to facilitate the fusion of vehicles with street scenes. Initially, vehicles and street scenes are sampled independently, followed by sorting and combining them according to depth in the real-world coordinate.

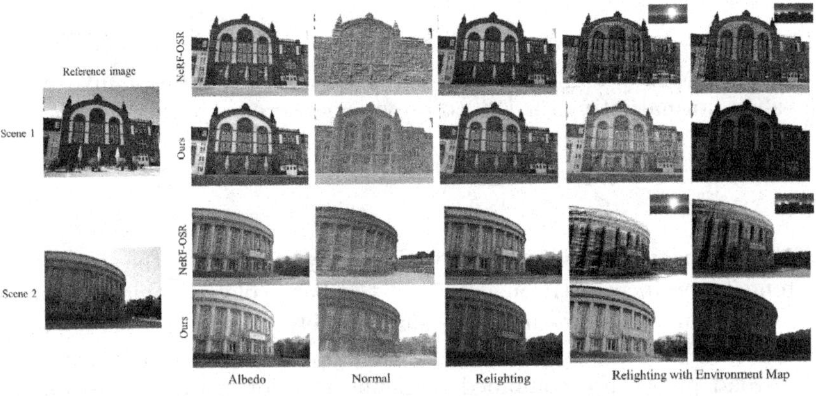

Fig. 4. Qualitative results of relighting on NeRF-OSR dataset [28]. Our method reconstructs normals with less noise and restores realistic shadows, enabling high-quality relighting results under various lighting conditions.

in the real-world coordinate. These additional points are specifically designated for the sampling of vehicles.

After the vehicle and the scene are both sampled, we sort and combine the sampling points based on their depth in the real-world coordinate, resulting in the combined sampling points $\{X\}$. Then, let $\mathbf{r} = \mathbf{o} + t\mathbf{d}$ denote the camera ray with origin $\mathbf{o}$ and direction $\mathbf{d}$, we obtain color and intrinsic properties of the vehicle and scene fusion by performing standard volume rendering:

$$c(\mathbf{r}) = \sum_{X_i} T_i \alpha_i \mathbf{c}_i, \quad T_i = \exp(-\sum_{k=1}^{i-1} \sigma_k \delta_k) \tag{12}$$

$$\boldsymbol{n}(\mathbf{r}) = \sum_{X_i} T_i \alpha_i \boldsymbol{n}_i, \quad \boldsymbol{k}(\mathbf{r}) = \sum_{X_i} T_i \alpha_i \boldsymbol{k}_i \tag{13}$$

where $\alpha_i = 1 - \exp(-\sigma_i \delta_i)$, $\delta_i = t_{i+1} - t_i$. Similar to [34], we truncate the density of scene sampling points within the vehicle's bounding box to prevent confusion between the scene and the vehicle. We can extract the mask of the foreground vehicle by comparing the depth of the vehicle and the scene, serving as the input for our FusionNet.

Fusion Network: Our Street Scene Renderer has achieved a photo-realistic composition of the appearance and shadow of foreground vehicles and street scenes. However, there are still some differences between the virtual vehicle models and the real vehicles, such as dust on the car body and fading of the car paint, etc. Therefore, we propose FusionNet to further eliminate the differences between virtual and real. We segmented vehicles from the street scene dataset and disturbed the clarity and color saturation of the vehicles to simulate the differences

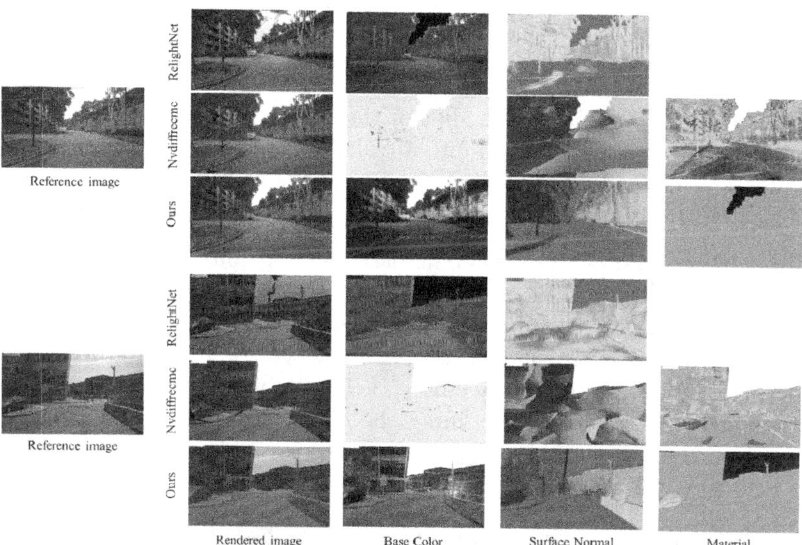

Fig. 5. Qualitative comparison of intrinsic properties decomposition. Our method reconstructs the true base color without any residual shadows. Furthermore, we achieve smoother normals and consistent materials, enhancing rendering quality.

Fig. 6. Qualitative comparison of foreground vehicle fusion. Our method harmoniously blends vehicles with street scenes, accurately restoring details such as highlights and shadows on the vehicles.

between virtual and real. We used a network structure similar to [39], with the addition of a vehicle segmentation mask as input, and an extra supervision item for color to ensure that the appearance of vehicles stays unchanged.

4   Experiments

4.1   Datasets and Baselines

NuScenes Dataset: NuScenes dataset [3] provides comprehensive resources for autonomous driving research. To simulate an urban environment, we select data from two distinct scenes from nuScenes dataset. Each scene's data encapsulates a video recorded by six strategically placed cameras on a moving vehicle. In our experiments, we used images from all six viewpoints.

NeRF-OSR Dataset: NeRF-OSR [28] dataset contains eight sites captured from 3240 viewpoints using a DSLR camera across 110 different recording sessions. For each site, a 360-degree shot of the environment map was also taken. Images from each site are divided into a training set and a test set with ground-truth illumination maps. We applied our Neural Street Field model to two scenes from NeRF-OSR dataset, and obtained quantitative evaluation results.

Baseline Methods: We use different baselines for different tasks for comparison. We establish Neural Street Field on NeRF-OSR dataset, and make qualitative and quantitative comparisons in terms of the quality of rendering and relighting. To decomposite intrinsic properties of scenes, we compare our method with RelightNet [40] and Nvdiffrecmc [11]. Lastly, we compare the methods of 2D illumination estimation, examining the effects of using generative network techniques in the integration of foreground vehicles and street scenes.

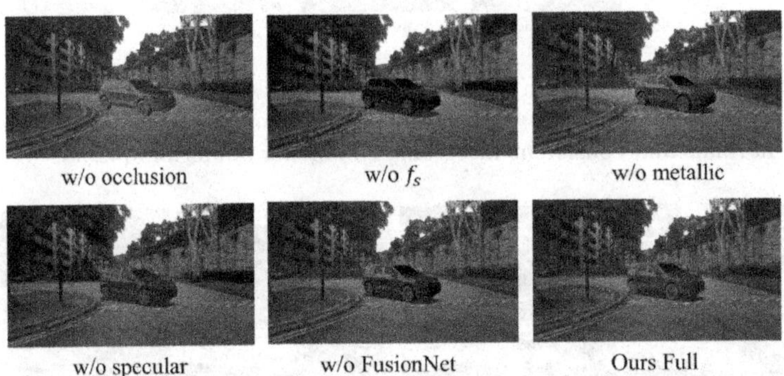

Fig. 7. Fusion results when removing or replacing a certain module of our proposed fusion process for ablation study.

Table 1. Quantitative Evaluation on NeRF-OSR [28] dataset

Method	Scene 1		Scene 2	
	PSNR	MSE	PSNR	MSE
NeRF-OSR [28]	15.86	0.026	15.97	0.025
Ours (w/o norm. reg.)	17.12	0.019	17.83	0.016
Ours (w/o shadow)	17.73	0.017	18.12	0.015
Ours	**17.95**	**0.016**	**18.56**	**0.014**

4.2   Rendering and Relighting Quality

Outdoor Relighting Quality: As shown in Fig. 4, We select two scenes from
NeRF-OSR dataset and build the Neural Street Field model to conduct relighting
evaluations for outdoor scenes. Our quantitative evaluation results are reported
in Table 1. Our method outperforms previous methods in PSNR and MSE by a
large margin, recovering more accurate albedo, normal, and shadow effects. We
also evaluate the relighting effect on two environment maps out of the dataset.
The spherical harmonics based lighting representation of NeRF-OSR cannot fully
capture the global lighting condition, and its bumpy normal also causes poor
relighting effects under bright lighting conditions. We comprehensively improve
the reconstruction quality of various intrinsic properties, thereby achieving much
better relighting effects than NeRF-OSR. The additional ablation study is conducted by: (a) Without normal regularization: We do not use additional normal
regularization terms (b) Without shadow: The shadow map is removed, which
means all surface points can be illuminated by the light source.

Intrinsic Decomposition Quality: Autonomous driving datasets commonly
provide multi-view videos in a center-outward manner, resulting in very sparse
viewing angles of the scene. Such images acquire limited 3D geometric information about the scene. Additionally, the images in the dataset commonly suffer
from motion blur and inconsistent brightness, which poses significant challenges
to accurately decomposing the intrinsic properties of the scene. As shown in
Fig. 5, We compare the quality of intrinsic decomposition with RelightNet [40]
and Nvdiffrecmc [11] RelightNet is unable to decompose material information
and shows significant errors in estimating base color and normal. Nvdiffrecmc
fails to reconstruct correct geometry and appearance from sparse observation
angles. Compared with these two methods, our method achieves high quality
intrinsic decomposition that can be directly used for render passes, producing
high-quality relighting results.

4.3   Vehicle Fusion Quality

Qualitative Comparison: As shown in Fig. 6, we compare the fusion effect of
our method with other methods. Hold-Geoffroy [12] first proposed a method for
estimating sky illumination, but they ignored the occlusion caused by objects in

the scene, leading to over bright appearance and incorrect shadows. Tang [29] considered both global and local lighting, which led to a more realistic appearance. However, due to the lack of 3D constraints, their results lacked consistency and were unable to bridge the gap between synthetic objects and real-world objects. Chen [5] tried using a generative network to modify the appearance of vehicles, but they could not restore details such as highlights and shadows. By contrast, our method synthesizes accurate highlights and shadow effects, leading to photo-realistic insertion of vehicles into the scene.

4.4   Ablation Study

To validate the effectiveness of the introduction of shadow map, Metal/roughness workflow and FusionNet, we conduct the experiments with the following settings:

- (a) Without occulusion: The shadow map is removed. All surface points are considered visible to the light source.
- (b) Without f_s: The specular component f_s of the BRDF is removed, retaining only the diffuse appearance.
- (c) Without metallic: The metalness of the vehicle is set to 0, modeling the vehicle as a completely non-metallic object.
- (d) Without specular: The metalness in specular component of f_s is set to 0, and the specular color turns white.
- (e) Without FusionNet: The FusionNet is removed from the fusion process.
- (f) Ours Full: Our method fuses a vehicle into a street scene.

The visualized results of the ablation study are shown in Fig. 7. We find that our full method achieves a photorealistic composition of the foreground vehicle and the street scene. Comparing (a) with (f), we can see that there are no shadow effects on the ground near the vehicle, and abnormal bright spots appear on the body of the car. This demonstrates the necessity of considering the shadows generated by objects within the scene and by the vehicle itself. Comparing (b) with (f), we find that the vehicle becomes a completely Lambertian surface, without any specular effects, failing to reflect the true appearance of the vehicle's material. This proves the advantages of the Metal/roughness workflow. Comparing (c) with (f), the appearance of the vehicle shifts towards white, this demonstrates the superiority of incorporating metalness into out Street Scene Renderer. Comparing (d) with (f), metalness makes the highlight display the correct color. Comparing (e) with (f), in the physically-based rendering results on virtual vehicles, there are issues of color inconsistency and over-saturation. FusionNet can eliminate the gap between virtual and real, yielding indistinguishably photorealistic results.

5   Conclusion

In this paper, we propose to synthesize photo-realistic street views with moving vehicles, editable illumination, and controllable viewpoints from sparse views of real-captured scenes and a synthetic dataset of 3D vehicles. A full radiance filed scheme is introduced for free-view synthesis of street scenes and vehicles, where illumination can be directly extracted from the real-captured scenes and transferred to the synthesized vehicle. Experiments show that our model can synthesize street view and vehicle images in free views, and significantly outperforms previous works in photo-realism and lighting modeling accuracy.

Limitations: Although our method achieves high-quality rendering of street scenes and fusion with foreground vehicles, it still has certain limitations. Our method remains reliant on some 2D priors, and additionally, we have not accounted for the effects of multiple light bounces. In our future work, we aim to extract more information from existing data to reduce our dependence on 2D priors. We plan to employ more sophisticated rendering techniques to minimize energy loss during the rendering process.

References

1. Abeysirigoonawardena, Y., Xie, K., Chen, C., Hosseini, S., Chen, R., Wang, R., Shkurti, F.: Generating transferable adversarial simulation scenarios for self-driving via neural rendering. arXiv preprint arXiv:2309.15770 (2023)
2. Agrawal, D., Xu, J., Mustikovela, S.K., Gkioulekas, I., Shrivastava, A., Chai, Y.: Nova: novel view augmentation for neural composition of dynamic objects. In: ICCV, pp. 4288–4292 (2023)
3. Caesar, H., et al.: nuscenes: A multimodal dataset for autonomous driving. In: Proceedings of the IEEE/CVF Conference on Computer Vision and Pattern Recognition, pp. 11621–11631 (2020)
4. Chen, Q., Koltun, V.: Photographic image synthesis with cascaded refinement networks. In: ICCV, pp. 1511–1520 (2017)
5. Chen, Y., et al.: GeoSim: realistic video simulation via geometry-aware composition for self-driving. In: CVPR, pp. 7230–7240 (2021)
6. Chen, Z., Zhu, L., Wan, L., Wang, S., Feng, W., Heng, P.A.: A multi-task mean teacher for semi-supervised shadow detection. In: CVPR (2020)
7. Dosovitskiy, A., Ros, G., Codevilla, F., Lopez, A., Koltun, V.: Carla: an open urban driving simulator. In: Conference on Robot Learning, pp. 1–16. PMLR (2017)
8. Fang, J., et al.: Augmented lidar simulator for autonomous driving. IEEE Robot. Autom. Lett. **5**(2), 1931–1938 (2020)
9. Gao, D., Li, X., Dong, Y., Peers, P., Xu, K., Tong, X.: Deep inverse rendering for high-resolution SVBRDF estimation from an arbitrary number of images. ToG **38**(4), 134-1 (2019)
10. Hasselgren, J., Hofmann, N., Munkberg, J.: Shape, light, and material decomposition from images using Monte Carlo rendering and denoising. NIPS **35**, 22856–22869 (2022)
11. Hasselgren, J., Hofmann, N., Munkberg, J.: Shape, light, and material decomposition from images using Monte Carlo rendering and denoising. arXiv:2206.03380 (2022)
12. Hold-Geoffroy, Y., Athawale, A., Lalonde, J.F.: Deep sky modeling for single image outdoor lighting estimation. In: CVPR, pp. 6927–6935 (2019)
13. Hold-Geoffroy, Y., Sunkavalli, K., Hadap, S., Gambaretto, E., Lalonde, J.F.: Deep outdoor illumination estimation. In: CVPR, pp. 7312–7321 (2017)

14. Hosek, L., Wilkie, A.: An analytic model for full spectral sky-dome radiance. ACM Trans. Graph. **31**(4) (2012). https://doi.org/10.1145/2185520.2185591
15. Isola, P., Zhu, J.Y., Zhou, T., Efros, A.A.: Image-to-image translation with conditional adversarial networks. In: CVPR, pp. 1125–1134 (2017)
16. Jain, J., Li, J., Chiu, M., Hassani, A., Orlov, N., Shi, H.: OneFormer: one transformer to rule universal image segmentation (2023)
17. Kajiya, J.T.: The rendering equation. In: 13th Annual Conference on Computer Graphics and Interactive Techniques, pp. 143–150 (1986)
18. Kar, O.F., Yeo, T., Atanov, A., Zamir, A.: 3D common corruptions and data augmentation. In: Proceedings of the IEEE/CVF Conference on Computer Vision and Pattern Recognition, pp. 18963–18974 (2022)
19. Lalonde, J.F., Matthews, I.: Lighting estimation in outdoor image collections. In: Proceedings of the 2014 2nd International Conference on 3D Vision - Volume 01. 3DV '14, pp. 131–138. IEEE Computer Society, USA (2014). https://doi.org/10.1109/3DV.2014.112
20. Li, L., Lian, Q., Wang, L., Ma, N., Chen, Y.C.: Lift3D: synthesize 3D training data by lifting 2D GAN to 3D generative radiance field. In: CVPR, pp. 332–341 (2023)
21. Li, W., et al.: AADS: augmented autonomous driving simulation using data-driven algorithms. Sci. Robot. **4**(28), eaaw0863 (2019)
22. Lin, Z.H., et al.: UrbanIR: large-scale urban scene inverse rendering from a single video. arXiv preprint arXiv:2306.09349 (2023)
23. Meder, J., Brüderlin, B.: Hemispherical gaussians for accurate light integration. In: ICCVG, pp. 3–15. Springer, Cham (2018)
24. Mildenhall, B., Srinivasan, P.P., Tancik, M., Barron, J.T., Ramamoorthi, R., Ng, R.: NERF: representing scenes as neural radiance fields for view synthesis. In: ECCV (2020)
25. Müller, T., Evans, A., Schied, C., Keller, A.: Instant neural graphics primitives with a multiresolution hash encoding. ToG **41**(4), 1–15 (2022)
26. Park, T., Liu, M.Y., Wang, T.C., Zhu, J.Y.: Semantic image synthesis with spatially-adaptive normalization. In: CVP, pp. 2337–2346 (2019)
27. Ros, G., Sellart, L., Materzynska, J., Vazquez, D., Lopez, A.M.: The synthia dataset: a large collection of synthetic images for semantic segmentation of urban scenes. In: CVPR, pp. 3234–3243 (2016)
28. Rudnev, V., Elgharib, M., Smith, W., Liu, L., Golyanik, V., Theobalt, C.: NERF for outdoor scene relighting. In: ECCV, pp. 615–631. Springer, Cham (2022)
29. Tang, J., Zhu, Y., Wang, H., Chan, J.H., Li, S., Shi, B.: Estimating spatially-varying lighting in urban scenes with disentangled representation. In: European Conference on Computer Vision, pp. 454–469. Springer, Cham (2022)
30. Wang, J., Ren, P., Gong, M., Snyder, J., Guo, B.: All-frequency rendering of dynamic, spatially-varying reflectance. In: SIGGRAPH Asia, pp. 1–10 (2009)
31. Wang, T.C., Liu, M.Y., Zhu, J.Y., Tao, A., Kautz, J., Catanzaro, B.: High-resolution image synthesis and semantic manipulation with conditional GANs. In: CVPR, pp. 8798–8807 (2018)
32. Wang, Z., Chen, W., Acuna, D., Kautz, J., Fidler, S.: Neural light field estimation for street scenes with differentiable virtual object insertion. In: ECCV, pp. 380–397. Springer, Cham (2022)
33. Wang, Z., et al.: Neural fields meet explicit geometric representations for inverse rendering of urban scenes. In: CVPR, pp. 8370–8380 (2023)
34. Wu, Z., et al.: Mars: an instance-aware, modular and realistic simulator for autonomous driving. In: CICAI (2023)

35. Xie, Z., Zhang, J., Li, W., Zhang, F., Zhang, L.: S-NERF: neural radiance fields for street views. In: ICLR (2023)
36. Yan, X., Zou, Z., Feng, S., Zhu, H., Sun, H., Liu, H.X.: Learning naturalistic driving environment with statistical realism. Nat. Commun. **14**(1), 2037 (2023)
37. Yang, Z., et al.: UniSim: a neural closed-loop sensor simulator. In: CVPR, pp. 1389–1399 (2023)
38. Zhu, Y., Zhang, Y., Li, S., Shi, B.: Spatially-varying outdoor lighting estimation from intrinsics. In: CVPR (2021)
39. Yu, J., Lin, Z., Yang, J., Shen, X., Lu, X., Huang, T.S.: Free-form image inpainting with gated convolution. In: Proceedings of the IEEE/CVF International Conference on Computer Vision, pp. 4471–4480 (2019)
40. Yu, Y., Meka, A., Elgharib, M., Seidel, H.P., Theobalt, C., Smith, W.A.P.: Self-supervised outdoor scene relighting. In: Proceedings of the European Conference on Computer Vision (ECCV) (2020)
41. Zhang, K., Luan, F., Wang, Q., Bala, K., Snavely, N.: PhySG: inverse rendering with spherical gaussians for physics-based material editing and relighting. In: CVPR, pp. 5453–5462 (2021)

Author Index

© The Editor(s) (if applicable) and The Author(s), under exclusive license
to Springer Nature Singapore Pte Ltd. 2026
M. Mahmud et al. (Eds.): ICONIP 2024, CCIS 2297, pp. 421–422, 2026.
https://doi.org/10.1007/978-981-96-7036-9

Made in the USA
Monee, IL
07 July 2026